Praise for the Revised Edition of *Wind Power*

"*Wind Power* is the most comprehensive, down-to-earth guide available. Paul Gipe, an acknowledged world expert, has produced an indispensable book for anyone considering purchase of a wind turbine. Whether discussing huge 2MW turbines or small home units, Gipe tells you what you want to know in straightforward, jargon free language, and with plenty of illustrations. Contrary to much trade literature, Gipe does not ignore the pitfalls of putting up a turbine. . . . *Wind Power* is a highly recommended book for anyone interested in the rapidly expanding world of wind energy."

Robert W. Righter, author of *Wind Energy In America: A History*

"If you only buy one book on wind energy, make it this one. . . . Paul Gipe has made a great book even better in his 2004 update of the classic *Wind Power for Home & Business*. I recommend this book to students, homeowners, and anyone who wants a clear picture of what they're getting into when they are hooked by the wind energy bug. The small wind energy world seems to have a disproportionate share of dreamers, cranks, and charlatans. Paul Gipe cuts through the fog and gives readers straight information on how to capture the energy in the wind. His clear explanation of terms, concepts, and recommended practices will save you many hours and many dollars."

Ian Woofenden, Senior Editor, *Home Power* magazine

"At long last, big as well as small wind machines are paying off. Grid-connected wind can compete! Paul Gipe, the internationally known wind expert, has captured the technology and the excitement of this fast-changing field. His explanations are clear and accessible. His anecdotes provide a human touch. If you're considering your own machine or investing in a wind company, or if you want to understand wind's rebirth, *Wind Power* is a must."

Paul P. Craig, Professor Emeritus of Engineering, University of California at Davis;
Chairman, Sierra Club National Energy Committee, 2000–2003

"Paul Gipe's new edition of *Wind Power* is exceptionally comprehensive and authoritative. Anyone interested in sustainable energy resources will find it informative. Those interested in wind machines for their home, farm, or business will find it essential."

Rich Ferguson, PhD, Research Director,
Center for Energy Efficiency and Renewable Technologies

"I have several hundred books on wind generators on my bookshelf. I've referred to Gipe's books repeatedly over the years. If I had to choose only two books out of all on the shelves, Paul Gipe's *Wind Power: Renewable Energy for Home, Farm, and Business* would definitely be a keeper."

Mick Sagrillo, Sagrillo Light & Power

"A truly comprehensive, beautifully illustrated book. . . . The new edition comes replete with useful extra features, a good range of cases studies and a sensible health warning about the need to take safety seriously. The only problem will be ensuring that eager colleagues do not make off with your copy!"

Professor David Elliott, Director, Energy and Environment Research Unit,
The Open University, Great Britain

WIND POWER

WIND POWER

Renewable Energy for Home, Farm, and Business

Paul Gipe

The Completely Revised and Expanded Edition

Chelsea Green Publishing Company
White River Junction, Vermont

Disclaimer: The installation and operation of a wind turbine entails a degree of risk. Always consult the manufacturer and applicable building, electrical, and safety codes before installing or operating your wind power system. For wind turbines interconnected with the electric utility, always consult with your local utility prior to installing your wind turbine. When in doubt, ask for advice. Recommendations in this book are not a substitute for the directives of wind turbine manufacturers or regulatory agencies.

The author assumes no liability for personal injury, property damage, or loss from using information contained in this book.

Designed by Peter Holm, Sterling Hill Productions
Printed in the United States
First printing, March 2004
10 9 8 7 6 5 4 3
Printed on acid-free, recycled paper

Library of Congress Cataloging-in-Publication Data
Gipe, Paul.
Wind power for home, farm, and business : renewable energy for the new millennium / Paul Gipe.
p. cm.
Completed rev. and expanded ed. of: Wind power for home and business.
White River Junction, VT : Chelsea Green Pub. Co., c1993.
ISBN 1-931498-14-8
1. Wind power. I. Gipe, Paul. Wind power for home & business. II. Title.
TJ820.G565 2003
621.4'5—dc22

2003019354

Chelsea Green Publishing Company
Post Office Box 428
White River Junction, VT 05001
(800) 639-4099
www.chelseagreen.com

Dedication

To my family and friends, whose encouragement has given me the freedom to choose my own path.

Disclosure

Paul Gipe has worked with ANZSES, APROMA, ASES, AusWEA, AWEA, BWEA, BWE, CanWea, CEC, DGW, EECA, GEO, KWEA, NASA, NREL, SEI, USDOE, Aerovironment, the Folkecenter, the Izaak Walton League, Microsoft, the Minnesota Project, PG&E, NRG Systems, SeaWest, the Sierra Club, the Rahus Institute, and Zond Systems, and has written for magazines in the United States, Canada, France, Denmark, and Germany.

Preface

Since the mid-1970s I've followed the development of wind energy around the globe. During this time I've been a proponent, participant, observer, and critic of the wind industry. As an observer I've traveled extensively, reporting on the technology and how it's being used. As a participant I've installed anemometers in Pennsylvania, hunted windchargers in Montana, and measured the performance of small wind turbines in California. As a proponent I've lectured about the promise of wind energy to groups from Vancouver to New Delhi, from Punta Arenas, Chile, to Hannover, Germany. And as a critic I've taken some wind companies to task when their environmental practices were no better than the technologies they intended to supplant.

In the early 1980s I prepared a daylong seminar on the prospects and pitfalls of wind energy. An earlier version of this book, published in 1983 under the title *Wind Energy: How to Use It,* grew out of the course notes for these seminars.

At that time there was a chasm between the books written for backyard tinkerers who wanted to build their own wind turbines, and those books surveying the entire field of wind energy. There was no book that answered the questions people raised in my seminars about how they could obtain a working wind system and not an experimenter's toy. *Wind Energy* was written to meet that need. The book was unique because it didn't simply look at the technology. It gathered tips and advice from leaders in the field and offered practical guidance on how to select, buy, and install wind turbines—and how to do so safely.

After extensive revision *Wind Energy* was reissued in 1993 by Chelsea Green Publishing as *Wind Power for Home & Business.* Since then the book has become a staple of both homeowners and professionals interested in the subject.

Today wind energy is a booming worldwide industry. The technology has truly come of age, and with today's heightened concern about our environment, this resurgence of interest is here to stay. Despite wind energy's success, there remains a need for a frank discourse on how to wisely use the technology. For this reason, I have written a comprehensive revision of my earlier books. Though *Wind Power: Renewable Energy for Home, Farm, and Business* follows the outline of the previous editions, this version includes considerable new material as well as a new, easier-to-use format.

In 1983 I sought to help newcomers to wind energy avoid the mistakes that I and others had made, and to spur development of this renewable resource. *Wind Power: Renewable Energy for Home, Farm, and Business* seeks the same end.

Bon Vent! (Good Wind!)

Acknowledgments

No one can write a book on a subject that crosses as many disciplines as wind energy without the help of numerous contributors. I am especially indebted to Vaughn Nelson at West Texas A&M University's Alternative Energy Institute. Vaughn first taught me the importance of swept area and how to quickly cut through the hype that often surrounds wind turbines. AEI's Ken Starcher has been invaluable for his technical expertise as well as his old-fashioned common sense. Nelson and Starcher have produced one of the most comprehensible texts on wind energy that I've found anywhere. The CD-ROM version was particularly helpful.

My thanks to Mick Sagrillo, Sagrillo Light & Power, and Hugh Piggott, Scoraig Wind Electric, for answering my many questions on battery-charging wind systems. Both Mick and Hugh are fonts of practical, hands-on knowledge of small wind turbine design. Their observations are peppered throughout the text.

Jim Salmon, Zephyr North, and Jack Kline, R.A.M. Design, were instrumental in the expanded chapter on wind resources, as were Dave Blittersdorf, NRG Systems, and Ken Cohn, Second Wind.

Small wind turbine manufacturers worldwide deserve a note of appreciation for responding to my frequent queries about their products. Over the years Mike and Karl Bergey, Bergey Windpower Company, have been notably forthcoming.

I extend my heartfelt gratitude to the Folkecenter for Renewable Energy and the people of Denmark for a fellowship to study the distributed use of wind energy in northwest Jutland. It was at the Folkecenter that I first learned how to use a griphoist to install small turbines.

My appreciation also to Bill Hopwood and Dennis Elliott for their contributions on siting. David Suey, Ken O'Brock, and Alan Wyatt for their help with mechanical wind pumps. Michael Klemen, Eric Eggleston, Claus Nybroe, and Jason Edworthy for their insightful comments on small wind turbine design. Capitola Reece, Gene Heisey, Art and Maxine Cook, Phil Littler, Sister Paula Larson, Eli Walter, and Bill Young for sharing their experiences. Gil Morrissey for his tutelage to a sometimes dim-witted apprentice electrician. Ed Butler for advice on how to do the job right. Klaus Kaiser, Christoph Stork, Bernard Saulnier, Charles Dugué, and Charles Rosseel for their help with the lexicon. Heiner Dörner for historical background on FLAIR. Susan Nelson for her faith in the future. And Nancy Nies for both her encouragement and her unstinting aid.

Contents

1

Introduction

Wind works. It's reliable. It's economical. It makes environmental sense. And it's here now. Wind machines are not tomorrow's technology. Whether it's on a giant wind farm in Minnesota, in a small village in Morocco, or in the backyard of a German farmer, wind energy works today in a variety of applications around the world. You, too, can put this renewable resource to work. The following chapters explain how to go about doing just that.

Wind technology has come a long way since the early 1970s, when the only wind turbines available were 1930s-era machines salvaged from ranches on the Great Plains. During the past three decades wind energy has truly come of age. But as you'll see in the chapters ahead, wind machines are not for everyone.

To use the wind successfully, you must have a good site and select the right machine. You also need something else—courage. A wind turbine is not cheap. It represents an investment in the future. And whether you install it yourself or contract a dealer to do it, the installation of a wind machine—small or large—is an undertaking fraught with risk and uncertainty. At some point, after considering all the pros and cons, only you can make the decision to proceed. You must weigh your choices, then act. The people who use wind energy are prudent, but they're doers.

During the past three decades wind energy has truly come of age.

People use wind machines for many reasons: economic, environmental, and philosophical. The knowledge that you're saving money—or earning it—is often sufficient reward for plunging into wind energy. Yet for many people there's more to it than that. Windmills have fascinated us for centuries and will continue to do so. Like campfires or falling water, they're mesmerizing; indeed, entrancing. People respond almost instinctively. Few escape feeling excitement at seeing a sleek turbine whirring in the wind.

Working with the wind is more than just a means to cheap electricity. It becomes a way of life, a way of living in closer harmony with the world around us. Harnessing the wind enables us to regain some sense of responsibility for meeting our own needs and for reducing our impact on the environment. By generating our own electricity cleanly and with a renewable resource, we can reduce the need for distant power plants and their attendant ills. Wind energy can and does make a difference.

If you're fortunate enough to install a wind machine with your own hands, you'll experience sensations few others can share. You'll know what it's like to gaze from the top of your tower at the countryside spread out before you. You'll know the feeling of seeing your wind machine spinning overhead for the first time. You'll rediscover the sense of accomplishment received from a job well done. You and your

Wind Energy Workshops

Mick Sagrillo, a grizzled veteran of the wind wars, organizes hands-on workshops annually through the Midwest Renewable Energy Association and Solar Energy International. His multi-day program often includes assembly and installation of a small wind turbine. Sagrillo (bottom) is an energetic and entertaining speaker, regaling his students with amusing anecdotes and homespun advice.

Hugh Piggott. Hugh displays one of the typical rotors he assembles in his workshops.

Hugh Piggott, a designer of small wind turbines, is a frequent contributor to Internet news groups on wind energy. He's unstinting with his time in explaining, to novices and pros alike, the arcane workings of small wind turbine generators. Piggot (top) offers several workshops per year on how to build your own wind turbine.

Several organizations, such as ECN in the Netherlands, offer workshops for professionals on commercial applications of wind energy. See the Workshops section in appendix (p. 422) for further details.

Mick Sagrillo. Mick lectures to a "hands-on" class in Amherst, Wisconsin, before installing a Proven wind turbine.

friends will discover the hearty, backslapping camaraderie that grows among people after several arduous days of working together. There's nothing quite like it. Installing a wind machine, however, will never be risk-free. Generating electricity isn't easy; it never has been. Your utility may make it look simple, but it's a tough and sometimes dangerous job. This book is designed to help you minimize the risk, to ensure—as much as possible—that you'll succeed in erecting a wind turbine that works reliably and safely.

Wind Power: Renewable Energy for Home, Farm, and Business was written for those who ask questions, who want to know what's going on around them, who want to do what they can themselves. It's for those who want to make a difference. Yet this isn't another how-to-build-your-own-windmill book, though much of what it contains is essential for building and installing one safely. Instead *Wind Power* gives you what you need to make intelligent choices (whether or not to build one yourself, for example). It's about how to proceed in a logical and methodical manner to determine if you can use wind energy and, if so, the kind of wind turbine that's right for you.

This book differs from others on the subject by both describing the technology and explaining how to evaluate what's important and what's not.

Before we plunge into the text, there are a few preliminaries we need to deal with first: the organization of the chapters ahead and the nomenclature that will be used.

How This Book Is Organized

Chapter 2 is an introduction to the varied applications for wind energy. It briefly discusses how wind energy is used today.

The wind itself is the subject of chapter 3. You'll learn the importance of wind speed and how to find out what you have. You'll also learn when it's necessary to measure the wind at your site—and when it's not.

In chapter 4 you'll find techniques for estimating the amount of energy that a wind turbine may capture, as well as how to interpret the information published by wind turbine manufacturers. With estimates of annual production in hand, you're then ready to evaluate the economics of your wind investment in chapter 5.

Chapter 6 explores wind technology: where we've been, where we are today, and where we're headed. The advantages and disadvantages of various technologies are explained. You'll also get a feel for what's on the market and how to evaluate it.

Towers, necessary companions to wind turbines, are the subject of chapter 7.

Cutting costs without cutting corners is covered in chapter 8. If you think building the wind machine yourself is the answer, this chapter asks you some hard questions. If you're a tinkerer, it suggests finding a used wind turbine. If you're looking for a trouble-free small wind turbine, this chapter narrows your options to installing a wind system yourself, from a kit.

Once you've decided what you need, chapter 9 explains how to put it all together: how to evaluate a manufacturer, what to include in a contract, and what to expect from a dealer.

You'll learn how you can sell power back to the utility company in chapter 10. And you'll find out why the utility may have some legitimate concerns about your doing so.

If, instead of generating electricity in parallel with the utility, you prefer to declare your energy independence, chapter 11 examines the components you'll need. This chapter explains why hybrid power systems using both wind and solar make more sense for remote sites than either of those technologies by itself.

Chapter 12 puts a modern twist on one of the oldest applications for wind energy: pumping water.

Chapter 13 examines where you can and can't install a wind turbine, and why. You'll also learn about potential land-use conflicts, how to avoid them, and—where you can't—how to deal with them.

Chapter 14 is designed for those who want to install a wind turbine themselves. This extensive chapter offers tips on installation, anchoring, and wiring.

Once your wind machine is installed you'll need to operate it properly, in order to ensure that it serves you well for many years to come. Chapter 15 reviews some simple start-up procedures and suggests how to operate, maintain, and monitor the performance of your wind system.

> Don't be fooled by pompous buzzwords. Pretentious terms don't make a wind turbine more reliable or more productive.

In chapter 16 we take a close look at a taboo subject not found in other books on wind technology—safety. Chapter 17 completes *Wind Power: Renewable Energy for Home, Farm, and Business* by looking at the future and where the technology may be headed.

Nomenclature—What Are We Talking About?

They've been called many things, some unprintable. Most know them as windmills. Whether we call them wind machines, wind turbines, or just windmills, the subjects of this book are kinetic devices intended to capture the wind and put it to work.

Wind Machines

The terms *wind machine* and *wind turbine* are simple and unpretentious. They do the job with the least fuss and are used interchangeably in the following pages. You'll find no reference to pompous buzzwords like *SWECS* (small wind energy conversion systems) or *WTGs* (wind turbine generators) here. Few people who work with wind energy for a living use such jargon. Those who do frequently do so merely to impress or, worse, to mislead. Don't be fooled. Pretentious terms don't make a wind turbine more reliable or more productive.

The term *wind machine,* as used here, shouldn't be confused with the huge electric fans found in orchards. These fans go by the same name but are used to stir still air during cold winter nights to protect valuable fruit. They move air; they're not moved by the air, as is a wind turbine.

When we're referring to the wind machine, tower, and ancillary equipment as a whole, *wind system* works well. The term *windmill* is reserved for the multiblade, water-pumping wind machine (American farm windmill) and for the European wind machine (Dutch windmill), because that's what nearly everyone calls them. Conventional wind turbines comprise three essential components: rotor, nacelle, and tower. The rotor, the spinning part of a wind turbine, is the most important. Because the rotor determines how much work a wind turbine can perform, the size of a wind turbine will often be described by the diameter of its rotor. The rotor attaches to the wind turbine's drive train, which typically sits atop the tower housed inside the nacelle. The tower simply raises the whole works above the ground, where it can better catch the wind. (See figures 1-1, Small wind turbine nomenclature, and 1-2, Medium-size wind turbine nomenclature.)

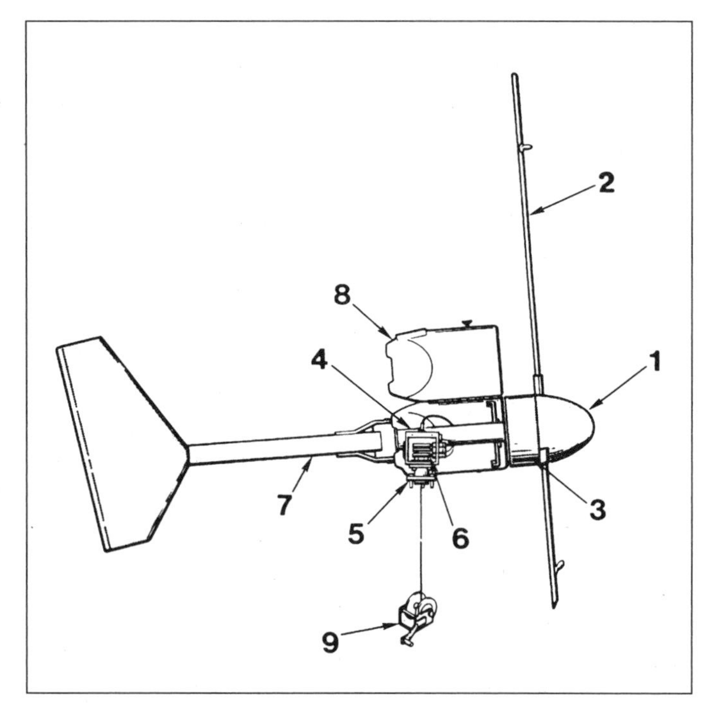

Figure 1-1. Small wind turbine nomenclature. (1) Spinner or nose cone. (2) Rotor blades. (3) Direct-drive alternator. (4) Mainframe. (5) Yaw assembly. (6) Slip rings and brushes. (7) Tail vane. (8) Nacelle cover. (9) Winch for furling the rotor out of the wind. (Bergey Windpower)

Power and Energy—There Is a Difference

In casual conversation, we use these terms interchangeably. But there is a difference. When we describe what a wind machine can do for us—what it can produce—we must be more specific. *Power* and *energy,* though closely related, have separate and distinct meanings. They are both related to work.

In the technical sense, work is performed when a force acts through some distance. When you push a stalled car, for example, you are applying a force. For work to be done, something must be accomplished: an object moved, lifted, or turned. If the object does not move, no work is accomplished. If the car did not move, no matter how hard you grunted and groaned to move it, no work was performed.

Energy is defined as the ability to do work or the amount of work actually performed. Both use the same units and are given in the same terms. When the wind strikes the blades of a wind turbine, it imparts a thrust or force that turns the rotor. A finite amount of energy in the flowing wind has been converted to rotational energy in the spinning rotor. When a force does work on an object, energy is transferred from one form to another.

Now couple that spinning rotor to a generator. Work is accomplished when electrons flow from the generator to a load, such as heating a wire in a toaster or turning a motor in a fan. Now take a closer look at the electric motor. The flow of electrons, what we call electricity, transfers the rotational energy imparted by the wind to the wind turbine rotor through wires or power lines to spin the shaft of the motor, giving us, once again, rotational energy. Because the motor spins a fan, work is accomplished. We've gotten something out of it.

The conversion from one form of energy to another is never 100 percent efficient; that is, you can't get out as much as you put in. There's always some energy lost in the process. Friction is the chief culprit. Though no one has ever built a machine that can convert 100 percent of the energy in one form to another, people keep trying. Their perpetual motion machines appear periodically as a "startling discovery" in the popular press. Wind turbines are a favorite target of this breed. Such machines never operate as claimed, nor do they usually produce useful work.

As someone who uses energy, you're concerned about the actual work accomplished and the amount of energy transferred. It's the bottom line of the technical balance sheet. You measure the performance of a wind machine by what it does for you, the work it performs, the energy it transfers and puts to use.

Where does power come in? *Power* is the rate at which work is performed, the rate at which energy is transferred (changed or released), the rate at which the energy in the wind passes through a unit of area. Power is given in watts (W), or in units of 1,000 watts, 1 kilowatt (kW).

Consumers demand power, and wind turbines can provide it. But what's most important isn't power, it's energy. Homeowners, for example, seldom pay for power. They pay for energy, given in units of kilowatt-hours (kWh). When homeowners buy a kilowatt-hour of electricity, they're paying for the energy delivered, not the power.

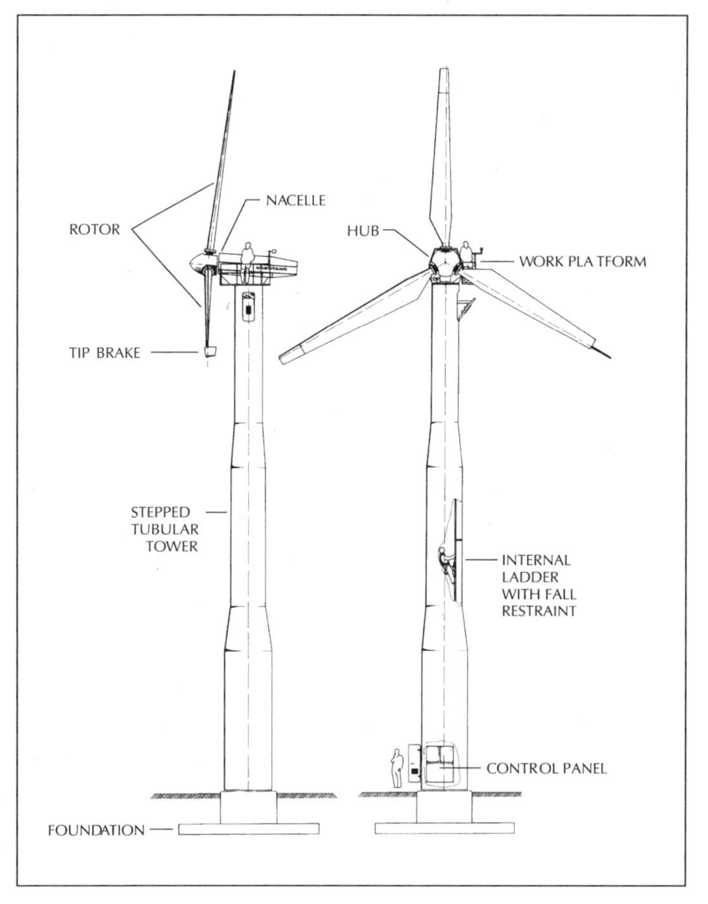

Figure 1-2. Medium-size wind turbine nomenclature. Typical first-generation (55 to 75 kW) Danish wind turbine. Note that even in such early designs, provisions were made to safely service the turbine by the inclusion of a fall-arresting system (on the ladder inside the tower) and a work platform on the nacelle. (NEG-Micon)

Watt's More

In an electrical circuit, current is the flow of electrons. It's given in units of amperes or amps. We perceive voltage as the pressure trying to force the electrons to flow through a circuit. But no flow takes place and no work is done unless a load, like a toaster, is attached to complete the circuit. (No electricity flows out of a receptacle until an appliance is plugged into it.)

Power in watts is the product of voltage times current. This represents the instantaneous rate of work being done when a force (voltage) moves electrons some distance through a wire (current). A toaster, for example, will operate at 110 volts (V) and will draw 10 amps (A), or:

$$\text{Power} = 110\,\text{V} \times 10\,\text{A} = 1{,}100\,\text{W} = 1.1\,\text{kW}$$

The toaster's demand for power is equivalent to eleven 100 W lightbulbs. Yet in the average household a toaster uses much less energy (power × time) than lighting, because it's used so infrequently—only a few minutes every morning. Lights, on the other hand, are used for hours on end. One 100 W lightbulb operating for 11 hours will burn as much electricity—use as much energy—as a toaster operated for 1 hour:

100 W Lightbulb:

100 W x 11 hours = 1,100 Wh = 1.1 kWh

Toaster:

1,100 W x 1 hour = 1,100 Wh = 1.1 kWh

The distinction between kilowatts and kilowatt-hours is critically important. Knowing the difference can keep you from being confused by a wind turbine's size, in kilowatts, and how much energy, in kilowatt-hours, it will actually produce.

Levers, Moment Arms, and Torque

Archimedes said that if you could get him a long enough pole, he could lift the earth.

Whenever you struggle with loosening a nut on a rusty bolt, you are faced with two choices: increase the force applied or increase the length of your tool. We are only so strong and our backs will only take so much strain before we begin to look at our options. We could stand on the wrench, thus applying our full weight (increasing the force), or we could go for a longer wrench. That's what most of us do—find a tool with a longer handle. This gives us a mechanical advantage (the use of less force to do the same job) by increasing the leverage we exert on the nut.

Be on guard for indicators of false precision.

The span from the nut to the end of the lever is the moment arm. A longer moment arm must act through a greater distance to turn the nut the same amount as a short moment arm or short handle. We must push the end of the wrench farther than before. This illustrates an important principle: Where we gain a mechanical advantage, we must make up for it by increasing the distance through which it acts. (It also demonstrates a concept summed up by Barry Commoner: There is no such thing as a free lunch. We gain by giving up something. Our hope is to come out ahead—on balance.) The work we wish to perform remains the same. We have only altered the relationship between force and distance—decreasing the one, increasing the other in direct proportion. This principle remains the same, whether we are talking about the length of a handle on a wrench, transmission ratios, or the number of pulleys in a block and tackle.

Torque is the product of the force applied to the lever and the span from the nut to where the force is applied. An understanding of torque and moment arms is helpful when considering what wind machine designers face when trying to prevent a tall tower from buckling in high winds. The wind's thrust on the wind turbine at the top of the tower is the force applied, and the span from the ground to the top of the tower is the moment arm through which it acts. Another example is the torque applied to the shaft attaching a wind turbine blade to the rotor. More torque can be delivered by a long blade than a short one, where all other conditions are the same. If the blade moves through the same arc, more energy can be delivered by a long blade than by a short one, with the same force applied. This affects design by requiring that a heavier (stronger) shaft be used, but this disadvantage is offset by the greater amount of energy delivered to the generator.

Equations—They're Informative

There are numerous tables and more than a few equations used in *Wind Power.* They're here for a reason. They express the relationships among quantities. The equation for the power in the wind succinctly explains why wind speed plays such a critical role. Wherever an equation appears in the text, there'll be an accompanying table to summarize the results. Tables are helpful because they're easier to read than equations, but trends are hard to spot. Where necessary, graphs are used to illustrate trends and general relationships, such as in the distribution of wind speeds over time.

False Precision—How to Avoid It

A long string of digits in a calculation gives a sense of precision that may or may not exist. Don't be deceived. Estimating the annual energy production of a wind turbine is an inexact science. The art of number crunching for technologies dependent on a natural resource—whether it's wind farming, mining, or drilling for oil—is finding the best approximation.

In the real world, natural phenomena seldom follow the orderly relationships shown in textbooks. If you plotted a graph of the relationship between wind speed and the instantaneous power generated by a wind turbine, you'd find that the points representing your measurements are scattered. When your job is to interpret this information, you draw a line that best fits the measured data. The resulting line is an approximation of what happened; it doesn't say exactly what happened.

Knowing this, you should be on guard for indicators of false precision. Take the potential performance of a wind turbine as an example. It's absurd to say that a wind machine will generate 495 kilowatt-hours per month when the data used in the estimate suggests the wind turbine could deliver from 450 to 550 kilowatt-hours per month. Why not simply say 500 kilowatt-hours per month and be done with it? By rounding off the calculation, you indicate uncertainty. Better yet, present the estimate as a range of values from 450 to 550 kilowatt-hours. Then you know that the estimates are only an approximation of what could occur.

What's New in This Version of *Wind Power*

- An expanded section on wind resources, including more detail on how to account for changes in elevation and temperature.
- An expanded section on evaluating power curves, including actual measured performance from several micro and mini wind turbines.
- An expanded section on siting wind turbines, including actual measured noise emissions from micro wind turbines.
- More wind turbine manufacturers worldwide.
- More information on medium-size or commercial wind turbines.
- More examples and case studies of those successfully using the wind.
- An expanded section on used wind turbines that now includes used medium-size turbines such as those once used on wind farms.
- Expanded appendixes that include more sources of information, more manufacturers of both small and medium-size wind turbines, a new glossary and lexicon, sources for used and reconditioned turbines, an expanded bibliography, more conversion factors, and a new feature to aid in constructing your own wind power spreadsheets.

For updates, changes, and corrections to this version of *Wind Power,* visit the Chelsea Green Web site at www.chelseagreen.com or the author's Web site at www.wind-works.org.

Since the advent of pocket calculators in the 1970s, and now the universal availability of personal computer spreadsheets, the results of calculations are often presented in meaningless detail. In the days of the slide rule (yes, this does date me), every student learned that a calculation was only as accurate as the divisions on the rule—as accurate as the least accurate number in the calculation. You couldn't carry a number out to 10 decimal places, even if you wanted to. Though slide rules have gone the way of the dodo, the concept remains valid:

The results of a calculation are only as accurate as the least accurate value used.

Consider average wind speed. It's normally presented to three significant figures. For example, 10.5 mph has two figures to the left and one to the right of the decimal; 8.25 m/s has two figures to the right of the decimal. If we were to use these average speeds in a calculation to estimate the energy production of a wind machine, the results should likewise be presented to no more than three significant figures. Say the calculation resulted in 22,525.49 kilowatt-hours on your calculator. Only the first three figures are valid, not the seven indicated. The result (22,500 kilowatt-hours) is more realistic, considering the accuracy of the numbers used to derive it. Ignoring the concept of significant figures leads to false precision.

In wind energy, size, especially rotor diameter, matters.

Most scientists and engineers are accustomed to approximate arithmetic. They round off numbers every chance they get. It makes their work easier. More importantly, it allows them to quickly get to the heart of a calculation without wasting time on needless detail. The use of approximate arithmetic is how an engineer can take a seemingly complex problem, such as estimating a wind turbine's potential energy production, and solve it mentally or scratch it out on the back of an envelope. Where appropriate, the calculations in *Wind Power* have been liberally rounded off.

Units—Yes, Metrics, Too

Purists about either the metric or English system of measurement should be forewarned. The wind industry in North America uses an unholy mix of both. Towers installed in the United States are sold in feet. For this reason, tower height will generally be given in feet. Wind turbines, on the other hand, are produced around the world. The convention among wind turbine manufacturers worldwide is to use the metric system to describe the diameter of their turbines. Consequently, the size of a wind turbine's rotor will often be given in meters, not feet.

The description of wind speeds isn't as clear-cut. In general, metric measurements in meters per second (m/s) are preferred everywhere but the United States, where miles per hour (mph) is still used. Because the wind industry in the United States is gradually adopting the metric system for wind speed, metric units frequently appear in the text.

If you have an aversion to either system, don't panic. In *Wind Power* the English or metric equivalent will often be included within parentheses. Conversion tables for common metric and English units can also be found in the appendix.

Size—It's All Relative

In wind energy, size, especially rotor diameter, matters. Today's wind turbines range in size from the minuscule Marlec 500 with a rotor only 0.5 meter (1.7 ft) in diameter to giant machines whose rotors reach nearly 100 meters (330 ft) in diameter. The Marlec can generate 20 watts in a good stiff wind. The large machines are capable of as much as 3 MW or 3,000,000 watts!

Size designations in *Wind Power* are somewhat arbitrary. The Marlec 500 is small; the 3-megawatt (3 MW or 3,000 kW) turbine is clearly not (see figure 1-3, Relative size).

Small wind turbines are often subdivided into micro, mini, and household-size machines. Micro wind turbines, such as the Marlec 500, are the smallest of the small, and are typically less than 1.25 meters (4 ft) in diameter. This class includes Southwest Windpower's popular Air series (see figure 1-4, Micro wind turbine).

Mini or cabin-size wind turbines are intermediate between the micro turbines and the

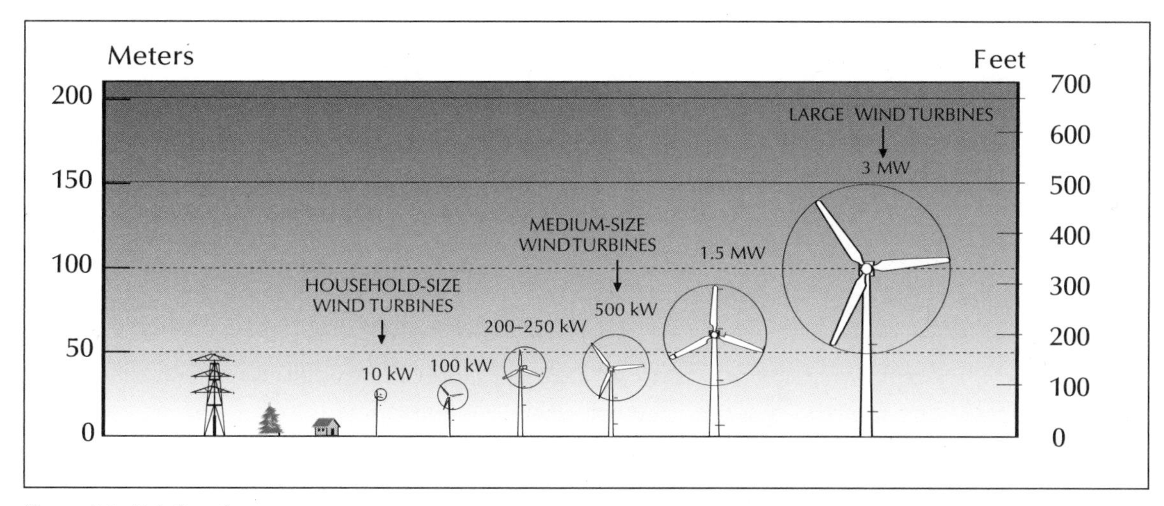

Figure 1-3. Relative size.

much larger household-size machines. They are less than 3 meters (10 ft) in diameter and include Bergey Windpower's XL1 and Proven Engineering's WT600 (see figure 1-5, Mini wind turbine).

Household-size wind turbines (from the Danish term *hustandmølle*) are the largest of the small wind turbine family. As you would expect from the broad range of home sizes available, wind turbines in this class span a wide spectrum. They include turbines as small as Vergnet's GEV 4.1 with a rotor 4 meters (13 ft) in diameter; Bergey's Excel, a turbine weighing more than 1,000 pounds (463 kg); and the 29-foot (8.8 m) diameter Wind Turbine Industries model, which weighs a hefty 2,300 pounds (1,045 kg) (see figure 1-6, Household-size wind turbine).

Medium-size turbines are larger still. They have steadily grown from the commercial turbines of the early 1980s with rotors from 10 to 15 meters (30 to 50 ft) in diameter to machines approaching the megawatt size, from 50 to 60 meters (150 to 200 ft) in diameter (see figure 1-7, Medium-size wind turbine).

There's only one word to describe today's multimegawatt turbines—*giant.* When the diameter of the rotor approaches the length of a football field, it's a large wind turbine. The poor performance and noise from experimental behemoths in the early 1980s gave

Figure 1-4. Micro wind turbine. LVM micro turbine, shown here on a Danish summer cottage, uses a six-blade rotor slightly less than 3 feet (0.9 meter) in diameter and is capable of 0.05 kW. The building is occupied only during vacations.

Figure 1-5. Mini wind turbine. Foreground: Bergey 850 in operation at the Wulf Test Field in California's Tehachapi Pass. This cabin-size turbine uses a three-blade rotor 8 feet (2.4 meters) in diameter and is capable of 0.9 kW. This model has been superseded by Bergey's XL1. Background: Medium-size wind turbines at one of the world's largest wind power plants.

Figure 1-6. Household-size wind turbine. Bergey Excel in operation at a home near Tehachapi, California. The Excel uses a rotor 7 meters (23 ft) in diameter and is capable of 7 kW.

Figure 1-7. Medium-size wind turbine. Vestas V39 atop Elliott's Hill in Northern Ireland. As its name implies, the V39 uses a rotor 39 meters (128 ft) in diameter and can generate 500 kW.

wind energy a bad name that took years to erase. The success of medium-size turbines in California and northern Europe spelled the death knell for the giants until the introduction of modern megawatt-size turbines in the late 1990s (see figure 1-8, Large wind turbine).

In the next chapter we'll look at how wind energy is used today.

One last word of advice before we begin. Pay attention, ask questions, work safely, and, most importantly of all, don't believe everything you read in sales brochures or on the Internet.

Figure 1-8. Large wind turbine. Germany's experimental 3 MW Growian (Grosse Wind Energie Anlage), with a rotor 100 meters (330 ft) in diameter, was the world's largest wind turbine when it was built in the 1980s. Growian operated only a few hours at the Kaiser-Wilhelm-Koog test field before it was dismantled. Commercial turbines have now reached an equivalent size.

2

Applications—How to Use the Wind

Tout sur terre appartient aux princes, hors le vent.
(Everything on Earth belongs to princes, except the wind.)
—Victor Hugo, *La Rose de l'Infante*

Look around you. How do you currently use energy? To power your appliances, to heat and light your home, to run your factory? In almost every case, a wind turbine can be used to supplement or replace your consumption of conventional energy. Mongolian nomads use modern micro turbines to boil water for tea, the Dutch use medium-size turbines to drive their lock gates, and midwesterners use commercial wind turbines to light their schools. The possibilities are limited only by the imagination. In this chapter we'll glance at some of the different ways wind energy is being used today.

Applications for wind power can be divided into several broad categories: generating electricity at remote sites, producing electricity in parallel with the electric utility, heating, and pumping water.

Generating Power at Remote Sites

Next to their reputation for mechanically pumping water and grinding grain, wind turbines are best known for their ability to generate power off the grid at remote sites. They've distinguished themselves in this role for decades. During the 1930s, when only 10 percent of North American farms were served by electricity, literally thousands of small wind turbines were in use, primarily on the American Great Plains. These home light plants provided the only source of electricity to homesteaders in the days before rural electrification brought electricity to all.

That's not true everywhere. There may be as many as 100,000 small wind turbines in use by nomadic herdsmen in northwestern China. These small turbines (so small they can be carted on horseback from one encampment to another) are the sole source of power available on the Asian Great Plains that stretch from China to the Soviet Union (see figure 2-1, Nomadic micro turbine).

What's a remote site? There's no hard-and-fast rule. Utilities will build a power line almost anywhere if someone will pay for it. As a rule, anyone more than 0.5 mile (1 km) from an existing utility service will find it cheaper to install an independent power system than to bring in utility service.

Even today three-fourths of all small wind turbines built are destined for standalone power systems at remote sites. Some find their way to backcountry homesteads in Canada and Alaska, far from the nearest village. Others serve mountaintop

Figure 2-1. Nomadic micro turbine. Marlec's Rutland 910 at a yurt on the Mongolian steppe. The Rutland 910 is so small it can be carried on horseback. (Peter Fraenkel)

Figure 2-2. Remote telecommunications. A hybrid wind and solar power system powering a telecommunications tower on the frontier between Argentina and Chile. The 2 kW AgroLuz turbine is modeled after the 1930s-era Jacobs windcharger and uses a pitch-regulated rotor 4.1 meters (13 ft) in diameter.

telecommunications sites where utility power could seldom be justified (see figure 2-2, Remote telecommunications).

Surprisingly, an increasing number are being put to use by homeowners determined to produce their own power even though they could just as easily buy their electricity from the local utility. Bruce Simson, for example, is a vice president for strategic planning at British Columbia Hydro, a Canadian electric utility. Yet Simson has lived off the grid since the early 1990s. When the lights in nearby Vancouver went dark, he jokingly asked the BC Hydro's operations manager why he couldn't make the utility more reliable.

Because the wind is an intermittent resource, remote applications generally require some form of storage. For remote homes, where a fairly steady supply of electricity is expected, battery storage becomes a necessity. Batteries store surplus energy during windy days for later use during extended calms. In battery-charging wind systems, direct current (DC) from the batteries can be used as is, in DC appliances or inverted to alternating current (AC) like that from the utility.

At many remote sites, small wind turbines produce power at less cost than gasoline or diesel generators. At such sites, says Mike

Bergey of Bergey Windpower, wind systems are also more cost-effective than solar photovoltaics alone. If these are your only other sources of power, and if you have the wind, a battery-charging wind system may be for you.

Hybrids

In the early days of the wind revival, solar and wind proponents wore blinders. If you wanted a wind system for a remote site, the dealer would happily oblige, sizing the wind turbine and the batteries to carry your entire load. If you'd asked solar dealers to do the same, they would have covered your roof with solar cells (PV modules). Never the twain would meet.

Fortunately, that's not so today. All now agree that hybrids using both wind and solar capitalize on the other technology's assets. In many locales, wind and solar resources complement each other. Strong winds in winter are balanced by long sunny days in summer, thus enabling designers to reduce the size of each component. They've found that these hybrids perform even better when coupled with small backup generators to reduce the amount of battery storage needed.

Low-Power Applications

There are numerous applications for low-power, off-the-grid systems where battery storage isn't required. One classic application is the cathodic protection of pipelines, where a small wind turbine provides an electric charge to the surface of the metal pipe. The charge counteracts galvanic corrosion in highly reactive soils. Storage isn't needed during calm winds because corrosion is a slow process that occurs over long periods. Eventually the wind returns and again protects the exposed metal. All cathodic protection in rural areas was, at one time, provided by wind turbines. Today pipelines primarily use small PV modules for cathodic protection, but wind machines are making a comeback (see figure 2-3, Cathodic protection).

In other cases, small batteries may be necessary for proper operation of the wind-charger and for storing small amounts of charge for windless periods. Two examples are the powering of walklights and the charging of electric fences (see figure 2-4, Electric fence charging). One popular low-power application is the charging of batteries on sailboats. In yacht harbors around the world, multi-blade micro turbines such as Ampair and Marlec are a common sight. Mounted on the decks or in the rigging, the turbines maintain the boats' batteries while in dock (see figure 2-5, Onboard battery charging).

Figure 2-3. Cathodic protection. Rafael Oliva's Bergey 1500 powers a cathodic protection system for a natural gas well on the windswept Patagonian steppes near Rio Gallegos, Argentina.

Figure 2-4. Electric fence charging. This Marlec 910F is charging an electric fence at the Folkecenter for Renewable Energy in Denmark. The Marlec's battery is hidden in the grass at left.

Figure 2-5. Onboard battery charging. An Ampair 100 on the deck of a sailboat in Copenhagen's inner harbor. Thousands of micro turbines are used in marine applications worldwide.

Village Electrification

One-third of the world's people live without electricity. In China alone half the population lives without access to utility power. Many Third World nations are scrambling to expand their power systems to meet the demand for rural electrification. Most are following the pattern set by the developed world: Build new power plants and extend power lines from the cities to rural areas. With the advent of reliable hybrid power systems using wind and solar energy, however, this approach to rural electrification doesn't make as much sense today as in the past.

Developing nations will find it more cost-effective, says Mike Bergey, to install hybrid power systems rather than to stretch heavily loaded, and often unreliable, central-station power from the large cities. Though these hybrid systems generate little power in comparison to central power plants, Third World villages need little power. One kilowatt-hour of electricity provides 10 times more services in India than it does in Indiana.

Hybrid power systems featuring small wind turbines, because of their relative low cost, enable strapped governments to get power into villages quickly. As the central power system expands to these villages, the hybrid systems can be removed and sent on to even more remote villages.

The strategy works in the developed world as well. Rather than pay for line extensions to remote farms in the foothills of the Pyrenees, France pays for installation of hybrid wind and solar systems.

Interconnecting with the Utility

Some wind turbines generate electricity identical to that supplied by the electric utility: constant-voltage, constant-frequency alternating current. Since the late 1970s homeowners, farmers, and businesses in nearly all countries have been permitted to connect these wind turbines with the utility network. Literally thousands of wind turbines have been interconnected with local electric lines around the globe. For example, Bill Hopwood's 20 kW Jacobs wind turbine has been in near-continuous operation for 20 years.

Of course, commercial wind farms produce bulk electricity for the utility network in many countries. By 2003 there were more than 60,000 medium-size wind turbines in operation worldwide. Collectively they were generating more than 40 terawatt-hours (40,000 million kWh) of electricity annually.

Though the concept is technically simple, and the law affecting interconnection in many countries favorable, there are numerous obstacles. First, you must deal with the utility company and its network, a task that too often could try the patience of Job. Second, if there's a power outage, your wind machine will typically be idled as well, though there are ways around this (see chapter 11). And because your wind turbine will be competing with bulk electricity supplied by the utility, the economic requirements are more stringent than for an off-the-grid system. Electric utilities have had more than 100 years of experience producing electricity. They've had ample time to learn how to do it cheaply.

Lighting the Way

Princeton, Massachusetts, is a New England village of 3,500 souls 50 miles (80 km) west of Boston. Since 1984 Princeton's Municipal Light Department has operated a small wind farm on the southern flank of Mount Wachusett. When neighboring communities were putting their money into the ill-fated Seabrook nuclear reactor, the small public utility invested in eight Enertech E44s through a $550,000, 10-year bond. While their neighbors are still paying for their mistake, the Princeton utility is debt-free. Today the town gets 1 percent of its electricity from what has become the longest-running municipal experiment with wind energy in North America. The 40 kW Enertechs were large turbines for the time, and they have not been without problems. Nevertheless, the wind farm has generated 3 million kWh through the year 2000, and has become such an integral part of the community that logos on the sides of the light department's trucks proudly depict the town's wind turbines.

How It Works

There are two types of wind turbines suitable for producing utility-compatible power: those that use induction (or asynchronous) generators, and those that use inverters. (These will be discussed more fully in chapter 8.) Regardless of which is used, their interconnection with the utility is the same. Cable from the wind turbine is connected to terminals within your utility service panel or connected directly to the utility's transformer. In effect, the wind turbine becomes a part of your home's electrical circuit not unlike that of any large electric appliance.

If the wind turbine is on your side of the utility's kilowatt-hour meter, the wind system reduces your consumption of utility-supplied electricity, regardless of how you use electricity. The wind turbine will power your clock, your stereo, your refrigerator, or your lights. If you're a farmer, it will run your milkers or your feeders. In short, you'll use wind-generated electricity wherever you presently use utility power.

Commercial wind farms produce bulk electricity for the utility network in countries around the globe.

When the wind is blowing, the wind turbine produces electricity that flows to your service panel. From there it seeks out those circuits where electricity is being consumed. With electricity it's first come, first served.

Public-Spirited

One advantage of wind energy is its accessibility to those in all walks of life. Installing a wind turbine isn't limited to large corporations with access to Wall Street's financial clout. Individuals and small groups can make a difference, both in their communities and in the greater world beyond, by working with wind energy.

Public-spirited. Two of three wind turbines donated to the community of Nevada, Iowa. Shown here are Wind World 200 kW turbines installed near the high school's football field.

Maybe it's their Scandinavian ancestry, or maybe it's the connectedness of small-town life; whatever the reason, midwestern citizens have pioneered the use of wind turbines at public institutions such as schools and hospitals.

There's no better example of this than Nevada, Iowa. Located north of Des Moines, and not far from the university town of Ames, Nevada is a bucolic village of 6,000. Here Harold and Marjorie Fawcett and Harold's sister Josephine Tope donated three wind turbines to the community. As a result of their gift, two 200 kW Wind World turbines were installed at the local high school, and one Vestas V27 was installed at the town's sewage treatment plant to produce electricity for the community hospital.

Playing with wind. Nordex 600 kW turbine on the playground at the public middle school in Forest City, Iowa. This turbine uses a rotor 43 meters (140 ft) in diameter.

Scattered across the Midwest are a number of other wind turbines installed at public schools. Several are in Iowa, including Spirit Lake, Akron-Westfield, and Forest City. The turbines were installed as part of a program to help schools cut their utility bills. They have done that, but they have also taught a new generation that wind energy works and can be a part of the community.

Electricity will flow to the first circuit where it's needed. If more energy is being generated than can be used by the first circuit, it will flow to the next, and so on. When the wind machine can't deliver as much energy as is needed, the utility makes up the difference. There are no fancy electronics controlling which circuit gets what. It's all accomplished silently and effortlessly. That's the beauty of electricity.

If you're not using electricity when the wind machine is operating, or if you're not using as much as it's generating, the excess flows from your service panel through the utility's kilowatt-hour meter and out to the utility's lines. In some cases the utility will permit you to run your kilowatt-hour meter backward. It seems mysterious, but it works. And it does so neatly, cleanly, and without fuss.

Farming the Wind

Under the right conditions, for example, when the utility pays a sufficient price for wind-generated electricity, you may be able to farm the wind for profit. You can do so by selling surplus generation from a wind turbine connected to your farm, home, or business. Or you could install a group of wind turbines intended solely for generating electricity for sale to the utility. The latter has become the most successful use of wind energy in history (see figure 2-6, Wind power plant).

A commercial wind farm, or wind power plant, is nothing more than a large-scale version of a wind turbine interconnected with an electric utility on the customer's side of the kilowatt-hour meter. But rather than meeting the domestic demand of a home or business, all the electricity generated by these wind power plants is delivered to the utility.

In the early 1980s some small wind turbines originally designed for farms or small businesses found their way into California's mushrooming wind industry. Literally thou-

Figure 2-6. Wind power plant. A linear array of modern wind turbines along a canal in the Netherlands. The wind turbines supply bulk electricity to the utility network served by the giant gas-fired power plant in the background.

sands of such wind machines were installed on California wind farms (see figure 2-7, The dawn of modern wind power).

These were wind turbines that just a few years before were being installed in backyards in the United States and Denmark. The biggest machines of that era produced about 50 kW, with rotors from 13 to 15 meters (44 to 50 ft) in diameter. Today commercial wind turbines are 10 to 30 times more powerful.

While most wind-electric generation in North America is produced by giant wind power plants, this is not the case in northern Europe. Denmark's and Germany's experiences with wind energy are vastly different from that of North America.

Two-thirds of the wind-generated electricity in Denmark and Germany is produced by wind turbines in small groups or clusters. Many homeowners, farmers, and small businesses in northern Europe operate their own medium-size wind turbine or share in the operation of a cooperatively owned group of turbines.

> Many homeowners, farmers, and small businesses in northern Europe operate their own medium-size wind turbine or share in the operation of a cooperatively owned group of turbines.

There are other differences as well. While North Americans were erecting 5 kW and 10 kW wind turbines in their backyards during the early 1980s, the Danes were installing 55-kilowatt machines in theirs. Even today Danish or German farmers will install a 900 kW turbine on their land, while their Yankee counterparts will be limited to a wind turbine of less than 100 kW, often less than

Figure 2-7. The dawn of modern wind power. A Bonus 55 kW on the shores of Denmark's Limfjord in northwest Jutland. One of the Danish manufacturer's first production units, this turbine was still operating after more than 20 years in service.

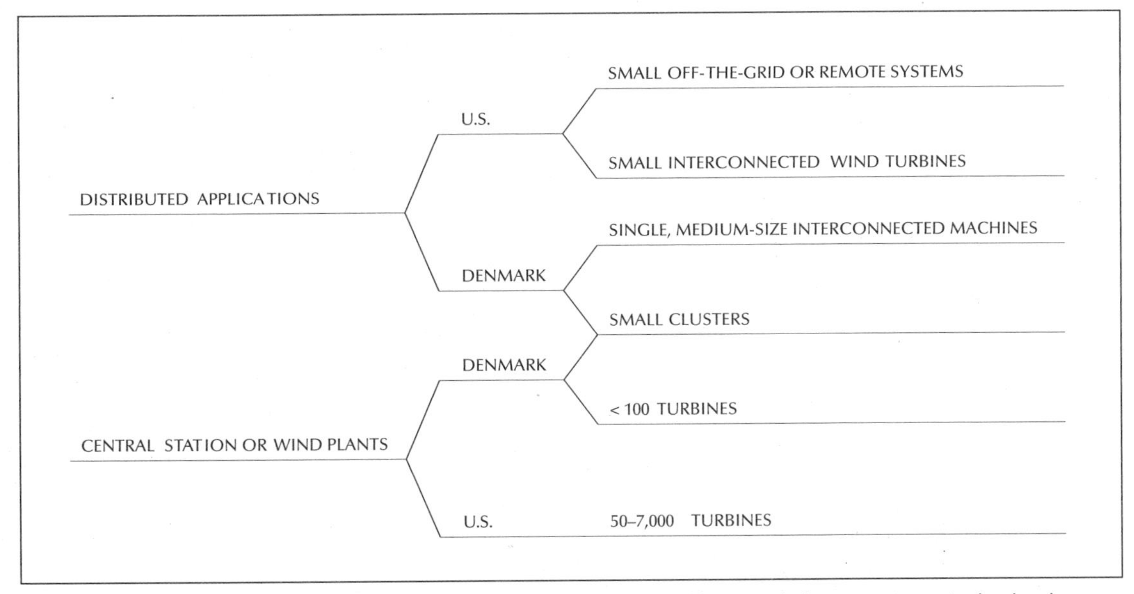

Figure 2-8. Models of wind development. Europeans have followed a different path than Americans in the development of wind energy. Most commercial wind turbines in northern Europe are installed in small clusters, compared to the massive wind farms found in North America.

10 kW in size (see figure 2-8, Models of wind development).

As Denmark and Germany have decisively shown, medium-size wind turbines need not be limited solely to commercial wind power plants; they can be successfully used to power farms, homes, and businesses.

Heating

In temperate climates heating makes up most of a home's energy demand. In many areas there is a good correlation between the availability of wind energy and the demand for heat. It's not surprising, then, that there have been numerous attempts to use wind machines strictly for heating. Advocates of the wind furnace concept believed that the same winter winds that rob a house of its warmth could be used for heating. They argued that generating low-grade energy with either an alternator or a mechanical churn would be less costly than producing the same amount of utility-compatible electricity. It's simpler, proponents said, than trying to produce the high-grade electricity demanded by the utility.

Unfortunately it never worked out that way. Both Danish and American companies tried to commercialize the concept during the 1970s and early 1980s, and again in the 1990s. None has proven popular. Experience has shown that in most cases it's cheaper and easier to interconnect the wind system directly with the utility than to generate heat and store the surplus for windless periods. Often it's more cost-effective to produce a high-grade form of electricity that can be used for all purposes, including home heating if desired, than to build a wind turbine that can only be used for one function (see figure 2-9, UMass Wind Furnace).

There are several examples of household-size turbines being used to provide supplemental heating. For years Bergey Windpower has marketed its Excel model to midwesterners in "all-electric" homes, where the wind turbine will provide some or all of the heating load. But the Excel produces utility-compatible electricity that can be used anywhere in the home—or sold back to the utility. Similarly, French manufacturer Vergnet has had a long-running installation on Britain's windswept Orkney Islands, generating electricity that is used for heating. Like Bergey's Excel, Vergnet's turbines

Figure 2-9. UMass Wind Furnace. A circa-1980 experimental wind turbine at the University of Massachusetts designed for heating the experimental building nearby. Several firms tried to commercialize the concept, with little success.

Figure 2-10. Wind pumping. *Foreground*: An American farm windmill on a Colorado ranch near the Wyoming state line. *Background*: 700 kW NEG-Micon turbine—part of the Ponnequin wind power plant.

can also be used to produce utility-compatible power as well as electricity for heating.

Pumping Water

Wind machines have historically been used to pump water, and pumping water remains an important application of wind energy today in both the developed and the developing worlds. The American farm windmill, known as the Chicago mill in some parts of the world, dependably pumps low volumes of water from shallow wells. These multiblade wind pumps are still extensively used for watering remote stock tanks on North America's Great Plains, Argentina's pampas, Australia's outback, and South Africa's veldt. There are probably more than a million of these wind pumps still in use worldwide (see figure 2-10, Wind pumping).

Researchers at West Texas A&M University's Alternative Energy Institute and the U.S. Department of Agriculture's Agricultural Research Service have made major advances in water-pumping technology, first with wind-assisted irrigation in the 1980s and then with wind-electric pumping systems in the 1990s. In cooperation with small wind turbine manufacturers, these researchers have developed pumping systems that couple modern electronics to small windchargers that eliminate the need for cumbersome batteries. Under certain conditions, these wind–electric pumping systems will deliver more water at lower cost than the traditional farm windmill.

Whether you want to pump water, heat your house, or sell power to the utility, the wind speeds at your site will largely determine how successful you are at harnessing the wind. The speed of the wind and its relationship to the energy that a wind turbine can capture are the subjects of the next two chapters.

3

Measuring the Wind

The way of the wind is a strange, wild way.
—Ingram Crockett, *The Wind*

"Hi, I want a windmill."

"You sure?

"Yeah, I can't wait to tell the utility to go you-know-where."

"Hmm. I'll bet they'll be glad to hear that. How much wind do you have?"

"Wind? Oh, we've gots lots of wind. It's always windy here."

More than once, new owners of wind systems—as well as not a few would-be wind farmers—have learned an expensive lesson, and one that may seem patently obvious. Wind generators with little wind are like dams with little water. They don't work, or at least they don't work very well. There's wind everywhere, but not everywhere has enough. Ample wind is a prerequisite for successfully siting a wind machine, and the more the better. But just how much is enough? How do you know whether the wind over your site is sufficient? If you're living on the west coast of Denmark or in the Texas Panhandle, you probably have enough wind.

Cup anemometer. Knowing the wind is essential to working with wind energy. Here a cup anemometer is used to measure wind speed in 1980. In the background is the 640 kW Nibe B turbine in Denmark, one of the world's most successful early research turbines.

Unfortunately few people live where the wind has such a well-deserved reputation for ferocity. Most of us need a better description of the wind than "It's always windy here."

In this chapter we'll discuss the wind, what it is, how local climate and terrain affect it, and how it changes over time. We'll explore the meaning of wind power and how wind speed and power increase with height. You'll learn where to find wind information for your area and how to determine the winds at your site. For those with an aversion to math, easy-to-use charts or tables are presented wherever a formula appears. As mentioned earlier, our objective is to determine if we have enough wind to put wind energy to work.

Wind—What Is It?

The atmosphere is a huge, solar-fired engine that transfers heat from one part of the globe to another. Large-scale convective currents set in motion by the sun's rays carry heat from lower latitudes to northern climes. The rivers of air that pour across the surface of the earth in response to this global circulation are what we call wind, the working fluid in the atmospheric heat engine.

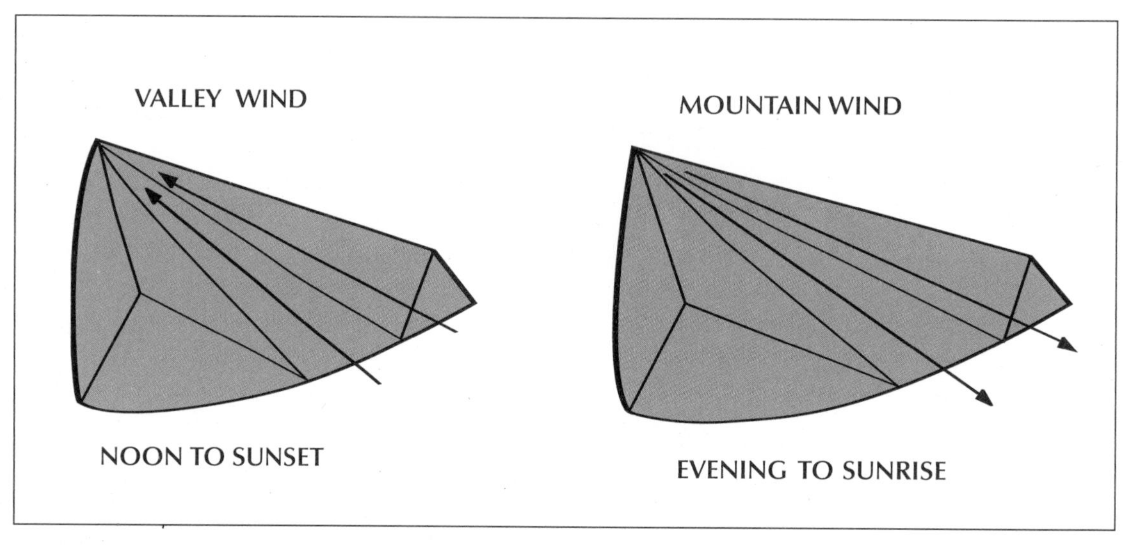

Figure 3-1. Mountain–valley winds. Warm air rises up the valley during the afternoon and descends during the evening. (Battelle PNL)

When the sun strikes the earth, it heats the soil near the surface. In turn, the soil warms the air lying above it. Warm air is less dense than cool air, and, like a helium-filled balloon, it rises. Cool air flows in to take its place and is itself heated. The rising warm air eventually cools and falls back to earth, completing the convection cell. This cycle is repeated over and over again, rotating like the crankshaft in a car, as long as the solar engine driving it is in the sky. Thermals, rising currents of warm air that boil up over land during bright daylight hours, are as much sought after for soaring by humans as by hawks. The cumulus clouds of summer are a sign of the convective circulation that causes winds to strengthen in late afternoon. If you're a pilot, you probably prefer to fly in the early-morning hours when winds are light. On the other hand, if you're making a trip to inspect a wind machine, the midafternoon—when you're more likely to find it running—is better.

Winds are also stronger and more frequent along the shores of large lakes and along the coasts because of differential heating between the land and the water. During the day, the sun warms the land much quicker than it does the surface of the water. (Water has a higher specific heat and can store more energy without a change in temperature than can soil.) The air above the land is once again warmed and rises. Cool air flows landward, replacing the warm air, creating a large convection cell. At night the flow reverses as the land cools more quickly than the water. In late afternoon, when the sea breezes are strongest, winds can reach 10 to 15 mph (5 to 7 m/s) on an otherwise calm day. Land–sea breezes are most pronounced when winds due to large-scale weather systems are light. The influence of land–sea breezes diminishes rapidly inland and is insignificant more than 2 miles (3 km) from the beach.

The winds along the shore are also higher because of the long unobstructed path (fetch) that the wind travels over the water. Hills, trees, and buildings block the wind on land. The shores of the Great Lakes and the cost of northern Europe have average wind speeds approaching 12 to 15 mph (6 to 7 m/s), partly due to these effects.

The mountain–valley breeze is another example of local winds caused by differential heating (see figure 3-1, Mountain–valley winds). Mountain–valley breezes occur when the prevailing wind over a mountainous region is weak and there's marked heating and cooling. These breezes are found principally in the summer months when solar radiation is the strongest. During the day, the sun heats the floor and sides of the valley. The warm air rises up the slopes and moves upstream. Cooler air is drawn up from the plains below, causing a valley breeze. At night the situation reverses and the mountains cool more quickly than the lowlands. Cool air cascades down the slopes and is channeled through the valley to the plains. Nighttime mountain breezes are generally stronger than valley breezes, with winds reaching speeds of 25 mph (11 m/s) in valleys with steeply sloping floors located between high ridges or mountain passes.

Mountain–valley breezes can be reinforced by the prevailing winds when the two flow in the same direction. One place where the effects of channeling are pronounced is on North America's Pacific coast, where onshore winds are funneled into narrow gorges through the mountains. Convective flow can reinforce the funneling effect when heating on interior deserts causes large temperature differences between the coast and the interior. Average wind speeds of more than 20 mph (9 m/s) are typical, with winds in some seasons averaging well above that.

Many of the windy passes through mountain chains on the West Coast of the United States lie east to west. The tremendous wind potential in the Columbia River Gorge, east of Portland, Oregon, is one such example. Another is the Altamont Pass, east of San Francisco, where cool air off the Pacific Ocean rushes over the low pass into the hot interior of the San Joaquin Valley. The San Gorgonio Pass near Palm Springs, California, is similar. This sea-level pass through the San Bernardino Mountains channels cool coastal air onto the blazing Sonoran Desert. Similar flows occur from the San Joaquin Valley across the Tehachapi Mountains onto the Mojave Desert. Winds through the Tehachapi Pass are more subject to the passage of cold fronts than are California's other passes. The storms associated with the cold fronts push masses of dense air through the Tehachapi Pass, sometimes causing winds up to 100 mph (50 m/s).

Long ridges across the path of the wind, like Cameron Ridge in the Tehachapi Mountains, also enhance the flow over the summit (see figure 3-2, Increase in wind speed over a ridge). Wind speeds may double as the flow accelerates up the gradual slope of a long ridge. As many California wind farmers have learned the hard way, this enhancement occurs only in the last third of the slope near the crest. Turbines lower down the slope perform dismally, as do those on the leeward side.

A similar terrain enhancement of the wind has been found on the island of Oahu in Hawaii. There the northeasterly winds sweep around the end of a long ridge that runs the length of the island from north to south. The winds are accelerated as they pass over the

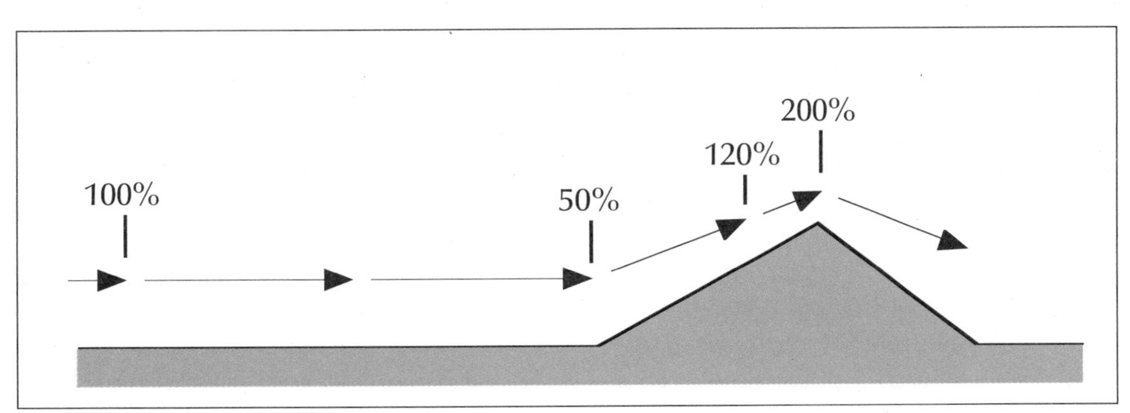

Figure 3-2. Increase in wind speed over a ridge. Wind speed increases near the summit of a long ridge lying across the wind's path. Though there can be some acceleration of the flow on the flanks, wind speeds typically are lower at the foot of the ridge. (Battelle PNL)

ridge, but more so when they pass around the ends. Kahuku Point, at the northernmost end of the ridge, has long been a prime site for wind development because of this local effect.

Mount Washington, in the White Mountains of New Hampshire, is one of the few areas in the eastern United States where terrain enhancement is well known. In fact, this phenomenon was first observed on Mount Washington and the Green Mountains of Vermont. Mount Washington has an average speed of approximately 38 mph (17 m/s), the highest average wind speed on the East Coast. On April 12, 1934, an observatory on the summit recorded the highest wind speed ever measured, 231 mph (103 m/s).

The effect of terrain on wind speeds within the region has been known for some time and has been used to advantage in the past. The Green Mountains of Vermont form a long northeasterly ridge that sits astride the prevailing winds just east of the long unobstructed fetch across Lake Champlain. Because of the potential speed enhancement, Palmer Putnam chose the Green Mountains as the site for the experimental Smith-Putnam wind turbine during the 1940s. Unfortunately the same terrain features that enhance wind speeds also create turbulence that can wreak havoc on wind machines. The 1,500-kilowatt Smith-Putnam turbine eventually succumbed to just such turbulence atop Grandpa's Knob after only a few years of operation.

Mountains and ridges offer higher winds for reasons other than channeling. Prominent peaks often pierce temperature inversions that can blanket valleys and low-lying plains. Temperature inversions cause a stratification of the atmosphere near the surface. Above the inversion layer normal air flow prevails, but below the winds are stagnant. The air beneath an inversion layer may be completely cut off from the air circulation of the weather system moving through the region. Temperature inversions are common in hilly or mountainous terrain such as southern California and western Pennsylvania, both areas notorious for their air pollution episodes. In northern latitudes inversions are common during the fall and winter.

The inversion layer itself may accelerate the wind. The wind above a temperature inversion, essentially a giant lake of stagnant air, blows across the surface of the inversion unimpeded by the hills, trees, buildings, and other features that often hinder the wind. Ridgetops may possess not only more frequent winds, but also stronger winds, because they may break through this inversion layer. The wind can skip across the inversion layer as though across the surface of a lake until it strikes an exposed mountaintop.

There are numerous regional winds around the globe that can also have a powerful influence on successfully siting a wind turbine. These winds, like the powerful chinook that roars down the east side of the Rocky Mountains, are due to infrequent local meteorological and geographic anomalies. The Santa Ana in southern California, for example, results from high pressure systems that occasionally move over the Basin and Range province of the American Southwest. But whether it's the föhn in the Alps, the sirocco sweeping across the Mediterranean from North Africa, the legendary mistral of southern France, or the tramontana howling out of the eastern Pyrenees, these winds are a force to be reckoned with. They can power wind turbines or destroy them. In the early days of the U.S. Department of Energy's wind program, chinooks with winds sometimes above 100 mph (50 m/s) wreaked havoc on the flimsy experimental turbines at the Rocky Flats test center near Denver, Colorado. Today wind farmers in Tehachapi harness the once feared Santa Ana winds. And in southern France operators of small wind plants eagerly look forward to the tramontana that sends tourists scurrying for cover at nearby resorts.

Wind Speed and Time

The wind is an intermittent resource: calm one day, howling the next. Wind speed and direction vary widely over almost all measuring periods. Because wind speed fluctuates, it becomes necessary to average wind speed over a period of time. That most commonly used is the average speed over an entire year.

The average annual wind speed itself is not constant. It varies from year to year. The average speed can change as much as 25 percent from one year to the next. This can amount to more than 2 mph (1 m/s) in a moderate wind regime where an average of 10 mph (5 m/s) is the norm.

Average wind speeds vary by season and by month. "March roars in like a lion and goes out like a lamb" is a popular adage signifying that early spring is windy while summer is not. For much of North America's interior, winds are light during summer and fall and increase during winter, reaching their maximum in spring (see figure 3-3, Historical monthly average wind speed).

When we looked at the differential heating of the earth's surface and its effects on local winds, we saw that wind speeds often increased during late afternoon after convective circulation had been set in motion. This tells us that wind speeds vary by time of day, not only because of changing weather, but also because of convective heating. The effects of local convective winds are greater during summer when winds are light and solar heating is strong, but they also occur throughout the year. Convective circulation leads to a dramatic difference between wind speeds during daylight hours and those at

Wind Speed Units

There are several scales of wind speed in common use. The old imperial scale of miles per hour (mph) is still used in the United States. Most meteorologists use meters per second (m/s). Many sailors continue to use nautical miles per hour (knots). Some national weather services use kilometers per hour (km/h). While conversions from one unit of measurement to another are given in the appendixes, there are some simple rules of thumb that can help you keep them straight. There are slightly more than 2 mph for every m/s. Similarly, there is slightly more than 1 mph for every knot, and slightly more than 0.5 mph for every km/h.

1 knot	= 1.15 mph
1 m/s	= 2.24 mph
1 km/h	= 0.621 mph

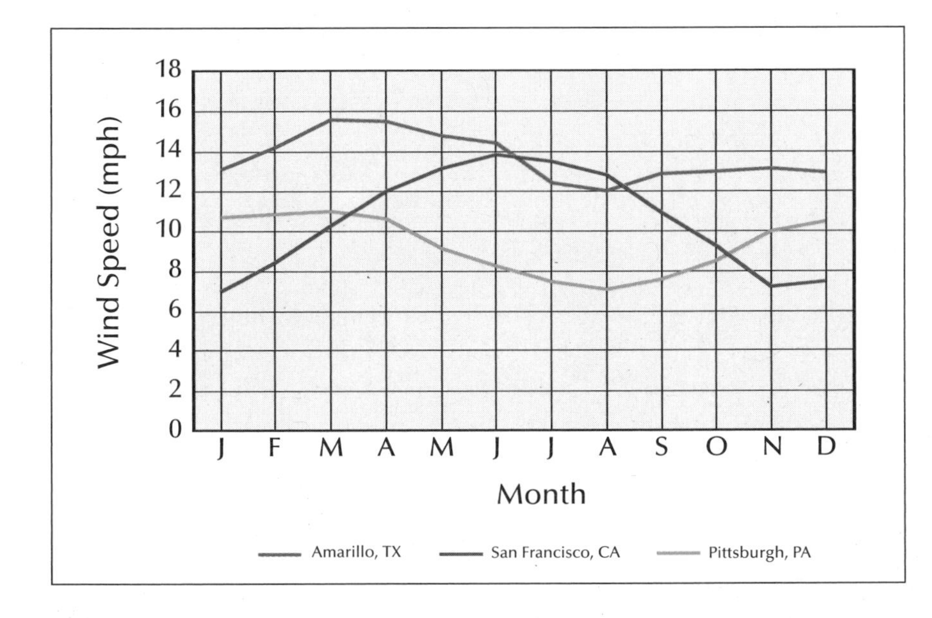

Figure 3-3. Historical monthly average wind speed. For the interior sites of Pittsburgh, Pennsylvania, and Amarillo, Texas, wind speeds increase during winter and spring. Wind speeds are strongest near San Francisco in summer, when seasonal flows of cool marine air rush toward the blistering San Joaquin Valley. It's this seasonal flow that drives the wind power plants in the Altamont Pass east of San Francisco. The annual average wind speed at Amarillo is 13.7 mph (6.1 m/s); at Pittsburgh, 9.3 mph (4.2 m/s); and at San Francisco, 10.5 mph (4.7 m/s).

The Beaufort Scale

One of the more unusual measures of the wind is the Beaufort scale of wind force. Originally developed for use at sea, the scale didn't initially refer to wind speed—only to an arbitrary scale of "force." Since its introduction, the scale has been adapted for use on land and to reflect a range of wind speeds in various units. It's a handy tool for sharpening your understanding of the wind. The scale was devised during the Napoleonic wars by British admiral Sir Francis Beaufort. The scale is seldom used by landlubbers in North America, but it is still common today to hear British weather forecasts in terms of wind force, such as "a Force 10 gale will strike the coast of Scotland." The beauty of the Beaufort scale is its simplicity. You don't need sophisticated instruments to gauge the strength or "force" of the wind because it uses observations of common objects such as smoke

Beaufort Scale of Wind Force

Strength or "Force"	Mean Speed (knots)	(m/s)	Range of Speed (knots)	(mph)	(m/s)	Wind Pressure (lb/ft²)	(N/m²)
0	0	0	<1	<1	0–0.2	0.00	0
1	2	1	1–3	1–3	0.3–1.5	0.01	1
2	5	3	4–6	4–7	1.6–3.3	0.08	4
3	9	5	7–10	8–12	3.4–5.4	0.27	13
4	13	7	11–16	13–18	5.5–7.9	0.57	27
5	19	10	17–21	19–24	8–10.7	1.2	58
6	24	12	22–27	25–31	10.8–13.8	2.0	93
7	30	15	28–33	32–38	13.9–17.1	3.0	146
8	37	19	34–40	39–46	17.2–20.7	4.6	222
9	44	23	41–47	47–54	20.8–24.4	6.6	313
10	52	27	48–55	55–63	24.5–28.4	9.2	438
11	60	31	56–63	64–72	28.5–32.6	12	583
12	68	35	64–71	73–82	32.7–36.9	16	748
13	76	39	72–80	83–92	37–41.4	20	935
14	85	44	81–89	92–103	41.5–46.1	24	1,169
15	94	48	90–99	104–114	46.2–50.9	30	1,430
16	104	53	100–108	115–125	51–56	37	1,750
17	114	59	109–118	126–136	56.1–61.2	44	2,103

Source: Adapted from *The Generation of Electricity by Wind Power* by E.W. Golding.

rising from a chimney or the swaying of trees. Beaufort devised the scale from 0 to 12, but it was extended in 1955 by the U.S. Weather Bureau to Force 17 . The Beaufort scale of wind force is also presented in the appendix with common descriptions in several languages.

General Description	
Description	On Land
Calm	Smoke rises vertically, flags hang limp
Light air	Smoke drift indicates direction, wind vanes don't move, but flags begin to unfurl
Light breeze	Wind felt on face, leaves rustle, wind vanes begin to move, flags unfurl and begin to extend
Gentle breeze	Leaves and small twigs in constant motion, light flags extended
Moderate breeze	Raises dust, leaves, and loose paper; small branches move, dry sand begins to drift
Fresh breeze	Small trees in leaf begin to sway, crested wavelets form on inland waters
Strong breeze	Large branches in motion, overhead wires begin to whistle; umbrellas used with difficulty
Moderate gale	Whole trees in motion; resistance felt when walking against the wind
Fresh gale	Twigs break off trees, wind generally impedes progress when walking
Strong gale	Slight damage to roof and homes, chimney pots and slates damaged
Whole gale	Trees uprooted, considerable structural damage
Storm	Widespread damage
Hurricane	Devastation

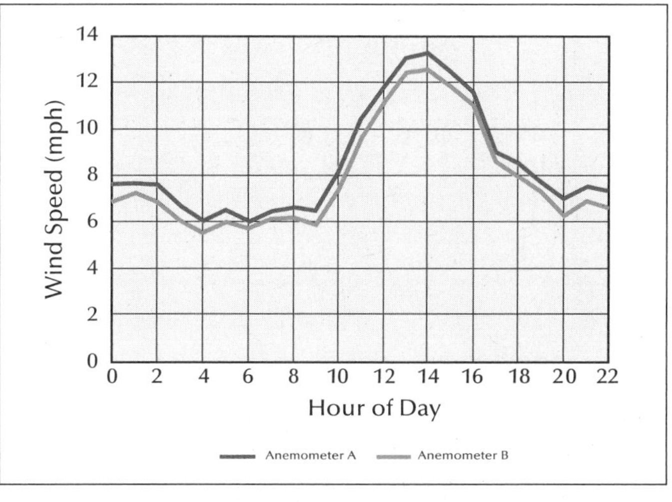

Figure 3-4. Diurnal hourly wind speed for Wulf Test Field. Measurements from two anemometers during December 1999 exhibit a typical diurnal pattern: Wind speeds increase in the afternoon due to convective heating. This pattern is evident even at a windy site in California's Tehachapi Pass. The speed at anemometer A is slightly more than 0.5 mph greater than that at anemometer B.

night (see figure 3-4, Diurnal hourly wind speed for Wulf Test Field).

The diurnal difference in wind speeds is less marked in winter because there's less convective circulation. During winter and spring, winds are dominated by storm systems. It's the recurring storms of winter that push up the average wind speeds across the Midwest and northeastern United States. Storms formed in the Gulf of Alaska are the source for the winds that funnel through the Tehachapi Pass in California.

Power in the Wind

One of the most important tools in working with the wind, whether designing a wind turbine or using one, is a firm understanding of what factors influence the power in the wind. For the sake of thoroughness, we'll start right at the beginning. As E. W. Golding said so succinctly in his classic *The Generation of Electricity by Wind Power,* "Wind is merely air in motion." Good so far. The air about us has mass (think of it as weight if you're unfamiliar with the term). Though extremely light, it has

Wind Speed Notation

In many previous works, wind speed has been indicated variously as (V) for velocity and, more recently, as (S) for speed. In this book (V) is used for wind speed. You will, however, find (U) as well as the lowercase (v) in books for professional engineers and meteorologists.

substance. A bucket of air is similar to a bucket of water, but the bucket of air is lighter. It has less mass than does the water, because air is less dense. It's more diffuse. Like any other moving substance, whether it's a bucket of water plummeting over Niagara Falls or a car speeding down the autobahn, this moving air contains kinetic energy. This energy of motion gives the wind its ability to perform work.

When the wind strikes an object, it exerts a force while attempting to move it out of the way. Some of the wind's kinetic energy is given up or transferred, causing the object to move. When it does, we say the wind has performed work. We can see this when leaves skid across the ground, trees sway, or the blades of a wind turbine move through the air.

The amount of energy in the wind is a function of its speed and mass. At higher speeds more energy is available, in much the same way that a car on the highway contains more energy than a car of equal size it passes. It takes more effort—energy—to stop a car driven at 70 mph than it does one at 50 mph. Likewise, heavy cars contain more energy than light cars traveling at the same speed. This relationship among mass, speed, and energy is given by the equation for kinetic energy where (m) represents the air's mass and (V) is its velocity, or speed in common parlance.

$$\text{Kinetic Energy} = 1/2\, mV^2$$

The air's mass can be derived from the product of its density (ρ) and its volume. Because the air is constantly in motion, the volume must be found by multiplying the wind's speed (V) by the area (A) through which it passes during a given period of time (t).

$$m = \rho AVt$$

When we substitute this value for mass into the earlier equation, we can find the kinetic energy in the wind:

$$\text{Wind Energy} = 1/2\, \rho AVtV^2 = 1/2\, \rho AtV^3$$

We've gone through this derivation for a reason. Equations are the language of science, and in this terse, compact script the fundamentals of wind energy are precisely stated. But before we go over each of them, let's complete one more step.

Power, as you may remember, is the rate at which energy is available, or the rate at which energy passes through an area per unit of time:

$$P = 1/2\, \rho AV^3$$

Power (P), we've now learned, is dependent on air density, the area intercepting the wind, and wind speed. Increase any one of these and you increase the power available from the wind. But most importantly, slight changes in wind speed produce significant effects on the power available.

Power Density

At sea level with a temperature of 15°C (59°F), where air density is 1.225 kg/m^3, then power density (P/A) in watts per area of the wind stream is:

$P/A = 0.6125\,V^3$, where P is in W, A in m^2, and V is in m/s

$P/A = 0.05472\,V^3$, where P is in W, A in m^2, and V is in mph

$P/A = 0.00508\,V^3$, where P is in W, A is in ft^2, and V is in mph

Air Density

The wind is a diffuse source of power because air is less dense than most common substances. Water, for example, is 800 times more dense than air. Don't be misled, though; the wind can pack quite a punch, as anyone who has survived a hurricane will tell you.

International Standard Atmosphere

The following are standard atmospheric conditions used internationally to define the reference conditions for wind turbine performance.

Sea-level pressure (p)	= 29.92 in Hg
	= 760 mm Hg
	= 1013.25 mb, or hPa
	= 1.01325×10^5 N/m^2 or Pa
Sea-level temperature (T)	= 69°F
	= 15°C
	= 288.15 K
Air density (ρ)	= 1.225 kg/m^3
	= 0.07651 lbs/ft^3

Air density decreases with increasing temperature. Air is less dense in summer than in winter, varying 10 to 15 percent from one season to the next. On an annual average, seasonal changes in temperature have only a modest influence on the power in the wind. Changes in elevation, however, can produce severe changes in air density.

For our performance calculations we'll assume that we're in an area where conditions approximate those near sea level and the temperature is about 59°F (15°C). Of course, if you plan to install a wind machine at the North Pole or on top of a mountain these assumptions don't apply.

Although air density is one of the critical factors influencing the power available in the wind, the assumption of standard sea-level conditions is sufficient for most applications. Still, overlooking the effect that temperature and elevation have on air density can lead to unpleasant surprises. When Rick Solinsky installed his new 2 kW Proven wind turbine in the Sierra Nevada near Lake Tahoe, he was disappointed in its performance. The turbine was significantly underperforming, and he wanted to know why. Was it the inverters? Was it something he'd done improperly during installation? The answer was all around him. The air density at Solinsky's 6,000-foot (1,800 m) site was much lower than that at sea level. There was simply less for a wind turbine to work with for a given wind speed at his elevation than at sea level.

Air density is inversely related to elevation and temperature: It decreases with increasing elevation or increasing temperature. This simple statement obscures the exponential relationship between air density and elevation. People who live within a few hundred feet or a few hundred meters of sea level can often safely ignore elevation. This is true for most of North America east of the Rocky Mountains as well as the lowlands of northern Europe. But once we move into the Rocky Mountains of western North America or, say, the Pyrenees Mountains between France and Spain, elevation has a profound effect on air density and wind turbine performance. For example, the air density at 5,000 feet (1,500 meters) atop Cameron Ridge in California's Tehachapi Pass is about 15 percent less than that at sea level.

Changes in temperature produce a smaller effect on air density than elevation, yet they are not insignificant. For example, increasing the temperature from 59°F (15°C)—the standard temperature used when estimating wind turbine performance—to 86°F (30°C) decreases air density 5 percent, a difference that could be critical in commercial applications. Canadian meteorologist Jim Salmon adds that temperature extremes from - 60°C (- 76°F) in the Arctic north to 40°C (104°F) in the prairie provinces are not unheard of. Changes in air density under such conditions are not merely academic questions. Some generators in wind turbines installed in the far north of Canada have burned out because the air can be as much as 27 percent more dense than at the standard temperatures for which the turbine was designed.

Thus it behooves us to carefully examine air density whenever conditions at our site vary much from those of the international standard atmosphere.

Air Density

For the sake of simplicity, we'll use metric units to find air density in kg/m^3. While the following calculations are somewhat involved, they are presented graphically in the accompanying figures. The calculations are also presented in tabular form in the appendixes. The tables in the appendixes include the equations in a form that can be easily entered into a personal computer spreadsheet.

Air density, as a function of changes in temperature and pressure, can be found using the gas law:

$$\rho = p/RT$$

where p is air pressure in N/m^2 or Pascal, R is the gas constant, 287.04 J/kgK, and T is temperature in Kelvin. For example, the air density at the standard temperature of 15°C (59°F) at sea level is:

$$\rho = 101{,}325 \,/\, [287.04 \times (273.15 + 15)]$$

$$\rho = 1.225 \text{ kg/m}^3$$

Estimating the effect that changes in temperature produce on air density is complicated by the normal decrease in temperature with elevation. Temperature typically decreases 6.5°C per 1,000-meter increase in elevation. This is the normal or environmental lapse rate (Γ).

From the hydrostatic equation we can find the air pressure (p) at a given elevation. Once pressure is known, we can find air density by incorporating the change in temperature caused by the lapse rate:

$$p = p_o \, [(T_o - \Gamma Z)/T_o]^{g/R\Gamma}$$

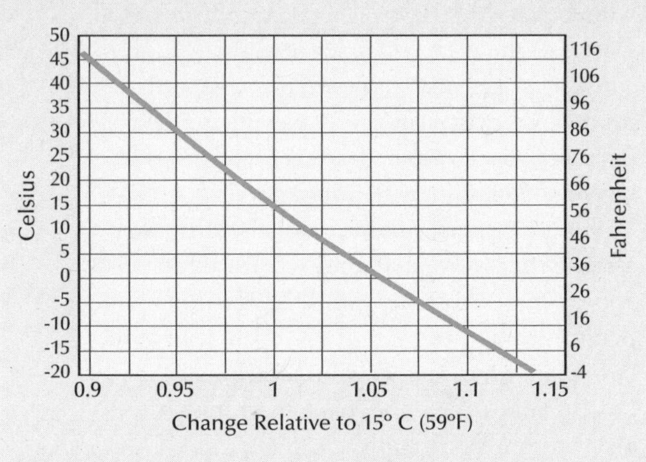

Change in air density with temperature. Percent change relative to conditions at sea level and 15°C (59°F).

where p_0 is the sea level air pressure, T_o is 288.15 K, Z is the elevation above sea level, and g is the acceleration due to gravity of 9.807 m/s^2.

Air pressure is often given in millibars (mb), 1/100 of a Pascal. Therefore at an elevation of 1,500 meters in the Tehachapi Pass, the pressure in mb is

$$p = 1013.25 \times \{[288.15 - (6.5/1{,}000 \times 1{,}500)]/288.15\}^{[9.807/(287.04 \times 6.5/1000)]}$$

$$p = 845.55 \text{ mb}$$

The normal temperature at an elevation of 1,500 meters (T_Z) is:

Swept Area

Power is directly related to the area intercepting the wind. Wind turbines with large rotors intercept more wind than those with smaller rotors and, consequently, capture more power. Doubling the area swept by a wind turbine rotor, for example, will double the power available to it. This principle is fundamental to understanding wind turbine design. Knowing this, you can quickly size up any wind machine by noting the dimensions of its rotor (see figure 3-5, Relative size of small wind turbines).

Consider a conventional wind turbine whose blades spin about a horizontal axis. The rotor sweeps a disk the area of a circle

$$A = \pi r^2$$

where (A) is the area and (r) is the radius of the rotor (approximately the length of one blade). This formula gives us the area of the wind stream swept by the rotor of a conventional wind turbine. Swept area is proportional to the square of the rotor's radius (or diameter). Relatively small increases in blade

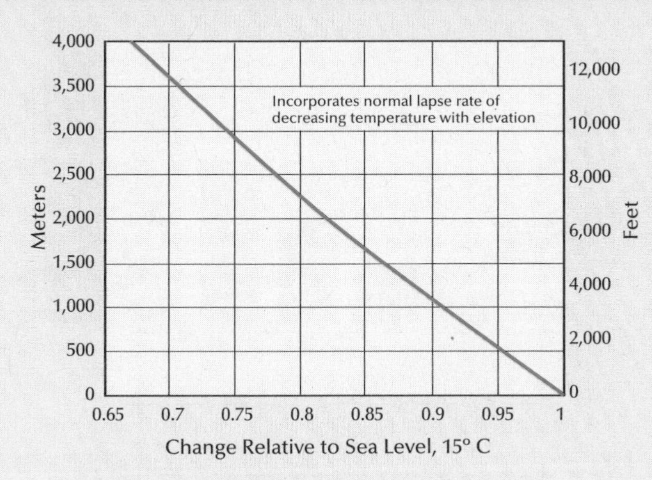

Change in air density with altitude. Percent change relative to conditions at sea level and 15°C (59°F). The chart incorporates the normal lapse rate of 6.5°C per 1,000-meter increase in elevation.

$$T_z = T_o - \Gamma Z$$
$$T_z = 288.15 - (6.5/1{,}000 \times 1{,}500)$$
$$T_z = 278.40\ \mathrm{K}$$

Thus air density is:

$$\rho = p/RT$$
$$\rho = [845.55/(287.04 \times 278.40)] \times 100$$
$$\rho = 1.058\ \mathrm{kg/m^3}$$

or some 14 percent less than that at sea level and 15°C.

Because the normal lapse rate doesn't always reflect actual temperature changes with elevation, it is sometimes necessary to find the air density for a specific temperature and elevation. For example, if Rick Solinsky wanted to know how well his Proven wind turbine would perform at his site, he would need to find the temperature at his site during his observations.

Again let's use Cameron Ridge in the Tehachapi Pass, at 1,500 meters above sea level, as an example. We'll assume it's summer and the temperature is 90°F (32°C). To determine air density, we need to first find air pressure. We'll use the hypsometric equation to find pressure changes with changes in elevation where (exp) is the base e exponential and approximately 2.71828:

$$p = p_o \exp(-Zg/RT)$$
$$p = 1013.25 \times \exp\{(-1{,}500 \times 9.807)/[287.04 \times (273.15 + 32)]\}$$
$$p = 857\ \mathrm{mb}$$

We now substitute this value into the equation for air density:

$$\rho = \{857/[287.04 \times (273.15 + 32)]\} \times 100$$
$$\rho = 0.977\ \mathrm{kg/m^3}$$

or some 20 percent less than air density at sea level and 15°C (59°F). Thus when conditions vary significantly from those at sea level, we ignore changes in air density at our peril.

length produce a correspondingly large increase in swept area and, thus, in power. Compare the area swept by one wind turbine with a rotor diameter of 10 and that of another with a rotor diameter of 12 (the units are not important here):

$$A_2 = (r_2/r_1)^2 A_1$$
$$A_2 = (12/10)^2 A_1$$
$$A_2 = 1.44 A_1$$

Increasing the rotor diameter by 20 percent (from 10 to 12) increases the capture area by 44 percent. Now consider the effect doubling the length of each blade—doubling the rotor diameter—has on the wind turbine's swept area:

$$A_2 = (r_2/r_1)^2 A_1$$
$$A_2 = (2/1)^2 A_1$$
$$A_2 = 4 A_1$$

Doubling the diameter increases the swept area four times. This exponential relationship between swept area and the power available

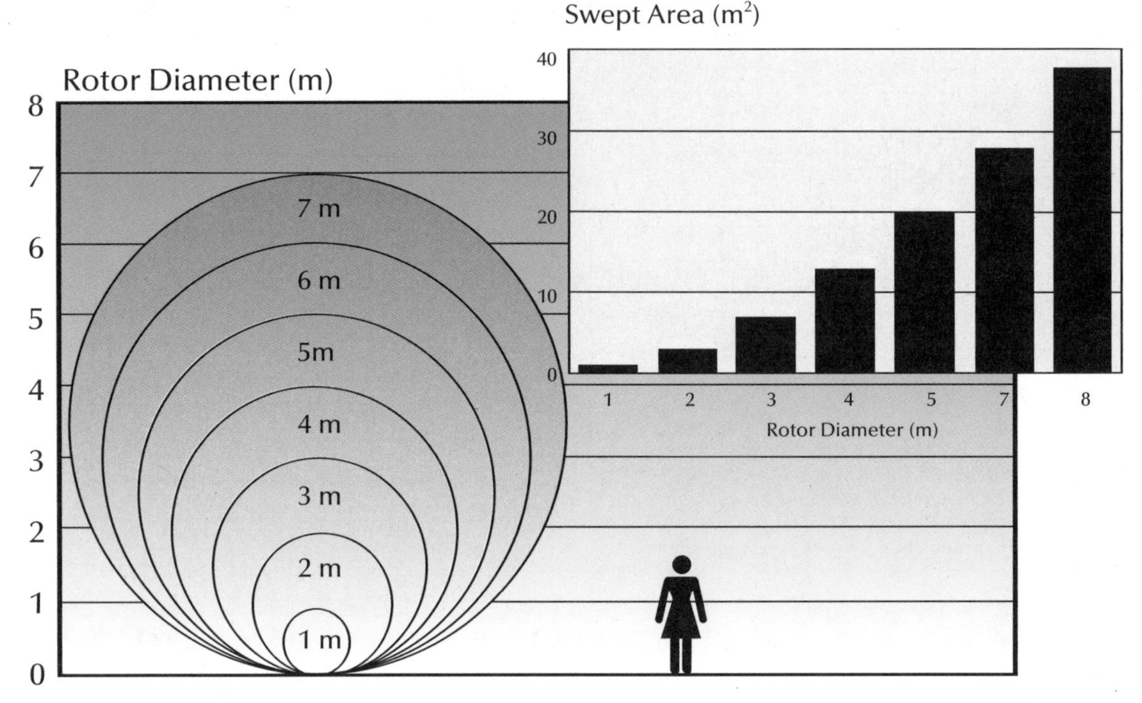

Figure 3-5. Relative size of small wind turbines. Marlec Engineering's Rutland 910 wind turbine uses a molded plastic rotor about 1 meter (3 ft) in diameter. It intercepts about 2/3 m^2. Southwest Windpower's Air 403 uses a rotor 1.2 meters (3.8 ft) in diameter that sweeps about 1 m^2 (12 ft^2). World Power Technologies' H40 uses a rotor slightly more than 2 meters (7 ft) in diameter. It sweeps more than five times the area of the Rutland and three times that of the Air. World Power's H80 uses a 3-meter (10 ft) rotor. J. Bornay's Inclin 3000 sweeps almost 13 m^2 (135 ft^2) with a rotor 4 meters (13 ft) in diameter. Vergnet's GEV 5/5 uses a 5-meter (16 ft) rotor. Bergey Windpower's Excel uses a 7-meter (23 ft) rotor and sweeps nearly 39 m^2 (410 ft^2).

to a wind turbine explains a crucial wind energy axiom: Nothing tells you more about a wind turbine's potential than the area swept by its rotor—nothing. The wind turbine with the bigger rotor will almost invariably generate more electricity or pump more water than a turbine with a smaller rotor. *Rotor diameter* is the best shorthand for "swept area." That's why in this book wind turbines are often referred to by their rotor diameter. It's a shorthand for the area of the wind they intercept.

Wind Speed

No other factor is more important to the amount of wind power available to a wind turbine than the speed of the wind. Because the power in the wind is a cubic function of wind speed, changes in speed produce a profound effect on power. Consider the power available at one site with a wind speed of 10 (again, the units are not important) and another site with a wind speed of 12. From an earlier equation we learned that power is proportional to the cube of wind speed:

$$P_2/P_1 = (V_2/V_1)^3$$
$$P_2 = (12/10)^3 P_1$$
$$P_2 = 1.73 P_1$$

Although there's only a 20 percent difference between the wind speeds at the two sites (12/10 = 1.2), there's 73 percent more power

Nothing tells you more about a wind turbine's potential than the area swept by its rotor—nothing.

available at the windier location. This is why there's such a fuss concerning the proper siting of a wind machine: Small differences in wind speed caused by bordering trees or buildings can drastically reduce the power a wind turbine can potentially capture. To grasp the full effect, consider what happens when the wind speed doubles from one site to the next. Doubling wind speed does not simply double the power available. Instead, power increases a whopping eight times:

$$P_2 = (20/10)^3\ P_1$$
$$P_2 = 2^3\ P_1$$
$$P_2 = 8\ P_1$$

We can summarize the power equation with these general rules:

- Power can be affected by changes in air density when sites differ markedly from those at standard sea-level conditions.
- Power is proportional to the area intercepted by the wind turbine. Double the area intercepting the wind and you double the power available.
- Power is a cubic function of wind speed. Double the speed, and power increases eight times.

At this point an important question arises. What wind speed are we talking about? The average wind speed? If so, what average wind speed? The annual average? Whichever we use determines the results we get.

Using the average annual wind speed alone in the power equation would not give us the right results; our calculation would differ from the actual power in the wind by a factor of two or more. To understand why, remember that wind speeds vary over time. The average speed is composed of winds above and below the average.

To illustrate, let's calculate the power density (P/A), the rate at which energy passes through a unit of area, for an annual average wind speed of 15. *Power density* is a term frequently used by wind energy experts because it's a convenient shorthand for how energetic the winds are during a period of time, typically a year. Power density is normally given in units of watts per square meter (W/m^2) but we don't need the units just yet.

$$P/A_1 = 1/2\ \rho V^3$$
$$= 1/2\ \rho \times 15^3$$
$$= 1/2\ \rho \times 3{,}375$$

Now, what happens if the wind blows half the time at 10 and half the time at 20? The average speed remains 15.

$$(10 + 15)/2 = 15$$

Yes, but watch what happens to the average power density using these two wind speeds.

$$P/A_2 = 1/2\ \rho \times [(10^3 + 20^3)/2]$$
$$= 1/2\ \rho \times [(1{,}000 + 8{,}000)/2]$$
$$= 1/2\ \rho \times 4{,}500$$
$$(P/A_2)/(P/A_1) = 4{,}500/3{,}375$$
$$P/A_2 = 1.33\ P/A_1$$

How can this be? Both have the same average speed. The answer rests with the cubic relationship between power and speed.

Grab a cup of coffee, sit back, and ponder this statement by Jack Park, one of the pioneers of America's wind power revival during the 1970s: The average of the cube of many different wind speeds will always be greater than the cube of the average speed. Or stated another way, the average of the cubes is greater than the cube of the average. In this case, the average of the cube for two wind speeds (10 and 20) is 1.33 times the cube of the average.

The reason for this paradox is that the single number representing the average speed

Power is a cubic function of wind speed.
Double the speed, and power increases eight times.

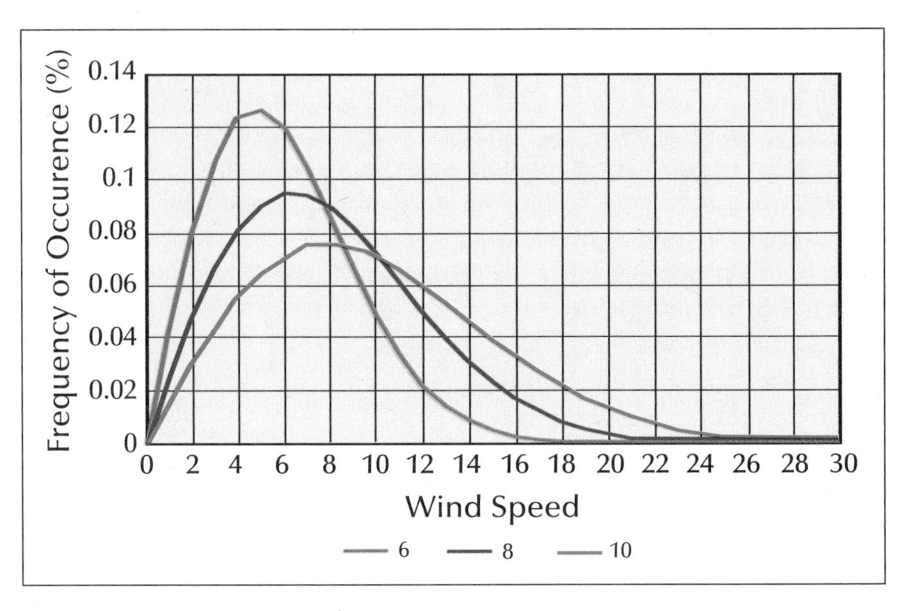

Figure 3-6. Rayleigh wind speed distribution. The average wind speed is all that's needed to define the shape of the Rayleigh distribution. As the average wind speed increases, the curve shifts toward the higher wind speeds on the right of the chart.

ignores the amount of wind above as well as below the average. It's the wind speeds above the average that contribute most of the power.

Speed Distributions

If we plotted a graph of the number of times, or frequency, with which winds occur at various speeds throughout the year, we'd find that there are few occurrences of no wind and few occurrences of winds above hurricane force. Most of the time wind speeds fall somewhere in between these extremes.

The occurrence of winds at various speeds differs from one site to the next, but in general follows a bell-shaped curve (see figure 3-6, Rayleigh wind speed distribution). These distributions of wind speeds can be described mathematically in an attempt to approximate the real world. Meteorologists use the Weibull distribution and its companion, the Rayleigh distribution, to characterize wind resources when the actual distribution of wind speeds over time is unavailable.

As we've seen from the previous example, summing the power contributed by a range of wind speeds rather than simply calculating the power from the average wind speed more accurately reflects the power available. For example, the power density calculated from the Rayleigh distribution for a given average wind speed is almost twice that derived from the average wind speed alone. This relationship holds for many sites with moderate to strong average annual wind speeds, but it will underestimate the potential at some and overestimate the potential at others. The real world is never as tidy as the mathematical models portray it.

Sites in the trade winds of the Caribbean often have high average wind speeds. But the winds are steady. They have few occurrences of extremely high winds. At trade wind sites, the Rayleigh distribution overestimates potential generation.

The relationship between the power density derived from the average speed alone and that from a speed distribution, whether an actual distribution measured at a site or a mathematical distribution, is what Jack Park called the cube factor or Golding's Energy Pattern Factor. For example, the cube factor for the Rayleigh distribution is 1.91.

The power in the wind at three different sites with exactly the same average wind speed illustrates the importance of Park's "cube factor" (see table 3-1, Effect of Speed Distribution on Wind Power Density for Sites

Frequency Distributions

While most of those who use wind energy seldom need to employ the following equations for speed distributions, it's important to know when and how they apply. For example, most manufacturer estimates of annual energy output include a brief notation that the estimate derives from a Rayleigh distribution. Some manufacturers substitute the expression "Weibull k = 2"—an arcane way of stating the same condition. In other words, performance of the wind turbine will vary if actual conditions differ from these standard speed distributions. The Weibull wind speed distribution is a mathematical idealization of the distribution of wind speeds over time. This distribution is determined by two parameters: C, the scale factor that represents wind speed, and k, the shape factor that describes the form of the distribution. A typical shape factor for midlatitude sites is about 2. Sites in the trade-wind belt have much higher values for k. The Weibull distribution can be found from

$$f(V) = k/C\ (V/C)^{k-1}\ \exp[-(V/C)^k]$$

where f(V) is the frequency of occurrence for the wind speed (V) in a frequency distribution, (exp) is the base e exponential function, C is the empirical Weibull scale factor in m/s, and k is the empirical Weibull shape factor.

The Rayleigh distribution is a special case of the Weibull function where k, the shape factor, is 2. Thus, to use the Rayleigh distribution you need only know the average wind speed. The Rayleigh distribution can be found from

$$f(V_R) = dV\ (\pi/2)\ (V/V_{avg}{}^2)\ \exp[-\pi/4\ (V/V_{avg})^2]$$

where dV is the width of the wind speed bin, V is the speed of the wind speed bin, and V_{avg} is the average wind speed.

Consider two sites where the Weibull parameters are known: Tera Kora, and Helgoland. Tera Kora is on the northeast coast of Curaçao, an island in the Caribbean. In 1993 Kodela, the island utility, installed a small wind plant there of twelve 250 kW turbines. The turbines have been successful partly because the site has a high average wind speed and partly because it's in the trade-wind belt. With a shape factor of 4.5, there are few occurrences of very high winds that can damage the turbines. Helgoland is an island off the coast of Germany in the North Sea. The distribution of wind over Helgoland has a shape factor of 2.09. Even though both sites have about the same average wind speed of slightly more than 7 m/s, there is more energy in the wind on Helgoland than at Tera Kora. The shape of the wind distribution on Helgoland indicates that there are more occurrences of high winds there than over Tera Kora—and because of the cubic relationship between wind speed and power, these stronger winds contribute significantly to the overall energy available. This is reflected in the greater power density on Helgoland (400 W/m^2) than that on Curaçao (280 W/m^2).

Site Comparison Between Tera Kora, Curaçao and Helgoland, Germany

	Tera Kora	Helgoland
Avg. Speed, Vavg (m/s)	7.3	7.1
Shape factor, k	4.5	2.09
Scale Factor, C (m/s)	8	8
Power Density (W/m^2)	280	400
Energy Pattern Factor	1.19	1.83

The shape of the distribution also affects how

Tera Kora. One of the Caribbean's most successful wind plants, provides about 1 percent of Curaçao's electricity.

well various wind turbines will perform. Most medium-size wind turbines, such as those installed at Tera Kora, are designed for sites with Rayleigh distributions and have high-rated wind speeds. To improve performance in the Caribbean, however—where the trade winds dominate—the turbines should be optimized for the wind regime by designing for a lower-rated wind speed than elsewhere to better capture the lower-speed winds that dominate the energy distribution. The equations for the Weibull and Rayleigh distributions in speadsheet format can be found in the appendixes. Note that most spreadsheet software contains a function command for the Weibull distribution that greatly simplifies using this equation.

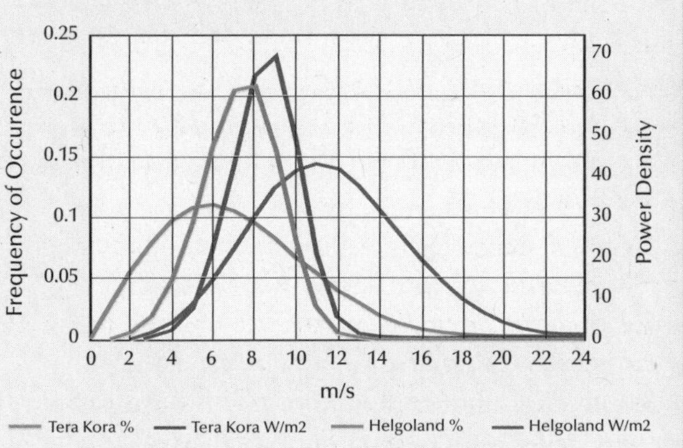

Weibull distribution. Distribution of wind speeds and power density at two sites with Weibull distributions.

Table 3-1
Effect of Speed Distribution on Wind Power Density for Sites with Same Average Speed

Site	Annual Average Wind Speed (m/s)	(mph)	Wind Power Density (W/m²)	Energy Pattern Factor or Cube (Factor)	Wind Power Class (at 10 m)
Culebra, Puerto Rico	6.3	14	220	1.4	4
Tiana Beach, New York	6.3	14	285	1.9	5
San Gorgonio, California	6.3	14	365	2.4	6

Source: Battelle, *PNL Wind Energy Resource Atlas,* 1986.

with Same Average Speed). Though a site in New York experiences the same average wind speed as one in Puerto Rico, the Caribbean island lies in the trade-wind belt and has more constant winds. These steady winds produce less power over time than a temperate wind regime like that of New York. In contrast, the blustery winds that rush through the San Gorgonio Pass contain 66 percent more power than the gentler winds bathing Puerto Rico.

Meteorologists have characterized the distribution of wind speeds for many of the world's wind regimes. For temperate climates such as that of the continental United States or Europe, the Rayleigh wind speed distribution offers a good approximation.

In a study of small wind turbine performance, Wisconsin Power & Light found that the Rayleigh distribution produced reasonably good estimates when used to project annual electricity generation. Its estimates were only about 5 percent less than the power actually produced. Monthly estimates were less reliable.

Because the speed distribution plays such an important role in determining power, it's always preferable to use an actual measured distribution whenever possible. Battelle Pacific Northwest Laboratory notes that the measured distribution for one site near Ellensburg, Washington, produces a power density of 320 W/m^2, twice that from a Rayleigh distribution (160 W/m^2) for the same average speed of 5.3 m/s (12 mph).

Despite its limitations, the Rayleigh distribution remains a useful tool for most sites. Meteorologists often use a more flexible mathematical formula, the Weibull distribution, which can more closely model the wind speeds at a wide range of sites than the Rayleigh distribution. The Rayleigh distribution is a member of the Weibull family of speed distributions.

Where do we stand now? We can calculate power density in two ways. We can sum a series of power density calculations for each wind speed and its frequency of occurrence (the number of hours per year the wind blows at that speed) for the site's distribution of wind speeds. Or we can use the average wind speed and the appropriate cube factor.

At a sea-level site with a temperature of 15°C (59°F), air density is 1.225 kg/m^3. If the site has a Rayleigh distribution, and thus the Energy Pattern Factor (EPF) is 1.91, then the power density in W/m^2 is:

Average Annual P/A = 0.05472 V^3
x EPF W/m^2, where V is in mph
Average Annual P/A = 0.6125 V^3
x EPF W/m^2, where V is in m/s

For those wanting to work completely in English units, power density in W/ft^2 is:

Average Annual P/A = 0.00508 V^3
x EPF W/ft^2, where V is in mph,
and A is area in square feet

If the site has an average annual wind speed of 4 m/s (9 mph), the annual power density is:

$$P/A = 0.6125 \times 4^3 \times 1.91 \text{ W/m}^2 = 75 \text{ W/m}^2$$

And once we know the annual power density at a site, we can quickly estimate the annual wind energy density in kilowatt-hours per year per square meter of the wind stream (kWh/yr/m^2):

$$E/A = (P/A) \times (8{,}760 \text{ h/yr}) \times (1 \text{ kW}/1{,}000 \text{ W}) = (75 \text{ W/m}^2) \times (8{,}760 \text{ h/yr}) \times (1 \text{ kW}/1{,}000 \text{ W}) = 656 \text{ kWh/m}^2$$

Table 3-2
Annual Wind Power and Energy Density for Rayleigh Distribution

Annual Average Wind Speed (m/s)	(mph)	Annual Power Density (W/m^2)	Annual Energy Density (kWh/m^2)
4	9.0	75	656
5	11.2	146	1,281
6	13.4	253	2,214
7	15.7	401	3,515
8	17.9	599	5,247
9	20.2	853	7,471

Estimating the amount of energy available annually to the wind turbine becomes simply the product of energy density (E/A) and the turbine's swept area (A) in square meters (see table 3-2, Annual Wind Power and Energy Density for Rayleigh Distribution).

$$E = (E/A) \times A$$

For example, Southwest Windpower's Air 403 intercepts about 1 m^2 (12 ft^2) of the wind stream. At a site with a 4 m/s average wind speed it will intercept about 650 kWh per year. It won't actually capture that much, and we'll discuss why in the next chapter. We're not quite ready to estimate how much of this energy a wind machine is capable of capturing. There's one step remaining.

Because the wind turbine will be mounted atop a tower that's typically two to three times taller than the tower used to measure wind speed, we need to estimate the wind speed at the top of the tower at the proposed wind turbine's hub height. How wind speed and power change with height is the subject of the next section.

Wind Speed, Power, and Height

Wind speed, and hence power, varies with height above the ground. Wind moving across the earth's surface encounters friction caused by the turbulent flow over and around mountains,

Logarithmic Model of Wind Shear

The logarithmic extrapolation of wind speed with height is

$$V = \ln(H/z_0)/\ln(H_o/z_0) V_o$$

where (V_O) is the wind speed at the original height, (V) is the wind speed at the new height, (H_O) is the original height, (H) is the new height, and (z_0) is the roughness length.

The European approach has worked well along the coastlines of the North German Plain, particularly in Denmark. In open areas with few windbreaks, such as coastal sites (roughness class 1 in the European system), the logarithmic model produces a result similar to that of the 1/7 power law. Farther inland, results from the two methods diverge. For inland sites, the logarithmic model finds more energy in the wind than does the 1/7 power law.

The effect of height on wind speed is so great that it often necessitates measuring actual wind speeds at hub height rather than relying on estimates produced by either system. Yet in the absence of actual measurements, the 1/7 power law is a reasonable, if sometimes conservative, approximation.

hills, trees, buildings, and other obstructions in its path. These effects decrease with increasing height above the surface until unhindered air flow is restored. Consequently, as friction and turbulence decrease, wind speed increases.

As you can imagine, frictional effects differ from one surface to another, depending upon its roughness. Friction is higher around trees and buildings than it is over the smooth surface of a lake. In the same manner, the rate at which wind speed increases with height varies with the degree of surface roughness. Wind speeds increase with height at the greatest rate over hilly or mountainous terrain, and at the least rate over smooth terrain like that of the Great Plains. Because of this, the benefits of using a tall tower are often greater when siting in hilly terrain than they are on the Llano Estacado (the staked plains) of the Texas Panhandle.

At low wind speeds, the change in speed with height or wind shear is less pronounced and more erratic. In light or calm winds, as may be encountered during a temperature inversion, wind speeds may increase slightly between the ground and a certain height, and then begin to decrease. Real-world experience has shown that changes in wind speed with height are not constant. In the Altamont Pass east of San Francisco, Pacific Gas & Electric Company found that above 200 feet (60 m), wind speeds decreased with increasing height. The utility found this effect after it installed a Boeing Mod-2, the rotor of which was 300 feet (91 m) in diameter. At times the uppermost part of the rotor would extend above the layer of fast-moving air. On average, however, wind shear is positive, and wind speed increases with height.

This effect is so important that data on wind speeds will always include the height at which the wind was measured. If the height is not specifically mentioned, it is usually assumed to be about 10 meters (33 ft) above the ground, though many measurements at airports in North America were made at 15 to 20 feet (4 to 6 m). Most wind turbines will be installed on towers much taller than this to take advantage of the stronger, less turbulent winds aloft. To elevate a Bergey Excel above the trees in a shelter belt surrounding one Alberta home, Nor'Wester Energy Systems' Jason Edworthy used a tower 100 feet (30 m) tall. Commercial wind turbines in Europe and North America are now being installed on towers 50 to 70 meters (160 to 230 ft) tall. Some of the bigger wind turbines in Germany have been installed on even taller towers!

The easiest way to calculate the increase in wind speed with height is to use the power law method. Another approach using logarithmic extrapolation is common in Europe. Logarithmic extrapolation is mathematically derived from a theoretical understanding of how the wind moves across the surface of the earth. In contrast, the power law equation is derived empirically from actual measurements. The power law equation may be less scientific, but it works well and is more conservative than the logarithmic method.

The following equation illustrates how to use the power law method where V_o is the wind speed at the original height, V is the wind speed at the new height, H_o is the original height, H is the new height, and α is the wind shear exponent:

$$V/V_o = (H/H_o)^{\alpha}$$
$$V = (H/H_o)^{\alpha} V_o$$

For example, consider the increase in wind speed when doubling tower height from 10 to 20 or from 30 to 60 (the units are unimportant; it's the ratio that counts):

$$V = (20/10)^{0.14} V_o$$
$$V = 2^{0.14} V_o$$
$$V = 1.1 V_o$$

Both systems require the user to estimate surface roughness. Where the rate of increase in wind speed with height is unspecified, it is commonly assumed in North America that the "1/7 power law" applies—that is, the surface roughness exponent is 0.014 representing open plains. Empirical results indicate that the 1/7 power law fits many, though certainly not all, North American sites (see table 3-3, Changes in Wind Speed and Power with Height for Selected DoE-Candidate Wind Turbine Sites in the United States).

On terrain where the 1/7 power law applies, doubling tower height increases wind speed by 10 percent. Increasing tower height five times—say, from 10 to 50, or from 30 to 150—may increase wind speed as much as 25 percent (see figure 3-7, Increase in wind speed with height). Once again, the 1/7 power law is just a guide.

Although wind shear often follows the 1/7 law, it doesn't always. Obstructions significantly reduce wind speeds near the ground. Over row crops such as corn, or over hedges and a few scattered trees, wind speed increases

The Wind Shear Exponent α

The wind shear exponent α varies with the time of day, season, terrain, and stability of the atmosphere. Shear is low where there is minimum surface roughness and high where there are numerous objects to disturb the flow. German engineer Jens-Peter Molly in his book *Wind Energie* presents a simple formula for calculating *a* from a measure of the surface roughness:

$$\alpha = 1/\ln(z/z_0)$$

where z_0 is the surface roughness length in meters and z is the reference height. For example, when the surface roughness length is 0.01 m, and the reference height is 10 m (33 ft), then

$$\alpha = 1/\ln(10 \text{ m}/0.01 \text{ m})$$
$$\alpha = 1/\ln(1{,}000)$$
$$\alpha = 0.144$$

or about that of the 1/7 power law.

Surface Roughness Lengths and the Wind Shear Exponent α

Terrain	Surface Roughness Length z_0 (m)	Wind Shear Exponent α
Ice	0.00001	0.07
Snow on flat ground	0.0001	0.09
Calm sea	0.0001	0.09
Coast with onshore winds	0.001	0.11
Snow-covered crop stubble	0.002	0.12
Cut grass	0.007	0.14
Short-grass prairie	0.02	0.16
Crops, tall-grass prairie	0.05	0.19
Hedges	0.085	0.21
Scattered trees and hedges	0.15	0.24
Trees, hedges, a few buildings	0.3	0.29
Suburbs	0.4	0.31
Woodlands	1	0.43

Note: Relative to a reference height of 10 m (33 ft)

Source: Adapted from *Characteristics of the Wind by Walter Frost and Carl Aspliden in Wind Turbine Technology*, and *Windenergie: Theorie, Anwendung, Messung* by Jens-Peter Molly.

Figure 3-7. Increase in wind speed with height. A simple chart for estimating the increase in wind speed as a function of relative tower height. To use the chart, find the ratio of tower height (H) to anemometer height (H_o), then find the intercept with the line representing the wind shear exponent. The intercept indicates the relative increase in wind speed at the tower height. For example, if the hub height of the wind turbine will be five times taller than the anemometer height and the 1/7 power law applies, find 5 on the horizontal axis, move vertically until you intercept the curve labeled 0.14, then proceed horizontally to the intercept with the vertical axis.

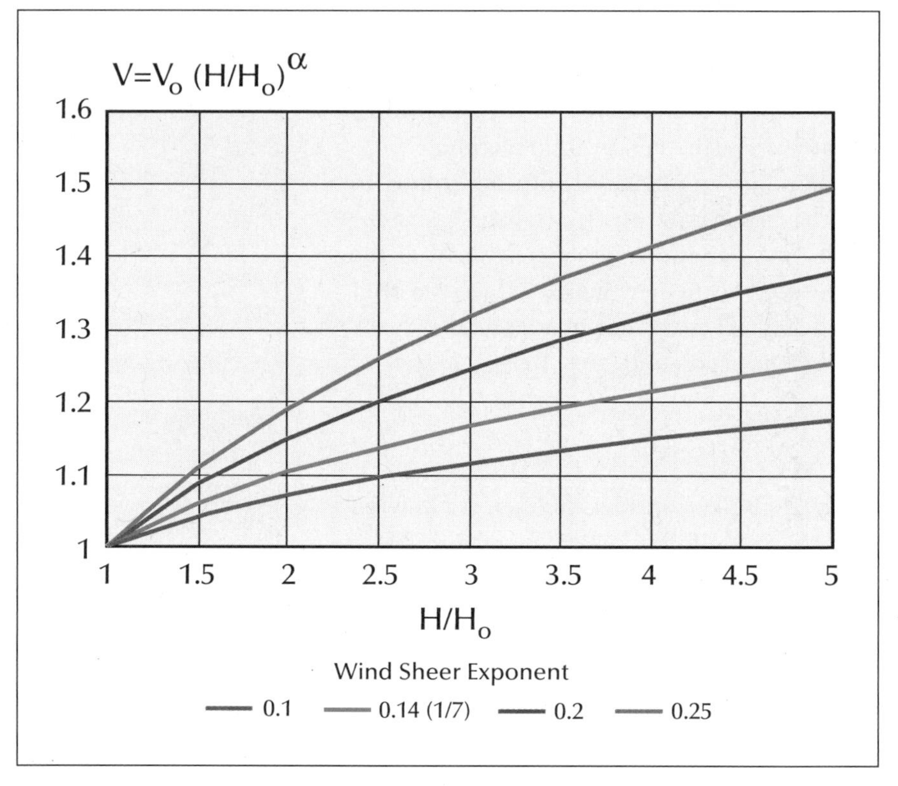

more dramatically with height than that predicted by the 1/7 law. Wind shear, α, rises to one-fifth (0.20). When the surface is rougher still, say with more trees and a few buildings, α increases further to one-fourth (0.25). The speed profile becomes even steeper over woods and clusters of buildings.

The increase in wind speed with height only holds true for the height above the effective ground level. The wind rushing over a field of corn sees the top of the corn, not the soil on which it grows, as the effective ground level. For a woodlot this is the uppermost point where the branches of the trees are touching, not necessarily the tops of the trees.

> Increasing tower height fivefold increases the power available in the wind nearly twofold at sites where the 1/7 power law applies.

Let's see what all this means by examining the effect a tall tower will have on the average wind speed at a site in Kansas where the 1/7 power law applies. In our example the wind speed was measured at 30 feet (10 m), H_O, and we want to install our wind turbine on a 150-foot (45 m) tower, H. If you don't like working with formulas, all is not lost. This calculation and others have already been made in figure 3-7 for a range of tower heights and wind shear.

$$V = (H/H_o)^{\alpha} V_o$$
$$V = (150/30)^{1/7} V_o$$
$$V = 5^{1/7} V_o$$
$$V = 1.26 V_o$$

The wind speed at the height of the wind turbine will increase 26 percent by installing it on a 150-foot tower.

But we can't stop here. Power is a cubic function of speed. Where P_O is the power at the original height of 30 feet and P is the power at the new height, the increase in power on the 150-foot tower in our example is given in the following formula. As before, there's no need to calculate the change in power density with height if you're averse to mathematics—just use figure 3-8 with the wind shear exponent appropriate for your site. This brings us to another rule of thumb:

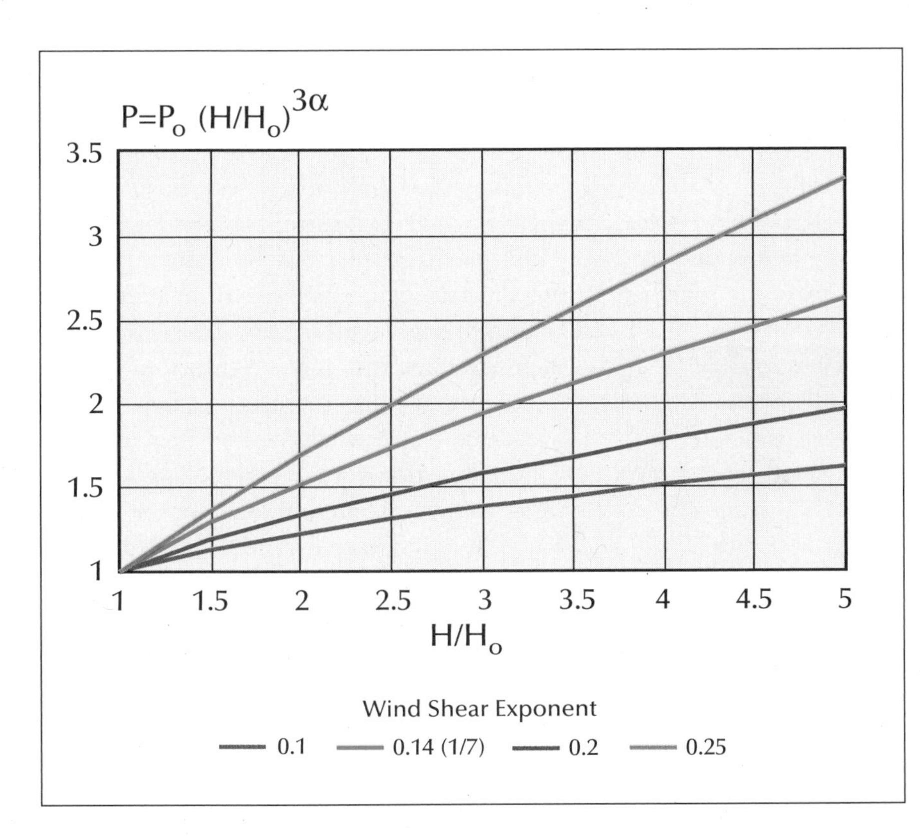

Figure 3-8. Increase in wind power with height. A simple chart for estimating the increase in wind power as a function of relative tower height. To use the chart, find the ratio of tower height (H) to anemometer height (H_o), then find the intercept with the line representing the wind shear exponent. The intercept indicates the relative increase in wind power at the new height. For a fivefold increase in tower height, power increases almost two times when the 1/7 power law applies.

Increasing tower height fivefold increases the power available in the wind nearly twofold at sites where the 1/7 power law applies.

$$P = (H/H_o)^{3\alpha} P_o$$
$$P = 5^{[3 \times (1/7)]} P_o$$
$$P = 1.99 P_o$$

Like all rules of thumb, this is only an approximation of the real world. In table 3-3 the increase in power with height doesn't exactly follow this rule. The reason? The distributions of wind speeds change slightly at new heights. For example, the wind shear exponent

Table 3-3
Changes in Wind Speed and Power with Height for Selected DoE-Candidate Wind Turbine Sites in the United States

Site	Wind Speed (m/s) at Height 9.1 m 30 ft	Wind Speed (m/s) at Height 45.7 m 150 ft	Speed Increase	Wind Shear Exponent α	Power Increase	Wind Shear Exponent α
Finley, North Dakota	6.1	9.1	1.49	0.25	3.15	0.24
Block Island, Rhode Island	5.0	7.4	1.48	0.24	3.06	0.23
Boardman, Oregon	3.8	5.5	1.45	0.23	2.73	0.21
Huron, South Dakota	4.7	6.8	1.45	0.23	2.53	0.19
Russel, Kansas	5.3	7.3	1.38	0.20	2.16	0.16
Clayton, New Mexico	5.4	7.3	1.35	0.19	2.06	0.15
Minot, North Dakota	6.5	8.4	1.29	0.16	1.97	0.14
Amarillo, Texas	6.3	8.1	1.29	0.16	2.04	0.15
San Gorgonio Pass	6.2	7.7	1.24	0.13	2.03	0.15
Livingston, Montana	6.8	8.4	1.24	0.13	1.74	0.11
Kingsley Dam, Nebraska	5.3	6.5	1.23	0.13	1.79	0.12
Bridger Butte, Wyoming	7.0	8.4	1.20	0.11	1.59	0.10

Source: Battelle PNL, *Wind Energy Resource Atlas,* 1986.

for wind speed at Clayton, New Mexico, is 0.19, but the shear exponent decreases slightly to 0.15 when considering the increase in wind power density at the new height.

Clearly you can't talk about wind speed without also referring to the height at which it's measured. The two always go together, though sometimes the height is assumed. In general the average wind speed at a specific site refers to the speed at the height of the anemometer. Confusion arises when you start talking to wind turbine manufacturers. The wind speeds they use may be either at the hub height of the wind machine or at some other height. It makes a big difference. Most rate their products at hub height; some don't. Reputable manufacturers and their dealers will clearly state which method they use.

After two decades of working with the wind, we've learned an important lesson—the hard way. No amount of historical wind data can substitute for knowing the wind at the specific site where you want to put your wind turbine. This includes measuring the wind at the proposed height of the turbine to avoid extrapolations that may or may not reflect actual conditions. More on this later. Next we will look at what historical wind data is available.

The Nocturnal Jet

High wind shear may be a regional phenomenon of North America's Great Plains. It's characteristic of Buffalo Ridge in southwestern Minnesota, according to meteorologist Ron Nierenberg. This contrasts with California's Tehachapi Pass, where shear on Cameron Ridge is near zero. At exposed sites in Minnesota, says Nierenberg, wind shear is often double that of the 1/7 power law, from 0.2 to 0.3. It's similar in Iowa and Wisconsin. Trees, he notes, are the cause. For comparison, Cameron Ridge in California is virtually treeless.

During summer months when wind speeds are typically low in continental wind regimes, a nocturnal jet may occur at a certain height above ground where the wind shear exponent can reach 0.4. This "jet" has nothing to do with the jet stream, Nierenberg explains; it's simply a layer of fast-moving air. "There are lots of places in the world where there's a localized zone of high winds, a so-called jet. Buffalo Ridge, at 600 meters [2,000 ft] above sea level, is possibly just high enough to pierce it." There's a similar jet about 300 meters (1,000 ft) above California's San Joaquin Valley. In part it is this jet that produces the winds in the Tehachapi Pass.

Measurements on Buffalo Ridge near Chandler, Minnesota, found wind shear in the 30- to 50-meter (100- to 160-foot) range typical of the Great Plains. But at 50- to 70-meter (160- to 230-foot) heights, the shear exponent jumped to 0.42. "The resource is very strong," says Rory Artig, an engineer with Minnesota's Department of Public Service. "It's quite different from that in California." Artig found that the two-and-a-half-year average annual wind speed at Chandler was 6.9 m/s (16 mph) at 30 meters (100 ft), 7.6 m/s (17 mph) at 60 meters (200 ft), and 8.2 m/s (18 mph) at 90 meters (300 ft) above ground level. While the 20 percent increase in wind speed may seem modest, it produces a 67 percent increase in the power available.

Published Wind Data

In general, wind data in all countries has been gathered near centers of population. People congregate in areas sheltered from storms and severe weather. (Cities are built more often in valleys than on windswept mountaintops.) Consequently, data from airports, military bases, and weather stations may not reflect the winds that exist at more exposed sites. Using data from these stations alone may lead to underestimating the potential wind power in an area. On the other hand, airports offer a vast clearing with minimal trees and buildings. In heavily wooded or developed areas, the open expanse at an airport may offset any sheltering effects of its location.

The data may be unreliable for other reasons as well: The wind-measuring instruments may have been inaccurate or poorly placed. The instruments at many airports were not properly maintained, and frequently were located on or adjacent to the terminal building. At these stations, the data better reflects the turbulence around the building than anything else.

Data from remote sites is even more problematic. In some cases wind data was collected only during daylight hours or for a few hours during the summer months. In either case, the data doesn't represent what could be expected throughout the day or throughout the year. At some stations, so few observations of wind speed were recorded that the observer was literally noting whether it was windy or not.

Historical average speeds are also available from air quality monitoring stations at both conventional and nuclear plants and from some industries. The limits on data from these sources are the same as those on data from airports. We measure air quality where wind speeds are low and where pollutants concentrate, such as in urban canyons and narrow mountain valleys. These are less-than-ideal locations for a wind machine, and the wind speeds are unrepresentative of better upland sites. For example, wind speeds measured in the deeply incised valley of the Rhine have little bearing on the Eifel Mountains, a plateau of low hills near Germany's border with Belgium where numerous medium-size wind turbines have been successfully installed.

There are also numerous sources of short-term wind data: government energy offices, universities, and nonprofit organizations. One or more of them might have collected wind data in your area. Check around and find out what work has been done and who has the results.

Wind data is not always presented in the form of annual average wind speeds. In both the United States and Europe professional meteorologists have categorized wind resources into a series of wind classes.

In the United States, Battelle Pacific Northwest Laboratories mapped average annual wind power rather than simply wind speed. Battelle devised a numerical rating that corresponds to one of seven wind power classes. Each class represents a range of power densities. For example, class 4 represents wind power density from 200 to 250 W/m^2, or wind speeds from 5.5 to 6 m/s (12.3 to 13.4 mph).

Battelle derived the maps from computer modeling of historical wind data, terrain, and regional weather patterns. The values shown represent only those sites such as hilltops, ridge crests, and mountain summits that are free of obstructions and are well exposed to strong prevailing winds. By giving a range of possible values rather than a single number, Battelle doesn't lure users into the mistaken notion that they can estimate wind speed with precision.

Denmark's Risø National Laboratory has done similar work in Europe. Its European Wind Atlas provides a detailed look at the wind resources of the 12 countries in the

Calculating the Wind Shear Exponent (α)

If you're measuring wind speeds at different heights, or if you've found data for wind speeds at different heights, you can calculate the wind shear exponent (α) for your conditions from

$$\ln(V/V_o)/\ln(H/H_o)$$

where V is wind speed at the upper anemometer, V_o is the wind speed at the lower anemometer, H is the height of the upper anemometer, and H_o is the height of the lower anemometer.

For example, consider the previous example of the diurnal wind speeds measured at the Wulf Test Field. The average wind speed at anemometer A was 8.5 mph, and that at anemometer B was 7.9 mph. Anemometer A is mounted at a height of 56 feet, anemometer B at a height of 36 feet. (The ratio of height A:B is 1.6.)

$$\alpha = \ln(8.5/7.9)/\ln(56/36)$$

$$\alpha = \ln(1.076)/\ln(1.556)$$

$$\alpha = 0.073/0.44$$

$$\alpha = 0.17$$

For the period measured, the wind shear approximates that typical for a short-grass prairie, which accurately characterizes the site.

Sources of Wind Data in the United States

The National Weather Service (NWS) and Federal Aviation Administration (FAA) have collected the United States' most extensive records. This data is stored at the National Climatic Data Center (NCDC) in Asheville, North Carolina. The NCDC also stores wind data from other federal agencies, including the Civil Aeronautics Administration, the USDA Forest Service, and the Department of Defense. At some of the stations, wind speed and direction have been recorded for more than 30 years. Periodically this information has been tabulated into a "Summary of Hourly Observations" that provides the long-term average wind speed, as well as the speed distribution. These summaries can be helpful when evaluating the winds at your site. You can find this data at http://lwf.ncdc.noaa.gov/oa/ncdc.html. The NCDC data, however, isn't always accurate; much of it has been collected at major airports, which are often sheltered from the wind.

Much of the work of collecting U.S. wind data has already been done by Battelle Pacific Northwest Laboratory. Battelle studied the country's official summarized data. It has also analyzed data from numerous new sources, as well as data from the many short-term recording stations installed specifically to measure wind energy. It carefully reviewed the records, then presented its results as maps of average annual wind power density. Battelle's maps and new high-resolution digital maps from the National Renewable Energy Laboratory can be found on the Web at www.nrel.gov/wind/wind_map.html.

European Community: Ireland, Great Britain, Denmark, Germany, France, Belgium, Luxembourg, Spain, Italy, Portugal, Greece, and the Netherlands. Like Battelle, Risø presents the data as a range of values. But Risø goes a step farther and suggests the wind speeds likely at sites with differing surface roughness, such as along coastlines. The European data is also available as a computer model for professional meteorologists.

Medium-size wind turbines in commercial applications are most commonly found in areas with power densities greater than 200 W/m^2 or average wind speeds above 5.5 m/s (12 mph) at 10 meters (33 ft) above ground level. This corresponds to the wind resource designated class 4 by Battelle. This resource is equivalent to a power density of 300 to 400 W/m^2, or an average wind speed greater than 6.5 m/s (15 mph) at 30 meters (100 ft) above the ground, the typical wind turbine height of the late 1980s. At tower heights of 50 meters (164 ft), at which most new medium-size wind turbines will be operating, the same resource is equivalent to a power density of 400 to 500 W/m^2, or an average annual wind speed greater than 7 m/s (>16 mph).

Most of the wind development in California has occurred on windier sites. Wind speeds on the crests of ridges in the Tehachapi Pass average 18 to 19 mph (8 to 8.5 m/s). Sites atop Altamont Pass's rolling hills are less windy than those in Tehachapi: 13 to 18 mph (6 to 8 m/s)

Extensive wind development has taken place in Europe along the North Sea coast. At hub height along the coast of the Netherlands, for example, wind speed averages about 7.5 m/s (17 mph). At less well-exposed sites nearby in the German state of Niedersachsen (Lower Saxony), wind speeds at hub height average 6.8 m/s (15 mph).

One of the lessons learned from California's wind rush during the early 1980s was the necessity of understanding the wind resource in complex terrain. (This was a commercial version of "It's always windy here, we don't need to measure the wind.") It was common then to monitor the wind with one anemometer for every 150 to 350 proposed turbines. As a result, many operators greatly overestimated the amount of wind their turbines would intercept. On average California wind turbine operators projected twice the amount of electricity the turbines actually produced. They were off by 100 percent! By the late 1990s California wind developers had seen the light and were monitoring the winds with one anemometer for every two to three turbines.

Surveying Your Site

To evaluate the potential at your site, begin by asking yourself what it is that you want. Is it the instantaneous wind speed, the average annual speed, or the distribution of wind speeds? If you want to know when it's too windy to go sailing, instantaneous wind speed will suffice. If you want to estimate the annual or monthly energy output from a wind machine, then at least the average wind speed is necessary. Preferably you'll want the wind speed distribution as well. If you plan to use the wind turbine as your sole source of power at a mountaintop retreat, for example, then more detailed information may be required, such as the number of calm days and the time between them.

In most cases you'll be interested in how much energy a wind machine can produce in your area; more specifically, at your site. To estimate annual energy output, the speed distribution is preferred. The speed distribution gives you the most accurate results, but it isn't always necessary. Average speed will often suffice, especially if you have some measure of the Energy Pattern Factor, too.

Now that you know what's needed, find out if someone has already done the work for you. Start by locating anyone nearby who has installed a wind turbine. "Talk to old-timers," says Mick Sagrillo. They may remember someone who once used a windcharger. Pay them a visit.

Or find folks who have an anemometer. What have they found? Before you take any of this data to heart, determine if your site and the others are comparable. Is the data typical of what you could expect? If so, you've saved a lot of time and effort. Normally you won't be so lucky and you'll have to make your own survey if you live in North America. In northern Europe, on the other hand, the meteorological network is so dense and the prices paid for wind-generated electricity so attractive that very rarely is measuring the wind justified for any but the largest wind farms. In either case, a site survey need not be elaborate. It can be as simple as looking at nearby trees.

Data collection. Measuring the wind can be as simple as pushing a button and writing down the data.

When data isn't available, start by looking at the vegetation. Trees and shrubs are frequently touted as a good qualitative indicator of wind speed (see figure 3-9, Biological indicators). High winds and a harsh environment of ice and snow will deform them. The severity of the deformation, whether the tree is slightly flagged or completely bent to the ground, can be used as a rough gauge of wind speed. The types of deformity are:

- **Brushing:** Branches and twigs sweep downwind. This can be observed on both conifers and deciduous trees.
- **Flagging:** Branches sweep downwind; upwind branches are cropped short.
- **Throwing:** Trunk sweeps downwind.

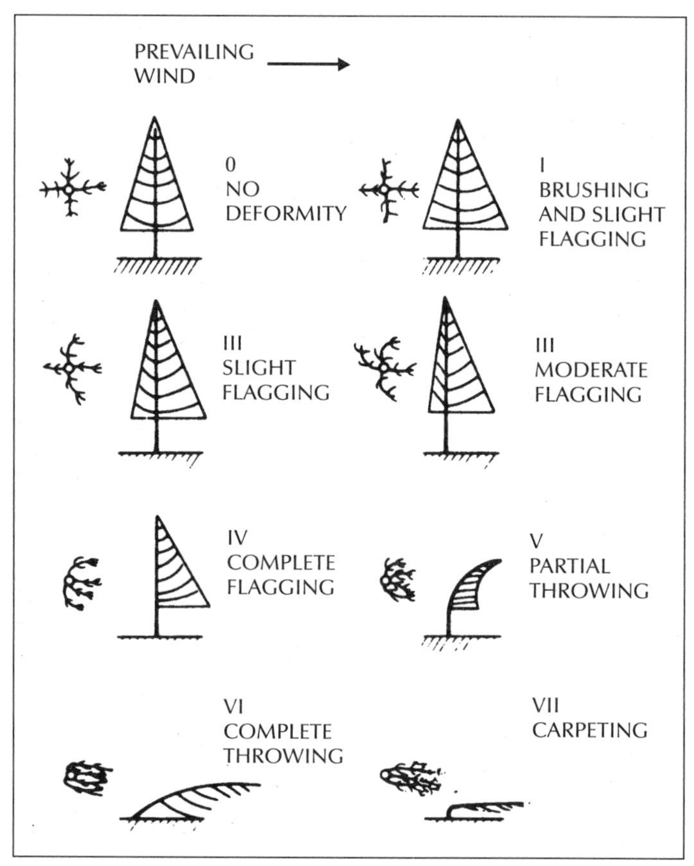

Figure 3-9. Biological indicators. The degree to which conifers have been deformed by the wind can be used as a rough gauge of average annual wind speed (see table 3-4, Griggs-Putnam Index of Deformity). (Battelle PNL)

- **Carpeting:** Trunk bends to the ground. Found frequently in alpine or severe environments where trees grow only a few feet above the ground.

Use the Griggs-Putnam scale of deformity to find the range of wind speeds represented. For example, the divi-divi tree is a popular icon on the island of Curaçao off the coast of Venezuela. The heavily flagged and partially thrown tree symbolizes the island as much as its famous orange liqueur. In the Griggs-Putnam index this tree would suggest average wind speeds of 7 to 8 m/s (15 to 18 mph). In fact, exposed sites on the island experience average wind speeds slightly more than 7 m/s.

There are limitations to this technique. First, don't get excited by one or two odd-shaped trees. One flagged tree is insufficient to make a judgment. There are many other causes for tree deformity besides the wind. You'll need to note several of the same species with an equal amount of deformation to determine if the wind is the cause. (Each species varies in its susceptibility to flagging.) Conifers, especially pine and fir trees, are more reliable indicators of wind strength than are deciduous trees. Moreover, deformation is more obvious where freezing salt spray or ice frequently accompanies high winds. Such conditions are often found along coastlines and on mountaintops.

If you can't find any deformation, don't despair. The absence of flagging doesn't necessarily indicate a low speed. Too often the value of trees as a wind speed indicator is overplayed. At best, it gives only a crude range of possible wind speeds, and even then it's most useful only where conifers dominate.

Next, find the nearest airport or other station where long-term records have been kept. Convert the wind data to power density and compare that with wind atlases for your area. Are they similar? Is the airport representative of your site, or is it better exposed or more sheltered? If the airport lies down on the valley floor and your site is on a plateau overlooking the valley, your site may experience much higher winds.

Now put the pieces together. Estimate what you think is your average wind speed

Table 3-4
Griggs-Putnam Index of Deformity

Wind Speed	Index I	II	III	IV	V	VI	VII
(mph)	7-9	9-11	11-13	13-16	15-18	16-21	22+
(m/s)	3-4	4-5	5-6	6-7	7-8	8-9	10

and power. Give yourself room for error. Avoid pinning your hopes on one number alone and instead use a range of values. Be conservative. Most people, including professional wind farmers, overestimate the amount of wind available.

With these numbers in hand, use the techniques in the next section to estimate the output from several different wind machines. Choose the one that delivers the amount of energy you need and determine its cost. Call the manufacturer or wind system dealer for realistic estimates. Look at the economics.

Next, ask yourself how much risk you're willing to assume. Even though you may have done an admirable job of estimating the wind regime in your area from published sources, obstructions and terrain features can greatly reduce the actual wind energy available. If you don't mind this uncertainty, or if you're on the Great Plains or the steppes of Central Asia and there isn't a tree for miles, then there's little need to go to the trouble of conducting a full-fledged wind speed survey. Measuring the wind at your site only becomes necessary when you're unwilling to take the risk that there's sufficient wind to produce what you expect.

For commercial wind power plants the economic risk is often too great to proceed without on-site measurements. Professional wind developers outside northern Europe nearly always measure the wind first. But for residential and small commercial users, this isn't always necessary. In Denmark, Germany, and the Netherlands the wind resource is well known, and seldom do those installing individual machines measure the wind first. In North America and elsewhere around the world where there are vast differences in terrain from one locale to the other, however, measuring the wind may make sense.

Measurements, if they are to be made at all, should be taken at the intended location of the wind machine and at its proposed height. This is particularly important in rough terrain or where there are obstructions. Reliable measurements can only be made when the anemometer extends well above nearby trees and buildings. Even tall grain crops and low-lying shrubs raise the effective ground level, severely reducing the wind speed measured by anemometers on towers less than 10 meters (33 ft) tall. If you need to measure the winds at your site, take a survey of the trees and buildings nearby and estimate their heights. You may find that the anemometer, and eventually the wind turbine, should be erected elsewhere.

Estimating the Height of Obstructions

Remember those comic pictures of a ragged artist thrusting his thumb at the world? He had a good reason for doing that. Artists use the technique to gauge proportions. You can also use it to estimate the height of nearby trees and buildings. A pencil works better, though. Here's how to use it.

Identify an object of known height at about the same distance from you as the tree or building you wish to measure. TV antennas, telephone poles, and houses work well. Hold the pencil at arm's length and sight along it. Line up the top of the pencil with, for example, the top of a tree. Slide your thumb down the pencil until it lines up with the bottom of the tree. Now turn to the object of known height and again sight along the pencil. While keeping the pencil at arm's length, move your arm up or down until your thumb lines up with the bottom of the object. Judge the proportions by noting how much of the pencil extends above the object. Is it twice the height, one-third greater, or the same?

In Wisconsin, Mick Sagrillo uses a similar technique that compares the shadow from an object of known height to the shadow of the object in question. Simply measure the shadow from the object of known height—for example, a fence post—and the shadow from the object of unknown height. The relationship between the shadow of the fence post and its height is the same as the relationship between the shadow of the obstruction and its height.

Another technique is to use a so-called tangent height gauge commercially available from science supply houses. The inexpensive plastic device is easy to use.

Figure 3-10. Anemometer, mast, and data logger.
Top: Guyed anemometer masts of thin-walled tubing can be used to measure wind speed. Bird's-eye view.
Bottom: Electronic data recorders can be used to collect wind data or—with the appropriate sensors—to monitor small wind turbine performance. (NRG Systems)

Assume that the wind data you've examined is unconvincing. You want to measure the wind at your site to get a better picture of what's there. What next? Anemometers measure wind speed. Wind vanes indicate direction. That's simple enough. More complex is the kind of recording equipment you'll need. In the next section we'll go over the equipment that's on the market and discuss what probably meets your needs best.

Measuring Instruments

To perform a wind resource assessment you'll need an anemometer, mast, and recorder (see figure 3-10, Anemometer, mast, and data logger). Instruments designed for the wind turbine market are generally less expensive and easier to use than those designed for meteorological use; they also give you more of the information you seek.

A wind-measuring instrument is composed of two parts: the sensor (the anemometer head), and a means for displaying the data it measures. The sensor generates an electrical signal that's proportional to wind speed. Cup anemometers are the most common wind sensors. The spinning cups drive either a DC generator or AC alternator. With the advent of pocket calculators, digital displays have become popular. (The data is the same, it just looks different.) Better-quality instruments typically use AC alternators and measure frequency. These anemometers are more accurate than those using DC generators. The least expensive anemometers, widely used even by professionals, produce an electrical pulse, which is then counted over a period of time, often just a fraction of a second.

Whatever system is used, the sensor (the anemometer head) either drives a meter that displays instantaneous wind speed or feeds data to a recorder. Unlike the displays of cheap wind speed meters that indicate only instantaneous wind speed, recorders store information for future retrieval. At one time all meteorological data was recorded on strip charts, but wind prospecting and the boom in electronics have revolutionized wind speed measurement.

Before we go any farther, let's clear up a common misconception. Instantaneous wind speed indicators are useless for finding the average wind speed. They're fun to watch, but that's it. To be of value in a site survey you would have to check them every hour, 24 hours a day, every day for months on end. Instantaneous wind speed meters are useful only for developing a better understanding of the wind.

Strip-chart recorders are obsolete. An electronic odometer or accumulator is an overall better choice for an inexpensive means to log wind data.

Similar to the odometer on the dashboard of your car, an accumulator counts the miles or kilometers of wind that pass the anemometer. To estimate average wind speed, simply divide the distance the odometer records by the elapsed time between readings.

Early electronic odometers required the observer to keep track of the elapsed time. Today most instruments do that for you, as well as much more. As Wisconsin's Mick Sagrillo emphasizes, you need more than merely the average wind speed. You also need some measure of the distribution of wind speeds to accurately assess the potential of a site. Today many inexpensive instruments can produce both an average wind speed and a measure of the Energy Pattern Factor.

To record the actual speed distribution requires a significant jump in sophistication and cost. As the need for more sophisticated measurements has increased, accumulators have evolved into data loggers. In essence, data loggers consist of multiple accumulators that tally the data falling into each accumulator's domain. For example, each accumulator could represent a given wind speed range. Winds 0–1 would fall into the first accumulator, winds 2–3 would fall into the second, and so on. At the end of the observation period, the contents of each register can be used to plot the speed distribution. This distribution can then be used to calculate power density or can be compared to a wind turbine's power curve to project potential performance.

Most data loggers process some of the measured data and record the results. These smart recorders significantly reduce the time and expense of analyzing the data later. By reducing the amount of data that must be stored, they can record data for months or longer.

Anemometer Towers

Telescoping television masts are one option for an anemometer tower. Masts with four and five bays are available. The best choice today is hinged towers of lightweight tubing (see figure 3-11, Anemometer mast).

Some try to cut corners by installing the anemometer on the roof of a building. Unfortunately, turbulence around the building lowers wind speed dramatically.

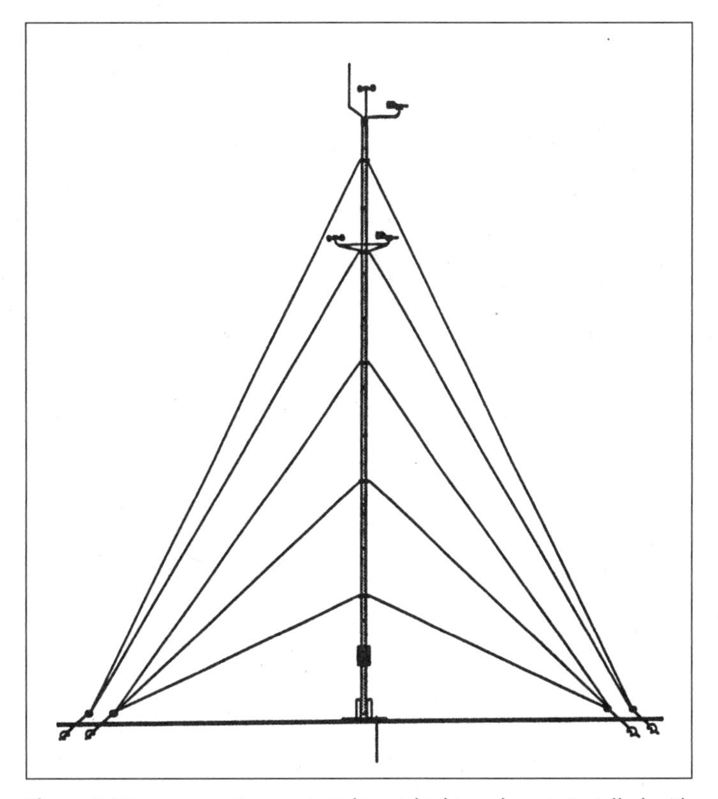

Figure 3-11. Anemometer mast. Lightweight, hinged masts installed with screw anchors greatly simplify wind measurements at remote sites. (NRG Systems)

Avoid this practice. Most of the time, mounting an anemometer (or a wind turbine, for that matter) on a building is impractical and a waste of time. Where do you attach the guy cables, for example? Will it be necessary to drill holes in the roof for anchors? If that doesn't stop you, imagine trying to erect a tall slender mast on a steeply pitched roof; it's an accident waiting to happen.

From bitter experience, wind prospectors have learned there's absolutely no substitute for measuring the wind at the height where you plan to install your wind turbine. Since even at the best sites the wind turbine will be installed on a tower at least 80 feet (24 m) tall, you'll need an anemometer mast of equivalent height. Fortunately there are mast systems designed specifically for this purpose. These masts can be erected by two people in a matter of hours. Masts are available up to 50 meters (164 ft) high.

Survey Duration

"Yeah, you got a fine site here," said the dealer as he installed the anemometer George had ordered. "I bet you've got 12 mph."

Two days later the dealer returned. After examining the anemometer he said, "Just as I thought, an easy 12 mph average." The dealer then persuaded George to buy his wind machine.

A wet finger in the air on the first visit would have been just as accurate. Maybe the dealer didn't know how to measure wind speed. Then again, maybe he was a con artist. It's hard to tell. The site was indeed a good one, and the wind turbine (made by a reputable manufacturer) was installed in a workmanlike manner. The site could have had a 12 mph average wind speed, but the dealer or the buyer wouldn't have known that after two days of measurement.

How long is enough? That's another tough question. Average wind speeds can vary as much as 25 percent from year to year. But it's obviously impractical to gather 10 years of data from a site before you decide whether or not to install a wind machine. Battelle suggests gathering one year of data. Even so, your site's average speed will be dependent on how normal the year has been with respect to the long-term average. Was it a typical year, or was it windier? Was it an El Niño year or a La Niña? To answer that question you must examine the wind data from the nearest airport or other long-term recording station and compare the test year's results with the station's historical average.

Try to establish a correlation between your site and the airport using the data summaries. If you're lucky you may find that a full year of measurements isn't necessary. But four months is a minimum. Anything less is guesswork. If you're not going to gather a full year of data, make sure that you at least capture the wind season, typically winter and spring months for much of the midlatitudes.

Data Analysis

Making sense out of the data from a site survey is more akin to alchemy than it is to science. Much is left to the judgment (or imagination) of the observer. You must determine whether the data is representative of the site and not surrounding obstructions, whether the year is normal, and whether or not there is a direct relationship between the site and, for example, a nearby airport with long-term records.

1. Calculate the ratio between your site's average speed and the airport's.
2. Adjust the airport's historical average by this ratio.

Step 1 establishes whether your site is windier or less windy than the airport. Step 2 normalizes the results for a typical year.

This approach assumes that a correlation exists between your site and the airport. There may not be one, particularly in rough or mountainous terrain. Try to use an airport in terrain similar to yours and with a similar exposure to the wind. The nearest airport may not always be the best choice.

Another method is to use linear regression analysis. This technique gives a measure of the ratio between the two sites, by testing the degree of correlation, and projects the site's average speed. Linear regression analysis is the same as graphically drawing the best-fitting line between two sets of data. Pocket calculators with engineering or statistical functions make the job a cinch, as does most spreadsheet software.

You'll find that hourly or daily wind speeds are often too erratic to establish a correlation between the two sites. Average weekly speeds are more stable. They tend to smooth out the passage of weather systems and local diurnal variations.

With your wind speed and power estimates in hand, you can now estimate how much a typical wind turbine will generate using the techniques explained in the next chapter.

Another Option—Install a Small Wind Turbine

If you don't want to go through the trouble and expense of performing a site survey, there's one avenue left. "Install a small wind machine," says Mick Sagrillo, a remanufacturer of small wind turbines in Wisconsin. The idea sounds crazy at first. But it makes sense the more you look at it (see figure 3-12, Anemometer mast as micro turbine tower).

You can install a micro wind turbine for about the same cost as an anemometer, mast, and logger. Agreed, this isn't a low-cost way to test the wind, but it works. What you get is an operating wind system. You gain hands-on experience and you learn exactly what you want to know: how well a wind turbine will perform at your site. Sagrillo points out that if the turbine doesn't work as well as you expected, it's far easier to resell a used wind turbine and tower than a used anemometer and mast. There are simply more people wanting small wind turbines than anemometers.

Like Sagrillo, Nor'Wester Energy Systems' Jason Edworthy says that small wind turbines seldom justify full-blown wind resource assessments, especially for off-the-grid use. In most stand-alone applications, the high value that a wind turbine adds to a hybrid wind and solar system warrants its use, often regardless of the wind resource. There are exceptions, of course. It makes no sense, for example, to install a wind turbine in a forest where trees will block the wind, just as it makes no sense to mount a solar panel under the shade of an awning.

Many purchasers of micro turbines have opted for the small machines over measuring the wind with a recording anemometer. Their reasoning is simple: "My photovoltaic panels don't give me enough power in winter, and I need a supplemental power source. A small wind turbine could be helpful, and I can buy a micro turbine for about what I'd pay for a recording anemometer. In the meantime I'll get something usable, kilowatt-hours, and I can track the turbine's actual energy production if I want to quantify the wind potential."

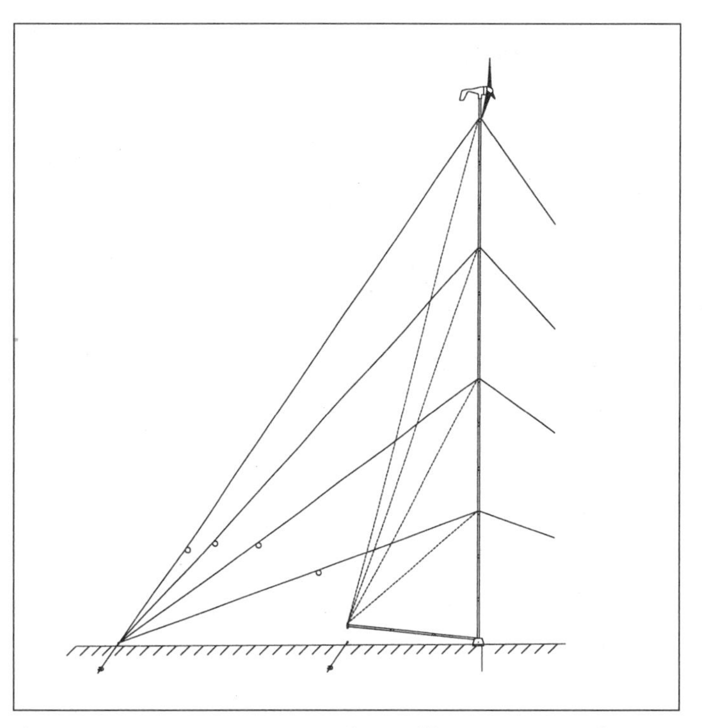

Figure 3-12. Anemometer mast as micro turbine tower. A guyed anemometer mast uses multiple sections of thin-walled tubing to support a micro wind turbine instead of an anemometer. (NRG Systems)

To take best advantage of this strategy, it helps if you want to eventually use a larger machine. Installing the micro turbine enables you to gradually phase in your wind generation as you prove its worth. For best results, you need to closely monitor the performance of the wind turbine just as you would record the wind speeds from an anemometer.

If you found, for example, that a wind turbine 7 meters (23 ft) in diameter meets your needs best, you could begin with a micro turbine about 1 meter (3 ft) in diameter. Then when you have successfully demonstrated that wind works at your site, you could graduate directly to the bigger turbine. If that's still too big a leap, you could install a turbine 3 meters (10 ft) in diameter instead.

There are a couple of ways to do this. You could install a foundation and anchors suitable for the 7-meter machine and erect a smaller turbine on a light-duty tower. Or you could install the small turbine on a heavy-duty foundation and tower. In the first case, the tower and turbine would be traded in for a heavy-duty tower and the 7-meter machine. In the second scenario you would trade in only the small turbine for the bigger one. This won't work with all wind systems, because towers and turbines may not be interchangeable, but it gives you much more flexibility than confronting an all-or-nothing decision.

This doesn't mean you shouldn't study the wind at your site. Monitoring the wind is instructive, says Edworthy. Hold a meter to the wind to develop a feel for its strength. Better yet, Edworthy adds, fly a kite. Attach streamers to the line and watch how those streamers near the ground roil and flap, and those higher up smooth out. It's easy and fun, too. Those streamers tell you about something that's invisible—the wind. And where those streamers fly smoothly is where you want your wind turbine.

In the next chapter we'll put this wind data to work as we estimate how much energy you can expect from typical small wind machines.

4

Estimating Output—How Much to Expect

Once we have decided where we want to put our wind machine and determined how much wind is available to it, we can proceed to the next step: estimating the amount of energy that typical wind machines can produce. With an estimate of the annual energy output in hand, we can examine the economics of various sizes and brands of wind machines in order to find the one that offers the most for our money or best fits our needs.

There are three methods you can use. The first is a back-of-the-envelope technique using the swept area of the wind turbine. With this method you can quickly evaluate the potential output of any wind machine. First find the average speed or power, as discussed in the previous chapter. Then simply calculate the swept area of the wind turbine's rotor. If you use it often enough, the technique will become so familiar that you'll be able to do it in your head. The second method is more involved and requires a wind speed distribution for your site and a power curve for each wind turbine you'd like to evaluate. The third approach uses manufacturers' published estimates for typical wind regimes.

As in the preceding chapter, formulas will be included to show precisely what we're doing and so you know where the numbers come from. They will also help explain important concepts. As before, formulas will be accompanied by tables or charts summarizing the results of calculations for a standard set of conditions.

Swept Area Method

Our first step is to find the power in the wind—the power density in W/m^2. Once we have determined power density, by whatever means necessary, we can easily estimate the potential power output from a wind machine. All we need is the area swept by the wind turbine's rotor.

Think about it for a moment. What is it that captures the wind in a wind machine? Is it the tower, the transmission, the generator? No, of course not. It's the spinning rotor. Yet this concept is difficult for many to grasp, even those working in the wind industry. The tower is important, as we have learned, and so is the generator. But they are not directly responsible for capturing the wind. Inevitably many people look at the size of the generator first. Yet the generator tells you very little about the size of a wind turbine. Rotor diameter says it all.

Let's assume we want to use a micro wind turbine with a rotor that intercepts

1 m^2 (slightly more than 10 ft^2) of the wind stream at a site with an average annual wind speed of 6 m/s (13.4 mph) at hub height.

From table 3-2 in the previous chapter, we found that the power density of the wind at a site with an average speed of 6 m/s is about 250 W/m^2. The power in the wind is:

$$P = P/A\ A$$
$$P = (250\ W/m^2) \times (1\ m^2)$$
$$P = 250\ W$$

What we are seeking, however, is not power but energy. When you pay your utility bill, you are primarily paying for the energy you've used. The amount of energy consumed is the product of power and time—how long the power was used. In the case of a wind machine, the energy it intercepts is a function of average power and how long it is available. In this example, we're using the average annual wind power. There are 8,760 hours in a year. Remembering that there are 1,000 watts per kilowatt, then:

The generator tells you very little about the size of a wind turbine. Rotor diameter says it all.

$$E = P(t)$$
$$E = (0.250\ kW) \times (8{,}760\ h/yr)$$
$$E = 2{,}190\ kWh/yr$$

Consequently this wind turbine will intercept about 2,200 kilowatt-hours (kWh) of energy passing through the rotor in one year.

That's not what the wind machine will pluck from the wind passing through the rotor, because we can't capture all the power available. If we could, the wind would come to a halt at the rotor and nothing further would happen. The maximum that we can capture at the rotor, the theoretical limit, was derived by the German aerodynamicist Albert Betz. The Betz limit is 59.3 percent (16/27) of the power in the wind available to the rotor.

In practice, wind turbine rotors deliver much less than the Betz limit. Optimally designed rotors reach levels above 40 percent. Usable energy is even less because energy is lost in converting the kinetic energy of the rotor to electrical energy. There are also losses due to rapid changes in wind speed and direction that are not accounted for in our simple formulas.

Well-designed drive trains operate consistently above 90 percent efficiency. The efficiency of generators, on the other hand, varies significantly depending on how they are loaded. When running at their rated output, generator efficiency can also be above 90 percent. But wind turbines infrequently drive their generators at the rated output. Much of the time the generator is partially loaded, and its efficiency suffers as a result. Power conditioning on some wind turbines interconnected with the utility can also contribute to significant losses.

Conventional wind turbines also miss some of the wind available. Unlike an anemometer, which readily measures gusts, a wind turbine does not respond to all gusts due to the rotor's inertia. The energy available in a gust as registered by an anemometer may not be extracted by the wind turbine. It may not "see" the gust.

Yawing or changing the direction of the turbine as the wind changes direction causes a similar problem for conventional wind machines. Cup anemometers capture wind from all directions. But a conventional turbine takes time to change its position and face fully into the wind; thus it misses a portion of the wind recorded by the anemometer. Vertical-axis wind turbines, because they are omnidirectional, capture the wind from all directions.

When you put all this together, a well-designed medium-size wind turbine can deliver about 30 percent of the overall energy available. This is what you can get out of the wind and actually put to use.

In practice, wind turbines capture from 12 to 40 percent of the annual energy in the wind, depending on the type of wind turbine

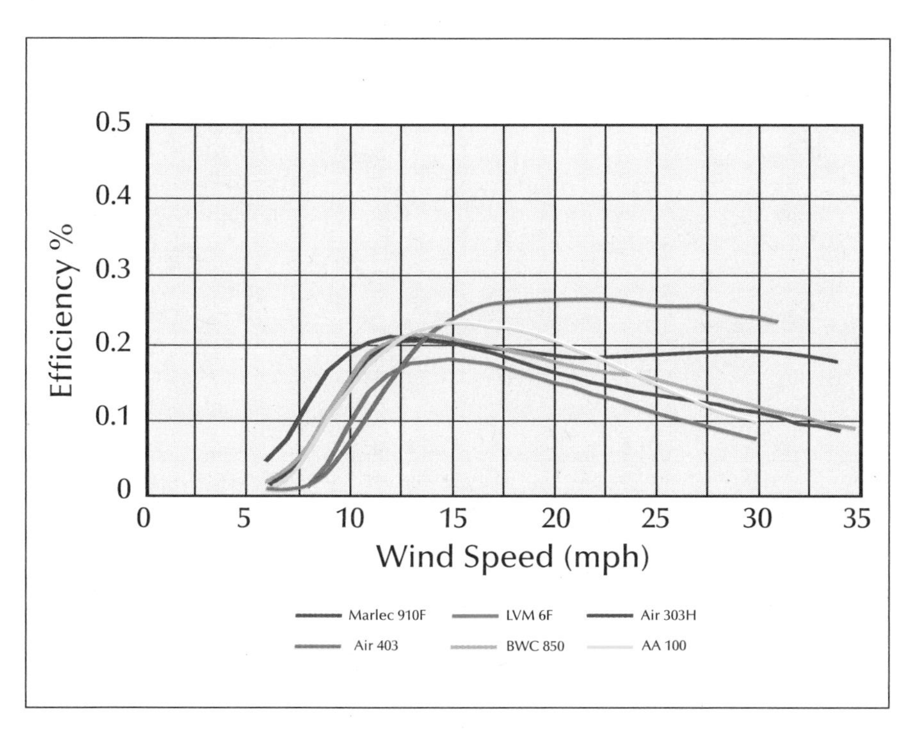

Figure 4-1. Micro and mini wind turbines' measured efficiency. The table shows the percentage of the power actually delivered to the batteries relative to the power in the wind. The data was derived from power measurements at the Wulf Test Field near Tehachapi, California, on Marlec's Rutland 910F, LVM's 6F, Southwest Windpower's Air 303H and Air 403, Bergey Windpower's 850 and Ampair's 100. The wind turbines were installed to each manufacturer's specifications, and the results are typical of what a consumer could expect. For four of the five turbines, peak efficiency of about 22 percent occurs around 13 mph (5.8 m/s). The Air 403 reached its peak performance of 27 percent at an unusually high wind speed of 20 mph (8.9 m/s). In three of the five designs efficiency gradually declines as wind speed increases. This is typical of most wind turbines both large and small, excepting those built by Southwest Windpower.

and the wind regime where it's operating. Wind turbines are designed for use in specific markets under specific wind conditions. Outside these conditions, the wind turbine performs less optimally. That's not to say they're less cost-effective, just that the overall system conversion efficiency may be less.

Small Wind Turbines

Small wind turbines are less efficient at capturing the energy in the wind than medium-size wind turbines. Despite the hype about new airfoils and breakthrough generator technology, small wind turbines seldom deliver more than 30 percent of the energy in the wind over any significant period of time. Most fall far short of that. In measurements at the U.S. Department of Agriculture's experiment station in Bushland, Texas, a Bergey 1500 converted a maximum of 23 percent of the energy in the wind to electricity.

One manufacturer of small battery-charging wind turbines crows that it uses "an airfoil so advanced it approaches the theoretical limits of efficiency." While this may be possible in a wind tunnel, it's extremely unlikely in the real world, where wind turbines must operate. Measurements of small wind turbines at the Wulf Test Field indicated that micro and mini wind turbines will deliver less than 23 percent of the power in the wind to the batteries in a stand-alone power system at most sites (see figure 4-1, Micro and mini wind turbines' measured efficiency).

If we use our example of a micro turbine with a swept area of 1 m^2 that intercepts 2,200 kWh of wind annually, and assume that it's capable of capturing and delivering to the load 20 percent of the energy available, the estimated annual energy output (or annual energy production) of the turbine is 440 kWh:

$$(2{,}200 \text{ kWh/yr}) \times 20\% = 440 \text{ kWh/m}^2\text{/yr}$$

Actual performance will probably be somewhat less, because few small wind turbines perform this well at such a windy site (see table 4-1, Annual Energy Output Estimates for Small and Medium-Size Wind Turbines).

Small wind turbines are typically designed to perform best in the low-wind regimes where

Table 4-1
Annual Energy Output Estimates for Small and Medium-Size Wind Turbines

				Small Turbines		Medium Turbines	
Annual Average Wind Speed (m/s)	Nominal (mph)	Annual Power Density (W/m²)	Annual Energy Density (kWh/m²)	Overall Conversion Efficiency (%)	Annual Energy Output (kWh/m²)	Overall Conversion Efficiency (%)	Annual Energy Output (kWh/m²)
4	9	75	656	0.20	130	0.36	240
5	11	146	1,281	0.20	260	0.35	450
6	13	253	2,214	0.19	410	0.33	720
7	16	401	3,515	0.16	570	0.29	1,000
8	18	599	5,247	0.15	770	0.26	1,340
9	20	853	7,471	0.14	1,020	0.23	1,720

Notes: Small wind turbine efficiency derived from measurements at the Wulf Test Field. One micro turbine tested delivered 23% efficiency in higher-speed wind regimes, but this is not the norm.
Medium-size wind turbine efficiency derived from tests by DEWI, Windtest KWK, and product literature.

people live, generally at sites with average wind speeds of 4 to 5 m/s (9 to 11 mph). In locales with higher average annual wind speeds, their performance drops off dramatically. At extremely windy sites, small wind turbines typically convert only 12 percent of the energy in the wind. This is a normal result of their basic design. Because of the cubic relationship of power to wind speed, there's so much energy available at windy sites that designers can afford to capture only a small part of it.

To summarize, the steps necessary to estimate the annual energy output (AEO) of any wind turbine are:

1. Find the wind power at the site and the height at which the wind machine will operate, and calculate the energy density.
2. Find the area swept by the wind turbine's rotor.
3. Assume a reasonable value for the overall conversion efficiency of the entire wind system.

Thus

$$\text{AEO} = (\text{P/A}) \times (\text{A}) \times (\%\ \text{efficiency}) \times (8{,}760\ \text{h/yr}) \times (1{,}000\ \text{W/kW})$$

When you look at the product literature describing a wind turbine, the swept area isn't always apparent. Most manufacturers, though, now list it along with other measures of performance. What is always obvious, or clearly stated in the literature, is rotor diameter. Given rotor diameter, you can calculate the area swept by the rotor.

You can get the feel of this technique by working through another example. Let's assume we want to use a mini wind turbine the size of a Bergey XL1 with a rotor 2.5 meters (8.2 ft) in diameter (D). The rotor intercepts:

$$\begin{aligned} A &= \pi\ (D/2)^2 \\ &= \pi \times (2.5/2)^2 \\ &= \pi \times 1.25^2 \\ &= \pi\ 1.6 \\ &= 4.9\ \text{m}^2\ \text{(or about 50 ft}^2\ \text{of the wind stream)} \end{aligned}$$

Let's say we plan to install the turbine in the Texas Panhandle, where the average annual wind speed at hub height is 12 mph (5.4 m/s) and the distribution of wind speeds approximates a Rayleigh function. Now let's see what we can expect using the formula for power density in W/m² when wind speed is in mph and area in m².

$$\text{AEO} = (0.05472\,V^3) \times (1.91\ \text{EPF}) \times A \times 20\% \times (8{,}760\ \text{h/yr}) \times (1\ \text{kW}/1{,}000\ \text{W})$$

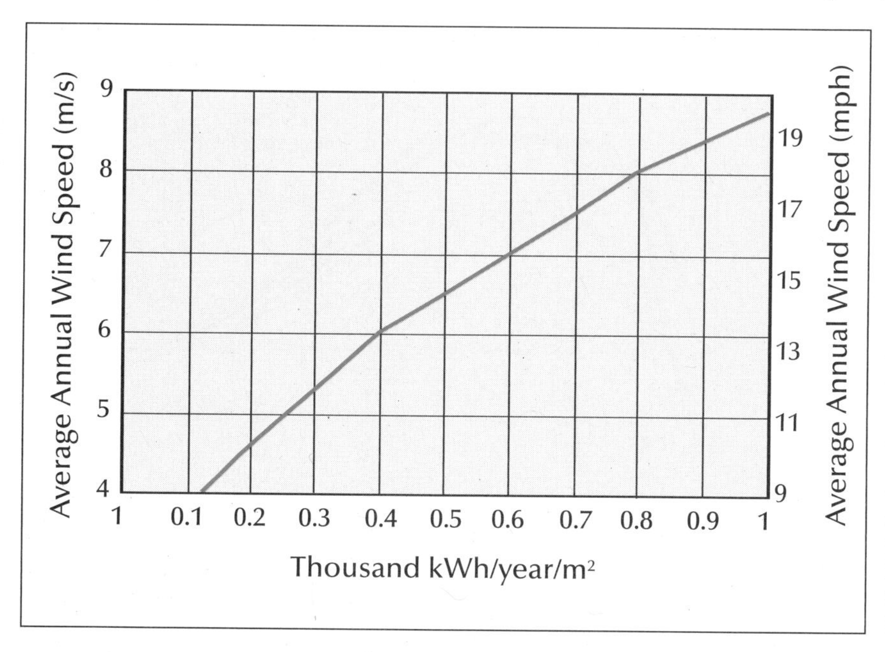

Figure 4-2. Annual energy output for small wind turbines per square meter of swept area. The chart allows you to estimate the amount of energy small wind turbines can capture relative to swept area as a function of hub-height annual wind speed. This chart is derived from measurements on four small wind turbines at the Wulf Test Field and from product literature. It is best applied to small wind turbines less than 7 meters (23 ft) in diameter. To use the chart, find the average annual wind speed at hub height for your site on the vertical axis. Follow the horizontal line until it intersects with the curve. Then proceed down the page to the horizontal axis. For example, a small wind turbine at a site with an average annual wind speed of 5 m/s should be capable of generating about 0.25 thousand kWh/yr/m^2 or 250 kWh/yr/m^2. In a battery-charging system, it's unlikely that all this energy will be delivered to useful loads.

$$= [0.05472 \times (12 \text{ mph})^3 \text{ W/m}^2] \times 1.91 \times 4.9 \text{ m}^2 \times 20\% \times (8{,}760 \text{ h/yr}) \times (1 \text{ kW}/1{,}000 \text{ W})$$
$$\sim 1{,}550 \text{ kWh/yr}$$

The chart in figure 4-2 summarizes the results for a range of average wind speeds relative to a swept area of one square meter. Simply find the area swept by the wind turbine in square meters and multiply it by the energy that's likely to be captured by a small wind turbine.

For example, at a site with a 5 m/s (11.2 mph) average wind speed, a typical small wind turbine captures about 250 kWh/yr/m^2. Our mini turbine from the previous example intercepts 4.9 m^2, therefore

$$AEO = (4.9 \text{ m}^2) \times (250 \text{ kWh/yr/m}^2) = 1{,}225 \text{ kWh/yr}$$

Now let's use the same turbine in another example. In this case, assume you don't know the average wind speed but you've identified the location on a wind power map and found it's a class 4 site with a wind power density of 200 to 250 W/m^2. You want to know what you can expect if you install the turbine on a 100-foot (33 m) tower.

Battelle's wind power classes are based on wind speeds at 10 meters (33 ft) above the ground. They've assumed that the increase in wind power with height for most sites corresponds to the 1/7 power law. The 100-foot tower is equivalent to the 30-meter height used by Battelle in its assessments of the United States' wind resource. At a height of 100 feet, there's 320–400 W/m^2 in the wind:

$$A = 4.9 \text{ m}^2$$
$$AEO = (P/A) \times (A) \times (8{,}760 \text{ h/yr}) \times 20\% \times (1 \text{ kW}/1{,}000 \text{ W})$$
$$= (320\text{–}400 \text{ W/m}^2) \times (4.9 \text{ m}^2) \times 20\% \times (8{,}760 \text{ h/yr}) \times (1 \text{ kW}/1{,}000 \text{ W})$$
$$\sim 2{,}750\text{–}3{,}400 \text{ kWh/yr}$$

Medium-Size Wind Turbines

Medium-size wind turbines are considerably more productive per square meter of rotor swept area than small wind turbines. The peak efficiency of most medium-size turbines is nearly double that of small turbines (see figure 4-3, Medium-size wind turbines' measured efficiency). Overall conversion efficiencies are proportionally greater, as well.

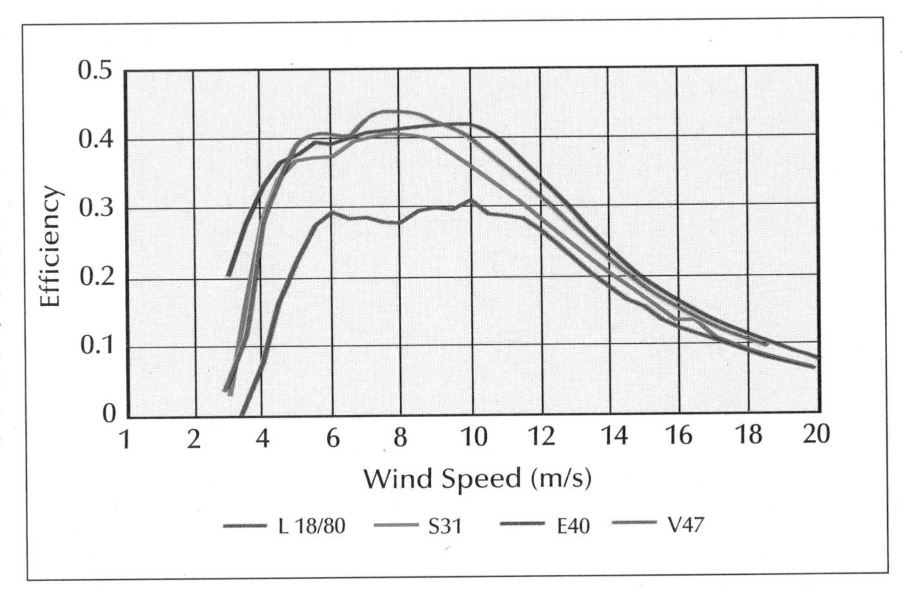

Figure 4-3. Medium-size wind turbines' measured efficiency. The overall conversion efficiency for four typical medium-size wind turbines: the Lagerwey 18/80, Südwind S31, Enercon E40, and Vestas V47. Efficiency is derived from power curves measured by the Deutches Windenergie Institut and Windtest Kaiser-Wilhelm-Koog. The Lagerwey is an 18-meter (59 ft) diameter 80 kW turbine; the Südwind is a 31-meter (102 ft), 300 kW turbine; the Enercon is a 40-meter (131 ft), 500 kW turbine; and the Vestas is a 47-meter (154 ft), 660 kW turbine.

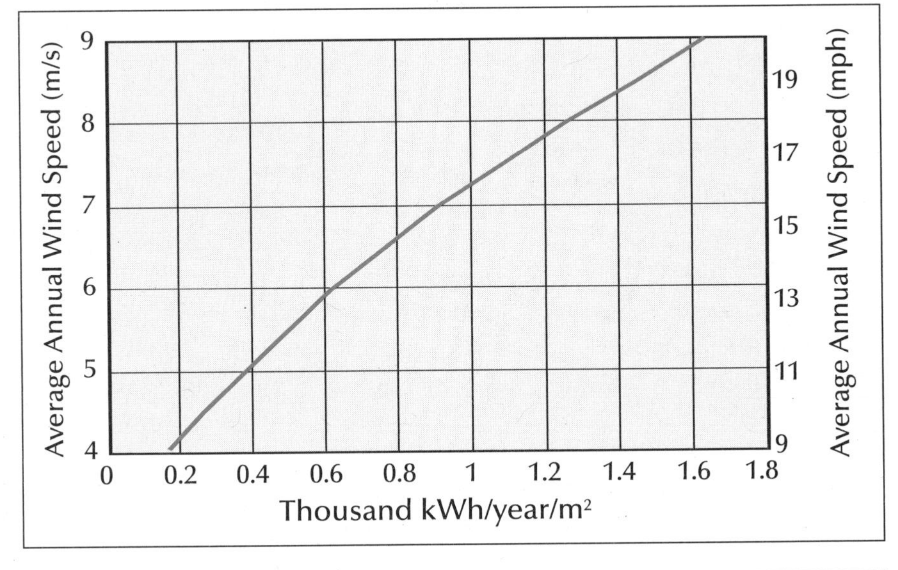

Figure 4-4. Annual energy output for medium-size wind turbines per square meter of swept area. This chart synthesizes annual production estimates from most of the medium-size wind turbines on the international market. Note that medium-size wind turbines are more efficient at capturing the energy in the wind than are small wind turbines. The wind speed on the vertical axis is for hub height.

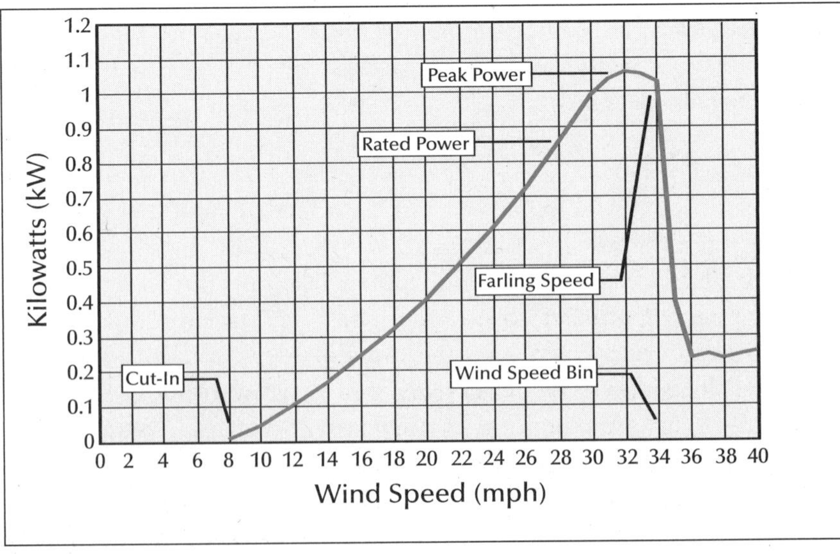

Figure 4-5. Sample small wind turbine power curve. The wind speed on the horizontal axis is for hub height. Note that the power on the vertical axis should represent the average power that the turbine will produce within each wind speed bin, not the maximum instantaneous power occasionally reached. Published power curves should denote the reference height for the wind speed measurements (usually hub height), whether the data was verified by an independent test field, and what averaging period was used (instantaneous, 1-minute, or 10-minute).

You can apply the same method used to estimate annual energy generation from small turbines for medium-size wind turbines. For example, at a site with a 6 m/s (13.4 mph) average annual wind speed, typical medium-size wind turbines will capture 33 percent of the energy in the wind.

Let's say you wanted to estimate the production from a 250 kW wind turbine with a rotor 25 meters (82 ft) in diameter.

$$A = \pi R^2$$
$$= \pi \times 12.5^2$$
$$\sim 500 \text{ m}^2$$

From the previous example, the power density for a 6 m/s site is 250 W/m^2. Consequently, the annual energy output is

$$AEO = (250 \text{ W/m}^2) \times (500 \text{ m}^2) \times 33\%$$
$$\times (8{,}760 \text{ h/yr}) \times (1 \text{ kW}/1{,}000 \text{ W})$$
$$\sim 350{,}000 \text{ kWh/yr}$$

You can find the same result by using the chart in figure 4-4. For a site with a 6 m/s average wind speed, the typical medium-size wind turbine captures about 700 kWh/yr/m^2. For a 25-meter wind turbine, then:

$$AEO = (700 \text{ kWh/yr/m}^2) \times (500 \text{ m}^2)$$
$$\sim 350{,}000 \text{ kWh/yr}$$

While the swept area technique is simple and straightforward, it encompasses all the critical factors affecting how much gross energy a wind generator will convert to electricity. More importantly, frequent use of this technique reinforces the significance of rotor diameter or swept area as opposed to generator size in gauging the relative size of wind turbines, both large and small.

Power Curve Method

Where you have access to the wind speed distribution for your site, or at least for the nearest long-term recording station, you can use a wind turbine's power curve to estimate the annual energy output. This is the method used by meteorologists when determining the potential generation from a wind machine in a commercial wind power plant. Essentially you match the speed distribution with the power curve to find the number of hours per year the wind turbine will be generating at various power levels (see figure 4-5, Sample small wind turbine power curve).

Calculating Swept Area

If the swept area of a wind turbine isn't provided, you can calculate it by using the formulas below. For a conventional wind turbine, use the formula for the area of a circle; for an H-rotor, use the formula for the area of a rectangle; and for Darrieus rotors, use the formula approximating the area of an ellipse.

Conventional rotor	$A = \pi R^2$
H-rotor	$A = DH$
Darrieus rotor	$A = 0.65\ DH$

where R is the radius of the rotor (half the diameter), D is the diameter, and H is the height of the blades on a vertical-axis wind turbine.

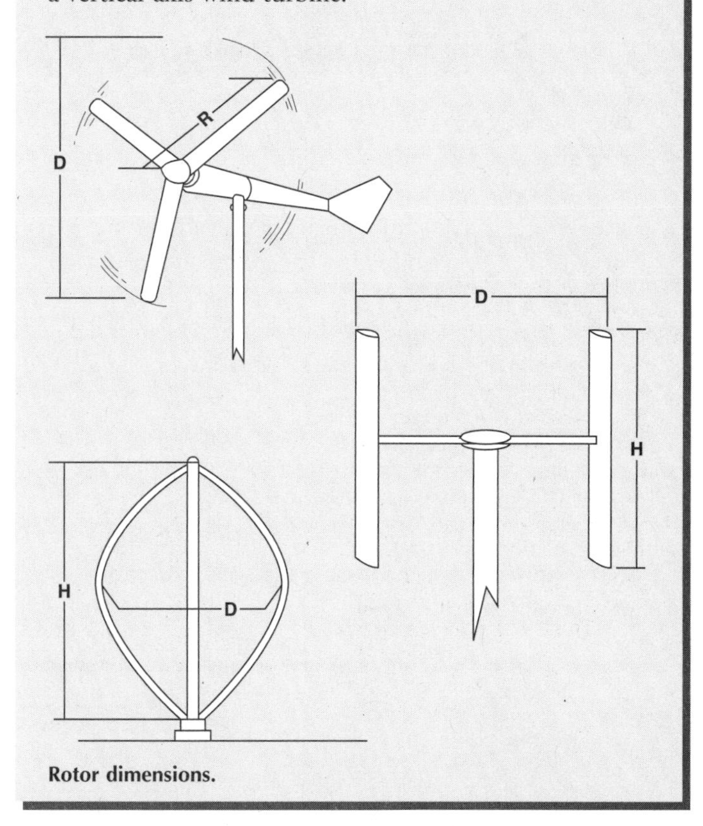

Rotor dimensions.

Swept Area Rules of Thumb

For the metrically challenged, a reasonable rule of thumb is that 1 square meter of rotor area is about equal to 10 square feet—10.8 ft^2 to be exact.

Swept Area, Rotor Diameter, and Nominal Power Rules of Thumb

Swept Area (m^2)	Nominal (ft^2)	Nominal Rotor Diameter (m)	(ft)	Nominal Power Rating (kW)
1	10	1.1	4	0.1
5	50	2.5	8	0.75
10	110	3.6	12	1
50	540	8	26	10–20
100	1,080	11	37	25–40
500	5,380	25	83	200–250
1,000	10,800	36	118	300–400
2,000	21,500	50	164	500–700
3,000	32,300	62	203	800–1,000
4,000	43,000	71	233	1,000–1,500
5,000	53,800	80	262	1,500–2,500

Avoid Average Speed Confusion

Newcomers to wind energy, in their zeal to estimate how much electricity they can generate with a particular wind turbine, sometimes confuse the power curve with graphs of annual energy output. Mike Bergey of Bergey Windpower has found that some customers erroneously apply the average annual wind speed at their site to the wind speed bin shown on the power curve. Unfortunately, this approach ignores the effect of the speed distribution on a wind turbine's production. If you want to use the average annual wind speed method, you must use tables or charts of the annual energy output. If you want to use the power curve method, you must use a wind speed frequency distribution, preferably a distribution for your specific site.

The power curves for most medium-size wind turbines have been independently measured by international testing laboratories because their customers demand them.

First, a word of caution. The power curves proffered by some manufacturers of small turbines can best be characterized as informed guesswork. View small wind turbine power curves and the energy calculations that result with a good dose of skepticism. There are no government agencies or independent testing laboratories ensuring the accuracy of published power curves for small wind turbines. In contrast, the power curves for most medium-size wind turbines have been independently measured by international testing laboratories because their customers demand them. The efficiency calculations on a sample of medium-size turbines in figure 4-3 used power curves verified by European test fields—not rosy projections provided by manufacturers.

Let's use the sample power curve in figure 4-5 to find the power the turbine will produce at wind speeds from cut-in through furling in a 12 mph (~5 m/s) wind regime with a Rayleigh distribution. From the distribution, we can calculate the length of time that the winds occur at each speed. We can exclude speeds below cut-in because no power is produced. Similarly, we can ignore wind speeds above 40 mph (18 m/s) because there are few occurrences of winds above this speed in this wind regime (see figure 4-6, Power curve method of calculating annual energy output).

For example, at a wind speed of 18 mph (8 m/s) this small turbine will produce somewhat more than 300 watts. Winds occur at this speed about 300 hours per year (see table 4-2, Estimating Annual Energy Output Using a Power Curve). In this one speed bin, the turbine produces:

(0.33 kW) x (300 h/yr) ~100 kWh/yr

Across the entire speed range, we find that our sample wind turbine will generate about 1,400 kWh per year at this site.

See the appendixes for an example of a spreadsheet calculating the annual energy output for a Bergey XL1.

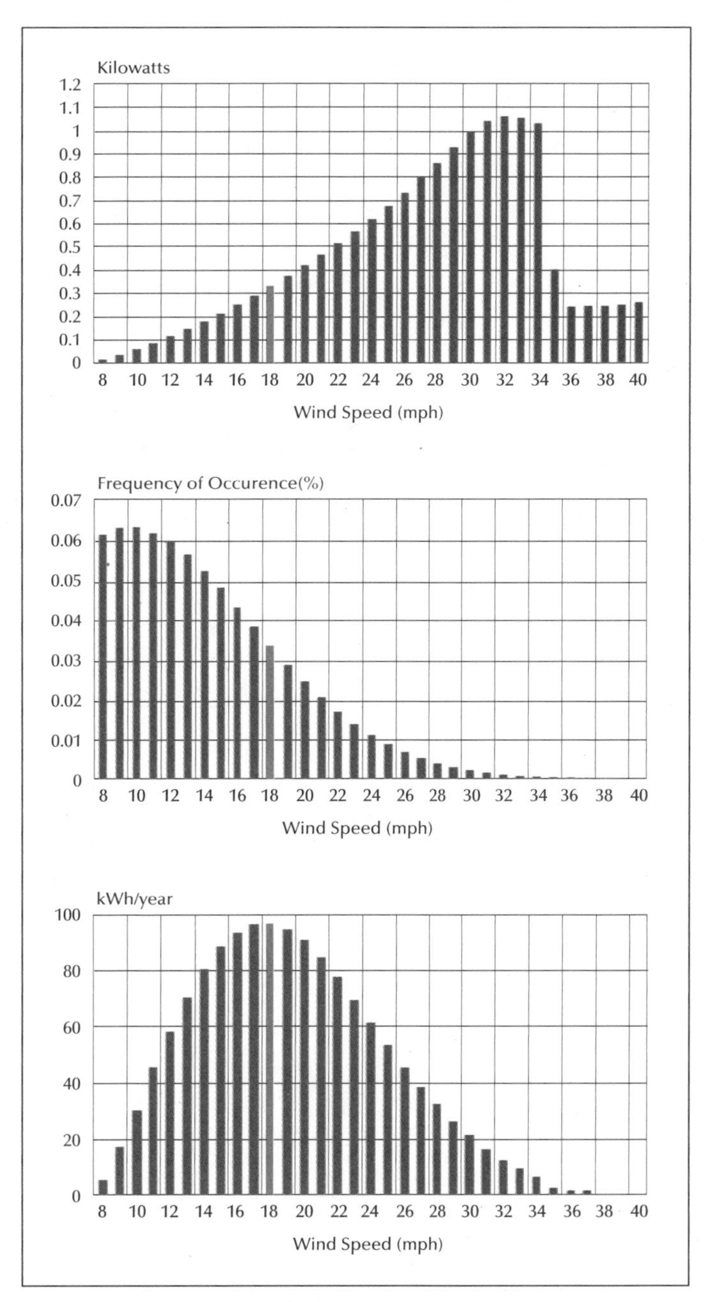

Figure 4-6. Power curve method of calculating annual energy output. Power is shown on the vertical axis, and wind speed at hub height on the horizontal axis. In this example the wind turbine produces slightly more than 300 watts at a wind speed of 18 mph (8 m/s). Winds of this speed occur 0.0335 percent of the time or about 300 hours per year. Consequently, this wind speed bin contributes almost 100 kWh per year of the turbines' total annual generation.

Manufacturers' Estimates

Most manufacturers provide estimates of the amount of energy they expect their turbines will capture under standard conditions: hub-height wind speeds, Rayleigh distribution (Weibull distribution with a shape factor of 2), sea level, and 15°C. The format varies. Some companies provide a chart of the estimated annual energy output (AEO in American jargon) at various annual average wind speeds (see figure 4-7, Sample curve of annual energy output). Others provide the same data in tabular form (see table 4-3, Manufacturer's Estimate of Annual Energy Output for Vergnet GEV 4/2).

Suppliers of battery-charging wind turbines to the off-the-grid market often present this data as kWh per month rather than kWh per year. Southwest Windpower, for example, estimates that its Whisper H40, which uses a 7-foot (2.1 m) diameter rotor, will produce about 100 kWh per month at a site with a 12 mph (~5 m/s) average wind speed, or about 1,200 kWh per year.

No matter what technique is used, these projections represent the gross amount of energy that a wind turbine can be expected to produce. For a host of reasons, rarely is all this energy put to use. For example, energy is lost in the cables connecting the wind turbine to the loads.

For medium-size wind turbines that are connected to the grid, it's not uncommon that up to 5 percent of the gross energy captured by the wind turbine is lost. This is

Table 4-2
Estimating Annual Energy Output Using a Power Curve

Wind Speed Bin (mph)	Instantaneous Power (kW)	Rayleigh Frequency Distribution (% Occurrence)	(h/yr)	Energy (kWh/yr)
8	0.010	0.0616	539	5
9	0.030	0.0631	553	17
10	0.055	0.0632	554	30
11	0.082	0.0620	543	45
12	0.111	0.0597	523	58
13	0.142	0.0564	494	70
14	0.175	0.0524	459	80
15	0.210	0.0480	420	88
16	0.247	0.0432	378	93
17	0.286	0.0383	336	96
18	0.327	0.0335	294	96
19	0.370	0.0289	253	94
20	0.415	0.0246	216	90
33	1.055	0.0009	8	9
34	1.030	0.0007	6	6
35	0.400	0.0005	4	2
36	0.240	0.0003	3	1
37	0.250	0.0002	2	1
38	0.240	0.0002	1	0
39	0.250	0.0001	1	0
40	0.260	0.0001	1	0
Annual Energy Output (AEO)				1,416

Note: Sample small wind turbine 2.4 meters (8 ft) in diameter, Rayleigh distribution average annual wind speed = 12 mph

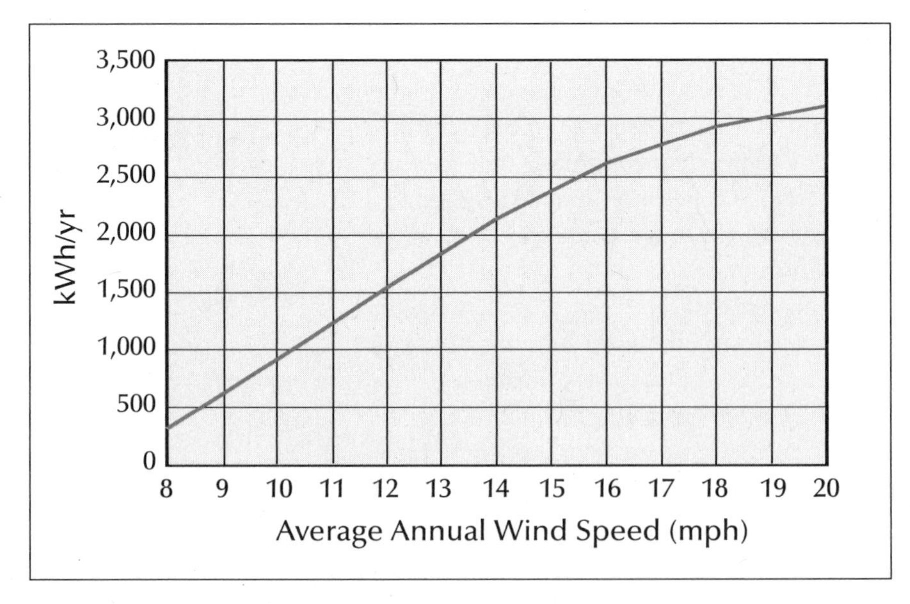

Figure 4-7. Sample curve of annual energy output. This chart presents an estimate of the gross annual generation of electricity for a small wind turbine 2.4 meters (8 ft) in diameter relative to annual average wind speed at hub height. The estimates assume a Rayleigh speed distribution (Weibull shape factor of 2). The same data is also frequently presented in tabular form. Most manufacturers note that "Your performance may vary."

Table 4-3
Manufacturer's Estimate of Annual Energy Output for Vergnet's GEV 4/2

Annual Average Wind Speed (m/s)	Annual Energy Output (kWh/yr)
5	3,500
7	6,900
9	9,300

Notes: Rotor diameter = 4m; rated power = 2 kW
Source: Vergnet SA

partly due to the wind turbine's operating less than 100 percent of the time (the turbine must be periodically stopped for maintenance) and partly due to losses in the cables and transformers. There are further losses when wind turbines are clustered together in a wind farm, because machines downwind produce less than those upwind.

In battery-charging systems, batteries that are fully charged may not be able to take more energy when, in a good stiff wind, the turbine is churning it out. Moreover, some of the energy that is eventually stored in the batteries is lost due to the inherent inefficiency of battery storage. Additional losses are incurred when using an inverter to convert the DC stored in the batteries to run AC appliances.

The Method of Bins

Power curves, like many other aspects of wind energy, are just approximations of what happens in a complex environment. The smooth line seen in sales brochures belies a complexity that frequently confuses consumers. Often the new owner of a small wind turbine will become disenchanted after watching the power being produced at various wind speeds.

In actual use the instantaneous power from a wind turbine varies for a particular wind speed, depending on whether the rotor was coasting from a previous gust at the time the measurement was made (in which case power would be greater than average) or was coming up to speed and the anemometer registered the gust but the rotor did not (in which case power would be less than the norm). It's very difficult for an observer to monitor simultaneously both fluctuating power and wind speed. There's a great deal of scatter in power measurements (see figure 4-8, Small wind turbine power curve scatter plot).

Because of the wide fluctuations in power and wind speed measurements, manufacturers and testing laboratories have agreed to average the measurements over periods of

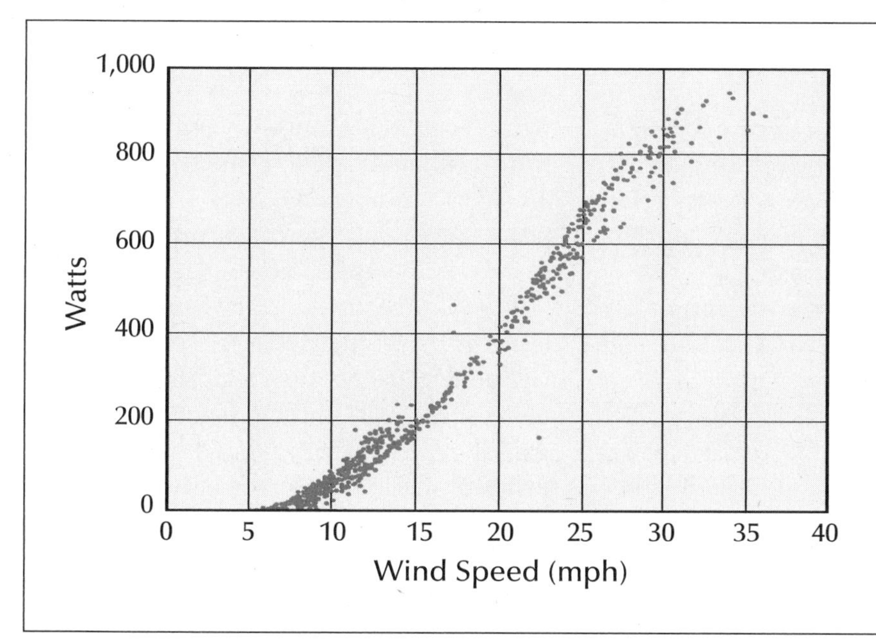

Figure 4-8. Small wind turbine power curve scatter plot. Measurements of power and hub-height wind speed collected on an 8-foot (2.4 m) diameter wind turbine at the Wulf Test Field during a two-week period. Data on wind speed and power was averaged every 10 minutes and recorded. Note that there is a bifurcation in the power curve in winds from 10 to 15 mph (5–7 m/s). This is typical of the kind of problem that can arise when measuring the power from wind turbines of any size. High winds at this site are consistently from the northwest, but low winds are both from the northwest and also from the southeast, where they pass over a nearby row of trees. The trees reduce the wind speed seen by the anemometer but have little effect on the turbine. Thus, the power curve shifts vertically, giving a misleading indication of low-wind performance.

Power Curve Nomenclature

Start-up speed. The wind speed at which the rotor first begins to turn after being at rest. Once spinning, wind turbine rotors can coast to lower wind speeds than those necessary to start the rotor revolving.

Cut-in speed. The wind speed at which the generator first begins to produce power. If the wind turbine in a battery-charging system uses a permanent-magnet alternator, voltage is generated whenever the rotor is turning, but current doesn't begin to flow until the voltage exceeds that in the batteries. For wind turbines interconnected with the utility network, cut-in occurs when the generator is connected to the grid.

Rated speed and power. The wind speed at which the generator produces the advertised power. Though frequently used as a reference for the size of a wind machine, rated speed and power have little utility. Most small wind machines and many medium-size wind turbines will produce more than their rated power.

Peak power. The maximum power the wind generator is capable of producing.

Cut-out speed. The wind speed at which the wind generator stops producing power or activates some means of high wind speed protection. This can be accomplished by applying a brake or other mechanism for physically stopping the rotor, or feathering the blades. Most medium-size wind turbines turn themselves off in winds above 50 mph (25 m/s) to reduce wear and tear and protect the turbine from damage. Small wind turbines using permanent-magnet alternators typically have no cut-out speed. They continue producing power in high winds. Most, however, do have a furling speed.

Furling speed. The wind speed at which a wind machine using a tail vane begins to fold or furl toward the tail. This reduces the area of the rotor intercepting the wind, protecting the turbine in high winds.

Wind speed bin. Though wind speed varies over a continuum, measurements used to develop power curves group power measurements into discrete registers or bins. The 5 mph bin represents winds from 4.5 to 5.5 mph. For wind speeds in m/s, each bin is typically 0.5 m/s wide. The 5 m/s bin, for example, would represent winds from 4.75 to 5.25 m/s.

time sufficiently long for the averages to become repeatable from one test to the next. Professionally produced power curves result from extensive averaging using the "method of bins." This technique sorts a series of power measurements into a corresponding interval or register of wind speed. For example, a series of power measurements averaged in the wind speed interval from 9.5 to 10.5 mph represents the average power produced in the 10 mph (~4.5 m/s) "bin." For medium-size wind turbines, the international standard is to average a minimum of thirty 10-minute samples (five hours of measurements for each bin) to accurately characterize the power at any one wind speed. Small wind turbine manufacturers may sometimes average a minimum of thirty 1-minute samples to derive their power curves (30 minutes of measurements for each bin).

Power curves published by reputable manufacturers depict the average power the turbine will produce within each wind speed interval. Some small wind turbine manufacturers still publish power curves produced from measurements of maximum instantaneous power, not the average power. Power curves derived in such a manner are misleading, and seldom will the consumer see the same results as those published by the manufacturer.

Performance Ratings

The practice in several countries has been to describe a wind machine's size by referring to its generator capacity in watts or kilowatts at some rated wind speed. This power rating is then used extensively in product promotion. For a moment, though, let's take an excursion. Let's go to Denmark—Roskilde, to be specific—to the Danish Test Center for Wind Turbines.

"Helge, that's a beaut, what size is it?"

"Ten meters."

"What? No, I meant how big is it; my Danish isn't too good."

"It's a 10-meter, but there's a bigger 12-meter down the road. Would you like to see it?"

"We're not going another step until you tell me how big that is!"

"You Americans are so demanding."

"All right, one more time: How big a generator does it have?"

"Which one?" Helge asked quizzically.

Boy, what a case of jet lag, the American said to himself. "I want to know how big the generator is. You know, the thing that generates the electricity, the guts of the machine."

Helge's patience was beginning to wear thin. "It has two generators, the largest of which is 30 kilowatts, but it has a 10-meter rotor driving it."

"Phew, I thought I'd never get it out of you," said the exasperated Yankee.

"*Tak* [thanks]," replied Helge.

Such an exchange has probably taken place more than once. We find ourselves in this predicament—identifying wind machines by their generator size—because many of the early pioneers in wind technology came from the electric utility industry or were designing wind turbines for the utilities. In common parlance we refer to power plants by the combined size of their generators. For example, an Inuit village runs a 50 kW diesel generator; it is the plume from a 500 MW coal-fired power plant that clouds the valley; and it was the 900 MW Unit 2 reactor that was damaged at Three Mile Island.

Utilities try to run their plants as close to full load as possible when they are in operation so the generators perform as efficiently as possible. As a result, a 500 MW generator normally produces 500 MW. Engineers understandably use this power rating when they talk to one another.

Wind Power Plant Losses

It's fairly straightforward to estimate how much any individual wind turbine will produce using the techniques described in this chapter. There are, however, numerous losses not accounted for by these simple methods, especially when groups of turbines are clustered together. These include losses for availability, electrical resistance, and array interference. Together, these can be significant.

No wind turbine operates 100 percent of the time. All wind turbines must be periodically inspected, if not maintained. When a wind turbine is stopped for maintenance, it's no longer "available" for operation—the wind industry's term of art for how much of the time the machine is in service.

Most medium-size and larger turbines are available for operation more than 98 percent of the time. That is, the wind turbine is stopped for reasons other than a lack of wind less than one week of the year. High availability became such an expected part of wind turbine operations by the mid-1990s that trade publications found newsworthy any hint that a company's availability had fallen to less than 97 percent. Electrical losses in large projects with long cable runs can amount to 3 percent.

More substantial losses can accrue from the interference of one wind turbine with the next in a large array. Interference can cut production 5 to 10 percent. As seen in some California wind farms, poorly designed arrays, where the turbines are too close together, losses can be even greater.

Accounting for availability, electrical losses, and array interference, actual electricity delivered may be 85 to 90 percent of that derived from a simple estimate of what a wind turbine can produce.

Wind machines are different because the wind is an intermittent resource. Fluctuations in wind speed cause generator output to vary, as well. Seldom does a wind generator produce its rated output for any extended period of time. Moreover, optimal rotor and generator combinations depend on the wind regime. A wind turbine with a 10-meter rotor may, for instance, perform most efficiently (deliver the most energy) matched with a 10-kilowatt generator in regions with low average wind speeds, but the same rotor may work better with a 25-kilowatt generator where it's windier.

Rotor diameter and swept area are better measures of a wind turbine's capability than its generator rating because it's the rotor and not the generator that captures the wind and converts it to a useful form. The generator comes later in the conversion process. Because nearly all wind turbines today use conventional rotors that sweep a circle, rotor diameter becomes a ready shorthand for the area swept by the rotor.

Consider Ian Woofenden's experience with an African Wind Power 3.6 and an older World Power Technologies Whisper 1000 at his home in Washington State's San Juan Islands. Peak power for the AWP 3.6 he tested was 900 watts, while the Whisper was rated at 1,000 watts. After comparing the performance of both turbines, Woofenden, a *Home Power* magazine associate editor, exclaimed, "Swept area rules!" The AWP 3.6 intercepts almost twice as much wind as the Whisper 1000 and the average power from the turbine he tested reflected this, "putting the lie to the standard rating system we've been working with," he says. The AWP machine produced "just over double the energy of the Whisper," even though the Whisper had a higher power "rating" than the AWP 3.6.

Most Europeans, like Helge in our fictional anecdote, refer to the size of their wind machines by rotor diameter. For example, Danish manufacturer Vestas has consistently designated its various models by rotor diameter in meters. Vestas shipped hundreds of its V15s to California during the state's great wind rush. In the late 1990s hundreds more of its V27s were installed. At the dawn of the new millennium, it was selling its V47 and the even larger V66 and V80. In Germany, Enercon has sold thousands of the wind turbine it dubs the E40 and its brawnier sibling, the E60.

The awareness of rotor diameter's importance in designating a wind turbine's size, coupled with a reluctance to part completely with traditional generator ratings, has led to hybrid designations that use both rotor diameter and generator capacity. In the early 1980s, for example, Enertech introduced its model 44-40, which used a rotor 44 feet (13 m) in diameter to drive a 40 kW generator. Lagerwey has sold hundreds of its 18/80 (18-meter diameter, 80 kW) model to Dutch farmers. In the late 1990s Südwind was marketing its N4660 (46-meter diameter, 600 kW) in Germany.

The rated power system is not only confusing but can also be misleading. First, there's no reference speed to compare one turbine to another: Rated speeds range from 22 mph (10 m/s) to more than 30 mph (15 m/s). Second, some manufacturers rate their machines at peak power output, and others don't. Medium-size machines, especially those using aerodynamic stall to regulate peak power, will often exceed their rated capacity, sometimes by up to 30 percent.

A few, shall we say, less-than-reputable manufacturers have taken advantage of the emphasis on generator size by adding large generators to relatively small rotors. By using this rating system, it's possible to slap a 6-foot plank on the shaft of a 25 kW generator and call it a 25-kilowatt wind turbine. In one particularly notorious case, Fayette Manufacturing built a 10-meter turbine and saddled it with a 95 kW generator. Most other manufacturers would have rated a turbine of this size at 25 to 35 kW. Years of results from some 1,000 of these machines in California proved that they performed no better and often much worse than other turbines with rotors of similar swept area driving only 25 kW generators. After years of poor performance, the Fayette machines were eventually scrapped.

Because of questionable power ratings by some manufacturers and general confusion by consumers as to what the numbers mean, the American Wind Energy Association (AWEA) attempted to clarify performance ratings in the 1980s by calling for a standard list of parameters. These included maximum power (not rated) and, most importantly, the annual energy output at various average speeds. AWEA hoped to replace the rated power at rated speed nomenclature with values that made more sense. Unfortunately the association never enforced the standard, and the same problems exist in the 21st century.

Efficiency

Now that you've estimated what a wind machine will produce at your site, you may conclude that it isn't enough. If that's the case, there are three ways to increase output:

1. Increase wind speed—that is, find a better site or use a taller tower.
2. Increase the swept area of the rotor by finding a wind turbine with a larger rotor diameter.
3. Improve the wind turbine's conversion efficiency—that is, find another wind turbine.

Efficiency is placed last for a reason. Many people have a disturbing fondness for the word *efficiency.* Invariably they will say, "Yeah, all those calculations are fine. But what's the most efficient turbine built today?" Several wind turbine manufacturers cater to this obsession by hyping their wind turbines as the "most efficient" on the market. What many overlook when they focus on efficiency is that energy output is more sensitive to wind speed (because of the cube law) and swept area (because of the square of the rotor's radius) than efficiency.

Wind turbines must first be reliable; second, they must be cost-effective. Efficiency is important, but it's not the sole criterion for judging the performance of a wind machine.

To gauge the potential of a wind turbine, ignore the size of the generator or its purported efficiency and get right to what matters most: the monthly or annual energy output for your site. But if the AEO is not available, remember that nothing outside the wind itself, no other single parameter, is more important in determining a wind machine's capability of capturing the energy in the wind than the area swept by the rotor.

Investors, not just homeowners but also savvy executives who should have known better, have lost staggering amounts of money, because they didn't grasp this concept. Even after years of experience in operating wind machines, and reams of technical documents, there's always someone who has made the startling discovery of a way to beat the Betz limit. (Most of them have never heard of Betz, unfortunately.) These wonderful new devices are not only more efficient than any previous wind machines, but will also produce more energy than theoretically possible. That's not to say that it can't be done. The Betz limit, after all, is only a theory. But no one has done it yet, and plenty have said they could. And as sure as the rain will fall, there'll be more people willing to part with their savings before punching a few numbers into a calculator or scribbling on the back of an envelope.

> A few, shall we say, less-than-reputable manufacturers have taken advantage of the emphasis on generator size by adding large generators to relatively small rotors.

Now that we have the tools to estimate the gross amount of energy that a wind turbine will produce, we can proceed to using this information to examine cost-effectiveness.

5

Economics—Does Wind Pay?

Ill blows the wind that profits nobody.
—William Shakespeare, *Henry VI, Part III,* Act II

Will a wind machine pay for itself? Will it be a sound investment? Or, more simply, is it worth the trouble? The answers to these frequently asked questions are elusive. They depend on a number of speculative variables not subject to precise calculation, such as inflation, interest rates, and the desired rate of return. Nor is there just one straightforward way to look at the economics.

A better question may be: Does a turbine have to pay for itself? Eric Eggleston, a wind energy engineer, notes that for those living off the grid it's a question of the least costly way of getting electricity to the site. It's not about payback at all. Certainly, commercial-size wind machines intended to produce bulk power in competition with conventional fuels must be cost-effective. But small wind turbines need not pay for themselves overnight, or even within 10 years, to prove beneficial. As long as they pay for themselves within their expected lifetimes, they're economical in the strict sense of the term.

Yet even this dictum may be too restrictive when the value of wind energy is compared to other products. Consumers often buy items of equivalent cost that provide no monetary return whatsoever. Small wind turbines, which are relatively more expensive than medium-size wind turbines, can be justified as easily as any other consumer product on this basis. For this reason Mike Mangin, a community wind advocate in Wisconsin, terms any wind turbine less than 20 kW that's interconnected with the utility "recreational wind." But wind machines are not luxuries. Unlike a swimming pool or a rack of snowmobiles in the backyard, wind machines save or even earn their owners money. And they do more. They generate electricity cleanly. That has value, as well, a value that's not currently incorporated in the price of utility-supplied power.

All too often consumers look only at the initial cost. They contrast this with what they are accustomed to paying their utility and throw their hands up in despair. There's no contest! "The wind may be free," they might be overheard lamenting, "but it sure costs a lot to catch it." They're right: Wind turbines are expensive. But there are always two sides to every equation, and the other side of the wind system equation is the revenue it earns or the money it saves.

Buying a wind turbine is a lot like buying a house. You can always rent at a lower initial cost, but you will almost always spend more over the long term. Mike Bergey of Bergey Windpower likes to point out that you're paying for a wind machine, whether you want to or not, every time you mail your check to the local utility. Although, as with death and taxes, you may not be able to entirely avoid dealing with the utility, a wind machine can at least help you reduce those monthly payments. In the process, you assume responsibility for reducing some of the environmental impact from your energy consumption.

A wind machine need not meet your entire electrical load in order to be economical. Many mistakenly think that the wind system is worthwhile only if it eliminates their entire electric bill. Sometimes this may be true. But a wind machine may be a good buy even when it provides only a small portion of your electricity. In fact, for many applications interconnected with the utility in North America, it's better to select a wind system that will produce less energy than currently used, because utility buyback rates are often less than retail rates. It doesn't make any sense to sell the utility energy for 3 cents per kilowatt-hour or less when the the utility will turn around and sell it back to you for 8 cents per kilowatt-hour.

Cost of Energy and Payback

There are several methods for evaluating the economics of wind energy. Determining the cost of energy (COE, for those with a bent for acronyms) produced by a wind system is one approach popular with government agencies. The COE accounts for initial cost, maintenance, interest rates, and performance over the life of the wind system. This life-cycle method produces an estimate of cost, in cents per kilowatt-hour, for the wind system over its life span. The results can be compared to today's cost of energy from new conventional sources.

Yet COE has limited value to individuals and businesses because it reveals only whether the wind system's generation will cost more or less than that from other sources. It doesn't tell you how much of a bargain—or cost—the wind system might be. The COE, consequently, cannot be used to judge whether or not your money would be more productive in an interest-bearing account at the local bank or some other investment.

Finding the payback, or the time it will take for the investment to pay for itself, is an easy way to gauge an investment's worth. Simply divide the wind system's cost by its projected revenue. If the time to payback is less than the life of the wind turbine, the turbine has paid for itself. But there may be better ways to use the money. Payback is related to return on investment. A short payback offers higher returns than a long payback. To maximize the return—to get the most for your money—you want as short a payback as possible.

Nevertheless, simple techniques such as payback are quick but misleading. They don't account for effects that take place after payback occurs. Payback is well suited for low-cost items such as storm doors, weather stripping, and added insulation, but not for costly long-term investments.

Payback gives no indication of the earnings a wind turbine will produce after it has paid for itself. Since wind generators are designed to last 20 years or more, much of their return takes place in later years after the intial cost has been recouped.

Most people are overly concerned about payback. If the wind system doesn't pay for itself within five years, they quickly lose interest. They fail to realize that some wind machines will pay for themselves many times over, even though they may have paybacks greater than five years. Wind systems are long-term investments and should be treated as such.

There's no better way to gain a sense of how a wind system will perform financially than to estimate the cash flow from one year to the next. Constructing a cash flow table is the only way to examine the economics over the long term with any degree of realism. It's what businesses use when they're considering investments in new equipment.

Let's examine the factors that affect a wind system's economics.

Economic Factors

The installed cost of a wind system is simply the cost of the wind turbine, tower, wiring, and installation, less any state or federal tax credits. Where they exist, tax credits reduce an individual's tax liability, instead of merely reducing taxable income, as with tax deductions. Tax credits effectively reduce the initial cost. In the United States, production tax credits offered during the 1990s applied only to sales of electricity, not to savings. Thus you had to dedicate the wind turbine's generation for sale to the utility, or to a third party, to qualify for the credit. Grants or subsidies based on the capital cost of the wind turbine also reduce the total installed cost. For up-to-date information on incentive programs, including those that effectively reduce installed cost, contact your national trade association or your accountant.

Maintenance costs are expenses for servicing or repairing the wind system. They can be expressed in cents per kilowatt-hour or as a percentage of the initial cost. The vast amount of experience from commercial wind turbines indicates that the cost of operating and maintaining medium-size turbines is about 1 cent per kilowatt-hour. The maintenance cost for small wind turbines is less well defined. During the 1980s Wisconsin Power & Light found, after monitoring a Bergey Excel for three and a half years, that the owner paid less than 0.3 cent per kilowatt-hour for the entire period. However, the utility found the cost of maintaining other brands sold during the early 1980s to be prohibitive. Nevertheless, there's a consensus that maintenance will cost about 1 cent per kWh, certainly for household-size and the larger medium-size turbines.

The cost of financing the purchase of a wind system can add significantly to overall costs. You immediately become aware of financing costs if you have considered using a loan to buy your wind system. It's much like building an addition onto your house with a loan from the bank. You pay for the use of the bank's money. You can't avoid financing costs simply by paying with cash. The cash invested in the wind system could have been earning interest at the bank. For homeowners and farmers, the installation of a wind turbine can be financed by increasing the mortgage on the property instead of by taking out a short-term loan. The financing or interest cost of the loan will then reflect the current mortgage rate.

Insurance is an often overlooked cost. There should be insurance not only for the wind system itself, but also for any accidents caused by the wind machine's operation. Since most wind systems require an extended period to pay for themselves, an owner would be foolhardy to operate an uninsured machine. An unexpected event could wipe out an expensive wind turbine before the investment had been recouped. Wind systems are also a potential hazard. Personal injuries and property damage can occur in a multitude of ways: The tower can fall over or (the most likely form of accident) someone can fall off the tower. Many insurance companies have little or no experience with small wind systems, and their coverage varies from a simple inclusion under a homeowner's existing policy, at no cost, to a policy written specifically for the wind turbine with an attendant premium. Businesses and farms can find a wealth of information on insurance for medium-size wind turbines. Schafer Systems' Phil Littler paid about $3,000 per year or about 1 percent of the turbine's installed cost on his Vestas V27 in Adair, Iowa, during the late 1990s. Paul White of Project Resources in Minneapolis, Minnesota, says insurance for a 660 kW Vestas V47 should cost about $5,000 per year, or somewhat less than 1 percent of the installed cost in the year 2000.

Taxes have a profound effect on the economics of any investment, largely on the cost side of the ledger. For homeowners in the United States, the costs of financing a wind system through a home mortgage are tax deductible. Because the interest costs are high in the early years of a loan, this deduction dramatically improves the attractiveness of small wind turbines to homeowners over what would be apparent from a simple payback calculation. The tax bonus is even more important for businesses, because they can deduct the cost of maintenance and other expenses.

The tax bracket of the investor determines the value of any deductions. Those in higher tax brackets save more by reducing their tax burdens than those who pay a lower tax rate. If you're in the 30 percent tax bracket, you save 30 cents for every dollar in tax deductions.

Similarly, homeowners receive more than 1 cent in value when they offset 1 cent's worth of electricity with a wind turbine because such savings are not taxed. (No one taxes you for using less electricity.) If homeowners offset 1 cent's worth of electricity purchases, they create 1.43 cents in value after taxes for someone in the 30 percent tax bracket [1/(100% – 30%)].

For those in lesser tax brackets the benefits are not as great, but they're still considerable. Conversely, where taxes are higher, the benefits are greater.

There are some hidden costs, as well. A wind turbine could potentially increase property taxes, though in some areas they are tax exempt.

Now for the income side of the account. The income derived from the wind system is a product of the annual energy output (AEO) and the value of the electricity. Like costs, the value of the electricity generated depends on a number of factors, the most important of which is the retail rate, or tariff for electricity. If we assume that whatever energy the wind machine generates will displace electricity that otherwise would be bought from the utility, the value of this energy (per kilowatt-hour) equals the utility's retail rate.

For a homeowner, the cost of electricity per kilowatt-hour is easy enough to estimate. To get an average cost in cents per kilowatt-hour, simply pull out the last 12 months' worth of your electric bills, sum them, and divide by the total amount of electricity consumed. For businesses, it may not be so clear.

Commercial customers pay not only for the energy they use, but also for the demand they put on the utility system. This charge is based on the maximum power drawn during the billing period in relation to their total energy consumption. This compensates the utility for maintaining generators online to provide power at the demand of the customer. By installing a wind generator a business lowers its total consumption while hardly affecting its peak demand. Thus the wind system could actually increase the demand charge while lowering costs for the energy consumed. This isn't a problem with all utilities, but farmers and businesses need to watch for it.

Once the future value of wind-generated energy in cents per kilowatt-hour has been found, the annual future value of electricity produced can be calculated. Multiply the retail rate in cents per kilowatt-hour by the amount of energy (kilowatt-hours) consumed annually. The result is the gross proceeds from offsetting the consumption of utility-supplied electricity.

One of wind energy's chief advantages over generating electricity by conventional means is that the fuel (the wind) is free. The bulk of the cost for a wind system occurs all at once. Once paid for, the energy produced costs little over the remaining life of the wind system. Conventional generation, on the other hand, consumes nonrenewable fuels whose costs continue to escalate. Thus, our analysis would be incomplete if we didn't account for the rising price of electricity. Like other aspects of energy production, the rate at which utility prices will rise over time is hotly debated.

Utility rates escalated sharply during the 1970s and early 1980s. These price hikes were due in part to the rapid rise in the cost of oil caused by the two oil embargoes, and in part due to completion of expensive new power plants. Oil-dependent utilities were hit the hardest, and their rates jumped dramatically. Energy costs at other utilities eventually also rose because coal and gas prices often track price increases in oil. Further, many utilities in the 1970s committed themselves to massive construction programs to meet expected growth in electricity consumption. These plants, both coal-fired and nuclear-powered, were completed during the early 1980s and were enormously expensive, particularly the nuclear reactors.

After these price shocks, electricity prices stabilized. Rate increases were much less severe during the late 1980s. In real terms, after accounting for inflation, the rates at some utilities declined. In the 1990s electric utility rates rose modestly, even though the world price of oil—the benchmark of energy costs—remained low. But as the war in Kuwait illustrated, future conflicts in the Middle East could once again cause shortages and higher prices.

The most important aspect of utility rate escalation is its relationship to inflation. Critics of utilities, or those hawking energy-saving devices, tend to overemphasize the effect that rising utility rates have over time. They fail to mention, or they conveniently ignore, the effect of inflation on rising rates in real terms. If utility rates were rising 10 percent per year, an inflation rate of 5 percent erodes the potential value of each kilowatt-hour.

At the turn of the new millennium, the utility industry was undergoing radical change worldwide. No one can predict what effect these changes will have on utility rates in real terms during the years to come. During the year 2000 the price of electricity in parts of the United States rocketed to heights once considered unimaginable, illustrating that electricity prices, like prices in the broader energy market, will remain highly volatile.

Residential Economics

Let's put all this together and construct a cash flow table of costs and income over the life of a typical household-size wind turbine. Assume that you plan to install a 7-meter Bergey Excel on a 100-foot-tall guyed tower costing about $35,000 in total. Your site at sea level has an average wind speed at hub height of 13 mph (about 6 m/s). Bergey Windpower estimates that the Excel S will generate about 14,000 kWh per year under these conditions.

From your observation of the local utility, you expect rates to rise 5 percent per year for the life of the wind system, with inflation of 3 percent. (This represents a real price increase of only 2 percent.) You'll pay 20 percent of the project's cost and borrow the rest. Fortunately you have convinced your banker to take out a second mortgage on your house at 8 percent interest for 10 years to pay for the wind system. Your accountant says you're in the 34 percent tax bracket. After explaining that future revenues are worth less than revenue today due to inflation, your accountant agrees to calculate the net present value for your results.

To calculate the net present value, your accountant will need to know the discount rate—the interest rate used to discount future cash flows. At a minimum the discount rate on a 20-year income stream will reflect the rate on a 20-year mortgage, and that's the value used in the following analyses. For businesses, however, the discount rate is equal to the businesses' cost of capital, or the cost of "lost opportunities," and can be significantly greater than the rate on long-term debt. Natural Resources Canada's renewable energy project analysis software, RETScreen, explains that in 2000 the discount rate for North American electric utilities varied from 6 to 11 percent.

We'll also assume that all the electricity will be used in your home at a retail rate of 10 cents per kilowatt-hour. Of course, this assumes that this is an appropriate-size wind turbine for your application and you can consume all the electricity on site. No tax credits or other subsidies apply, and because you'll use the wind turbine for your personal consumption, not for business, there are no deductions for operation and maintenance (O&M), insurance, or depreciation. Interest payments on the mortgage, however, are deductible. There is also considerable value in investing in energy savings, instead of paying for electricity consumption with taxable earnings over a long period (see table 5-1, Cash Flow for Single Household-Size Wind Turbine in U.S. Residential Use). In this example the turbine produces a net positive cash flow early in the 15th year.

Farm and Business Economics

Let's try another example, using similar conditions except that we'll install the wind turbine at a dairy. We'll assume that the farm consumes nearly 500,000 kWh per year and is capable of using 90 percent of the generation on site from a Vestas V29, a wind turbine 29 meters (95 ft) in diameter and capable of generating 225 kW.

The cost of maintenance, insurance, interest payments on a loan to buy the turbine, and the depreciation of the turbine are deductible business expenses in the United States. This reduces the farm's after-tax cost of wind-generated electricity. Depreciation deductions offer substantial tax benefits to users of capital-intensive equipment such as wind turbines. A business in the 30 percent tax bracket saves 30 cents in

Table 5-1
Cash Flow for Single Household-Size Wind Turbine in U.S. Residential Use

Year	Gross Revenue	O&M	Insurance	Loan Interest	Loan Principal	Tax Value (Cost)	Revenue (Loss)	Cumulative Revenue
0	-7,000	0	0	0	0	0	-7,000	($7,000)
1	1,400	-140	-350	-2,240	-1,933	$2,938	-1,725	($8,725)
2	1,470	-144	-361	-2,085	-2,087	$2,991	-1,686	($13,349)
3	1,544	-149	-371	-1,918	-2,254	$3,046	-1,647	($14,996)
4	1,621	-153	-382	-1,738	-2,435	$3,102	-1,607	($16,602)
5	1,702	-158	-394	-1,543	-2,630	$3,158	-1,566	($18,168)
6	1,787	-162	-406	-1,333	-2,840	$3,215	-1,525	($19,694)
7	1,876	-167	-418	-1,106	-3,067	$3,273	-1,485	($21,178)
8	1,970	-172	-430	-860	-3,313	$3,332	-1,444	($22,622)
9	2,068	-177	-443	-595	-3,578	$3,391	-1,403	($24,025)
10	2,172	-183	-457	-309	-3,864	$3,450	-1,363	($25,388)
18	3,209	-231	-578	0	0	$4,937	4,127	$2,259
19	3,369	-238	-596	0	0	$5,183	4,349	$6,608
20	3,538	-245	-614	0	0	$5,443	4,583	$11,191
Total 20-year revenue							$14,129	
Net present value using a discount rate derived from a 20-year mortgage.							($4,000)	

Note: Generation used on site offsets consumption at retail rate. Tax credit applies only to sale of electricity.

Assumptions:

Rotor Diameter	7 m	retail rate	0.10 $/kWh	Utility Rate Escalation	5%
Avg. wind speed	6 m/s	Resale rate	0.02 $/kWh	Inflation Rate	3%
Yield	365 kWh/m^2/yr	% at retail rate	100%	Down Payment	20%
Installed Cost	$35,000	Tax credit rate	0.0165 $/kWh	Loan Term	10 yrs
O&M	0.01 $/kWh	%Tax credit used	100%	Loan Interest	8%
Insurance	1%	Tax bracket	35%	Discount Rate	6%

Results:

Swept area	38 m^2
Annual Energy Output	14,000 kWh/yr

Understanding Table 5-1

Gross revenue derive from offsetting the purchase of electricity from the utility with that from the wind turbine at the utility's retail rate. O&M is the cost of maintenance. The interest and principal portion of the loan are calculated to show how the debt is paid off. Tax value is largely due to the value of saving electricity instead of earning money to pay for it, and the avoidance of taxes that accrue to earnings. Tax value also derives in part from the interest deduction because interest payments are a deductible expense in the United States if part of a home's mortgage. Annual revenue is the sum of O&M, insurance, loan payments, and tax value. This table of cash flow is included with the formulas in critical cells in the appendixes.

taxes for every 100 cents in deductions. The farm's purchase of electricity from the utility is also a deductible expense. Thus there is an indirect cost in increased taxes for businesses saving electricity in the United States. To account for this, tax must be calculated on the gross revenue, or savings in electricity, produced by the wind turbine. In the United States tax credits or subsidies apply to each kilowatt-hour sold to the local utility (see table 5-2, Cash Flow for Single Medium-Size Wind Turbine [225–300 kW] in U.S. Business or Farm Application). The turbine generates a net positive cash flow in the fourth year.

Of course, your conditions may vary from those used here. Always consult your accountant or financial adviser before investing in wind energy.

Your conditions may vary from those used here. Always consult your accountant or financial adviser before investing in wind energy.

Risks

Analyzing the economics of a wind system is fraught with assumptions about the future. The assumptions you use may or may not reflect conditions over the turbine's design life. No one knows with certainty what the future will bring. There's a degree of risk associated with every investment. Consequently, there's no simple answer to the question, Is it a good deal?

When buying a wind system, you may be betting that utility costs will increase faster than inflation. That certainly was true during the 1970s and 1980s. But utility costs lagged behind inflation during much of the 1990s, and in some places actually declined. Interest rates and utility rate escalation have a profound effect on a wind turbine's economics, because they compound year after year. Slight changes in their relationship could change the attractiveness of wind energy as an investment.

Maintenance costs are equally important. Small wind systems are generally more expensive than larger wind turbines relative to the amount of energy they produce. For small wind turbines, maintenance costs can be a deciding factor at marginal sites. At low-wind sites, maintenance can consume much of a small wind turbine's revenue. Cutting annual maintenance costs by using an integrated, advanced small wind turbine that requires little or no maintenance can measurably improve the overall economics at such sites.

Implicit in these calculations has been the assumption that all the energy generated was used on site, offsetting energy that would have been bought from the utility. This may not be the case. Some of the energy may be generated at times when there is little or no need for it.

Table 5-2
Cash Flow for Single Medium-Size Wind Turbine (225-300 kW) in U.S. Business or Farm Application

Year	Gross Revenue	O&M	Insurance	Loan Interest	Loan Principal	Depreciation Deduction	Income Tax	Tax Credit	Revenue (Loss)	Cumulative Revenue
0	-60,000	0	0	0	0	0	0	0	-60,000	($60,000)
1	44,160	-4,800	-3,000	-19,200	-16,567	-60,000	14,994	792	16,379	($43,621)
2	46,368	-4,944	-3,090	-17,875	-17,892	-96,000	26,439	816	29,822	($13,799)
3	48,686	-5,092	-3,183	-16,443	-19,324	-57,600	11,771	840	17,256	$3,457
4	51,121	-5,245	-3,278	-14,897	-20,870	-34,560	2,401	865	10,097	$13,553
5	53,677	-5,402	-3,377	-13,228	-22,539	-34,560	1,011	891	11,034	$24,587
6	56,361	-5,565	-3,478	-11,425	-24,342	-17,280	-6,515	918	5,955	$30,541
7	59,179	-5,731	-3,582	-9,477	-26,290		-14,136	946	908	$31,449
8	62,138	-5,903	-3,690	-7,374	-28,393		-15,810	974	1,942	$33,391
9	65,244	-6,080	-3,800	-5,103	-30,665		-17,591	1,003	3,008	$36,400
10	68,507	-6,263	-3,914	-2,649	-33,118		-19,488	1,033	4,108	$40,507
18	101,216	-7,934	-4,959	0	0		-30,913		57,410	$426.394
19	106,276	-8,172	-5,107	0	0		-32,549		60,448	$486,842
20	111,590	-8,417	-5,261	0	0		-34,269		63,643	$550,486
Total 20-year revenue									$550,486	
Net present value using a discount rate derived from a 20-year mortgage.									$214,000	

Notes: Generation used on site offsets consumption at retail rate. Tax credit applies only to sale of electricity.

Assumptions:

Rotor Diameter	29 m	Retail rate	0.10 $/kWh	Utility Rate Escalation	5%
Avg. wind speed	6 m/s	Resale rate	0.02 $/kWh	Inflation Rate	3%
Yield	3720 kWh/m²/yr	% at retail rate	90%	Down Payment	20%
Installed Cost	$300,000	Tax credit rate	0.0165 $/kWh	Loan Term	10 yrs
O&M	0.01 $/kWh	%Tax credit used	100%	Loan Interest	8%
Insurance	1%	Tax bracket	35%	Discount Rate	6%

Results:

Swept area	661 m²
Annual Energy Output	480,000 kWh/yr

Understanding Table 5-2

Tax on income is the sum of the tax calculated on gross revenues (electricity savings), and deductions for maintenance (O&M), insurance, interest payments, and depreciation. Annual revenue is the sum of gross revenue minus operations and maintenance, insurance, loan payments, and income taxes, plus any tax credits. This table of cash flow is included with the formulas in critical cells in the appendixes.

The excess energy will then be sold back to the utility, and often at a price far lower than the purchase price or retail tariff. This reduces the wind system's overall savings. Where there's a demand for heat or hot water, such as at a dairy, the surplus generation can be dumped into a thermal storage system instead of selling it back to the utility at a reduced rate.

Cost-Effectiveness

Many people confuse the cost-effectiveness of wind turbines with their efficiency. The reason for installing wind machines is the generation of clean electricity. You could install the most inefficient wind machine ever built, the kind that causes an engineer to grimace, but if it

Taking a Gamble: Schafer Systems

Driving on I-80 west of Des Moines, Iowa, you can't miss Schafer Systems. If the huge American flag bordering the factory doesn't draw your eye, the gleaming Vestas V27 certainly will. Both are tucked between the I-80 right-of-way and the plant's south wall just outside Adair, Iowa.

The whirling 27-meter (90 ft) diameter landmark stands on the subtle drainage divide between the Mississippi River to the east and the Missouri River to the west. Adair—the second highest point in Iowa, says Schafer Systems' affable Phil Littler—is one of the state's best wind locations.

Littler, a solid midwesterner whose family has lived in Adair for more than 100 years, did his homework. He could teach a California wind farmer a lesson or two about how to measure the wind and what to look for in a turbine. He went so far as to measure the noise from an identical wind turbine himself before committing to the Danish manufacturer. Littler chose Vestas's white tubular tower over a cheaper lattice tower because of its clean aesthetic lines and the ability to work inside it in winter, protected from the elements.

Phil Littler and Schafer Systems V29 in Adair, Iowa.

The homework paid off. The Vestas V27 has been available for operation more than 99 percent of the time since installation in mid-1995. The 225 kW turbine consistently produces 500,000 kWh per year at what Littler estimates is a 13.1 mph (5.8 m/s) average wind speed.

As in most European wind turbines of this size, either there are two generators or the generator uses dual windings—as in Schafer's V27—to optimize performance in modest wind regimes such as the Midwest. An important 8 percent of the total electricity produced by the Schafer Systems V27 is from the generator's low-power windings.

Like many homeowners, Schafer Systems simply wanted a means to reduce its utility bill. Like farms and businesses throughout Europe, it chose wind energy.

Originally the turbine was sized to meet 75 percent of the company's demand. Because of rising consumption—strong demand by state lotteries for the point-of-purchase displays built by Schafer Systems—the turbine provides two-thirds of the factory's electricity. Rarely does the turbine produce more electricity than the plant consumes.

Because Schafer Systems does not sell the electricity to the utility, it didn't qualify for the federal tax credit. The subsidy primarily benefits large wind farm developers and not individuals, farms, or small businesses. Littler says the turbine is performing about 5 percent better than that projected by Vestas. Still, the project hasn't been risk-free. Lightning damaged the wind sensors atop the nacelle during the first year, and the cost of insurance is twice the $1,500 per year they had estimated. Maintenance costs, too, have been somewhat higher because of the high lightning incidence. As a result, Littler has extended the original estimate of payback from five years to between seven and eight years.

To avoid any complications in connecting the 480-volt wind turbine to the plant's electrical system, Littler hired the local utility to install the turbine's transformer and interconnection equipment. To everyone's surprise, the turbine has shut down more often from power quality problems on the utility side than from any problems with the wind turbine.

There was some concern about ice throw, and they can sometimes hear pieces falling on the roof during the winter, but Littler says that overall ice hasn't posed a problem.

Zoning posed no obstacles, either. Schafer easily rezoned the commercial property to allow the 200-foot (60 m) total height of the turbine and its blades. The company also notified the state department of transportation because of the turbine's proximity to the interstate highway.

Unlike the lottery players Schafer Systems serves, the small Iowa company didn't gamble on its future. It took a calculated risk—and made it pay.

Table 5-3
Relative Cost per kWh for Selected Wind Turbines

Rotor Diameter (m)	(ft)	(kW)	Swept Area (m^2)	Cost ($)	Specific Cost ($\$/m^2$)	AEO (kWh)	Relative Cost (kWh/yr)
1.2	3.8	0.4	1	$2,000	1,864	400	5.0
2.1	7.0	0.9	4	$3,300	923	1,300	2.5
2.5	8.2	1	5	$3,500	713	1,800	1.9
3.6	11.8	0.9	10	$4,700	462	3,800	1.2
29	95	225	661	$300,000	454	475,600	0.6

Note: AEO = 6 m/s (13.4 mph) average wind speed at hub height of 100 feet (33 m).

works well and is cheap enough, it could be more cost-effective than a modern engineering marvel that is unreliable or costs much more. If you deliver lower-cost electricity with an inefficient wind machine than with an efficient one, so be it.

There are several measures of cost-effectiveness in use: cost per kilowatt of installed capacity, cost per rotor swept area, and cost per kilowatt-hour of energy generated. The most frequently used measure of cost-effectiveness is cost per kilowatt. However, it's about as meaningful as the power rating in describing the size of a wind machine. This measure came into use the same way—utility engineers were accustomed to using it. Like power ratings, the cost per kilowatt works well for power plants that run at constant output. But for wind machines, the cost per kilowatt just confuses matters.

During the early 1980s a few American manufacturers took advantage of this situation. Because consumers were using cost per kilowatt to compare one wind machine to another, these manufacturers began offering products with higher generator ratings relative to the size of the rotor than was the norm. Their products weren't any better, nor would they produce more electricity than their competitors, but their cost per kilowatt was lower. One company, Fayette Manufacturing, rated its wind turbine nearly three times higher than other wind machines of comparable swept area. Even though its machine cost more than others, it appeared more cost-effective because of its higher generator rating. Fayette sold 1,600 wind turbines this way!

A truer measure of a wind machine's size is the area swept by its rotor, and a more useful measure of cost-effectiveness is the cost per swept area. The limitation on using cost per swept area is the assumption that all wind machines are equally efficient at converting the energy in the wind to electricity. They aren't. The cost per swept area is just a shortcut for what counts most: the relative cost per kilowatt-hour generated at your specific site.

This cost per kilowatt-hour isn't the same as the cost per kilowatt-hour you pay for electricity from the utility. The cost per kilowatt-hour measure should only be used for comparing one wind machine to another. It's not appropriate for comparing a wind turbine to other forms of energy because it doesn't account for all the costs and benefits from the wind turbine over its entire life cycle. It's merely a measure or "figure of merit" for comparison shopping—nothing more (see table 5-3, Relative Cost per kWh for Selected Wind Turbines).

In our first example, the Bergey Excel cost $35,000 and generated 14,000 kWh per year at our site. The cost per kilowatt-hour for one year of the Bergey at this site is $2.5 per annual kWh. The Vestas V29 cost $300,000, or 10 times the cost of the Bergey, but it generated 34 times more electricity. The relative cost of the V29 in our example is $0.6 per annual kWh.

It may seem arcane at first, but this is a technique used by professionals to compare the cost-effectiveness of various wind turbine models in different wind regimes. Cost-effective turbines at windy sites produce lower costs per kilowatt-hour. More costly wind turbines or those at less windy sites result in higher values.

The cost-effectiveness of any wind system is more sensitive to initial cost and annual energy output than to any other factor. That's why proper siting is so critical to maximizing electricity generation. Where finding a windier site to boost generation is impossible, using a taller tower may be your best alternative.

In the next chapter we can examine the technology available today.

6

Evaluating the Technology—What Works and What Doesn't

"Hey, that's a funny-lookin' windmill you got there. What is it?"

"A VAWT."

"Don't get smart with me, son. I asked you a simple question."

"Actually, it's an articulating, straight-blade VAWT."

"What was that? I don't work for the government, you know."

"Some call it a giromill."

"Well, that's better. Why didn't you say so in the first place? For a moment there I thought you were speaking in tongues."

As wind technology has grown, so has its vocabulary. At times it may seem as if the wind industry does speak in tongues. Nearly every conceivable wind turbine configuration has been tried at least once—most only once. Designs have run the gamut from the familiar farm windmill to contraptions such as the giromill. Despite the plethora of imaginative designs, only a few approaches have proven successful.

Wind energy has been used to generate electricity since the dawn of the electrical age. While Poul la Cour, the Danish Edison, was adapting European windmills to produce electricity in the 1890s, American Charles Brush was adapting the technology he knew best, the multiblade farm windmill, to perform the same task. Brush's 56-foot (17 m) diameter wind turbine generated 12 kW for charging batteries at Brush's Cleveland, Ohio, estate. It produced enough electricity for his domestic needs and rivaled that generated by fossil fuel for the New York mansion of banking magnate J. P. Morgan.

Wind-electric technology has developed sporadically since 1900, with significant advances in both the United States and Europe during the war years. The modern wind revival followed the energy crises of the 1970s, and despite setbacks the technology has made steady progress ever since. Modern wind turbines are capable of generating about 10 times more power than traditional windmills of equivalent size (see table 6-1, Comparison of Modern Wind Turbines with Traditional Windmills).

Wind machines designed for residential or remote uses, where simplicity is required, have evolved into highly integrated designs with few moving parts. These advanced small wind turbines typically use a rotor that spins about a horizontal axis upwind of the tower. Most of these designs use three slender blades and drive permanent-magnet alternators.

Similarly, most of today's medium-size wind turbines, like those used in wind power plants, use three blades oriented upwind of the tower and drive induction

Table 6-1
Comparison of Modern Wind Turbines with Traditional Windmills

	Rotor Diameter (m)	Rotor Diameter (ft)	Effective Capacity (kW)
American farm windmill	5	16	0.5
Modern household-size wind turbine	5	16	5–6
Brush multiblade	17	56	12
Modern wind turbine	17	56	75–100
Dutch or European windmill	25–30	80–100	25–30
Modern wind turbine	25–30	80–100	250–300

Table 6-2
Technology Development of Commerical Wind Turbines

Period	Nominal Rotor Diameter (m)	Nominal Rotor Diameter (ft)	Nominal Swept Area (m^2)	Nominal Capacity (kW*)
Late 1970s	12.5	40	100	40
Early 1980s	15	50	125	65
Mid 1980s	18	60	250	100
Late 1980s	25	80	500	250
Early 1990s	35	110	1,000	400
Mid 1990s	40	130	1,250	500
Late 1990s	50	160	2,000	750
Early 2000s	70	230	4,000	1,500
Mid 2000s	80	260	5,000	2,000

Note: *@ 60 Hz.

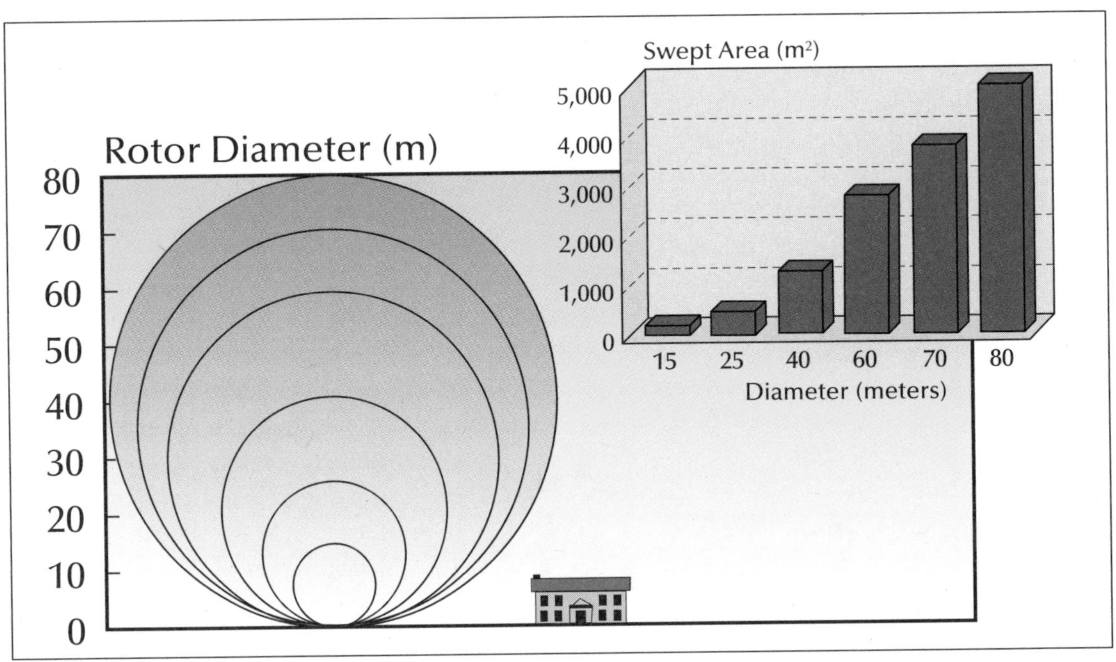

Figure 6-1. Medium-size wind turbines. The area swept by medium-size wind turbines has increased steadily since the early 1980s.

(asynchronous) generators. These commercial wind turbines have grown steadily in size from modest beginnings in the late 1970s and early 1980s (see figure 6-1, Medium-size wind turbines). Medium-size turbines in the early 2000s were typically 30 times the size of the commercial turbines installed in the early 1980s (see table 6-2, Technology Development of Commercial Wind Turbines).

In this chapter we'll look at where the technology stands today and why certain designs have become commonplace. We'll also look at the important difference between wind machines that use drag to drive their rotors and those that use lift, the reason modern wind machines use only two or three blades, the materials that are used to make these blades, the kinds of controls used to protect

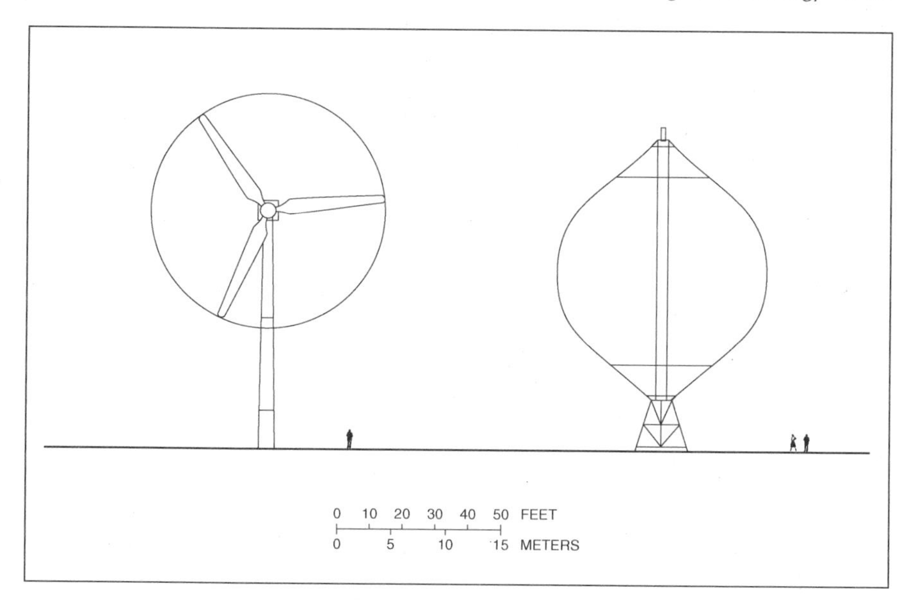

Figure 6-2. Horizontal- and vertical-axis wind turbines. Although the Darrieus or eggbeater turbine on the right spins about a vertical axis, it's equally as efficient at harnessing the energy in the wind as the conventional wind turbine on the left. Wind turbines of this size are capable of generating 200 kW. (Pacific Gas & Electric Co.)

the wind turbine, and the types of transmissions and generators now being used.

Orientation

There are two great classes of wind turbines: those whose rotors spin about a horizontal axis, and those whose rotors spin about a vertical axis (see figure 6-2, Horizontal- and vertical-axis wind turbines). Conventional wind turbines, like the Dutch windmill found throughout northern Europe and the American farm windmill, spin about a horizontal axis. As the name implies, vertical-axis wind turbines (VAWTs) operate much like a top or a toy gyroscope.

Vertical Axis

The principal advantage of modern vertical-axis wind machines over their conventional counterparts is that VAWTs are omnidirectional—they accept the wind from any direction. This simplifies their design and eliminates the problem imposed by gyroscopic forces on the rotor of a conventional machine as the turbine tracks the wind. The vertical axis of rotation also permits mounting the generator and drive train at ground level (see figure 6-3, Darrieus rotor with nomenclature).

Vertical-axis turbines can be divided into two major groups: those that use aerodynamic drag to extract power from the wind (for example, the cup anemometer) and those that use lift. We can further subdivide those VAWTs using airfoils into those with straight blades and those with curved blades. The simplest configuration uses two or more straight blades attached to the ends of a horizontal cross-arm. This gives the rotor the shape of a large H. Unfortunately this configuration permits centrifugal forces to induce severe bending stresses in the blades at their point of attachment.

During the 1920s French inventor D. G. M. Darrieus patented a wind machine that cleverly dealt with this limitation. Instead of using straight blades, he attached curved blades to the rotor. When the turbine was operating, the curved blades would take on the form of a spinning rope held at both ends. This troposkein shape directs centrifugal forces through the blade's length toward the points of attachment, thus creating tension, rather than bending, in the blades. Because materials are stronger in tension than in bending, the blades can be lighter for the same overall strength and operate at higher speeds than straight blades. Although the phi (ϕ) or

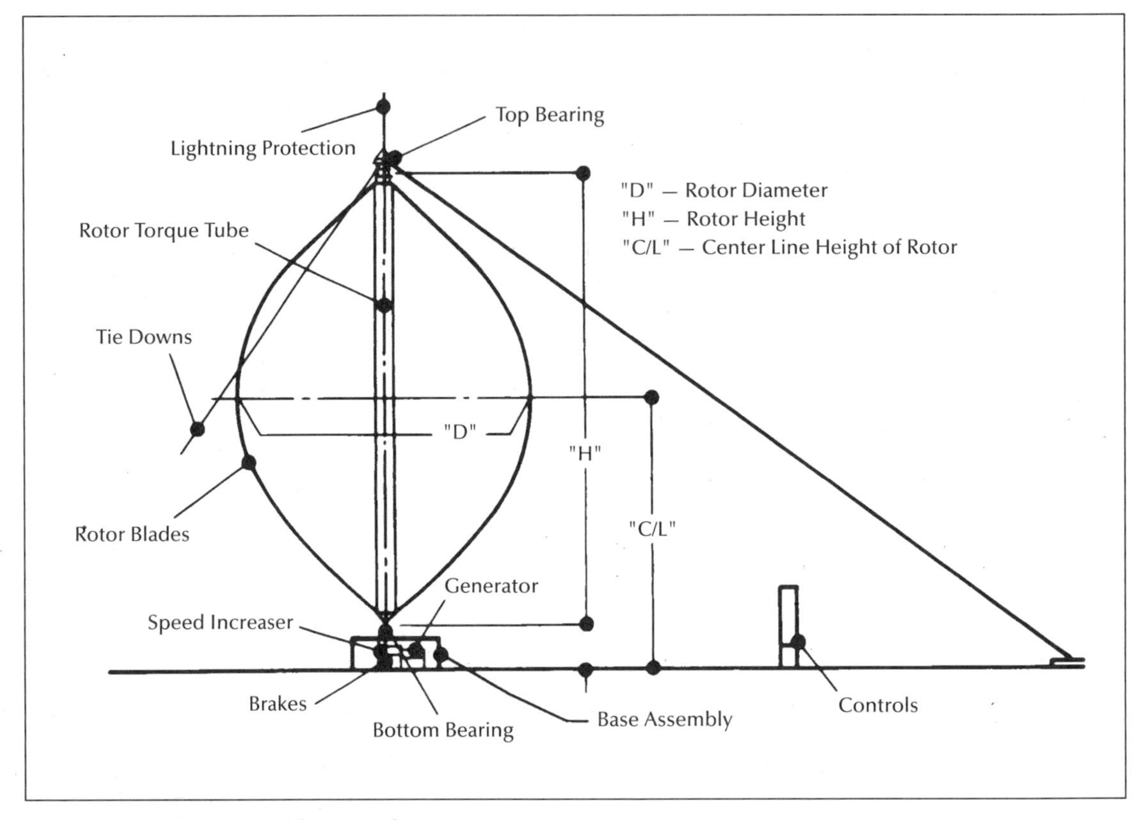

Figure 6-3. Darrieus rotor with nomenclature.

eggbeater configuration is the most common, Darrieus conceived several other versions, including Delta, Diamond, and Y. All have been tried at one time or another (see figure 6-4, Darrieus configurations). Some have likened the phi-configuration Darrieus turbine to a slice of onion spinning about a vertical axis.

Darrieus's concept eventually faded into obscurity. Canada's National Research Council reinvented the design in the mid-1960s, and subsequently Canadian wind research focused on Darrieus turbines (see figure 6-5, Novel Darrieus). In the United States, Sandia National Laboratories has also pursued the technology. Several firms in Europe and North America attempted to commercialize Darrieus technology but had little lasting success (see figure 6-6, Household-size Darrieus).

Work on the technology has practically ceased, and there are few Darrieus turbines still in service. Carl Brothers operates one of the last of the breed at Canada's Atlantic Wind Test Site on Prince Edward Island. The test site has operated a 35 kW DAF-Indal

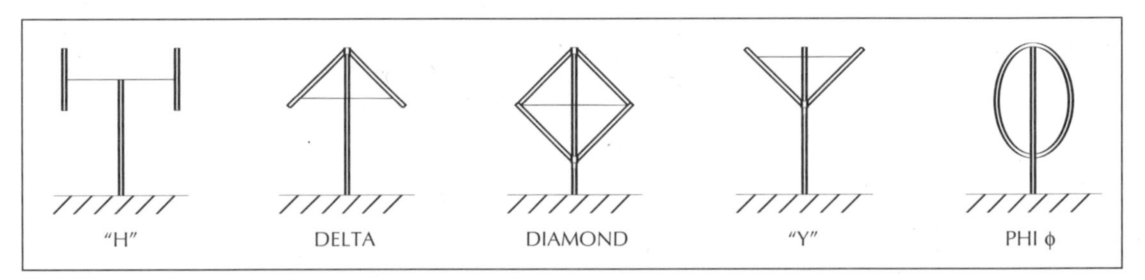

Figure 6-4. Darrieus configurations. There are several other Darrieus configurations besides the common eggbeater design.

Figure 6-5. Novel Darrieus. This experimental Darrieus design was under development near Pincher Creek in Alberta, Canada in the late 1990s. This unusual configuration eliminates the use of guy cables.

Figure 6-6. Household-size Darrieus. Depicted is the installation of a 10 kW Darrieus turbine in the early 1980s. Unlike most Darrieus turbines, this small Alcoa design used a cantilevered rotor, eliminating the need for guy cables.

design for more than 15 years, probably a world record for a Darrieus turbine. With the exception of some very small Savonius rotors, there are no VAWTs in widespread use.

Darrieus turbines were bedeviled by poor performance and poor reliability. The aluminum blades often fatigued and sometimes failed catastrophically. This was in part because the lift forces, which propel the blades, reverse direction every revolution, flexing their attachment to the torque tube or central mast. Another source of frequent flexing of the blades is inherent in the rotor's eggbeater shape. When the rotor is at rest, the blades sag due to their own weight, stressing the connection to the torque tube. Moreover, the presumed advantage of housing the drive train at ground level was offset by the large bearings and guy cables at the top of rotor.

Nor are Darrieus turbines reliably self-starting. Their fixed-pitch blades can't drive the rotor up to operating speed from a standstill unless the blades are parked in just the right position relative to the wind. While this isn't necessarily a serious limitation, it does require motoring the rotor up to speed.

To provide self-starting capability, several researchers in North America and Europe reverted to the H-rotor configuration. With straight blades, pitch can be varied as the blades orbit around the rotor's axis. Technocrats identify this kind of wind machine as an articulating, straight-blade, vertical-axis wind turbine.

Work on [Darrieus] technology has practically ceased.

Figure 6-7. Giromill or cycloturbine. Like all vertical-axis wind machines, this turbine can transmit mechanical power to ground level via a long shaft. *Top:* Blades "articulate" or change pitch relative to the wind. *Bottom:* The wind vane at the top of the rotor orients the blades with respect to the wind. Only one machine of this design by McDonnell Aircraft was ever built.

It's also known as a giromill or cycloturbine (see figure 6-7, Giromill or cycloturbine).

The H-rotor has one important advantage over the Darrieus design: It captures more wind. The intercept area of an H-rotor is a rectangle. For the same size wind machine—that is, where the height and diameter are the same—the H-rotor will sweep more area than an ellipse and should, theoretically, capture more wind energy.

Despite considerable research and development, especially in Great Britain, medium-size H-configuration VAWTs have never entered commercial production. Some small turbines were developed (for example, Pinson Cycloturbine), but none has succeeded commercially.

Horizontal Axis

Unlike Darrieus turbines, conventional wind turbines are not omnidirectional. As the wind

Figure 6-8. Fan tail. During the 18th and 19th centuries European windmills incorporated "fan tails" for automatically orienting the rotor towared the wind. This thatch-covered windmill is located near Pewsum in northwest Germany.

changes direction, horizontal-axis wind machines must change direction with it. They must have some means for orienting the rotor with respect to the wind. Traditionally the rotors of conventional wind turbines have been placed upwind of their towers and have incorporated some device for pointing the rotor into the wind.

With the Dutch windmill, for example, the miller had to constantly monitor the wind. When the wind changed direction, the miller laboriously pushed a long tail pole or turned a crank on the milling platform to face the windmill's massive rotor back into the wind. Later versions liberated millers from their labor by using fan tails that mechanically turned the rotor toward the wind (see figure 6-8, Fan tail).

On smaller wind turbines, such as the farm windmill, the task is much easier; a simple tail vane will do. The tail vane keeps the rotor pointed into the wind regardless of changes in wind direction (see figure 6-9, Tail vane).

Figure 6-9. Tail vane. The Dutch LMW turbine uses a tail vane to direct the rotor into the wind at the Folkecenter for Renewable Energy's test field in Denmark.

Passive Yaw

Both tail vanes and fan tails use the force of the wind itself to orient the rotor upwind of the tower. They passively change the orientation, or yaw, of the wind turbine with respect to changes in wind direction without the use of human or electrical power.

The rotor may also be placed downwind of

Tail Vanes

Tail vanes are deceptively simple. As the many designs on the market attest, as much art as engineering goes into the design of a tail vane for small wind turbines.

Scoraig Wind Electric's Hugh Piggott offers a rule of thumb: The tail vane's surface area should be greater than the square of the rotor diameter divided by 40. He also suggests that the tail boom should be similar in length to that of a rotor blade. The exception is for boating applications, where shorter tail vanes are advisable. Experience has also shown that tail vanes with a pronounced vertical shape are more effective than those with horizontal shapes.

Micro turbines often use fixed tail vanes that are part of the generator body. Maritime versions typically have a hole in the tail that enables grappling with a pole hook when it's necessary to take the turbine out of the wind. Britain's National Engineering Laboratory found that the yaw systems on micro turbines, while simple, were overactive in turbulent winds and could cut performance.

Tail vanes for mini and household-size turbines are typically hinged about a near-vertical axis. This allows the rotor to fold, or furl, toward the tail vane in high winds.

In the simplest versions, the tail vane is integrated with the tail boom, as in the old Bergey 850. For most designs, however, the tail boom carries the tail vane well beyond the nacelle. Manufacturers use booms made from pipe (SWP's Whisper series), box beams (Bergey), or truss assemblies (Enercon's experimental E12), or are cable-stayed, as in some Dutch designs. Dutch, German, and French manufacturers prefer tail booms that elevate the tail vane well above the rotor axis. Others simply extend the tail vane horizontally.

Where furling in used, the furling hinge is sometimes placed at the end of the tail boom (Marlec and some Dutch designs), but most designers place the hinge at the nacelle (Bergey, African Wind Power).

Although tail vanes are effective devices for passively controlling yaw, they become unwieldy as turbines increase in size. At the upper end of their range, tail vanes may require dampers or other devices to reduce the rate at which the wind turbine yaws into the wind.

While tail vanes have been used on American farm windmills up to 56 feet (17 m) in diameter, they are typically limited to much smaller wind machines. Among today's largest household-size turbine's using a tail vane is Vergnet's 10-meter (33 ft) diameter two-blade, upwind turbine. Above this size, electric yaw motors or fan tails are typically used to mechanically orient the rotor into the wind.

Novel twin tail. The University of New Castle's 5 kW experimental small wind turbine at Fort Scratchley on Australia's east coast. While unusual, the concept is not unique. During California's wind rush in the early 1980s, a San Diego company built an unsuccessful 40-foot (12 m) diameter upwind turbine using dual tail vanes.

the tower. Downwind rotors don't necessarily need tail vanes or fan tails. Instead, the blades are swept slightly downwind, giving the spinning rotor the shape of a shallow cone with its apex at the tower. This coning of the blades causes the rotor to inherently orient itself downwind (see figure 6-10, Horizontal-axis configurations). On heavy blades the coning angle may be only 1 to 2 degrees, says British aerodynamicist Andrew Garrad, while on lightweight blades it can be as much as 8 to 10 degrees.

Downwind machines are certainly sleeker than wind turbines using either tail vanes or fan tails (see figure 6-11, Downwind rotor). Some believe this gives downwind machines a more modern look. But they pay a price, say

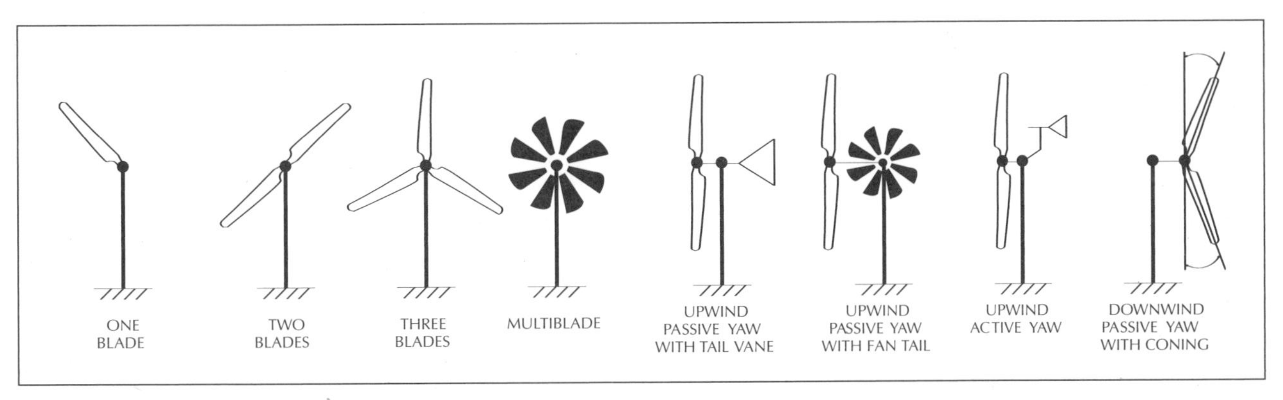

Figure 6-10. Horizontal-axis configurations. Upwind, downwind, one blade or two—it's all been tried at one time or another. (Adapted from J. W. Twidell and A. D. Weir, *Renewable Energy Resources*.

Figure 6-11. Downwind rotor. Proven Engineering's WT2200 on a headland in the Falkland Islands. Unlike most small wind turbines, the Scottish firm uses a self-orienting rotor downwind of the tower. The fairing helps keep the rotor downwind. The wind is from the right. (Proven Engineering)

proponents of upwind turbines. Downwind machines occasionally get caught upwind when winds are light and variable. Some downwind turbines would occasionally hunt the wind. Enertech and Storm Master turbines were notorious for "walking" around the tower in search of the wind. Critics add that the tower creates a shadow that disrupts the airflow as the blades pass behind the tower. This decreases performance, increases wear, and on two-blade designs emits a characteristic *whop-whop* sound that some find annoying.

West Texas A&M University's Vaughn Nelson asserts that upwind machines suffer a similar but less severe performance penalty. The wind piles up in front of a tower much like the small zone of turbulence just upstream from a stone in a swiftly flowing brook. Yet the stone creates a much bigger zone of disturbance or wake downstream.

One significant disadvantage of passive downwind machines is yaw control in high winds. A common method for protecting upwind turbines in high winds is to orient the

Ducted or Augmented Turbines

Just as some inventors are quick to grasp panemones (see figure 6-12) as a mysterious "newly discovered" technology for harnessing the wind, others envision wind turbines encased in shrouds or ducts. As with panemones and their paddles, shrouded turbines are easy to understand. Much like giant funnels, the shrouds concentrate or augment the flow across the wind turbine's rotor.

Unlike panemones, however, ducted turbines are wrapped in mysterious aerodynamic terms, such as *diffuser augmentation,* that are beyond the ken of most observers. Even supposedly sophisticated engineers have been snared by what at first appears to be a startling new technology "overlooked" by everyone else. As a result, augmented turbines have been plagued by hucksters and charlatans, the alchemists of wind energy.

Do augmented turbines work? Yes, of course. Panemones and cup anemometers work, too. Do augmented turbines produce the amount of energy promised at the cost promised? No, they have never fulfilled their often highly touted claims. Modern high-speed wind turbines, for all their limitations, reliably deliver quantities of electricity at increasingly competitive costs.

Unfortunately the U.S. Department of Energy funded development of the diffuser-augmented concept in the 1970s, lending credence to the design. In principle, an airfoil-shaped shroud surrounding a conventional wind turbine draws the wind through the rotor and augments—concentrates—the wind flowing across the blades. By doing so the diffuser-augmented turbine can achieve conversion efficiencies much higher than conventional wind turbines.

DoE abandoned the idea, along with many other "innovative" concepts. But the damage was done, and ducted turbines periodically reappear in the press as a promising new technology. Inventors often cite old DoE reports as proving the concept's merits.

One outspoken critic of diffuser augmentation is Heiner Dörner, a professor at the University of Stuttgart's Institute of Aircraft Design. Sure, Dörner says, wind tunnel tests show you that can double the wind speed across the rotor. But for this to happen, the wind must be homogeneous and flow directly into the concentrator—conditions available only in a wind tunnel. Rarely would such conditions exist in the real world, where such a turbine would operate. Yes, he says, the complete ducted assembly can yaw with changes in wind direction, but it can never perfectly follow the wind. Thus the concentrating effect will be difficult, if not impossible, to achieve consistently.

Indeed this is one of the results seen by the few ducted turbines that have been built. Engineers first noted in their reports that yes, they were able to augment the flow; they went on to say, however, that for various reasons the concentrating effect was much less than they had anticipated. Moreover, there were other problems. The shroud was more expensive and difficult to construct than they thought, yawing the turbine was difficult . . . and a veritable litany of excuses for why the design didn't work as promised was given.

rotor perpendicular to the wind. On small turbines with tail vanes, the rotor can be furled, taking the rotor out of the wind. Upwind turbines with active yaw controls mechanically direct the rotor out of the wind. Unless a downwind machine also has an active yaw drive, the rotor will always—well, nearly always—remain downwind of the tower.

Occasionally a downwind rotor without active yaw control can get caught upwind of the tower. This can prove disastrous unless the turbine's designers anticipated such an event. Storm Master and Carter turbines were plagued by this phenomenon. Because of these and other problems, today no medium-size turbines use downwind rotors with passive yaw control.

Active Yaw

Today all medium-size turbines use active yaw control to keep the rotor upwind of the tower. A wind vane mounted on top of the nacelle signals a hydraulic or electric motor to mechanically direct the rotor into the wind. As wind turbines have grown in size, so have the size, number, and sophistication of the yaw drives. One medium-size turbine today may incorporate several yaw motors in the nacelle, as well as yaw brakes to reduce the loads on the yaw gears.

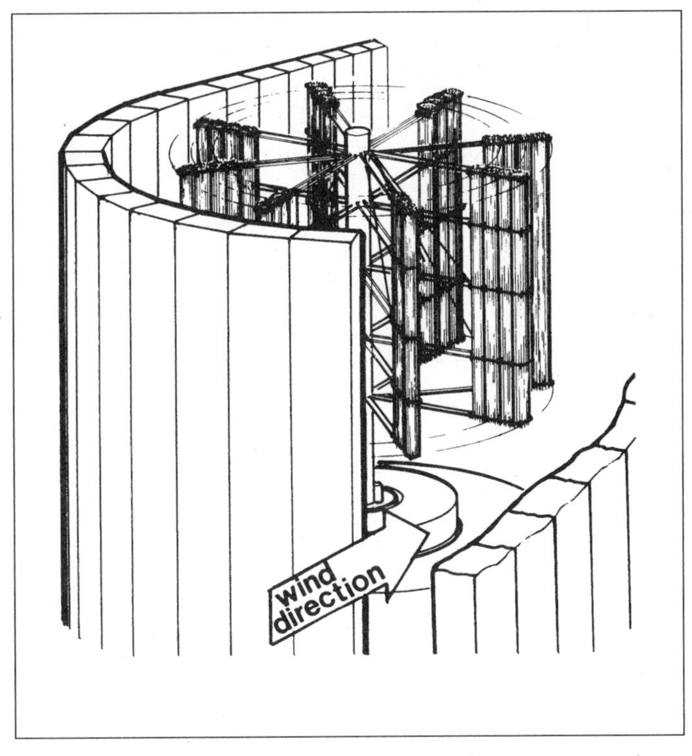

Figure 6-12. Panemone. A simple drag device used in ancient Persia for grinding grain. The vertically mounted blades were made by fastening bundles of reeds onto a wooden frame. The surrounding wall guides the prevailing wind onto the retreating blades. Many backyard inventors unfamiliar with aerodynamics construct similar ducted drag devices. (Sandia)

Lift and Drag

All modern medium-size and larger wind turbines use aerodynamic lift to drive their blades through the air. These machines have low solidity—that is, they use only two or three slender blades, and these blades operate at several times the speed of the wind that propels them. Modern wind turbine blades use airfoils or wings, much like the wing of an airplane. (In Danish a wind turbine blade is called a *vinge* or "wing.") These airfoils typically have a very high lift-to-drag ratio, a measure of their performance.

True drag devices are simple wind machines that use flat or cup-shaped blades to turn a rotor around a vertical axis (see figure 6-12, Panemone). In each, the wind merely pushes on the cup or blade, forcing it to move downwind; in doing so the rotor spins about its axis.

There are inherent drawbacks to drag devices that limit their use for generating electricity. At best only 15 percent (4/27) of the power in the wind can be captured by such machines. In comparison, the maximum possible for a lift device is 59 percent (16/27, the Betz limit). Drag devices also require more materials than comparable wind machines using lift.

Backyard tinkerers often turn their attention first to drag devices, because these machines are easier to understand and construct (see figure 6-13, Drag device). The

Figure 6-13. **Drag device.** Salesmen's model of a 1920s design using drag on the articulating flat panels to drive a water-pumping windmill. Though somewhat more sophisticated than a simple panemone, the aerodynamic performance of this approach is also limited.

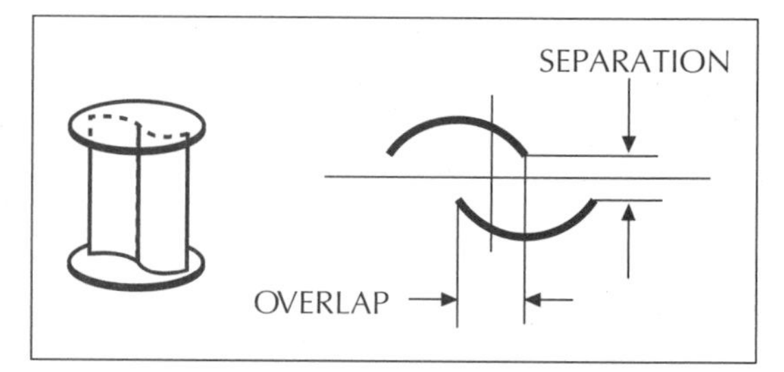

Figure 6-14. Savonius rotor. To achieve optimal performance, the two blades must be offset to permit some recirculation of flow. Because of its simplicity, a Savonius rotor is often the first choice of do-it-yourselfers.

wind pushes on a big wide blade, which moves with the wind. What could be simpler? Though inventors constantly create imaginative ways to use drag to power wind turbines, they always confront the physical limitations of drag propulsion: Lift devices are nearly four times more efficient at capturing the energy in the wind, relative to the area of the wind intercepted by the rotor. When examining the portion of the frontal area occupied by the wind turbine's blades, lift devices easily produce at least 50 times more power per unit of blade area than the same size turbine dependent upon drag, says Oregon State University's Robert Wilson, an authority on aerodynamics. The only way to improve the performance of drag devices is to incorporate some form of lift, as in some Savonius rotors, where there is recirculation of the airflow.

Lift devices easily produce at least 50 times more power per unit of blade area than those dependent upon drag.

Just to confuse matters, engineers occasionally refer to horizontal-axis wind turbines that use crude blades and operate at low rotor speeds as "drag devices." Examples are the farm windmill and the traditional European windmill. Technically these machines are not true drag devices, because they do use lift. Drag on the blades as they move through the air, however, is higher relative to the lift produced than is found on modern high-speed airfoils.

Figure 6-15. Modern S-rotor. Finns remain enamored with Savonius rotors. Windside Oy's S-rotor on display at a conference in the late 1990s.

In 1924 Finnish inventor Sigurd Savonius developed an S-shaped vertical-axis wind machine. Though his was principally a drag device, Savonius improved the rotor's performance by recirculating some of the airflow between the rotor's two halves (see figure 6-14, Savonius rotor). Air striking one blade is directed through the separation between the

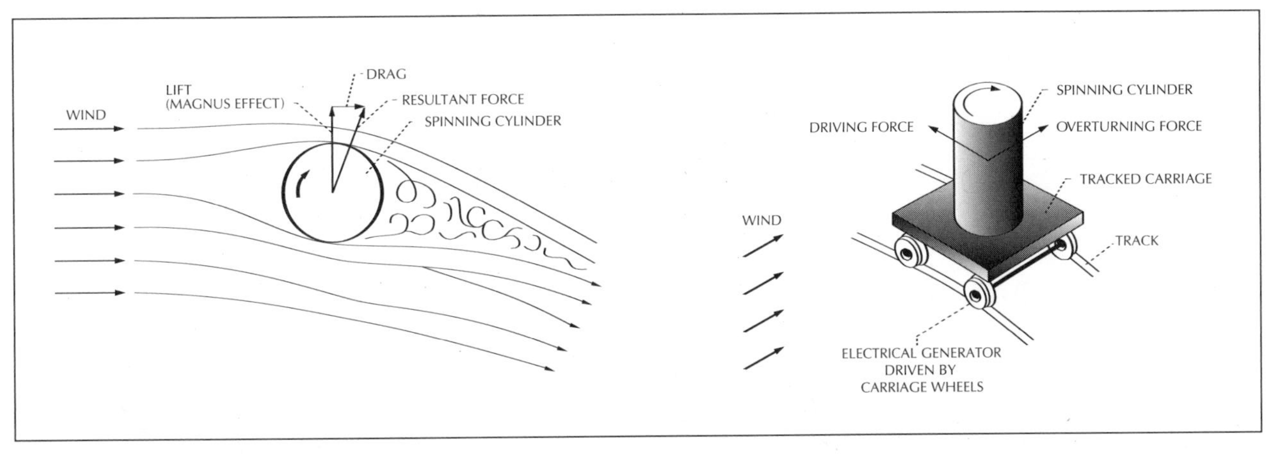

Figure 6-16. Magnus effect. A hybrid device that uses the lift created by a spinning cylinder demonstrates the Magnus effect. Spinning cylinders have been used to drive several kinds of wind machines, including a flat car on a track.

two halves of the S and onto the other blade. Researchers have measured conversion efficiencies of almost 30 percent under optimal conditions—considerably higher those of typical drag devices. In practice, however, S-rotors extract less than 15 percent of the power in the wind. Because of their poor performance relative to conventional wind turbines, Savonius rotors have never found broad commercial application. Today they're only found sitting in experimenters' garages or powering buoys (see figure 6-15, Modern S-rotor).

Another novel approach uses the Magnus effect. This is the lift or thrust produced when air moves over the surface of a spinning object (see figure 6-16, Magnus effect). It's what produces the curved flight in baseball's curveball. The pitcher imparts spin to the ball as it leaves his hand. Air rushing over the spinning ball forces the ball off its normal trajectory.

German engineer Anton Flettner built several such devices. Using the Magnus effect as a means of propulsion on two upright, spinning cylinders, he sailed the Atlantic to New York in 1925. The following year he built a horizontal-axis wind turbine 65 feet (20 m) in diameter in which he used four spinning cylinders to drive the rotor.

In 1933 J. Madaras constructed a 90-foot (28 m) cylinder 28 feet (9 m) in diameter at Burlington, New Jersey, in the hope of using this phenomenon to drive cars around a track. Like similar attempts to harness the Magnus effect, it was abandoned because the spinning cylinders were material-intensive. The same results could be obtained from true airfoils at less expense. Lift devices use airfoils with high lift-to-drag ratios to power the rotor. It seems mysterious that a wind machine with only a few blades can operate efficiently. But a modern wind turbine, because it uses lift, can capture the same amount of power with a smaller rotor, using fewer blades than a drag device. Modern wind turbines, those using lift, would make Buckminster Fuller proud: They "do more with less."

Why is this? Why do some wind machines, like the farm windmill, use multiple blades where others use only a few? Why do some blades taper from the root to the tip while others taper from the tip to the root? To understand the answers to these questions, we need to delve further into wind turbine aerodynamics.

Aerodynamics

Wringing the most energy from the wind striking a wind turbine rotor is exceedingly complex. Aerodynamicists Woody Stoddard and Dave Eggleston devote an entire book,

Figure 6-17. Taper. Wind turbine blades taper from root to tip. The saberlike shape minimizes solidity but also enables strengthening the root where the blade attaches to the rotor hub. Shown here is a blade for Vestas's V66, a 1.65 MW turbine. The blade is about 32 meters (105 ft) long.

Wind Turbine Engineering Design, to the subject. Suffice it to say that modern wind turbine rotors use very little material to capture the energy in the wind stream.

As contemporary wind energy pioneer Peter Musgrove has written, modern wind turbines present little solidity to the wind: The two or three slender blades occupy only 5 to 10 percent of the rotor disk. Power densities of 1,000 W/m^2 of blade area are typical. With so little material doing so much work, says Musgrove, the energy used to make the wind turbine is quickly recovered, often in less than one year.

The slender blades on these turbines typically taper and twist from hub to tip (see figure 6-17, Taper). For example, the blade of Vestas's V27 tapers from 1.3 meters (4.3 ft) in width at the hub to 0.5 meters (1.6 ft) in width at the tip. Why they do so is more difficult to explain.

It's intriguing that a sailboat can travel faster than the wind, and even more so when we learn that the boat sails faster across the wind than with it. Mariners discovered this fact centuries ago. Today we explain the paradox by speaking in terms of lift and drag. The blade of a modern wind turbine is much like the sail of a sailboat—both have good lift to drag, which allows them to travel faster than the wind.

To begin, let's look at the factors affecting the lift from an airfoil like that of a wind turbine blade. Air flowing over the blade causes both lift and drag. When you're driving down the highway and you stick your hand outside the window, lift from the air flowing over your hand (a crude airfoil shape) literally lifts your hand toward the roof. Drag pulls your hand toward the rear of the car. The sum of these two forces generates a thrust that pulls the blade on its journey through the air, much the way it pulls a sailboat through the water. This thrust is greatest when the sailboat is sailing across the wind or a blade is slicing through the wind, as on a horizontal-axis wind turbine

One measure engineers use to rate airfoil performance is the ratio of lift to drag. Designers of high-speed, high-performance wind turbines want a high lift-to-drag ratio. The lift-to-drag ratio is determined by the blade's angle of attack—the blade's angle with respect to the apparent wind, the blade's shape, and its aspect ratio.

One characteristic of high-performance airfoils is the shallow angle of attack at which they function best. Slight increases in the angle of attack—for example, from 0 to 15 degrees—produce increasing amounts of lift. A point is reached, however, where the flow over the blade separates from the airfoil and becomes turbulent. Lift then deteriorates rapidly and drag increases. At this critical juncture, the airfoil is said to stall.

Stall is a deadly condition in flight. Airplanes literally fall out of the sky when stall occurs, when there's no longer enough lift to support them. It's one of the leading causes of light-plane accidents. In a wind machine stall can be beneficial. We'll see why in a moment. But first consider the angle of attack: It's a function of the blade's angle to the plane of rotation—its pitch—and the apparent wind. For now assume the pitch is fixed, which it is on many wind turbines.

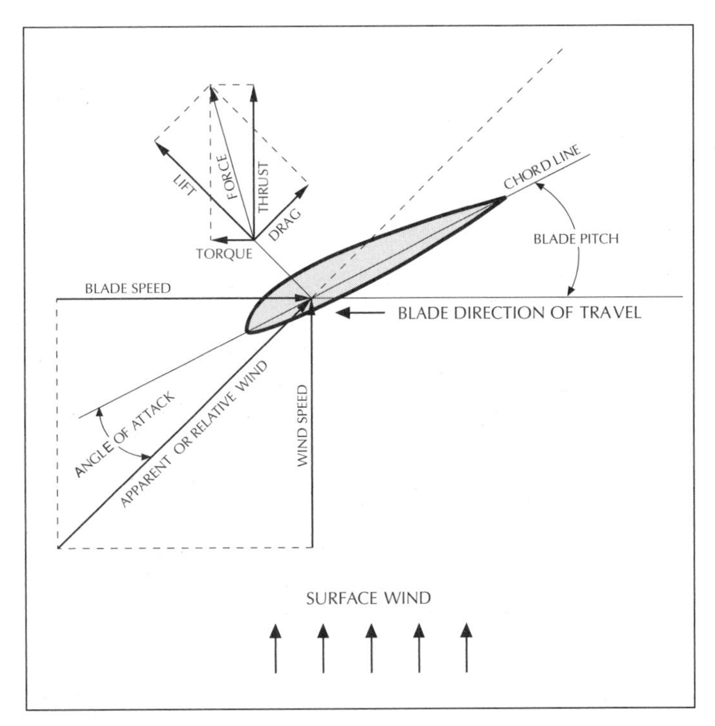

Figure 6-18. Apparent wind. Airfoil performance is gauged by the ratio of lift to drag. Lift is determined by the angle of attack. The pitch of the blade, the speed of the blade through the air, and the speed of the wind control the angle of attack and, consequently, lift.

Apparent Wind and the Angle of Attack

The apparent wind is the wind seen by the blade. It's the result of the airflow due to a combination of the blade's own motion and the wind across the ground. If you recall some of your high school physics, you'll note that the result is dependent on the relative strength of each. For example, if both were equal in speed and if they were acting at right angles to each other, the apparent wind would be acting at a 45-degree angle between the two (see figure 6-18, Apparent wind). On a sailboat under sail the wind you feel on your face is the apparent wind.

If wind speed increases while the blade's speed through the air remains constant, the position of the apparent wind swings toward the wind direction because it has become more influential. As the apparent wind changes position, it also changes the angle of attack. Reversing the process by increasing blade speed relative to wind speed causes the apparent wind to shift toward the direction of the blade's motion, decreasing the angle of attack. Wind turbine designers must decide how best to deal with this relationship for each airfoil, because there's an optimal angle of attack—a point where the lift-to-drag ratio is optimal and performance reaches a maximum.

To maintain an optimal angle of attack as wind speed increases, a fixed-pitch blade must increase its speed proportionally. Thus to maximize aerodynamic performance, the rotor must spin faster as the wind speed increases. Another way to say it is that the tip-speed ratio, the relationship between blade speed (at the tip of the blade) and wind speed, must remain constant in order to maintain optimal aerodynamic performance. Most designers of small wind turbines try to operate their wind turbines in this manner.

Note that blade pitch on nearly all medium-size wind turbines is fixed throughout the turbines' operating range. Blades on variable-pitch turbines are no exception. Contrary to popular belief, these turbines pitch their blades only to start the rotor spinning after the rotor has been at rest, and to reduce power when the turbines have reached the generator's rated capacity.

Operators of medium-size turbines, whether fixed pitch or variable pitch, have one significant advantage over users of small wind turbines. The pitch on these turbines can be adjusted to tailor the turbine's performance to the site, such as a ridgetop in the Tehachapi Mountains. On fixed-pitch turbines the preferred pitch can be set when the blade is first attached to the rotor, or it can, with more difficulty, be adjusted seasonally. One Danish manufacturer modified the hub on its fixed-pitch rotors so that the pitch could be adjusted mechanically just a few degrees to compensate for changes in air pressure. Operators of variable-pitch turbines can change the pitch-control algorithm, letting the rotor's pitch-control mechanism do the work.

In contrast, blade pitch is permanently fixed on nearly all small wind turbines. Seldom, for example, can the blade pitch of a small wind turbine rotor be adjusted to compensate for lower air density at higher elevations.

Unlike small wind turbines, many medium-size wind machines operate at a constant rotor speed. As wind speed changes, rotors on these turbines continue to spin at the same speed because of the induction generators they use. Consequently their rotors operate at varying tip-speed ratios. Designers are willing to sacrifice some performance for the simplicity of fixed-pitch blades driving constant-speed generators. On stall-regulated turbines, the airfoil begins to stall and performance erodes as wind speed increases. This reduces the rotor's power in high winds, making it easier for designers to build protective controls to keep the rotor from destroying itself. On pitch-regulated turbines, the blades change pitch in high winds, dumping excess power.

The amount of thrust driving the rotor is a function of not only the airfoil's lift coefficient, which depends on the blade's angle of attack, but also the area of the blade and its speed through the air.

Twist and Taper

For the sake of simplification, we've been looking at a blade as if the conditions it sees were constant along its entire length. That may be true for airplanes, but it's not so for wind turbines. Even when the pitch of the blade is fixed and rotor speed constant, the speed through the air of a point on the blade changes with its distance from the hub. On a conventional wind turbine, blade speed is higher at the tip than near the hub because it has more distance to cover in the same amount of time.

Because blade speed increases with distance from the hub, the apparent wind varies as well. The apparent wind increases in strength, and its position shifts toward the plane of rotation as you move out along the blade toward the tip. If the blade designer wants to maintain the angle of attack (to optimize performance) at the same time blade speed is increasing, the angle of attack must decrease toward the tip. As a result, wind turbine blades are twisted from root to tip (see figure 6-19, Twist). The twist reaches a maximum at the root and a minimum at the tip. Glance up at the next wind turbine you see and note that the tip of the blade is parallel with its direction of travel. On the Vestas V27 used in the previous example, the blades have a twist of 13 degrees at the hub and 0 degrees at the tip, ironically the same degree of twist as that of the Gedser mill, the distant forerunner of modern Danish turbines.

Figure 6-19. Twist. For most wind turbine airfoils, blade angle of attack varies from root to tip. But there's nothing new under the sun. Here the jalousie latticework of Germania Molen's blades twists about the stock or blade spar. Germania Molen, despite its name, is one of the Netherlands' nearly 1,000 functioning windmills. Note the airfoil-shaped leading edge. Jalousie slats are controlled by the spider linkage protruding from the main shaft. Yes, that's thatch covering the cap and smock. *Inset:* Vestas V39 blade illustrating taper and twist from root to tip. Note that the blade is much thicker near the hub than the tip. This allows the use of more strengthening materials where the loads are greatest. The blade is about 19 meters (62 ft) long. (Vestas Wind Systems)

Figure 6-20. Multiblade farm windmill. The characteristic curved metal blades of the American farm windmill resulted from the pioneering work of Thomas Perry. The Aermotor-brand windmill embodied the principles Perry learned after 5,000 tests on various "wheel" (rotor) designs in the late 19th century. Rotors using Perry's curved blades were almost twice as efficient as those using the wooden slats then common. His Aermotor also included "back gearing," which allowed the wheel to make several revolutions for each stroke. Perry's design has since been widely copied. Note the work platform for servicing the rotor. This is a common feature of farm windmill towers.

Solidity

Wind machine designers long ago learned that blade area (the number of blades, as well as their length and width) governs the amount of torque, or turning force, that a rotor can produce. The more blades or blade area of a given length for the wind to act on, the more torque it will produce. Greater solidity, the ratio of blade area to the area swept by the rotor, generates greater torque.

The American farm windmill uses multiple "sails" that cover nearly the entire rotor disk (80 percent solidity). It was designed to deliver high torque for pumping water in the low winds of late summer, and it does its job remarkably well (see figure 6-20, Multiblade farm windmill).

There's no better example of a high-solidity rotor fitting this description than the American farm windmill. It uses multiple sails that taper from the tip to the root. High torque assures you that the windmill will be able to lift the pump's piston (and the water with it) during late summer, when the need is greatest but the winds are lightest.

Yet the demands on electricity-generating wind turbines are different. Wind turbines are not necessary to generate electricity. We are not dependent upon them as the Amish are for water from their farm windmills. For us there are many other sources of electricity: the utility, photovoltaics, or a stand-alone generator.

We want power from a source and are willing to pay for it only when it's superior in some way to other sources—it's cheaper, cleaner, or provides some combination of benefits. To compete with these technologies, the wind machine must be designed to extract power from the wind in the most cost-effective and environmentally desirable manner possible. The farm windmill is too material-intensive for this task, even though its performance has greatly improved over the past century.

Early farm windmills, for example, used flat wooden slats for blades. In 1888 Aermotor introduced its "mathematical" windmill, which substituted sheet-metal blades for those of wood. Aermotor stamped a broad curve into the metal blade, and in doing so directed the air to flow over the back side of the following blade. Much like the slot effect in jib-rigged sailboats, this cascading of air over the following blades improved Aermotor's performance over that of its rivals. Unfortunately the "new and improved" farm windmills—all of which now use Aermotor's technique—still extract only 15 percent of the power in the wind.

Intuitively the multiblade farm windmill looks like it would capture more wind than a modern machine with only two or three blades. We sense that the rotor should have more blades to capture more wind. Consider what would happen if we carried this belief to its logical extreme, however. The optimal rotor would cover the entire swept area with blades, in effect producing a solid disk. No air would pass through. The wind would pile up in front of the rotor and flow around rather than through it. The wind speed behind the rotor would be zero. Instead of capturing more wind, we wouldn't capture any. There must be some air moving through the disk, and it must retain enough kinetic energy so it can keep moving to make way for the air behind.

The Betz Limit

We must strike a balance between a rotor that completely stops the wind and one that allows the wind to pass through unimpeded; between the amount of wind striking the rotor and the amount flowing through. German scientist Albert Betz demonstrated mathematically that this optimum is reached when the rotor reduces wind speed by one-third. By conserving the wind's momentum as it passes through the rotor, Betz calculated that the maximum a theoretical wind turbine could capture was 16/27 or 59.3 percent of the energy in the wind. "You can't beat Betz," says Julian Feuchtwang of the Centre for Renewable Energy Systems Technology in Great Britain, though many have tried.

Real rotors, says the Alternative Energy Institute's Vaughn Nelson, never achieve the Betz limit because of losses due to aerodynamic drag, losses around the blade tip, and losses due to rotation in the wake behind the rotor. As Nelson explains it, the wake spins opposite to the direction of the rotor as a result of conservation of angular momentum. The wind acts on the rotor and in doing so spins the rotor. In the process the wind itself is deflected and spirals downstream. To optimize rotor performance—to approach Betz's theoretical limit as nearly as possible—designers must opt for high rotor speeds and low torque, because high torque increases losses in the wake.

But, you may ask, if we were to lower the rotor's torque, wouldn't we be lowering the power it can produce, even if it's going to be more efficient at producing it? Yes, if we kept everything else the same. We don't. Power is a product of torque and rotor speed. To deliver the same amount of power as we decrease torque, we must increase rotor speed.

This strategy works up to a point, says Scoraig Wind Electric's Hugh Piggott. In high-performance rotors there are other important losses besides those caused by rotation of the wake, principally drag on the fast-moving blades. There are also tip losses from increased pressure around the end of the blade on rotors using only a few slender airfoils. This causes more air to flow around rather than over the blade—one reason for tip vanes on large aircraft.

Blade speed is a function of rotor diameter and rotor speed. Both are described by a single term, the ratio between the speed of the blade through the air at the tip and the wind speed: the *tip-speed ratio.* Tip speed increases either as rotor speed increases or as the length of the blade increases for a given rpm. Modern high-performance wind turbines operate at tip-speed ratios of 4 or more. In contrast, true drag devices operate at tip-speed ratios of less than unity.

Though efficiency improves with increasing rotor speed, there are practical limits. Increasing drag, which reduces the airfoils' effectiveness, is one. For small wind turbines, Wisconsin's Mick Sagrillo argues that durability is inversely proportional to tip-speed ratio. That is, small turbines that operate at high tip-speed ratios may wear out faster than those that operate at lower tip-speed ratios.

Scoraig Wind Electric's Piggott suggests that a tip-speed ratio of 5 is aerodynamically optimal for small wind turbines with slightly higher tip-speed ratios on rotors used in combination with direct-drive alternators. Medium-size turbines, such as the 660 kW Vestas V47 or the 500 kW Enercon E40, operate at tip-speed ratios of 4:1 to 6:1 at rated power (see table 6-3, Tip Speeds and Tip-Speed Ratios of Selected Wind Turbines).

Noise is another practical limit to higher tip-speed ratios. Noise is proportional to the speed of the blade tip. At higher tip speeds, vortices are shed from the tip, and it's these vortices that cause the swishing sound as the blades move through the air. For example, the tip speed of medium-size, constant-speed Danish wind turbines, at their rated wind speed, is about 60 m/s (130 mph). Higher tip speeds may increase blade noise. Some experimental wind turbines with unusually high tip-speed ratios near 10:1 were notoriously noisy.

Blade Number

Wind turbines need only one slender blade to capture the energy in the wind. To sweep the rotor disk effectively, a one-blade turbine must operate at higher rotor speeds than a two-blade turbine, thereby reducing the gear ratio required for the transmission, and hence the mass and cost of the gearbox. Since one blade should cost less than two or three, proponents argue that one slender blade delivers optimal engineering economy.

Table 6-3
Tip Speeds and Tip-Speed Ratios of Selected Wind Turbines

	Diameter (m)	(ft)	(rpm)	Selected Wind Speed (m/s)	Tip Speed (m/s)	Tip Speed Ratio[1]	Constant or Variable Speed
Farm windmill	3.05	10	78	6.7	12	1.9	v
Dutch windmill	25	82	25	12	33	2.7	v
Micro Turbines							
Ampair 100	0.91	3	750	10	36	3.6	v
Air 403	1.17	3.8	1,500	10	92	9.2	v
Household Turbines							
BWC Excel	7	23	310	14	114	8.2	v
Medium Turbines							
Enercon E40	40	131	36	13	75	5.8	v
Vestas V47	47	154	26	15	64	4.3	c/v
Bonus 1MW/54	54	177	22	15	62	4.1	c
Monopteros 50[2]	56	184	43	11	125	11	v

Notes: [1]At rated power.
[2]Experimental one-blade turbine.

One-Blade Wind Turbines

While never commercially successful, optimal material economy has for decades been a siren's song luring designers onto the rocks of flimsy, one-blade designs. German engineers have been particularly susceptible because of Ulrich Hütter's 1940s' doctoral thesis on the design of inexpensive, high-performance wind turbines.

Hütter taught for many years at the University of Stuttgart's Institute of Aircraft Design, alongside Professor Franz Wortman, himself well known for airfoil sections at the Institute of Aerodynamics. Wortman and his students pursued Hütter's minimalist design philosophy to its logical conclusion: the one-blade FLAIR or Flexible Autonomous 1-Bladed Rotor.

In the mid-1980s, Wortman installed a FLAIR prototype 8 meters (26 ft) in diameter at the university's test field near Schnittlingen, in southern Germany's Swabian Alps. The novel downwind rotor drove a 5.1 kW, four-pole induction generator—it, too, out of the ordinary. The generator on the grid-connected version incorporated high slip (14 percent), enabling the turbine to regulate rotor speed within a vary narrow range by using its Watt governor.

Originally developed for the German washing machine company Böwe, FLAIR was subsequently sold to aerospace giant Messerschmitt-Bölkow-Blohm. From it, MBB developed the Monopteros 15 series: turbines with rotors from 12.5 to 17 meters (40 to 55 ft) in diameter, driving generators from 15 to 30 kW (Monopteros 20). MBB also introduced a greatly scaled-up version of Wortman's FLAIR: the Monopteros 50 series, with rotor diameters from 47 to 56 meters (150 to 180 ft), rated from 550 kW to 1 megawatt. MBB's work on the turbine ceased in 1986 after Wortman's death.

Monopteros 20. The smallest entrant in Messerschmitt-Bölkow-Blohm's line of one-blade turbines was a direct descendant of Professor Franz Wortman's FLAIR. (MBB)

Riva Calzoni MP5. This MP5 was hardly out of production before it was displayed as a novelty in Milan's Museum of Science and Industry. (Nancy Nies)

Independently of Wortman's group, Riva Calzoni developed its own one-blade design. An Italian heavy-engineering company, Riva Calzoni built 25 of its MP5 and another 25 of its MP7 models by 1992, when it abandoned the small turbine market .

Both MBB and Riva Calzoni found that it was more profitable to build larger turbines for commercial wind farms than small turbines for rural residences. At one time MBB envisioned building monstrous 5 MW versions, dubbed Growian II (Grosse Wind Energie Anlage), along the German coast. But MBB completed only three 650 kW Monopteros 50 models near Wilhelmshaven

Never say die. As evidence of continuing interest in one-blade designs, a small German company was displaying a photo of its one-blade turbine on the Web in the late 1990s. (Schoder GbRmbH)

before abandoning the program. Of the two firms, Riva Calzoni was the more successful, eventually installing more than 100 of its 300 kW versions in Italy before discarding the one-blade approach entirely.

As late as the 1990s, the concept was still finding new adherents in Germany, though having little more success than their predecessors.

The giant German conglomerate Messerschmitt-Bölkow-Blohm built just such a series of one-blade wind turbines. But there are other, equally important design criteria besides lowest initial cost. Two blades are often used for reasons of static balance. Many modern wind turbines use three blades because they give greater dynamic stability than either two blades or one.

Rotors using two and three blades are also more efficient than rotors using only one due to aerodynamic losses at the tip of the blade. British engineer John Armstrong notes that, everything else being equal, one blade captures 10 percent less energy than two blades. And though one blade may be cheaper, engineers say the rotor it is attached to is just as heavy as one with two blades. First, the blade on a one-blade turbine must be stronger than a comparable blade on a two-blade turbine, because it must capture twice as much energy in the wind. Second, a one-blade rotor must compensate for the weight of the missing blade by using a massive counterweight. Because of its higher speed and greater aerodynamic loading, one blade will also emit more noise than two.

Ultimately, one-blade rotors provide no cost savings. Some manufacturers claim they can build three simple blades for the cost of one single high-performance blade and the sophisticated teetering hub required for a one-blade rotor.

The advantages of rotors with two blades over those with three are similar to those of rotors with one blade over those with two. The rotor is cheaper, lighter, and ideally operates at higher speeds, leading to lower-cost transmissions. Two blades are easier to install than three because they can be bolted to the hub on the ground, in the position they will assume on the assembled turbine. The disadvantages of two blades are similar, as well. Because of their higher speeds and greater rotor loading, they are often noisier than three blades.

Conventional wisdom holds that three-blade machines will deliver more energy and operate more smoothly than either one- or two-blade

Figure 6-21. Blade number. While most wind turbines use two or three blades, Windmission develops high-performance rotors that use several very slender blades. In the foreground is Windmission's Windflower at the Folkecenter for Renewable Energy's test field in Denmark.

turbines. They will also incur higher blade and transmission costs as a result of their lower rotor speed. British engineer Armstrong estimates that rotors with three blades can capture 5 percent more energy than two-blade turbines, while encountering fewer cyclical loads than either one- or two-blade turbines when reorienting the nacelle to changes in wind direction.

The dynamic or gyroscopic imbalance of two-blade turbines with changes in wind direction is a classic engineering challenge. The phenomenon can best be seen in the jerky or wobbly motion of small two-blade turbines as they yaw with the wind. When the rotor is vertical, there is little resistance to yaw, but as the rotor nears the horizontal, the inertia retarding the rotor from reorientating itself reaches a maximum. This occurs twice every revolution. This causes the turbine to yaw unevenly. On larger machines this effect is dampened with shock absorbers or by allowing the rotor to teeter.

Three blades minimize this dynamic problem and are preferred on small machines where yaw dampening or teetering hubs would be too costly. For this reason, Nolan Clark, who for several decades has overseen the U.S. Department of Agriculture's wind experiments, prefers three blades to two on small turbines.

According to California aerodynamicist Kevin Jackson, a two-blade rigid rotor encounters 10 times the force that a three-blade rotor sees. "This is why two-blade rotors are always teetered" on medium-size wind turbines, explains Jackson. With teetering, the two-blade rotor experiences even fewer cyclic loads than the common three-blade design.

Aerodynamics resemble other branches of engineering in that there are always trade-offs. The theoretically ideal rotor, says turbine designer Hugh Piggott, would use an infinite number of infinitely slender blades (see figure 6-21, Blade number).

Less prone to technical resolution is the perception that one- and two-blade turbines are less aesthetically pleasing than those with three blades.

Self-Starting

Low solidity reduces a rotor's ability to start spinning on its own. Remember that the apparent wind flowing over the blade is partly due to the blade's motion. When the rotor is stopped, the lift on the blades from the wind alone may not be enough to start the rotor into motion. One solution for rotors using fixed-pitch blades is to motor the rotor up to a speed where the aerodynamic forces are sufficient for the blades to drive the rotor. This is a common practice for Darrieus turbines. There were also a number of conventional wind turbines, mostly American designs, that required motoring the rotor up to speed, such as Enertech and ESI.

Don't be misled by glib talk of a "new" wind turbine that runs in low winds. Anybody can design a rotor to spin in light winds. But why bother?

Many designers are willing to sacrifice some performance to gain a self-starting capability. All wind turbines on the market in the early 2000s were self-starting. Typically their rotors would begin spinning in winds from 8 to 10 mph (4 to 5 m/s). British micro turbines built by Marlec, Ampair, and LVM use six-blade rotors to produce high starting torque and good low-wind performance, ideal for keeping sailboat batteries charged at port. The multi-blade rotors on these turbines are much quieter than their high-performance competition.

Don't be misled by glib talk of a "new" wind turbine that runs in low winds. Anybody can design a rotor to spin in light winds. But why bother? There's so little energy in winds below the start-up speed of conventional turbines that there's no economic justification for making the effort. The rotor may spin, but it won't produce enough electricity to make it worthwhile.

Darrieus turbines are typically not self-starting, though it's now known that they can self-start under the right wind conditions. These conditions—though infrequent—do occur. When the Darrieus rotor is at a standstill, only the wind across the ground acts on the blade. Because the pitch of the rotor is fixed, the blades stall and nothing happens. Normally the rotor must be motored up to speed. But on July 6, 1978, all that changed and a new corollary was added to Murphy's Law: "Wind turbines that won't self-start, will."

Canadian researchers were testing a 230 kW experimental Darrieus rotor on the Iles de la Madeleine off the coast of Quebec's Gaspé Peninsula. While repairs were under way, the brake was removed. Because it was thought that the rotor could not start itself, the turbine was left unattended overnight. During the night, the wind picked up. By the next morning the rotor was spinning out of control. Eventually the rotor spun off the tower and corkscrewed itself into the ground.

Darrieus rotors have also been plagued by a misperception that they're less efficient than conventional wind machines because the blades must run both with and against the wind. Even so, the blades on a Darrieus turbine produce lift for most of their orbit around the turbine's axis.

Sandia Laboratories has found that under ideal conditions, Darrieus turbines can extract more than 40 percent of the power in the wind. In other words, their performance is similar to that of conventional wind turbines; not any better but certainly no worse.

Developers of the Musgrove H-rotor found that a vertical-axis rotor with fixed-pitch blades can be made to start itself. (The Musgrove design, named for its British inventor Peter Musgrove, uses a novel means of protecting the rotor in high winds, as explained in the section on rotor controls.) The designers discovered that decreasing the blade's aspect ratio—its height to its width—by shortening and widening the blades created more lift while the rotor was at rest. The stubbier blades could start the rotor without robbing performance at operating speeds.

An H-rotor with articulating blades, such as a giromill or cycloturbine, is also self-starting. The pitch of each blade is set according to a predetermined schedule and the position of the blade relative to the wind. The blade's angle of attack is optimized at each position of its orbit around the rotor's axis. Controlling blade pitch with respect to the wind gives the rotor a reliable self-starting capability not found in the Darrieus rotor. It should also deliver better performance, because lift can be maximized regardless of whether the blade is advancing into, across, or with the wind. Giromills, however, have never lived up to expectations. They're also material-intensive.

Tip Vanes and Tip Torpedoes

When commercial aircraft began sporting winglets to reduce tip vortices, the use of tip vanes at the ends of wind turbine blades came to the fore. Wind turbine engineers freely admit that a portion of the air's flow over the blade, like that over an aircraft wing, is lost at the tip as the wind escapes around the end of the blade. Eliminating the lost lift by using a tip vane or winglet is nothing new. Aerovironment, the firm that built the Gossamer Condor and other aviation marvels, studied the question in the early 1980s. Due to the difficulties of constructing a tip vane, Aerovironment concluded that it was cheaper to simply extend the length of a conventional wind turbine blade than to add a winglet.

Some medium-size wind turbines, notably Bonus's 300 kW Combi model of the early 1990s, do incorporate unusual tip shapes. But instead of boosting performance, these tip designs are intended to reduce vorticity-induced noise. To cut noise, Bonus designer Henrik Stiesdal employed what he termed a tip torpedo.

Blades

Blades are one of the most critical and visible components of a wind turbine. Blades can be made from almost any material—and have been. European windmills used blades of wood and canvas, and this tradition survived well into the 19th century. In the 1970s researchers at Princeton University adapted the technology to build an experimental sail-wing turbine. Philosophy professor Gordon "Corky" Brittan used cloth sails on his Montana Windjammer, an ungainly turbine patterned after a Cretan windmill (see figure 6-22, Cloth blades).

Wood has always been popular. Early farm windmills used wooden slats, and wind-chargers of the 1930s used wood almost exclusively. Wood is still used on small wind machines. It's strong, readily available, easy to work with, relatively inexpensive, and has good fatigue characteristics. "Wood flexes for a living," explains Mick Sagrillo, who ran a small turbine repair shop in Wisconsin. "It works well in high-fatigue applications."

Wood

Wooden blades for small turbines are built either from single planks of Sitka spruce or from wood laminates. The blades are then machined into the desired shape and coated with a tough weather-resistant finish. The manufacturer then covers the leading edge with polyurethane tape to protect the blades from wind erosion and hail damage. This tape is the same as that used on the leading edges of helicopter blades. It's resistant to ultraviolet light and abrasion. Leading-edge tape is critically important on wood or wood-composite blades. "Without it you just have driftwood after a few years," says Sagrillo.

Few of those new to wind energy appreciate the wind's erosive force. If you need to be convinced of this, pay a visit to the Texas Panhandle or the Tehachapi Pass during the spring wind season. But don't forget to take your goggles. Sand and blowing grit scour anything in their path. This airborne sandpaper has deeply etched the galvanizing on the windward side of towers in the Tehachapi

Figure 6-22. Cloth blades. Partially reefed canvas sails on a Cretan sail windmill at the Internationales Muhlenmuseum, Gifhorn, Germany. Reefing allows the miller to control power in the rotor during high winds. Note the prow and stays, a distinguishing feature of Cretan sail windmills.

Pass. In areas prone to blowing sand, wind turbine blades have seen their leading edges damaged after only two years of use.

Though solid wood planks work well for small machines up to 5 meters (16 ft) in diameter, manufacturers prefer laminated wood (like that of a butcher's block) for larger turbines. Laminated wood offers better control over the blade's strength and stiffness, as well as limiting shrinkage and warpage.

In the laminating process, slabs of wood are bonded together with a resin. The resulting block can then be carved into the desired shape. By varying the types of wood, the direction of their grains, and the resin, a material can be produced that is stronger than any one part alone and stronger than a single plank of the same size. Laminated wood blades have been used on small wind turbines of all sizes.

Thinner slices of wood are also used to produce veneers. Layer upon layer of razor-thin slices are sandwiched together with a resin and molded into the airfoil shape. The process is widely used to build the hulls of high-performance sailboats and has been adapted successfully for wind turbine blades in both the United States and Europe.

Wood-composite blades fabricated by Michigan's Gougeon Brothers earned a reputation for strength and reliability in wind turbines up to 43 meters (142 ft) in diameter. Similarly, NEG-Micon used wood composites from its British affiliate on its 1.5 MW turbine.

Metal

In the late 19th century farm windmill manufacturers began replacing wooden blades with stamped, galvanized steel. Thin steel sheets have been used ever since. Steel is strong and well understood. That's why steel was chosen by Boeing engineers for the giant 300-foot (91 m) diameter Mod-2, and the subsequent 320-foot (98 m) diameter Mod-5B (abandoned by Boeing and the U.S. Department of Energy in Hawaii). The blades were constructed from structural steel—nothing fancy; the same steel used in bridges.

But steel is heavy. The hub, drive train, and tower must be more massive than on a wind machine with a lighter rotor. Both Boeing and Dutch manufacturers encountered numerous problems with steel rotors and the shafts supporting them.

Aluminum is lighter and, for its weight, stronger. For this reason, aluminum is used extensively in the aircraft industry. We can fabricate aluminum blades with the same techniques used to build the wings of airplanes: Form a rib and then stretch the aluminum skin over it. The blades on NASA's early Mod-0A were built in this way. On smaller machines a simpler method can be used by stamping a curve into the leading edge, folding the sheet metal over the spars, and then riveting it into place.

Aluminum can also be extruded, eliminating several fabrication steps. Various manufacturers once thought that blades could be mass-produced by extruding blades in the same way we manufacture gutters, drain spouts, window moldings, and ladder rails—by squeezing a hot piece of aluminum through a die.

> No manufacturer today builds wind turbines with metal blades.

Alcoa and Canadian fabricator DAF-Indal developed extruded aluminum blades for Darrieus turbines. They believed that the Darrieus rotor was ideally suited for their aluminum extrusions because the inertial forces on Darrieus blades are in tension. The blades endure less stress in the Darrieus rotor than they would in a conventional wind machine. They can also use a blade of a constant chord or width, such as those produced by extrusion.

Aluminum, unfortunately, has two weaknesses: It's expensive and it's subject to metal fatigue. Have you ever taken a piece of wire and broken it by flexing it back and forth a few times? That's metal fatigue, and it works

Figure 6-23. Aluminum blades. A 14-foot (4.3 m) diameter Wincharger on Mormon Row, Grand Teton National Park. The extruded aluminum blades on this turbine have endured for half a century.

the same way on the wing of an airplane or the blade of a wind turbine. Aluminum is a good material—when used within its limits. But on wind turbines, it hasn't been successful.

Most of the problems the hundreds of Darrieus turbines once operating in California encountered were due to metal fatigue where individual blade sections were joined together. The turbines were finally removed in the 1990s and sold for scrap. Imagine: An aluminum beer can on the shelf in your grocery could have once been part of a Darrieus turbine in California.

The only remotely successful use of extruded aluminum blades was for home light plants during the 1950s. At the time, Wincharger switched from wooden blades to extruded aluminum. Some Winchargers can still be found with their blades intact (see figure 6-23, Aluminum blades).

Metal blades, whether steel or aluminum, may also cause television and radio interference. Metal reflects television signals, and this can cause "ghost" images on nearby TV sets. This has proven to be far less of a problem than first thought, even among the 500 Darrieus turbines that once operated in California. No manufacturer today builds wind turbines with metal blades.

Fiberglass

Fiberglass (glass-reinforced polyester, or GRP to Europeans) or related plastic composites now dominate blade construction (see figure 6-24, Blade cross section). Like wood, fiberglass is strong, is relatively inexpensive, and has good fatigue characteristics. It also lends itself to a variety of designs and manufac-

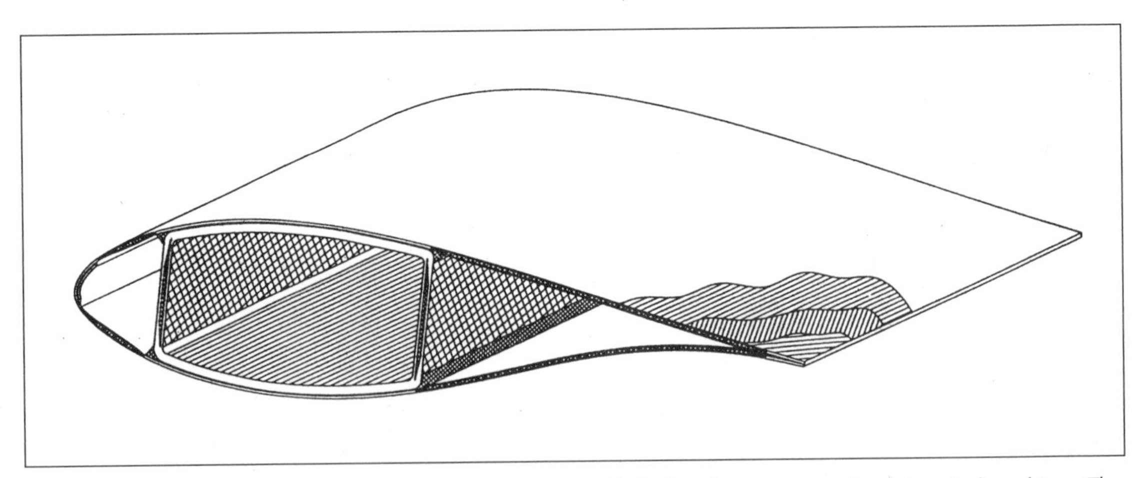

Figure 6-24. Blade cross section. Construction of the fiberglass blade found on many medium-size wind machines. The central section is the spar, which provides the blade's principal structural support. (Vestas Wind Systems)

turing processes. Fiberglass can be pultruded, for example. Instead of material being pushed through a die, as in extrusion, fiberglass cloth (like the cloth used in fiberglass auto body kits) is pulled through a vat of resin and then through a die. Pultrusion produces the side rails for fiberglass ladders and other consumer products. As with aluminum extrusions, pultruded fiberglass blades are recognizable by their constant chord (width). Bergey Windpower has used fiberglass pultrusions exclusively since abandoning aluminum blades in the early 1980s.

For sailors, fiberglass has become the material of choice. The same techniques used to build fiberglass boats have been successfully adapted by Danish, Dutch, and American companies to assemble wind turbine blades. These manufacturers place layer after layer of fiberglass cloth in half-shell molds of the blades. As they add each new layer, they coat the cloth with a polyester or epoxy resin. When the shells are complete, they are literally glued together to form the complete blade. Nearly all medium-size wind turbine blades incorporate some form of this process.

Filament winding has also been used to produce spars, the main structural members supporting some wind turbine blades. Fiberglass strands are pulled through a vat of resin and wound around a mandrel. The mandrel can be a simple shape like a tube or box beam, or it can be a more complex shape such as that of an airfoil. Originally developed for spinning missile cases, filament winding delivers high strength and flexibility. Though some blades have been made entirely from filament winding, the process is often used only for the blade spar. The blade is then assembled in a mold with a smooth fiberglass shell.

Small wind turbines use a variety of materials. Like Bergey Windpower, many use fiberglass. Southwest Windpower injection-molds blades on its Whisper H40 model with fiber-reinforced composites, but uses carbon-fiber reinforcing in its Air series (see figure 6-25, Carbon-fiber blade). Proven Engineering uses a rugged glass fiber-reinforced polypropylene on its novel downwind rotor.

Figure 6-25. Carbon-fiber blade. Handle with care. The trailing edge of Southwest Windpower's Air 403 blade is so sharp, the Arizona manufacturer ships it with a warning label. Note the shark-fin-like planform derived from the Glauert formula for high-performance airfoils.

Hubs

Like the spokes in a bicycle wheel, the blades become part of the rotor when attached to its hub. The hub holds everything together and transmits the transverse motion of the blades into torque. Three aspects of the hub are important: how the blades are attached,

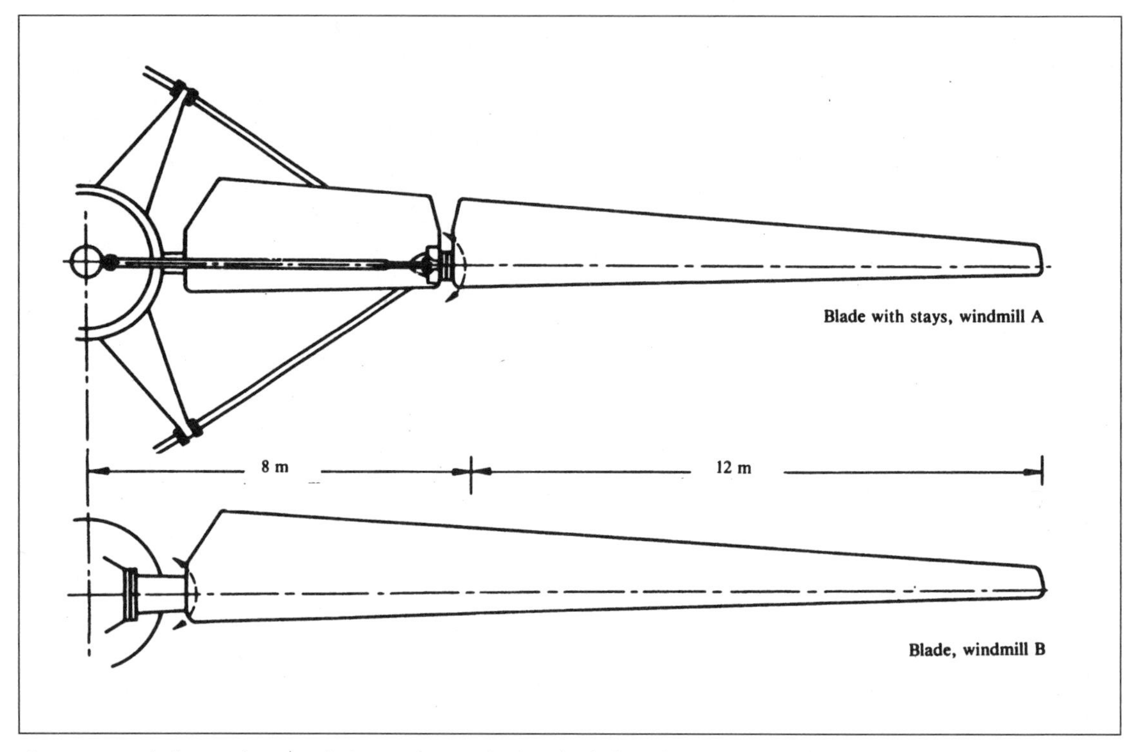

Figure 6-26. Blade attachment. Blades can be attached to the hub with stays, or cantilevered (attached at only one point). Early Danish wind machines used rotors braced with stays. Most contemporary designs use cantilevered blades, though a few small wind turbines use cable-stayed rotors. (Danish Ministry of Energy)

whether the pitch is fixed or variable, and whether or not the attachment is flexible.

Today, all medium-size wind turbines and nearly all small wind turbines use blades cantilevered from the hub—that is, they're supported only at the hub, just as the wing of a modern airplane is attached only at the fuselage (see figure 6-26, Blade attachment). During the late 1970s and early 1980s some European designs used struts and stays to brace the blades, following the pattern of the famous Danish wind machine at Gedser. Struts reduce bending on the root of the blade where it attaches to the hub, but they also increase the drag on the rotor. Consequently the spar, the main structural support of the blade, and its attachment to the hub need not be as massive as on a cantilevered blade. Struts and stays work fine on upwind machines, as long as the turbine stays upwind. They tend to fail when the turbine inadvertently yaws downwind. Early Danish designs, such as the Windmatic 14S, were susceptible to this weakness, which led to an industrywide abandonment of struts and stays for bracing the rotor (see figure 6-27, Struts and stays).

Most hubs are rigid: They don't allow the blades to flap back and forth in gusty winds. The blades may change pitch by turning about their long axis, but they don't change from the plane of rotation. For most of the wind turbines currently on the market, the blades bolt directly to a rigid hub. During the late 1980s and early 1990s several manufacturers of medium-size wind turbines reintroduced variable-pitch hubs, enabling more sophisticated control of the rotor in high winds. A growing number of manufacturers were incorporating pitchable hubs into their megawatt-size turbines in the early 2000s (see figure 6-28, Rotor hubs).

Several manufacturers have attempted to commercialize wind turbines using two-blade teetered rotors. The rotor on these

Figure 6-27. Struts and stays. Like its Danish predecessor at Gedser, the Windmatic 14S used struts and stays to brace the laminated wood blades. Note the fan tail for orienting the wind turbine into the wind. The blades used pop-up air brakes to limit rotor speed in emergencies.

machines would teeter or rock about the hub. As a unit, the rotor would swing in and out of the plane of rotation like a seesaw. This teetering action relieves stresses on the blade during gusty winds, when the turbine yaws to track changes in wind direction, and when the blade passes through the tower's shadow. Teetering hubs have been used on both upwind (WEG MS3) and downwind (ESI 54 and 80; Carter 25 and 300) turbines. Though engineers have long stressed teetering's advantages, no wind turbines using the technique have proven commercially successful.

Figure 6-28. Rotor hubs. *Top:* Cast steel hub for a 500 kW three-blade, fixed-pitch Danish wind turbine. The slotted bolt holes allow field adjustment of blade pitch. *Bottom:* Variable-pitch hub for 750 kW turbine. Note the pitch linkage mechanism and bolt holes for attaching the blade.

Even more complex hubs have been used on one-blade wind turbines, with a similar lack of commercial success (see figure 6-29, Teetering hub).

Figure 6-29. Teetering hub. One of MBB's Monopteros 50 wind turbines undergoing blade repair in late 1990 near Wilhelmshaven in northern Germany. Note the massive counterweight, teetering hinge, and work platform. The blade is about 25 meters (80 ft) long.

The hub attaches to the main shaft, which forms part of the turbine's drive train: the arrangement of shafts, gearboxes (where used), and generators that convert the motion of the spinning rotor into electricity.

Drive Trains

Over the decades several means have been used to transfer power from the rotor to the generator. Some turbines have driven the generator directly, others have used mechanical transmissions, and a few designs have attempted to use hydraulic or pneumatic transmissions. Where gearboxes have been used, some designs have integrated them into the structural support of the turbine, whereas most have mounted the transmission on the bed plate or frame of the turbine.

Driving the generator directly with the rotor eliminates the need for a transmission and reduces the complexity of the drive train. Direct drive should also offer slightly higher conversion efficiencies, since no power is lost in the gearbox. Direct drive, though, requires specially designed slow-speed generators that may be larger and consume greater amounts of expensive copper than conventional transmission-generator combinations.

Small Turbines

The most successful of the pre-REA (Rural Electrification Administration) windchargers, the Jacobs home light plant, used direct drive. Although the small turbine industry—Enertech, for example—flirted with gear-driven machines during the 1970s and early 1980s, most small wind turbines emulate the 1930s Jacobs and use direct drive (see figure 6-30, Direct-drive micro turbine).

Jacobs's chief competitor, Wincharger, used a simple gearbox. Wincharger used one large helical gear on the main shaft of the rotor to drive a small gear on the generator. During the late 1970s Jim Sencenbaugh produced his

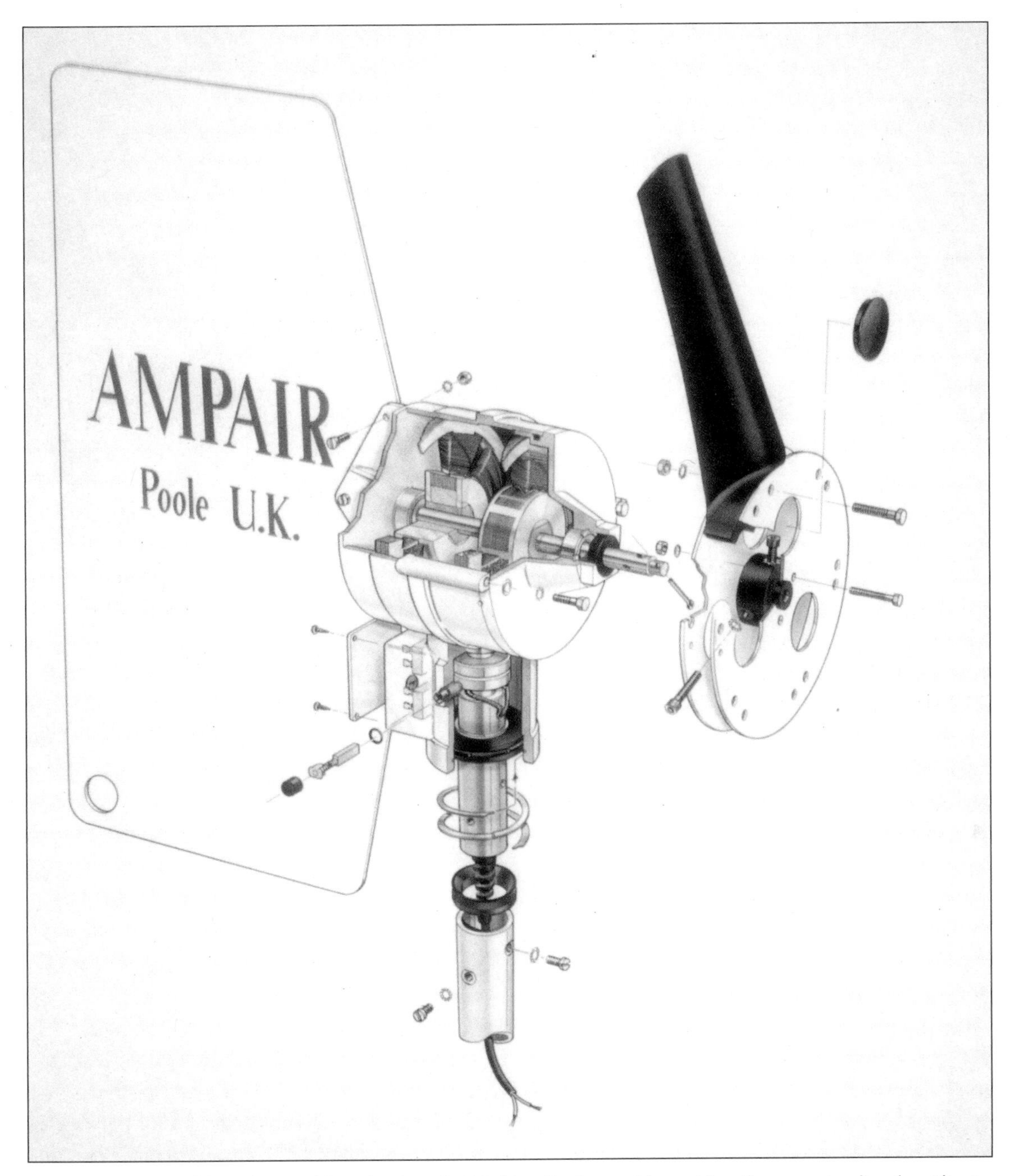

Figure 6-30. Direct-drive micro turbine. The Ampair 100, like all micro turbines, drives the generator directly without step-up gearing. Note the twin rotor and stator windings. (Ampair)

model 1000 using a similar approach. Sencenbaugh used the transmission to increase the 350 rpm speed of his rotor to the 1,100 rpm needed by the generator.

Other small turbines have used belts or chains. Most home-builts of the 1970s used this approach because belts, pulleys, and sprockets were cheap and readily available. Aeropower, for example, used cogged belts, like the timing belts on auto engines, on its small turbine. In practice, belts and chains have proven unreliable.

As wind turbines increase in size, the need for a transmission becomes more pressing

Figure 6-31. Integrated drive train. *Left:* A 15-meter (50 ft) diameter, 50 kW wind turbine patterned after the Enertech E44. The AOC 15/50 does not use a nacelle cover. From left to right: hub, main shaft bearing support and gearbox, generator, and brake. The generator hangs on the gearbox flange. *Right:* Integrated 2 kW turbine. Unlike other turbines in its size class, which use direct-drive alternators, Vergnet's model 4/2 uses a two-blade downwind rotor 4 meters (13 ft) in diameter to drive an integrated off-the-shelf induction generator.

because the speed of the main shaft decreases. For small machines, transmissions with only one or two stages of parallel shafts may suffice. But with medium-size wind turbines more stages may be necessary, or designers may even opt for planetary or epicyclic gearboxes.

Periodically, materials-conscious designers revisit integrated drive trains, where the transmission housing supports the rotor bearings, generator, and yaw assembly. Later Enertechs took this path, as did its design successor, Atlantic Orient (see figure 6-31, Integrated drive train).

Medium-Size Turbines

While several manufacturers of medium-size turbines have followed the integrated path blazed by Ulrich Hütter in Germany during the 1950s, most successful companies have mounted critical components independently on a structural frame. In the early 1980s the difference between integrated drive trains and traditional ones is what distinguished problem-prone American turbines from what became known as the Danish approach.

Rather than integrate the drive train, Danish manufacturers fabricated a metal frame or bed plate, to which they mounted the main shaft, transmission, generator, and other components (see figure 6-32, Danish drive train). The separate main shaft and support bearings on Danish machines allowed the transmission to be readily replaced without requiring removal of the rotor.

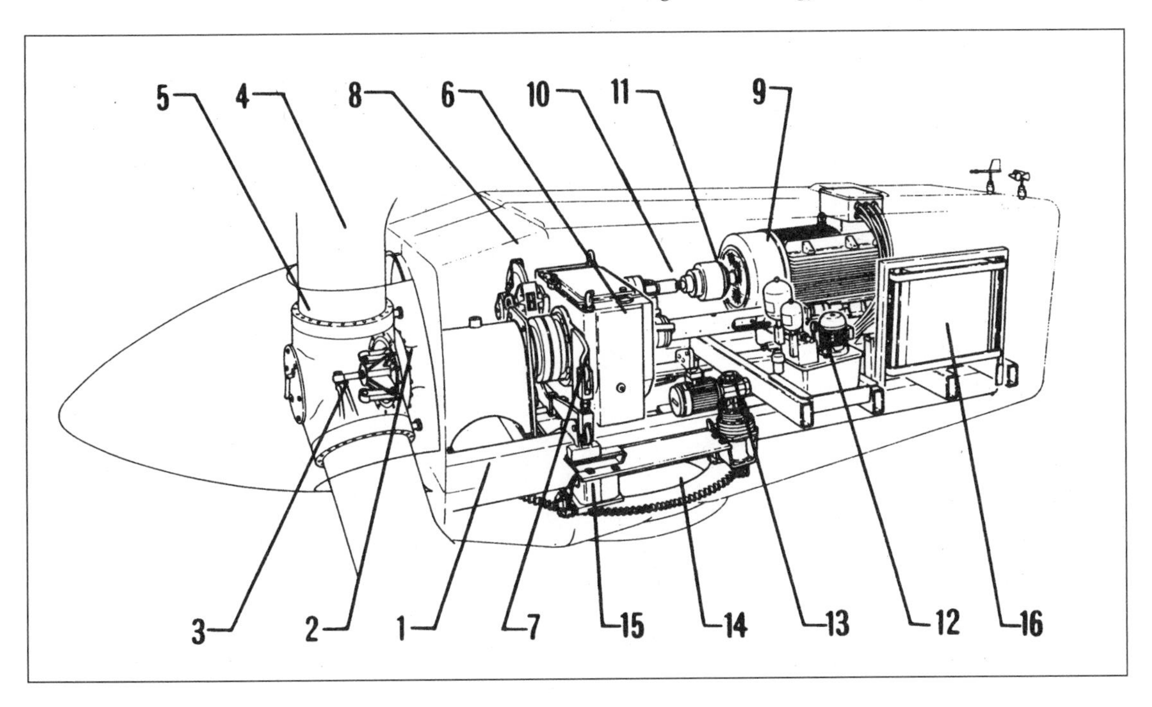

Figure 6-32. Danish drive train. *Top:* Vestas V27, a 225 kW variable-pitch wind turbine. (1) Bed plate (frame or strong-back). (2) Main shaft. Note that the bearings supporting the main shaft are independent of the transmission. On some wind machines the bearings in the transmission housing support the rotor. (3) Pitch linkage. (4) Blade root. (5) Hub. (6) Transmission or gearbox. (7) Transmission or gearbox torque relief. (8) Nacelle cover. (9) Induction or asynchronous generator. Older European wind turbines commonly used dual generators. This model switches poles on a single generator. (10) High-speed shaft. (11) Coupling. (12) Hydraulic pump and reservoir for pitch actuation. (13) Yaw or slewing drive for pointing the turbine into the wind. (14) Yaw or slewing ring. (15) Yaw brake. (16) Nacelle-control unit. (Vestas Wind Systems) *Bottom:* The Ecotecnia megawatt-class turbine uses a novel rotor support. (Ecotecnia)

Figure 6-33. Enercon direct drive. By directly driving a large-diameter ring generator, Enercon eliminates the need for a gearbox. *Top:* Part of ring generator being assembled at the Enercon plant in Aurich, Germany. *Bottom:* Cluster of Enercon E66s in northern Germany. The 66-meter (220 ft) diameter rotor drives a 1.8 MW ring generator. (Enercon)

American designers of integrated drive trains tended to abandon nacelle covers as an unnecessary amenity. This lack of foresight gave U.S.-designed turbines of the early 1980s an industrial appearance that certainly did little to further public acceptance of wind energy. And unlike Danish turbines, most U.S.-built machines of the period lacked work platforms for servicing the turbines.

Some European turbines used a variation on the integrated strategy. While conventional in most respects, these designs hung the rotor directly on the transmission's input shaft without relying on an independent main shaft. This produced a more compact nacelle but at a cost of requiring the transmission housing to withstand thrust loads on the rotor, as well as the torsional forces for which the transmission was designed.

As turbines have grown larger, engineers have sought novel ways to handle the wind's thrust on the rotor, the rotor's weight, and the

torque the rotor transmits to the transmission. In one version, much like the drive wheel on commercial vehicles, an extension of the nacelle's frame extends to support the rotor and main bearing while a small shaft transmits torque to the gearbox.

Designers specify gearboxes that use either parallel shafts, the conventional choice for medium-size wind turbines, or planetary transmissions. Parallel-shaft gearboxes occupy more space and weigh more than planetary transmissions, but are significantly quieter. Turbines in the 500 kW class and larger use a combination of planetary and parallel-shaft designs. The planetary stage is used on the low-speed shaft from the rotor where noise is less of a problem. The parallel stage is used on the high-speed side of the transmission to keep noise under control.

Transmissions on larger turbines are suspended on the main shaft, rather than mounted directly on the frame of the turbine. Dampening these shaft-mounted gearboxes allows the transmission to absorb fluctuations in torque caused by gusty winds, introducing needed compliance into the drive train of big turbines.

Gearboxes are not intrinsic to wind turbines. Designers use them only because they need to increase the speed of the slow-running main shaft to the speed required by mass-produced induction generators. If they choose, designers can opt for special-purpose, slow-speed generators and drive them directly without using a transmission.

Purpose-built, permanent-magnet alternators revolutionized the reliability and performance of small wind turbines by eliminating the need for a gearbox. Since 1994 German manufacturer Enercon has done the same for medium-size turbines. After the successful introduction of its 500 kW E40, Enercon grew rapidly to become Germany's largest manufacturer of wind turbines (see figure 6-33, Enercon direct drive).

Other Forms of Transmission

Hydraulics and pneumatics have been suggested for transmitting the rotor's torque to ground level. In theory this offers some advantages because a hydraulic drive, for example, can more easily be matched to the torque characteristics of a wind turbine rotor than can a mechanical transmission. Again, in principle, they should also be simpler. These advantages are offset, however, by inefficiencies and the complexity of making such a system actually work. The only large-scale test of hydraulic transmissions, the Bendix-Schachle turbine once owned by Southern California Edison Company, ended in ignominious failure as the turbine was plagued with hydraulic leaks. Eventually the utility cut the legs of the turbine with an acetylene torch and pulled the giant machine over with a crane. According to eyewitness Lloyd Herziger, the monster groaned in an agony of tearing metal as it crashed to the ground. No wind turbine using a hydraulic transmission to generate electricity has ever been either a technical or a commercial success.

Parallel-shaft gearboxes occupy more space and weigh more than planetary transmissions, but are significantly quieter.

Pneumatics have a proud pedigree. Jules Verne's last novel centered on wind turbines powering Paris with pneumatics. Bowjon in the United States and Koender in Canada successfully use wind-turbine-driven compressed air to pump water. Yet no one has successfully used pneumatics to produce electricity commercially.

Generators

Generators are not perpetual motion machines. They transfer power, but they don't create it. Power must be delivered to a generator before you can get power out of it. (In our case the prime mover, as it's called, is the rotor.) Nor are

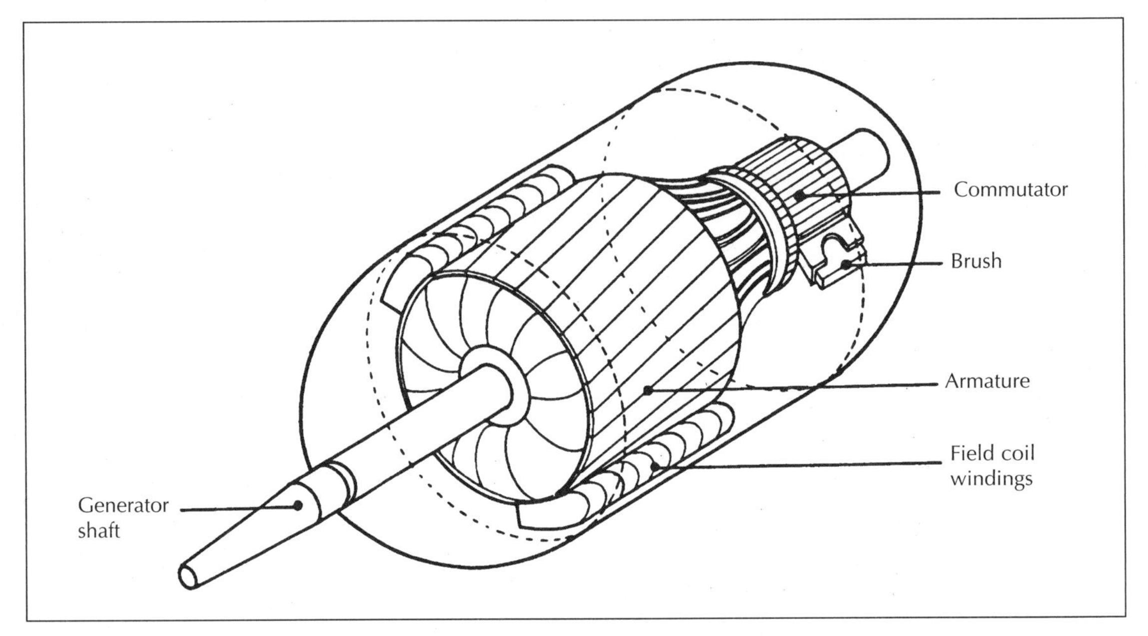

Figure 6-34. DC generator. Shown is a sketch of a direct-drive generator on 1930s-era Jacobs windchargers. Power is drawn off the spinning armature through brushes. Some of this power is used to energize the field coils. The commutator of a DC generator is simply a mechanical rectifier, picking off part of the alternator's AC waveform.

generators 100 percent efficient at transferring this power. The rotor will deliver more power to the generator than the generator produces as electricity. This leads us to a fundamental principle about the size of wind turbines. The size of a generator indicates only how much power the generator is capable of producing if the wind turbine's rotor is big enough, and if there's enough wind to drive the generator at the right speed. Thus we once again confront the fact that a wind turbine's size is primarily governed by the size of its rotor.

The generator converts the mechanical power of the spinning wind turbine rotor into electricity. In its simplest form a generator is nothing more than a coil of wire spinning within a magnetic field. Consequently, whether generating direct current (DC) or alternating current (AC), a generator must have:

1. Coils of wire in which the electricity is generated and through which it flows.
2. A magnetic field.
3. Relative motion between the coils of wire and the magnetic field.

By varying each of these conditions, you can design a generator of any size for any application.

Power in a simple DC circuit is the product of current and voltage. In a generator the armature is the coil of wire where output voltage is generated and through which current flows to the load (see figure 6-34, DC generator). The portion of the generator where the magnetic field is produced is the field. Relative motion between the two is obtained either by spinning the armature within the field or by spinning the field within the armature. As you would expect, the stationary part of the generator is the stator; the spinning part is the rotor.

The power produced by a generator depends on the diameter and length of the wires used in the armature, the strength of the magnetic field, and the rate of motion between them. Increase any one of these, and you increase the potential power of the generator. The size of the wire in the armature determines the maximum current that can be drawn from the generator before it overheats, melts its insulation, shorts out, or otherwise destroys itself. The heavier the wire, the more current it can carry. As long as the wind turbine's rotor continues to provide greater and greater amounts of power as wind speed increases, the generator will continue to produce more current until the generator overheats. To prevent such occurrences, generators usually employ a means for limiting current to a safe maximum.

Generators are rated in terms of the maximum current they can supply at a specified voltage and, for AC generators, at a specific frequency. This rating is given on the nameplate in amps and volts (and frequency, where appropriate), as kilowatts and volts, or as kilovolt-amperes (kVA). The generator may be rated for the current it can supply continuously, or the current it can supply for only a short period. If generator size is of concern to you, always check which rating is being used. Reputable manufacturers rate their generators for continuous, rather than intermittent, duty.

Let's turn to voltage, the other half of the power mix. Generated voltage depends on the rate at which magnetic lines of force are crossed by the wire loops in the armature. Designers alter voltage by changing the magnetic field, by changing the rate of motion between them, or both.

The generator's field is provided by magnets. With electromagnets, some power is used to "excite" or "energize" the field around the armature. The strength of this field is a function of the number of wire coils in the field windings and the current flowing through them. For example, if you double the number of coils in the windings, you double the strength of the field, thus doubling generated voltage.

Many of the windchargers built during the 1930s produced 32 volts. Resistance losses are proportionally higher when transmitting low-voltage power. Because of this, most reconditioned windchargers were rewound for 110 volts. The old wire was stripped off the generator and replaced with more turns of thinner wire. Less current could be drawn through the smaller wire than before, but the increased length of wire produced a stronger field, increasing the voltage. Generating capability was not affected and power from the generator remained the same, but the balance between the voltage and the current changed: The voltage increased, and the current decreased by an equivalent amount.

Permanent magnets can also provide the field. They don't require power for excitation because they're inherently magnetic. The principle means for increasing field strength with permanent magnets is to use magnets with greater magnetic density, such as by using neodymium-iron-boron or other rare-earth magnets.

The voltage can be increased by adding more field coils, by adding more permanent magnets, or by increasing the speed at which the armature windings pass through the field. This can be accomplished by increasing the diameter and length of the generator so there's room for more magnets, or by spinning the rotor faster.

Yes, all this does have some bearing on the design of wind-driven generators. To get a feel for how, let's examine two popular pre-REA wind machines. Both Jacobs and Wincharger used about the same size rotor (14 feet, or 4.3 meters); thus the power available to the generator and the speed of the rotors were

Permanent Permanent Magnets?

It can come as a surprise to learn that "permanent" magnets may not be permanently magnetized. Permanent magnets may lose some of their magnetization if overheated for extended periods. Southwest Windpower, for example, attributed some of the poor performance of its early Air series to "degaussing," or the weakening of the strength of the turbine's permanent magnets. This occurred, SWP explained, when heat wasn't dissipating fast enough from the Air's alternator. Later versions of the Air were redesigned to better conduct heat from the alternator to the micro turbine's cast-aluminum body.

roughly equivalent. Yet Jacobs chose to use a direct-drive generator, whereas Wincharger chose a transmission.

To produce the same power and voltage as Wincharger without a transmission, Jacobs had to design a generator that would operate at lower shaft speeds. Jacobs did so by increasing both the diameter and the length of its generator relative to that of Wincharger. This allowed the use of more field coils (six, to Wincharger's four). The coils were also larger.

The Jacobs generator's greater diameter also increased the speed at the periphery of the armature where it passed the field coils. Doubling the diameter doubles the rate at which the armature cuts through the field. The effect is the same as that of a 2:1 transmission, where the output shaft spins at twice the speed of the input shaft.

All in all, the Jacobs generator was considerably larger, and used much more copper and iron than Wincharger, to do the same job. But the Jacobs generator could do that job at a slower speed. Jacobs chose a low-speed generator for long bearing life and simplicity. It believed these advantages offset the generator's greater cost.

Barry Commoner's adage, "There's no such thing as a free lunch," puts it succinctly. Whether it's the design of generators or any other wind machine component, there's always a trade-off. You gain something only by giving up something else. You hope that what you gain is more valuable than what you've lost. That's as true today as it was during the 1930s. Designers of small wind turbines who stress long life and low maintenance choose lower generator speeds. The price they pay is a more expensive generator.

Table 6-4
Nominal Generator Speed in rpm for Induction (Asynchronous) Generators

	Europe 50 Hz	The Americas 60 Hz
4-pole	1,500	1,800
6-pole	1,000	1,200

Manufacturers of small wind turbines intended for remote sites in harsh environments may opt, as Jacobs did, for building slow-speed generators tailored to their wind turbine. That's what most of today's small wind turbine manufacturers have done. They build special-purpose, direct-drive, slow-speed alternators.

There are some exceptions. French small turbine manufacturer Vergnet uses off-the-shelf induction generators coupled to gearboxes (see figure 6-31b, Integrated drive train). Similarly, Wind Turbine Industries uses a conventional wound-field alternator but, unlike other companies, mounts the alternator vertically at the top of the tower, using a right-angle drive.

The trade-offs are also apparent in medium-size wind turbines. During the early 1980s many American-designed wind turbines that operated at high speeds were installed in California wind farms. These machines were not only noisy but also trouble-prone. Danish designs operating at much more modest speeds eventually won more than half the California market. Like Jacobs before them, the rugged Danish designs opted for lower speeds to reduce wear and tear. The Danish turbines typically drove a six-pole generator at 1,200 rpm, while their American competitors used cheaper four-pole generators running about 1,800 rpm. Today none of the early U.S. designs is still being built (see table 6-4, Nominal Generator Speed in rpm for Induction [Asynchronous] Generators).

Alternators

Windchargers of the 1930s produced DC by spinning the armature within the field. Power was drawn off the rotating armature through a commutator. During the 1960s the auto industry began replacing DC generators with alternators. Though alternators produce alternating current, they offer several advantages

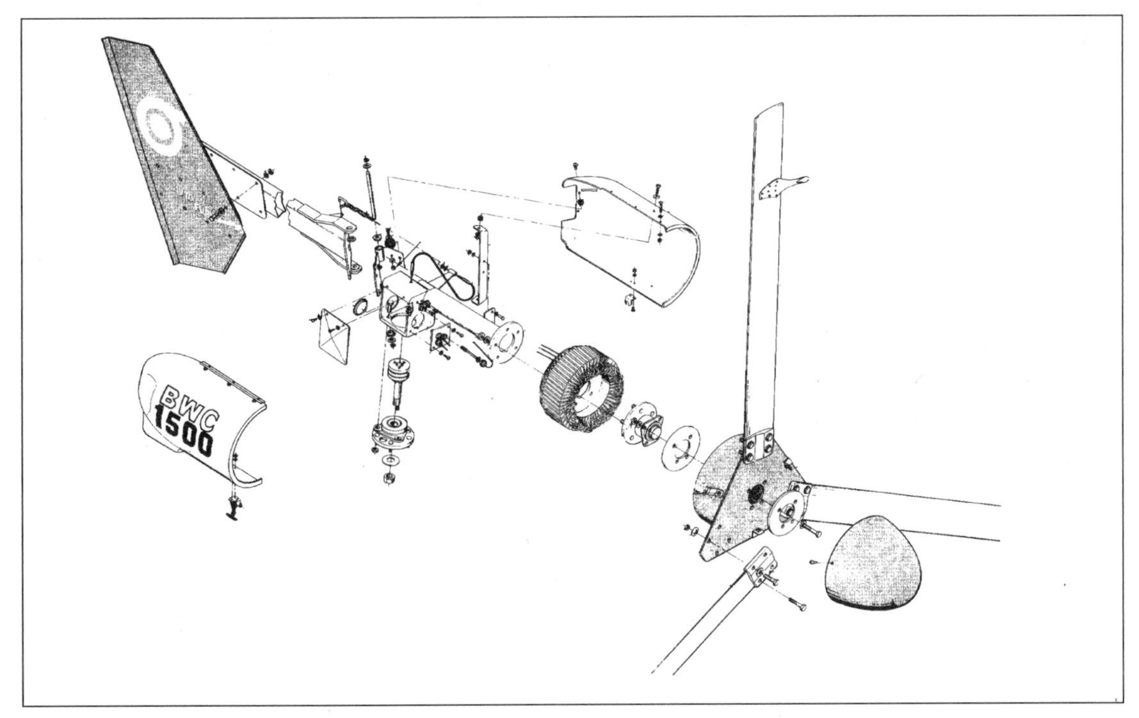

Figure 6-35. Inside out. Bergey Windpower places the permanent magnets inside what it calls the "magnet can" of its 1,500-watt model. Bergey then mounts the blades directly onto the end plate of the magnet can. Current is drawn off the stator coils (center), which are bolted to the mainframe flange (left). The rotor bearing (between the stator and magnet can) is an automotive wheel bearing. (Bergey Windpower)

over DC generators. For a given output, alternators cost less than generators. An alternator's slip rings are also more durable, because they don't carry the alternator's current, as do the brushes in a generator's commutator.

The battle between alternators and generators is far from over. Some mavericks, such as Mick Sagrillo, believe that DC generators still offer promise. Sagrillo, who rebuilds DC generators, argues that special-purpose generators, such as the Jacobs home light plant, use oversize brushes to ensure long life. These brushes don't wear out as quickly as many imagine, he says. Further, Sagrillo maintains that a generator gives better high-end performance than an alternator. Still, he concedes that alternators now dominate the market.

In today's alternator the field, rather than the armature, revolves. Power is drawn off the stator from fixed terminals. There are no brushes or commutators to wear out from the passage of high current. There's no arcing at the brushes. Excitation of the alternator's field is provided through slip rings on the rotor. But only enough power passes through the slip rings to excite the field (a small percentage of the alternator's output). Wear is negligible in comparison to the brushes on a DC generator.

There are no slip rings—no moving contacts—in a permanent-magnet alternator, since the field is permanently excited.

In a conventional alternator, the field revolves inside the stator. But Bergey Windpower and some other small turbine manufacturers spin the permanent-magnet field outside the stator. The case to which the magnets are attached rotates outside the stator. In this configuration the blades can be bolted directly to the case, and they often are. There is also another benefit: Centrifugal force presses the magnets against the wall of the magnet can (see figure 6-35, Inside out).

The magnets attached to the rotor of a

more conventional shaft-driven alternator, such as in Southwest Windpower's Air series, are thrown outward from the spinning shaft. Because of the high rotational speeds found in small wind turbines, especially when unloaded, designers must pay special attention to retaining the magnets. For example, in the Air series Southwest Windpower straps the magnets to the rotor with a metal band.

As the name implies, alternators generate alternating current. As the rotor spins, current rises and falls like waves on the ocean (electrons in the armature first are jostled in one direction, then alter course and are jostled the other). The alternator's frequency is the rate at which current rises and falls; it's given in cycles per second or hertz. The speed of the rotor and the number of poles or coils of wire determine the alternator's frequency. Drive the alternator faster and frequency increases; slow the rotor and frequency decreases. This explains why most small wind turbines generate variable-frequency AC. When wind speed rises, the turbine spins faster, increasing frequency (as well as voltage and current). When the wind subsides, frequency decreases.

In a simple alternator, the four coils are wired together in series as a single circuit producing single-phase AC. When three groups of coils are arranged symmetrically around the stator, the alternator produces three-phase AC, each phase one-third out of sync with the next. Most alternators used in wind systems produce three-phase AC. Three-phase alternators do more with less. The designer can more efficiently pack coils within the generator. Voltage is determined by the rate at which lines of force are cut by the armature. Thus we can increase power for the same current flow by increasing the number of coils and taking up all the available space within the generator.

Most small wind turbine alternators produce three-phase AC, to make the best use of the space inside the generator case. Some battery-charging models, such as Southwest Windpower's Air series, rectify the AC to DC at the generator; others, such as Bergey Windpower's XL1, rectify it at a controller that can be some distance from the generator.

If you've ever spun the shaft of a toy generator in your hand, you remember how it felt when the rotor would stick slightly as the coils in the armature aligned with the magnets in the field. As the coils passed by the magnets, the shaft would turn more easily. This same effect, cogging, occurs in large generators and motors. Cogging is of interest in wind machines because it can retard the start-up of the wind turbine in light winds when the poles are aligned. Increasing the num-

Air-Gap Generators

Air-gap or axial-flux generators are a popular generator design for small wind turbines in Britain. Both Marlec and Proven use this strategy by arraying the magnets and stator coils axially rather than in the more common radial pattern. In Marlec's 1700 and Proven's 2.5 kW model, the manufacturers use two sets of magnet rings, with stator coils sandwiched between them. To build bigger alternators, a designer can increase the alternator's diameter or add more disks. Jeumont Industrie did both in its 48-meter (160 ft) diameter, 750 kW wind turbines. Jeumont, a manufacturer of discoidal alternators for French nuclear submarines, adapted the technology to wind energy.

Proven air-gap alternator. *Hmmm, can I get this back together, once I get it apart?* thinks Mick Sagrillo as he struggles to peek inside Proven Engineering's permanent-magnet alternator.

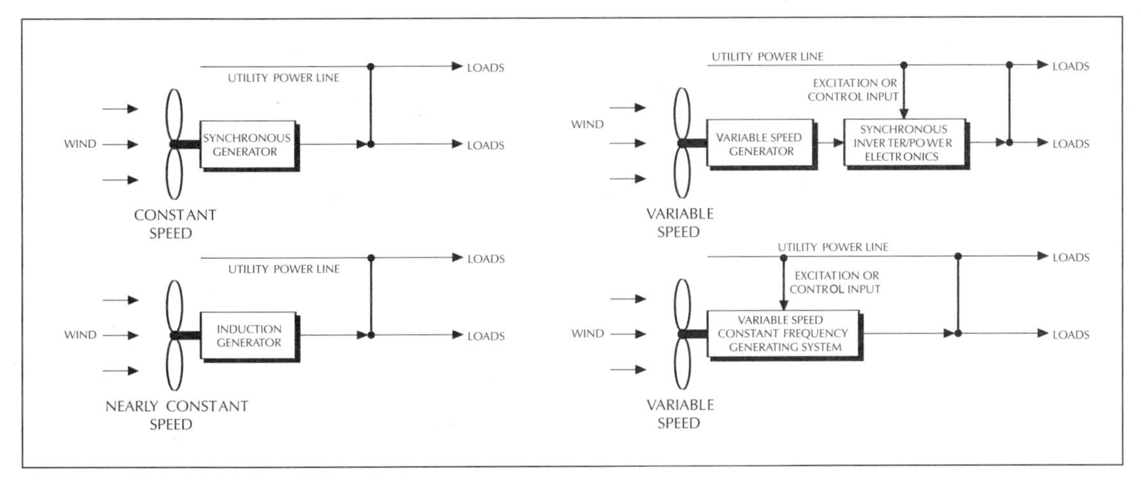

Figure 6-36. Utility-compatible wind machines. Techniques for generating utility-compatible electricity. Most medium-size wind turbines use induction generators. Most small wind turbines (and now some medium-size models, as well) use power electronics with a variable-speed generator. Very few wind turbines have ever been designed to drive true synchronous generators.

ber of poles reduces cogging, enabling the turbine to start more easily in light winds. Cogging is also a source of alternator whine, and by reducing cogging designers can reduce alternator noise. Scoraig Wind Electric's Hugh Piggott notes that skewing the slots in the laminations of the armature reduces cogging, and many small turbine manufacturers use this or similar techniques.

Variable- or Constant-Speed Operation

Wind machines driving electrical generators operate either at variable speed or at constant speed. In the first case, the speed of the wind turbine rotor varies with the speed of the wind. In the second, the speed of the wind turbine rotor remains relatively constant as wind speed fluctuates.

In nearly all small wind turbines, the speed of the rotor varies with wind speed. This simplifies the turbine's controls while improving aerodynamic performance. When such wind machines drive an alternator, both the voltage and frequency vary with wind speed. The electricity they produce isn't compatible with the constant-voltage, constant-frequency AC produced by the utility. If you used the output from these wind turbines directly, your clocks would gain and lose time, and your lights would brighten and dim as wind speeds fluctuated. Eventually you'd burn up every motor in the house. Unless you have a use for this low-grade electricity (heating, pumping water, and so on), the output from these wind machines must be treated or conditioned first, even if it's simply for charging batteries.

Because batteries can't use AC, the alternator's output must be converted to DC. As in your car alternator, diodes—electrical check valves that permit the current to flow in only one direction—rectify the AC output to DC, which is then used for battery charging.

To produce utility-grade electricity, either the alternator's AC or rectified DC can be treated with a synchronous inverter to produce constant-voltage, constant-frequency AC like the utility's (see figure 6-36, Utility-compatible wind machines). Most of these inverters, though not all, are line commutated. They must be interconnected with the utility to operate. The utility's alternating current provides a signal that triggers electronic switches within the inverter, which transfers the electricity at the right time to produce constant-frequency AC at the proper voltage. No utility power is consumed in the process. It's merely used as a signal to coordinate the switching.

In the past nearly all medium-size wind turbines, such as the thousands of machines installed in California during the early 1980s, operated at constant speed by driving standard, off-the-shelf induction generators. A number of manufacturers, however, have switched to variable-speed operation, especially on megawatt-size turbines. Many of these use a form of induction generators.

As in small turbines, operating the rotor of medium-size and larger turbines at variable speed theoretically improves aerodynamic performance. Still, data showing that any improvement is measurable for medium-size turbines is sketchy. More importantly, proponents of variable-speed operation argue, allowing the rotor to increase speed as a gust strikes the turbine reduces potentially damaging loads on the drive train. By absorbing these loads in the inertia of the spinning rotor, designers believe they can cut maintenance costs and extend the life of the turbine. One big advantage of variable speed is the reduction of the rotor's aerodynamic noise at low wind speeds relative to that of constant-speed machines.

Induction (Asynchronous) Generators

Induction or, as they're known in Europe, asynchronous generators became popular in medium-size and some household-size wind turbines for several reasons: They're readily available, they're inexpensive, and they can supply utility-compatible electricity without sophisticated electronic inverters. Induction generators are simply induction motors (like the motor in your refrigerator) in disguise.

An induction motor becomes a generator when driven above its synchronous speed. Plug a four-pole induction motor into an outlet in North America and the motor will turn slightly less than 1,800 rpm, consuming power. Leave it plugged in, but now drive the motor slightly faster. The motor will no longer consume power from the outlet. You're now supplying it. Spin the rotor just a little faster and the motor not only won't be consuming electricity, but will also be generating it. As you try to spin the motor faster, it gets harder to turn. The utility consumes the additional power as you produce it, without rotor speed appreciably increasing.

In a wind turbine driving an induction generator, as wind speed increases the load on the generator automatically increases as more torque (power) is delivered by the wind turbine's rotor. This continues until the generator reaches its limit and either breaks away from the grip of the utility or overheats and catches fire.

Induction generators are not true constant-speed or synchronous machines. The speed of induction generators varies slightly and is thus asynchronous. As the torque available from the wind turbine rotor increases, the generator speed slips by 2 to 5 percent, or 36 to 90 rpm on an 1,800-rpm generator. In an operating wind turbine this slip is imperceptible. But in a cluster of turbines the variation in slip from one to the next is detectable. In one moment the rotors of a small group of turbines will all be in sync—all spinning at exactly the same speed. But they will gradually fall out of sync, becoming more and more out of sync until the cycle is repeated.

Danish manufacturer Vestas exploits slip to advantage. Its V47 series Vestas uses electronic controls on the generator rotor to vary the slip by as much as 10 percent of nominal speed. This enables Vestas to enjoy some of the benefits of variable speed while avoiding the costs associated with true variable-speed operation.

So-called squirrel cage induction generators have proved extremely popular for wind turbines because they're widely available in a range of sizes from numerous manufacturers worldwide, and their interconnection with the utility is straightforward. You can literally plug it in and go. Early promotions for the defunct manufacturer Enertech depicted its wind machines being plugged into a wall socket.

Interconnection is a little more sophisticated today, but the principle remains the same.

A household-size turbine is wired to a dedicated circuit in your service panel. A medium-size machine is connected to the utility's nearest step-down transformer. Utilities are much more comfortable with induction generators than with synchronous inverters, because they understand them better. Synchronous inverters still remain a mystery to many utilities.

A more costly version of the induction generator uses a wound rotor. These doubly fed induction generators facilitate operating the wind turbine at variable speed, but require an AC-DC-AC inverter for producing utility-compatible electricity. Several wind turbine manufacturers have incorporated this technology (see figure 6-37, Doubly fed induction generator).

When you're looking at a wind machine's generator, there's no need to be dazzled by the technology employed. Your primary concern is what kind of power it produces. If you want a wind machine for charging batteries at a remote hunting cabin, you won't be able to use an induction generator. If you want utility-compatible power, then you can't use a permanent-magnet alternator that doesn't also include an inverter of some kind.

Don't be swayed by the size of the generator, either. It's only an indication of how much power the generator is capable of producing, not how much it will generate. In the 1980s, if you had asked Danish manufacturers what size generator they used in a given wind turbine they would look at you quizzically and ask, "Which generator?" Danish wind machines of the period often used two induction generators, one for low winds and another, much larger, generator for higher winds.

Figure 6-37. Doubly fed induction generator. A flange-mounted, variable-speed generator for the Zond 750 kW turbine in the background.

Dual Generators

Induction generators operate inefficiently at partial loads. For a wind machine with a generator designed to reach its rated output in a 25 to 35 mph (11 to 15 m/s) wind, the generator would operate at partial load much of the time. To improve performance, Danish designers would bring a smaller generator online in low winds so that it would operate efficiently at nearly full load. As wind speed increased, they would disengage the smaller generator while energizing the larger, or main generator. Thus both generators operated more efficiently than either did alone, and overall performance of the wind machine improved.

Essentially Danish designers gained the performance advantage of operating the rotor at variable speed by using two generators. The

Table 6-5
Typical Generating Hours for Raleigh Distribution

Wind Speed Average (m/s)	Wind Speed Nominal (mph)	Generating Hours/yr	% of Total Hours	Hours Below Rated Capacity	% of Generating Hours	Hours on Small Generator*	% of Generating Hours
5	11	5,970	68%	5,960	100%	4,480	75%
6	13	6,710	77%	6,670	99%	4,150	62%
7	16	7,200	82%	7,020	98%	3,650	51%
8	18	7,540	86%	7,090	94%	3,150	42%

Notes: *or low-speed windings.
Assumptions: cut in, 4 m/s; cut out, 28 m/s; small generator, 4–7 m/s; rated, 16 m/s.

use of dual generators permitted the wind turbine rotor to operate at two speeds. This enabled them to operate not only the generator at higher efficiencies but the rotor as well. Though they are not true variable-speed machines and can't take full advantage of the optimal tip-speed ratio, these turbines can bracket the optimal range. This is particularly useful in low winds where efficiency is most crucial.

Don't be swayed by the size of the generator, either. It's only an indication of how much power the generator is capable of producing, not how much it will generate.

In many parts of the world wind turbines would operate most of the time on the small generator of a typical Danish design (see table 6-5, Typical Generating Hours for Rayleigh Distribution). For example, in a windy region with a 7 m/s (16 mph) average wind speed, the wind turbine would be producing electricity more than 7,000 hours per year, nearly all of that below the turbine's "rated" capacity. More than half the time the turbine would be operating on the small generator. At Great Britain's first wind plant near Delabole, Cornwall, Peter Edwards found that his Danish turbines operated 77 percent of the time on the small generator.

The two generators may be in tandem and driven by the same shaft or they may be side by side, with the small generator being driven by belts from the main generator. Usually both generators are spun at the same time and are brought online not mechanically but by energizing the field electrically.

During the early 1990s Danish designers combined the two generators into one by using generators with dual windings and by switching the number of poles. Typically capacity of the small generator or low-power windings is about 20 to 25 percent of the main generator's capacity. For example, in Vestas's V27, the generator's low-wind windings are rated at 50 kW of the generator's full 225 kW. During low winds, the turbine's controller energizes six poles. Under these conditions the generator reaches its nominal synchronous speed at 1,200 rpm in the Americas (60 Hz) and at 1,000 rpm in Europe (50 Hz). In higher winds the controller switches to four poles and the generator reaches its synchronous speed at 1,800 rpm in the Americas and at 1,500 rpm in Europe.

Rotor Controls

The rotor is the single most critical element of any wind turbine. It's what confronts the elements and harnesses the wind. Because the blades of the rotor must be relatively large and operate at relatively high speed to capture the energy in the wind, they're the most prone to

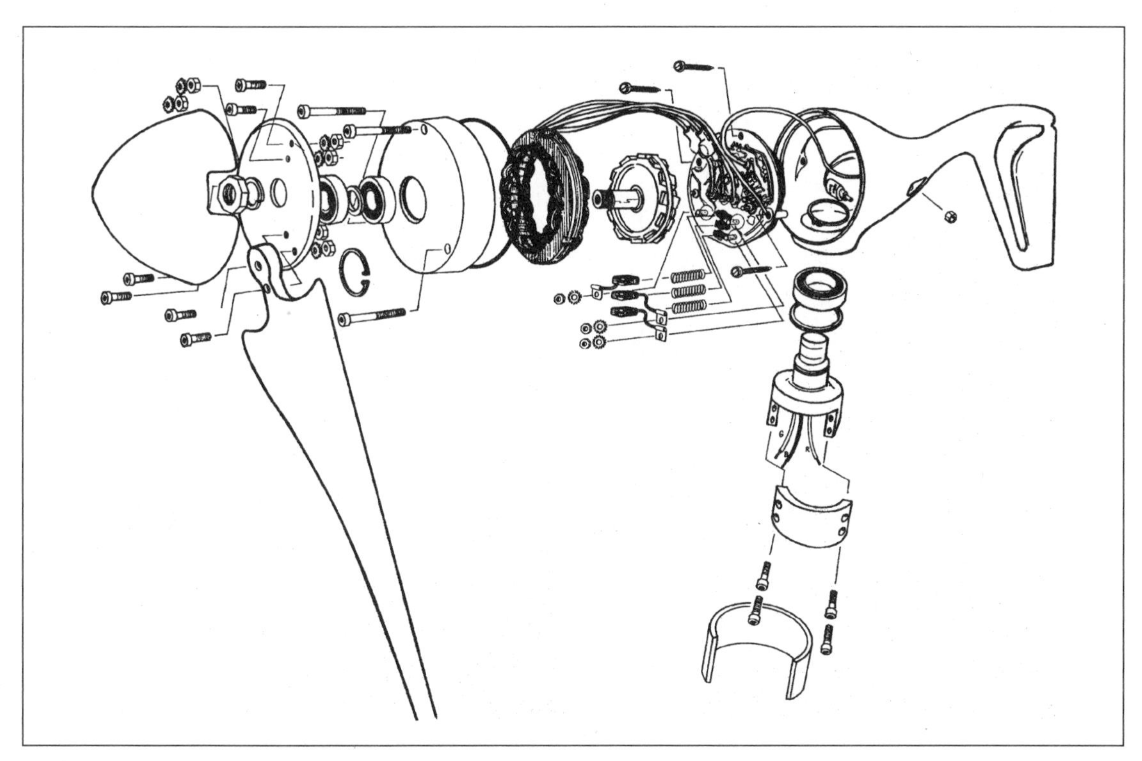

Figure 6-38. Flutter and dynamic braking. Southwest Windpower's Air series of micro turbines used blade flutter to limit the rotor's performance in high winds. Later models also used a form of dynamic braking controlled by the sophisticated circuit board mounted in the aluminum body. (Southwest Windpower)

catastrophic failure. How a wind turbine controls the forces acting on the rotor, particularly in high winds, is of the utmost importance to the long-term, reliable functioning of any wind turbine. Though there is some overlap, the technology chosen for small wind machines differs importantly from that of commercial-size turbines.

For the smallest micro turbines, those about 1 meter (3 ft) in diameter or smaller, the absence of controls may be acceptable under certain conditions. Multiblade turbines such as Marlec's Rutland, Ampair's 100, and LVM's Aero4gen rely on their relatively low rotor speed and rugged construction to survive high winds without any controls whatsoever. These turbines were designed for use on sailboats, to keep the batteries charged when the boat was moored in a protected harbor.

In contrast, Southwest Windpower relied on the controversial use of electronics and blade flutter to protect its lightweight but innovative Air series of micro turbines (see figure 6-38, Flutter and dynamic braking). Most wind turbine designers do everything they can to avoid blade flutter, because flutter can destroy both the blades and the wind turbine. "Flutter is like driving downhill with badly balanced wheels," says Scoraig Wind Electric's Hugh Piggott. Southwest Windpower, which originally designed its wind turbine for use on land, not at sea, used flutter to limit the aerodynamic performance of its sleek turbine. Early models were noisy and plagued with failures. Later models used an electronic controller built into the turbine to short the alternator's phases together to further control rotor speed, and reliability improved. Only time will tell whether this high-technology approach to controlling a micro turbine's rotor will be as successful as less prosaic but tried-and-true methods. However, even Marlec and LVM specify their furling models—for example, the letter *F* in

Figure 6-39. Decreasing frontal area. Mini wind turbines, such as LVM's 4-foot (1.2 m) diameter Aero6gen, reduce the rotor's frontal area to protect the turbine in high winds.

Figure 6-40. Halladay rosette or umbrella mill. Segments of the rotor furl in high winds by swinging out of the wind's path. The rotor also passively orients itself downwind of the tower. The large weight (left) counterbalances the weight of the downwind rotor.

Marlec's designation of the Rutland 913F—when the turbines will be used at exposed sites on land.

Furling remains the simplest and most foolproof method for controlling a small wind turbine rotor. Furling, in its various forms, decreases the frontal area of the turbine intercepting the wind as wind speeds exceed the turbine's operating range (see figure 6-39, Decreasing frontal area). As frontal area decreases, less wind acts on the blades. This reduces the rotor's torque, power, and speed. The thrust on the blades (the force trying to break the blades off the hub) and the thrust on the tower (the force trying to knock the tower over) are also reduced. This method of rotor control permits the use of lighter-weight and less expensive towers than on small wind machines where the rotor remains facing into the wind under all conditions.

Halladay's umbrella mill exemplifies the concept (see figure 6-40, Halladay rosette or umbrella mill). These 19th-century water-pumping wind machines automatically opened their segmented rotor into a hollow cylinder in high winds, letting the wind pass through unimpeded. Each segment was composed of several blades mounted on a shaft, allowing the segment to swing into and out of the wind. When the segments were closed, Halladay's windmill looked like any other water-pumping windmill from the period. But in high winds thrust on the segments would force them to flip open. This action was balanced by counterweights so the farmer could adjust the speed at which the windmill would open and close.

Horizontal Furling

Later inventors, such as the Reverend Leonard Wheeler, chose to use the same control concept (changing the area of the rotor intercepting the wind), but in a different manner. Rather than swinging segments of the rotor parallel to the wind, Wheeler thought it simpler to swing the entire rotor

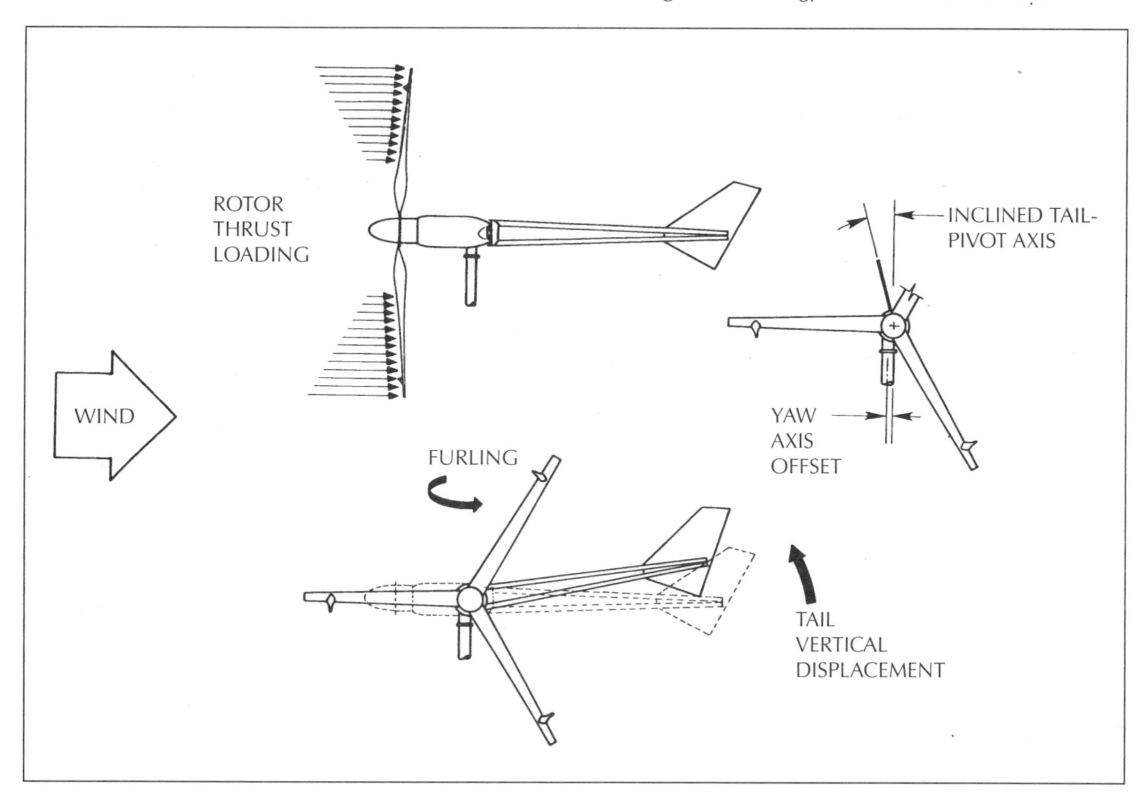

Figure 6-41. Horizontal furling. Bergey wind turbines furl in high winds by swinging the rotor toward the tail. Note that the yaw tube (the tube connecting the wind machine to the tower) pierces the nacelle off center. The rotor axis is offset from the yaw axis, causing the rotor to fold toward the tail in strong winds. The nacelle pivots about an inclined axis, enabling the weight of the raised tail vane to push the nacelle and rotor back into the wind. (Bergey Windpower)

out of the wind. He couldn't do this with the downwind rotor used by Halladay. Instead he placed his rotor upwind of the tower and used a tail vane to keep the rotor pointed into the wind.

Wheeler hinged the tail vane and rotor—or *windwheel,* to old-timers—to permit the rotor to swing or furl toward the tail. As it did, the rotor disk took the shape of a narrower and narrower ellipse, gradually decreasing the area exposed to the wind. The mechanism for executing this was the pilot vane, which extended just beyond and parallel to the rotor disk (see figure 12-1, Traditional farm windmill). Unlike the tail vane, the pilot vane was fixed in position relative to the rotor. Wind striking the pilot vane pushed the rotor toward the tail and out of the wind. In the folded position the rotor and pilot vane were parallel to the wind, like the segments of the Halladay rotor. The thrust on the pilot vane was counterbalanced with weights. As in the Halladay design, a farmer could determine the wind speed at which the rotor would begin to furl by adjusting the weights. The pilot vane went the way of hand cranks on cars as the American farm windmill evolved. Offsetting the axis of the rotor slightly from the axis about which the wind machine yaws around the top of the tower produced the same results: self-furling in high winds. When the rotor axis is offset from the yaw axis, the wind's thrust on the rotor creates a force acting on a small moment arm (lever), represented by the distance between the two axes. The wind's thrust pushes the rotor out of the wind toward the hinged tail vane (see figure 6-41, Horizontal furling).

On contemporary farm windmills, there are no weights and levers to counteract the furling thrust. Instead they use springs. By

Figure 6-42. Furling hinge. *Left:* African Wind Power's simple but rugged furling hinge smoothly controls the rotor in strong winds. *Right:* The 3.6-meter (12 ft) diameter AWP 36 is conservatively rated at 900 watts. Main frame, left; yaw assembly, center; angled hinge pin; tail boom, right. (Duncan Kerridge)

Figure 6-43. Vertical furling. In high winds Northern Power Systems furls its HR3 model by tilting the rotor skyward, following the example of a 1930s-era windcharger. A shock absorber dampens the rate at which the rotor returns to the running position. This design includes a winch and cable for manually furling the turbine, as shown here at the Centro de Estudio de los Recursos Energeticos in Punta Arenas, Chile.

adjusting the tension in the spring, the farmer controls the wind speed at which the rotor furls. To see this for yourself, find an operating farm windmill and watch it in high winds. It will constantly fold toward the tail and unfold without any intervention.

Self-furling is a marvel of simplicity. Millions of farm windmills using Wheeler's approach to overspeed control have been put into service around the world. It's what you might call a proven concept. And if it worked reliably for all those machines for all those years, it should still work today. It does. Many small wind turbines use furling of one form or another.

The Bergey series of small wind turbines carries simplicity even farther than does the farm windmill. Rather than using springs to control furling, Bergey uses gravity to return the rotor to its running position.

Bergey skews the hinge pin between the nacelle and the tail vane a few degrees from the vertical. As the rotor furls in high winds, the nacelle lifts the tail vane slightly. When the wind subsides, the weight of the tail pushes the nacelle back into the wind.

Wind machines that use furling, like the water-pumping windmills before them, can be controlled manually by furling the rotor with a winch and cable on models where such a feature is included (Bergey's Excel, for example). When furled, the rotor doesn't come to a complete stop—it will continue to spin—but this can make the difference between survival and failure in an emergency.

Furling design is as much art as science, and when done poorly can lead to wildly fluctuating power from the rotor as well as generating considerable noise. Passive furling is "crude but effective," says Scoraig Wind Electric's Hugh Piggott, "and simplicity pays off in a small wind turbine." Piggott, a master of the technique, designed the smoothly furling African Wind Power turbine (see figure 6-42, Furling hinge).

Vertical Furling

During the 1930s Parris-Dunn built a windcharger that used a variation on the furling theme. Rather than turning the rotor parallel to the tail vane, the company chose to tip the rotor up out of the wind. In high winds the turbine would take on the appearance of a helicopter (see figure 6-43, Vertical furling). As the winds subsided, the rotor would rock back toward the horizontal. Northern Power Systems and Juan Bornay are two manufacturers that have adopted this technology. It's a simple strategy that works well. Northern Power Systems' High Reliability (HR) 3 has survived winds in excess of 176 mph (79 m/s) in Antarctica.

Like the original Paris-Dunn, Northern Power uses a spring to control the wind speed at which the rotor begins to furl. The wind speed at which the rotor begins to pitch back is governed by the tension in the spring. Atlantis and other small German turbines accomplish the same effect with a counterbalancing weight. The principal difference between the two manufacturers is the means they use for dampening the action of the rotor and generator as it rocks back and forth. Gusty winds can cause the rotor to tip up, then quickly rock forward, dropping the rotor and generator onto the wind machine's frame and severely jarring the blades and the rotor's main shaft. Northern Power uses a shock absorber that dampens the return of the rotor to the running position. Others have followed the Parris-Dunn example, and simply use a rubber pad to cushion the blow.

In a variation on vertical furling, the rotor, nacelle, and tail vane act as one unit. As the rotor moves up out of the wind, the tail vane drops toward the tower. Spain's Juan Bornay produces a line of small turbines using this technique. World Power Technologies built several hundred turbines with this form of furling but found that it reduced directional stability and caused the turbine to rapidly yaw about the tower, so the firm abandoned it for horizontal furling.

Figure 6-44. Variable rotor area. Synergy's novel S3000 reduces frontal area in high winds by using the thrust on its unusual tail vane to lift the 2.7-meter (9 ft) downwind rotor skyward. (Synergy Power)

Figure 6-45. Downwind coning. Proven Engineering's flexible hinge, demonstrated here by Mick Sagrillo, allows the polypropylene blades to cone downwind of the tower to protect the rotor in high winds. As the rotor cones, the blades change pitch as a result of the complex hinge angle.

One of the most novel designs to use frontal-area control is that of the Chinese manufacturer Synergy (see figure 6-44, Variable rotor area). Synergy uses a direct-drive, downwind rotor and a large tail vane. The tail vane not only directs the rotor downwind of the tower, but also furls the rotor vertically in high winds.

Coning

Gordon Proven has taken a somewhat more straightforward approach in his downwind rotor. Proven reduces the intercept area in strong winds by allowing the blades to increasingly cone downwind of the tower, shedding loads as they do so. The glass fiber-reinforced polypropylene blades attach to a flexible hinge at the hub. Proven's cleverly designed hinge permits the blades not only to flex downwind, but also to change pitch as they flex (see figure 6-45, Downwind coning).

Longtime wind advocate Peter Musgrove's contribution to wind technology was designing a way to reduce the intercept area of a vertical-axis turbine (see figure 6-46, Variable-geometry H-rotor). The H-rotor configuration offers several advantages over a conventional Darrieus turbine. Its weakness is the tremendous forces trying to bend the blades at the juncture between the blades and the cross-arm. These bending forces can be reduced, and the speed of the rotor controlled, by hinging the blades.

In the Musgrove turbine, the blades are hinged to the cross-arm in such a manner that the portion of the blade above the cross-arm is not equal to the portion below. As the rotor spins, centrifugal force throws the heavier portion of the blade away from the vertical, varying the geometry of the rotor. The wind and rotor speed at which this occurs is determined by the mass distribution of the blade and the tension in a spring restraining the blades in the upright position. At high wind speeds the blades approach the horizontal, reducing the intercept area.

Figure 6-46. Variable-geometry H-rotor. In Peter Musgrove's ingenious design, the straight blades of the H-rotor are hinged so that they tilt toward the horizontal at high rotor speeds, reducing the rotor's intercept area. Shown here is an early attempt by a British firm to commercialize the concept: PI Specialist Engineers 6-meter (18 ft) diameter variable-geometry VAWT, built in 1979. The figure at the bottom of the machine is Barry Holmes, the company's technical director. (PI Specialists)

For many years Musgrove and his graduate student, Ian Mays, pursued development of the technology. Though there were several experimental versions built in Great Britain, including a massive 500 kW model in Wales, no wind turbine using Musgrove's "variable geometry" has succeeded commercially. Both Musgrove and Mays have gone on to become leaders in competing wind companies, building wind farms in North America and Europe using conventional horizontal-axis turbines, some of which use variable-pitch blades.

Changing Blade Pitch

When most people first consider the problem of controlling a rotor in high winds, they think immediately of changing blade pitch. This probably results from our exposure to propeller-driven airplanes. Indeed, a wind turbine rotor can be controlled much like the propeller of a commuter plane. Like changing intercept area, changing blade pitch affects the power available to the rotor. By increasing or decreasing blade pitch, we can control the amount of lift that the blade produces.

There are two directions in which the blades can be pitched: toward stall or toward feather. When the blade is turned until it's nearly parallel with its direction of travel (perpendicular to the wind), it stalls. Thus to stall the blade, we need turn it only a few degrees. To feather a blade, on the other hand, it's necessary to turn it at right angles to its direction of travel (90-degree pitch), or parallel to the wind. Feathering a blade requires it to be rotated farther about its long axis than stalling it does, causing the pitch mechanism to act through a much greater distance.

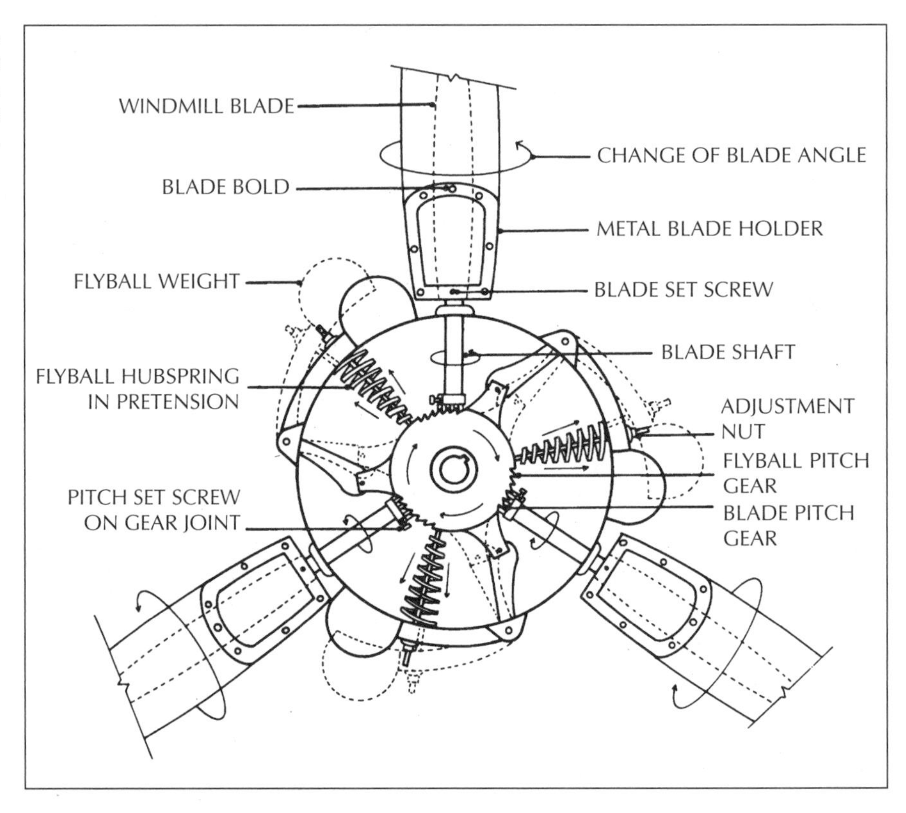

Figure 6-47. Flyball or Watt governor. Centrifugal force throws the weights away from the Jacobs governor, changing the pitch of the blades via a mechanical linkage.

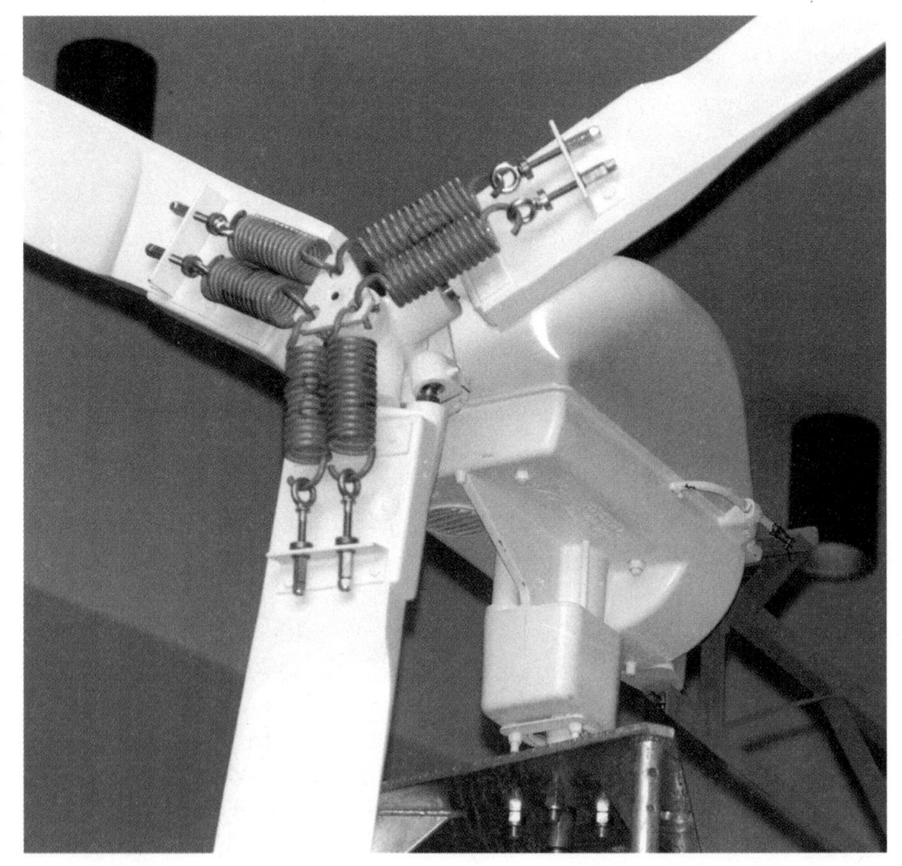

Figure 6-48. Blade-actuated governor. Many of the small wind turbines designed during the 1970s used this design, patterned after later versions of the Jacobs windcharger. The forces acting on the blades at high rotor speeds cause them to collectively change pitch.

Stall destroys the blades' lift, limiting the power and speed of the rotor, but it does nothing to reduce the thrust on the rotor or the tower. Though it is simpler to build a mechanism for stalling the blade than it is to build a feathering governor, the stalling technique is less reliable. On upwind machines, thrust on the blades bends them toward the tower.

Small wind turbine designs dependent on blade stall as the sole means of overspeed protection have a poor survival record. Historically the blades have had the nasty habit of striking the tower. Downwind turbines using stall regulation have had fewer problems, because the blades are forced to cone farther downwind and away from the tower. Still, they too have had a poor overall reliability record.

During the late 1990s several European manufacturers of medium-size turbines introduced variable-pitch rotors that pitched the blades to stall. These machines and their active controls are far more sophisticated than the passive pitch-to-stall systems sometimes used on small turbines. Where changing blade pitch is the primary means of control on an upwind rotor of a small wind turbine, experience has shown that the blade should rotate toward full feather. When it does, the drag on the blade is reduced to one-fifth that of a blade flatwise to the wind.

Governors for passively pitching the blades of small turbines appear in a variety of forms. During the 1930s, the Jacobs brothers popularized the flyball or Watt governor (see figure 6-47, Flyball or Watt governor). Above normal rotor speeds, the weights would feather all three blades simultaneously via a mechanical linkage.

It's relatively easy to design a mechanism on a small turbine that pitches each blade independently. Jacobs's innovation was to link blade pitching together; thus all blades changed pitch at the same rate. This avoids an aerodynamic imbalance that can damage the turbine.

When carefully adjusted, Jacobs's massive governor protected the 14-foot (4.3 m) rotor reliably. Jacobs introduced an even more resourceful solution on later models, called the Allied (after a windcharger on which it first appeared) or blade-actuated governor.

Why use weights when you don't have to? The blade-actuated governor uses the weight of the blades themselves to change pitch (see figure 6-48, Blade-actuated governor). Unlike the blades on the flyball governor, the blades not only turn on a shaft in the hub but also slide along the shaft. Each blade is connected to the hub through a knuckle and springs. The knuckle, in turn, is attached to a triangular spider. As the rotor spins, the blades are thrown away from the hub, causing them to slide along the blade shaft. When they do, the blades pull on the spider, which rotates all three blades together. The springs govern the rotor speed at which this occurs. Like the flyball governor, the blade-actuated governor works reliably when properly adjusted and built to the highest material standards.

In the late 1970s Marcellus Jacobs, the sole surviving founder of the original Jacobs Wind Electric Company, reentered the wind business. (The original firm ceased activity during the 1950s.) Along with his son Paul, Marcellus began manufacturing wind turbines patterned after his earlier models. His company briefly built wind turbines 7 to 8 meters (21 to 26 ft) in diameter. Jacobs's redesigned machine didn't depend solely on blade feathering to control rotor speed, since the blade-actuated governor was inadequate for a machine of this size. The new Jacobs turbine was also self-furling. The governor feathered the blades to limit power output to the alternator; overspeed protection was provided by furling the rotor toward the tail.

For many manufacturers of small wind turbines, mechanical governors have proven too costly and unreliable. Critics also charge that they are too maintenance-intensive for the modern wind turbine market. Still,

Figure 6-49. Pitch weights. Vergnet is one of the few small wind turbine manufacturers that uses pitch weights to protect the rotor in high winds. Contrast the distinctive tail vane on the upwind 10 kW model (left) to the 2 kW downwind model (right) at Vergnet's test field at Château Las Tours near Port-la-Nouvelle in the south of France.

French manufacturer Vergnet successfully uses pitch weights to govern its two-bladed rotors (see figure 6-49, Pitch weights). And there are hundreds of small windchargers operating in North America from the 1930s and from the Jacobs's revival in the 1970s that rely on mechanical governors. With proper maintenance and a supply of spare parts, these machines will last for several more decades. Owners of these turbines argue that mechanical governors provide better power regulation in high winds than does furling. In winds above the rated speed, power output drops sharply on some furling turbines, while small wind turbines using mechanical governors are able to maintain near-constant peak power. Though "there are substantially more moving parts in a pitching governor [than in a furling governor], I'd take a pitching governor every time," says Wisconsin's Mick Sagrillo.

It's possible to change blade pitch without using a mechanical governor. In the bearingless rotor concept, the blades are attached to the hub with a torsionally flexible spar. At high speeds the blades twist the spar toward zero pitch, stalling the rotor. Weights attached to the blades are sometimes used to provide the necessary force. There are no moving parts in the hub: no bearings, knuckles, or sliding shafts. Several have attempted to commercialize this technology on medium-size wind turbines. None succeeded, because they failed to master the complex rotor dynamics involved.

Carter wind turbines typified this overspeed control strategy. The filament-wound, fiberglass spar permitted the blade to twist torsionally. During high winds, small weights inside each blade would rotate the blade toward stall. The flexible spar also permitted the blade to cone progressively downwind of the tower in high winds, like the fronds of a palm tree during a hurricane. This design, though elegantly simple, was unreliable and fared poorly over time.

In medium-size turbines, the cost and complexity of variable-pitch hubs precluded their commercial use until well into the mid-1980s (see figure 6-28b, Rotor hubs). Part of the problem at the time was how to handle the large loads placed on the hub of a fully cantilevered blade. In the Danish experimental program at Nibe, Risø National Laboratory built two 600 kW turbines in the late 1970s. One Nibe turbine used struts and stays to reduce these cantilevered loads, allowing the turbine to vary the pitch of the outboard blade section only. The second turbine varied the pitch of the blade along its entire length (see figure 6-26, Blade attachment). The latter proved more successful. But it wasn't until the late 1980s that manufacturers of medium-size wind turbines began introducing full-span, variable-pitch rotors. (see figure 6-50, Full-span, variable-pitch rotors). As wind turbines continue to grow larger, the trend is toward full-span pitch control, especially in multimegawatt-size turbines where the higher costs and complexity of pitch control can be justified.

Figure 6-50. Full-span, variable-pitch rotors. In the late 1980s several manufacturers of commercial medium-size wind turbines, such as Britain's Wind Energy Group, introduced variable-pitch rotors. WEG's MS2 used wood-composite blades on a variable-pitch rotor 25 meters (82 ft) in diameter. Shown here is a cluster of the 250 kW turbines in California's Altamont Pass. The blades on the turbine in the foreground are feathered; the blades on the turbines in the background are in the operating position.

Aerodynamic Stall

Almost all wind machines without pitch control use aerodynamic stall to some extent for limiting the power of the rotor. This is particularly true of medium-size wind turbines using fixed-pitch rotors to drive induction generators. In winds above the rated speed, the tip-speed ratio for these turbines declines because the speed of the rotor remains constant. For wind turbines operating at constant speed, the angle of attack increases with increasing wind speed, reducing the aerodynamic performance of the blades.

Designers seldom rely on blade stall as the sole means of overspeed protection on wind turbines driving induction generators. Stall control is dependent on the generator keeping the rotor at a constant speed. Induction generators, in turn, are dependent on the utility for controlling the load. During a power outage, the generator immediately loses its magnetic field. The rotor, no longer restrained to run at constant speed, immediately accelerates. Stall now becomes ineffectual for regulating rotor speed until a new equilibrium is reached. Unfortunately this occurs at extremely high rotor speeds.

On an upwind machine with a tail vane, the rotor can protect itself from destruction by furling out of the wind. Since tail vanes are limited to small turbines, medium-size upwind machines—and all fixed-pitch downwind machines—must use a different strategy. Brakes are the most popular.

Once brakes have been selected as the means of limiting rotor speed during a loss-of-load emergency, they're also frequently used during normal operation. In a typical fixed-pitch wind machine, the brake is applied at its cut-out speed to stop the rotor. Wind turbines using this approach require extremely strong blades, in case the brake should fail and the rotor accelerate to destructive speeds.

Consider the case of a small downwind turbine driving an induction generator that was

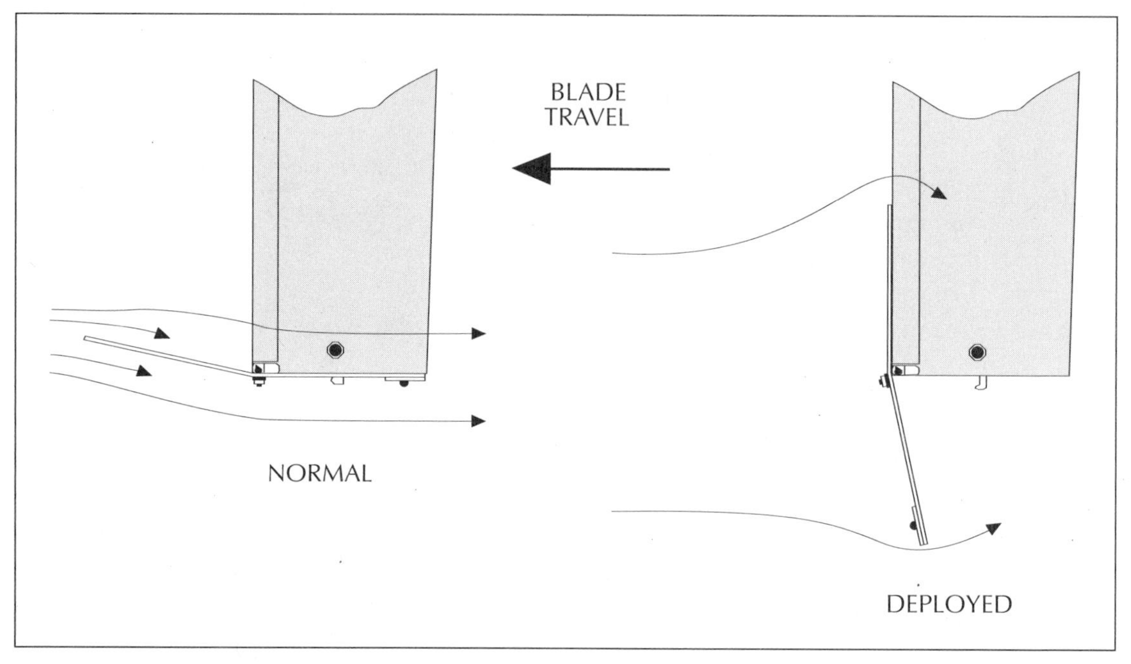

Figure 6-51. Tip brake. Enertech popularized the use of tip brakes for overspeed protection on downwind, induction wind machines. Tip brakes are noisy and rob power.

designed to brake to a halt at the cut-out speed. When the manufacturer, Enertech, first introduced the machine, the company stressed that the rotor was stall-regulated and that it could operate safely, if necessary, above the cut-out speed, without the brake. The rotor was braked, asserted Bob Sherwin, Enertech's vice president, only to minimize wear on the drive train at high wind speeds. The amount of energy in the wind at these higher speeds, he said, did not warrant the cost of capturing it. Mother Nature soon gave Sherwin's Enertechs ample opportunities to prove their mettle. The brake failed on several occasions. Rotors went into overspeed, and several Enertechs destroyed themselves, to Sherwin's chagrin. Stall alone wasn't enough to protect the rotor. Enertech later added tip brakes for such emergencies (see figure 6-51, Tip brake).

Mechanical Brakes

Brakes can be placed either on the main (slow-speed) shaft or on the high-speed shaft between the gearbox and the generator. Brakes on the high-speed shaft are the most common, because the brakes can be smaller and less expensive than the large disks and multiple calipers of those on the main shaft. When on the high-speed shaft, the brakes can be found between the transmission and the generator or on the tail end of the generator. In either arrangement, braking torque places heavy loads on the transmission and couplings between the transmission and generator. Moreover, should the transmission or high-speed shaft fail, the brake can no longer stop the rotor.

In general, brakes on fixed-pitch machines should be located on the main shaft, where they provide direct control over the rotor. (There's always a greater likelihood of transmission failure than failure of the main shaft.) But the lower shaft speeds require more braking pressure and greater braking area. As a result, the brakes are larger and more costly than those on the high-speed shaft.

Brakes can be applied mechanically, electrically, or hydraulically. Most operate in a fail-safe manner. In other words, it takes power to

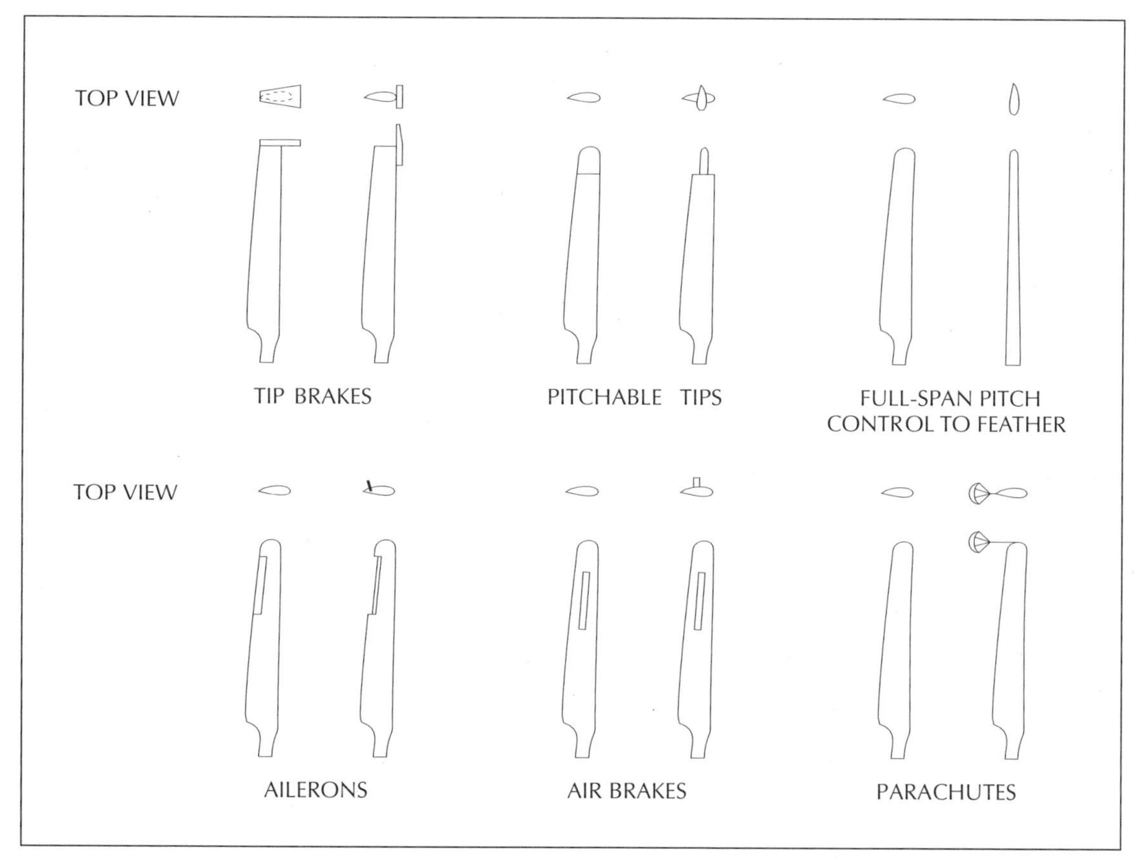

Figure 6-52. Overspeed controls. On medium-size wind turbines, the most common mechanism for protecting the rotor from overspeed is pitchable blade tips. Some turbines, such as the Windmatic 15S, use spoilers; others, like Enertechs, use tip brakes; a few have used parachutes. Large turbines and some medium-size designs use full-span pitch control.

release the brake. The brake automatically engages when the wind machine loses power. Springs provide the force in a mechanical brake, such as those on the defunct Enertech and ESI brands; a reservoir provides the pressure in the hydraulic brakes on Vestas turbines, for example.

The problem with brakes of any kind is that they can fail—not often, it's true, but once is enough. Brake pads require periodic replacement or adjustment. After extensive use, the calipers have to travel farther to reach the disk. If the brakes are spring-applied, pressure from the springs decreases, as does braking torque, as pad travel increases.

In one 40-foot (12 m) diameter model built briefly during the late 1970s, the undersize brake just wasn't up to the job. And tragically there were no backup devices to protect the turbine. Several machines ran to destruction, but some were brought under control when the designer, Terry Mehrkam, climbed the tower, wedged a crowbar into the brake, and manually forced the pads against the disk. This dangerous practice eventually cost Mehrkam his life.

Experience has taught wind turbine owners and designers alike that wherever a brake is used to control the rotor, there must be an aerodynamic means to limit rotor speed should the brake fail. There are three common choices for aerodynamic overspeed protection on wind machines without tail vanes and pitch controls: tip brakes, spoilers, and pitchable blade tips (see figure 6-52, Overspeed controls). These devices were frequently found on medium-size wind turbines

Figure 6-53. Air brake. Miller Cornelius Gerkes releases the air brake on Germania Molen in the Netherlands' Groningen province. Note the airfoil-shaped leading edge and the jalousie shutters.

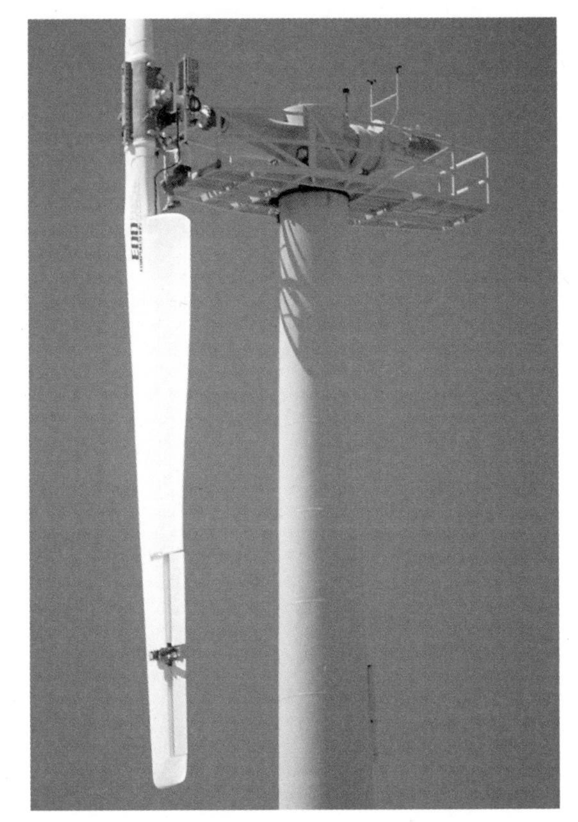

Figure 6-54. Ailerons. Northern Power Systems' experimental turbine unsuccessfully used ailerons (flaps) for overspeed control. This turbine was eventually abandoned.

using fixed-pitch rotors to drive induction generators in the 1980s.

Aerodynamic Brakes

Aerodynamic brakes are not a new idea. In the early part of the 20th century, during the heyday of pioneering work in aviation, some traditional Dutch windmills began to sport newfangled shapes on their wooden spars: jalousies, airfoil-shaped leading edges, and air brakes (see figure 6-53, Air brake).

Tip brakes are similar to air brakes. These plates attached to the end of each blade also use drag to slow the rotor (see figures 6-31a and 6-51). They're activated by centrifugal force once the rotor reaches excessive speed. They deploy a flat plate transverse to the direction of travel. They're simple, effective, and have saved many a fixed-pitch rotor from destruction. Tip brakes, however, have been likened to keeping your foot on the accelerator at the same time you're stepping on the brake. They keep the rotor from reaching destructive speeds, but do nothing to reduce the lift of the blade or the thrust on the wind turbine and tower. Tip brakes are also noisy, and reduce the performance of the rotor under operating conditions by increasing drag at the tip where blade speed is greatest.

Enertech eventually adopted tip brakes on its turbines after discovering that stall wasn't sufficient to protect the machine in high winds. When Bob Sherwin reintroduced the Enertech E44 design as the AOC 15/50 in the 1990s, he continued Enertech's use of tip brakes.

Somewhat more sophisticated was Northern Power Systems' experimental use of what Jito

Coleman called ailerons on the blade's trailing edge (see figure 6-54, Ailerons). Tehachapi sailplane pilot Kevin Cousineau says these devices should more correctly be labeled flaps. Regardless of what they're called, they've proven noisy and unreliable for use in wind turbines.

Most of the power in the wind is captured by the outer third of the rotor. Consequently it's not necessary to change the pitch of the entire blade to limit the rotor's power and speed. The performance of the blade in this region can be reduced by using spoilers or movable blade tips. In the Windmatic 15s, if there was a loss of load or the brake failed, centrifugal force activated spoilers along the back side of each blade. The spoilers popped out of the blade's surface and changed the shape of the airfoil, destroying its effectiveness.

The widespread success of wind energy in the 1980s is attributable to the "Danish" concept of rugged, three-blade upwind turbines driving induction generators. One of the principal reasons for their success was that nearly all of the stall-regulated Danish turbines of the period used pitchable blade tips for overspeed control. These passively activated tips were more effective than the tip brakes used on Enertechs, for example.

During a loss-of-load emergency, such as a power outage, higher-than-normal rotor speeds cause the blade tips to slide along a grooved shaft. As they move along the shaft, the tips pitch toward feather. This action decreases lift where it's greatest, while dramatically increasing drag. Both spoilers and pitchable blade tips have proven highly successful, though refinements have been necessary. Frequently, for example, only one or two of the blade tips would activate, rather than all three at the same time. While this was usually sufficient to protect the turbine, it caused many windsmiths sleepless nights.

As the turbines have grown larger, the tips are no longer passively activated or used only in emergencies. For example, both Boeing's

Figure 6-55. Pitchable blade tips. NEG-Micon's 1.5 MW turbine actively pitches the blade tips to control the rotor both in normal operation and in emergencies. Vestas V66 with feathered blades are in the background at the Tjæreborg test field near Esbjerg on the west coast of Denmark. The ungainly turbine on the concrete tower has since been removed.

experimental 2.5 MW Mod-2 and its 3.2 MW Mod-5B actively controlled the pitch of the outboard section of each blade. Though Boeing failed to commercialize its monstrous turbines, some Danish and German manufacturers have used partial-span pitch control, or active blade tip control, in multimegawatt-size turbines (see figure 6-55, Pitchable blade tips).

Putting It All Together

We'll now take a look at two classes of wind machines and how their manufacturers put all the pieces together. We'll also look at how they operate under normal and emergency conditions. The first group is the advanced small wind turbines designed specifically for high-reliability and low-maintenance applications, often off the grid. The second group is medium-size, or commercial, wind turbines such as those found in wind power plants around the world.

Figure 6-56. Minimalist small turbine. Southwest Windpower's Whisper 175 uses a 15-foot (4.6 m) diameter two-blade rotor to directly drive a permanent-magnet alternator. The rotor sweeps 175 square feet (16.4 m²), hence the designation.

Small Turbines

Small wind turbines should be designed for simplicity, ruggedness, and low maintenance. Most, but unfortunately not all, are. Karl Bergey liked to quote French pilot Antoine de Saint-Exupéry to express his design philosophy: "Perfection is achieved not when there is nothing more to add but when there is nothing more to take away."

Most of today's small wind turbines typically employ an upwind rotor and are passively directed into the wind by tail vanes, though some (Provens) are passively oriented downwind of the tower. Most use three blades, though some, such as Southwest Wind Power's Whisper 175, use only two (see figure 6-56, Minimalist small turbine). Nearly all micro and mini turbines drive special-purpose generators directly, without the use of step-up gearing. Among the larger household-size turbines are some, such as those of Wind Turbine Industries and Vergnet, that use off-the-shelf generators coupled to transmissions.

Nearly all small wind turbines are designed for stand-alone, battery-charging applications. Some, however, such as Bergey Windpower's XL1 and Excel, also work with special-purpose inverters for interconnection with the electric utility. A few household-size turbines, like those of Vergnet, drive induction generators and can be interconnected to the utility without an inverter.

With the exception of those driving induction generators, these turbines operate at variable speed. From start-up to rated wind speed, rotor rpm increases with increasing wind speed. Similarly, voltage and frequency increase as wind speed increases. Near the rated wind speed the blades begin to stall, reducing performance. As wind speed continues to increase, most small turbines begin to furl, swinging the rotor toward the tail vane, while in some designs the blades change pitch. Regardless of the technique, the turbines spill excess power to protect the rotor. When high winds subside, the turbine's rotor returns to its operating position automatically.

Most small turbines lack a mechanical brake, and many lack a furling cable for manually bringing the turbine under control during emergencies. Practically none include a parking brake for servicing the rotor, or a yaw brake to prevent the wind machine from yawing about the top of the tower. Both are important considerations if you plan to service the turbine atop the tower. For the most part, the designers of these machines stressed simplicity and ruggedness over greater control. Bergey Windpower and Southwest Windpower's Whisper line carry simplicity one step farther by integrating the hub and rotor housing into one assembly. All are designed for little or no maintenance, and with good reason. At windy sites, it's not uncommon for a wind machine to be in operation two-thirds of the time, or about 6,000 hours per year. At that rate a wind machine operates as many hours in the first six months of the year as an automobile driven 100,000 miles (170,000 km). Over a 20-year lifetime, a wind machine 3 meters (10 ft) in diameter will accumulate nearly 3,000 million revolutions, eight times that of a crankshaft in an automobile driven 100,000 miles!

In programs where performance has been monitored, wind machines built by both Northern Power Systems and Bergey Windpower have chalked up an impressive record of reliability. After two decades of development, these designs have proven more dependable in remote power systems than the conventional engine generators they were originally designed to supplement. The turbines did require regular service and occasional repairs, but they performed the job they were designed to do.

Though experimental wind turbines using integrated direct-drive, permanent-magnet alternators have been built as large as 600 kW in size, commercially successful designs have been limited to 30 kW or less.

Danish wind turbine owners learned in the 1970s that every wind turbine should include some form of aerodynamic overspeed control. From harsh experience, they knew that reliance on mechanical or electrical brakes was a recipe for disaster. While some manufacturers of micro turbines may argue that this caveat does not apply to their product, it clearly applies to all household-size turbines, as well as the much larger commercial wind turbines.

Medium-Size Turbines

In contrast to small wind machines, commercial wind turbines found in wind power plants use, for the most part, off-the-shelf induction generators, transmissions, brakes, yaw drives, electrical sensors, and controls.

The numerous components on a medium-size wind turbine require regular maintenance, which can be a function simplified by clustering the turbines together in one location, as in a wind farm, or within easy reach of the manufacturer's nearest service center.

This complexity, and the resulting need for maintenance, thwarted many American manufacturers who attempted to market household-size turbines for dispersed applications in the United States. These machines, best represented by Enertech, used a fixed-pitch rotor downwind of the tower to drive an induction generator. Enertech turbines, like contemporary medium-size wind turbines, used a brake to stop the rotor under normal and emergency conditions. If the brake failed, Enertech relied on tip brakes to protect the rotor from self-destruction.

Some of today's medium-size turbines still use fixed-pitch blades bolted rigidly to the hub. Like small wind turbines, most of these machines are self-starting (though some early models, like the Enertech, were motored up to their operating speed). From start-up to the cut-in wind speed, the rotor speed varies with wind speed until it reaches the speed at which the generator can be synchronized with the utility.

Danish wind turbine owners learned in the 1970s that every wind turbine should include some form of aerodynamic overspeed control.

From the cut-in to the rated wind speed, the rotor continues turning at the same speed while delivering more and more power. The wind machine may, from time to time, switch from one generator to another as wind speed varies, changing rotor speed like a driver shifting gears in a car. But overall the turbine operates at constant speed. (The rotor speed does vary slightly, but this is imperceptible to most observers.)

Above the rated wind speed the blades begin to stall, dumping excess power. When wind speeds exceed the turbine's cut-out speed, or with any abnormal occurrence such as excessive vibration, the brake is applied, bringing the rotor to a stop. In the typical Danish design, the turbine then yaws 90 degrees out of the wind to a parked position. When wind speeds fall back below the cut-out speed, or the emergency fault has been reset, the turbine yaws back into the wind and releases its brake. Soon after, the rotor accelerates to its operating speed.

On American downwind designs of the 1980s, such as on the Enertechs and Carters, there was no active yaw control. The rotor stayed downwind of the tower even when the rotor was parked due to coning of the rotor blades.

The brake on both types of machines is normally "on," or engaged. The brake can be released only when there is electrical power to the wind machine. Thus if there's a loss of power, the brake automatically returns to its engaged position, stopping the rotor.

On early Danish designs, such as those of the 1980s, if the brake was unable to prevent the rotor from reaching unsafe speeds, the backup aerodynamic controls were deployed. These then slowed the rotor to a safe speed, but they wouldn't stop the rotor. Once deployed, these aerodynamic controls would typically be reset manually to ensure at least a cursory inspection of the turbine.

Modern Danish and German turbines, because of their increased size, use aerodynamic means to bring the rotor under control. The brake is applied after the rotor has been slowed to a safe speed. Though the brake may be capable of stopping the rotor in an emergency, it isn't used during normal operation except to park the rotor.

There are important variations in the manner in which some medium-size turbines operate. Unlike most such turbines, Kenetech's model 56-100 employed a variable-pitch rotor downwind of the tower. More than 4,000 of these 18-meter, 100-kilowatt turbines had, at one time, been operating in California, the vast majority in the Altamont Pass. Even though it uses a variable-pitch rotor, the 56-100, and other turbines like it, maintains a constant pitch from cut-in to the turbine's rated output. Only in winds above its rated capacity does the rotor begin to pitch its blades. Thus the turbine uses only pitch to control peak power, and feathering the blades to protect itself from overspeed.

Several of the major European manufacturers have made the transition from stall control of fixed-pitch rotors to full-span pitch control, especially in megawatt-size turbines.

Another development in the mid-1990s was the large-scale introduction of full variable-speed operation in wind turbines from 250 to 500 kW using purpose-built direct-drive alternators. Since then the technology, used notably by German manufacturer Enercon and the Dutch manufacturer Lagerwey, has been applied to ever-larger turbines. Others, such as the German manufacturer DeWind, have incorporated doubly fed induction generators in a conventional medium-size turbine drive train to gain the advantages of variable speed operation by using mass-produced alternators.

Wind turbines, however, are only part of a wind system. Towers, the subject of the next chapter, are an integral part of any wind power system.

7

Towers

Towers are as integral to the performance of a wind system as the wind turbine itself. Without the proper tower, your wind machine isn't much more than an expensive lawn ornament, and could even become a hazard to all in the vicinity.

Towers, as a rule, are one of the few wind system components in which you have some choice. Unlike the selection of blades and generator, for example, which have been determined by the manufacturer, you may have several types of towers to choose from—at least for household-size and smaller wind machines. And for both small and as medium-size turbines, you can choose the height best suited for your site.

When considering tower options for small turbines, it's imperative to keep in mind that the tower be strong enough to withstand the thrust on the wind turbine (the force trying to knock the wind turbine off the tower) and the thrust on the tower (the force trying to knock the tower over). And unless it's a hinged tower that can be lowered to the ground, the tower must support not only the weight of the wind turbine, but also the weight of the people who will service it.

Towers for medium-size turbines are designed by the turbine manufacturer specifically for its turbine.

Foremost among the criteria for the correct tower is whether it's available in the height desired.

Height

As the wind industry has matured—and wind system users, as well—selecting a tower of the proper height has become increasingly important. In the early 1970s anything that would get the wind machine off the ground was acceptable.

Towers for wind machines used on the Great Plains during the 1930s were never very tall. The flat terrain and lack of obstructions didn't call for towers taller than 60 feet (18 m). Even so, by the late 1940s Wincharger was installing guyed towers 85 to 105 feet (25 to 30 m) in height, and Parris-Dunn was advising its customers that "the higher the tower the greater the power."

Parris-Dunn was advising its customers that "the higher the tower the greater the power."

Through painful experience, we've learned that economical power generation and good performance are obtained only on a tall tower. This has proven true for both small and commercial-size wind turbines.

We have known for some time that wind speed and power increase with height,

Stratospheric Heights

The tower heights of medium-size wind turbines are often in the ratio of 1:1 with the turbine's rotor diameter. In the early 1990s, for example, towers for medium-size wind turbines typically reached heights of 30 to 40 meters (100 to 130 ft) for wind turbines of 25 to 30 meters (80 to 100 ft) in diameter. By the mid-1990s towers were reaching heights of 40 to 50 meters (130 to 160 ft), and by the end of the decade towers 60 to 70 meters (200 to 230 ft) tall had become common. In wooded areas of Germany, a tower-height-to-diameter ratio of 1.5 has sometimes been used. Enercon, for example, was installing its giant E66 on towers more than 100 meters (330 ft) tall. The use of such tall towers may explain why wind turbines have become visible from commercial aircraft flying over Germany's central highlands. Tall towers permit the turbines to stand well above surrounding obstructions—trees and buildings—as well as the terrain. They also make the turbines far more visible—even from high-flying aircraft.

but the lesson didn't begin to sink in until wind systems were being installed in great numbers across the breadth of North America and Europe. We gained far more experience with power-robbing turbulence and what it can do to a wind machine's performance than we ever needed. As a result, recommended tower heights have gradually increased. The commercial success of medium-size wind turbines is, in large part, due to their use of increasingly tall towers.

This is a lesson that has been lost on many small wind turbine aficionados. Mick Sagrillo, never one to mince words when it comes to small wind turbines, is adamant about using as tall a tower as practical. He tells students in his workshops on how to install small turbines that the three most common mistakes people make with wind energy are using

1. too short a tower,
2. too short a tower, and
3. too short a tower.

Sagrillo likens using too short a tower to "putting solar panels on the north side of the roof so they won't get sunburned." He also advises his students to consider the height of mature trees when sizing towers. "Trees grow. Towers don't," he warns. Sagrillo's been working with small wind turbines long enough to know what happens when trees grow and eventually shelter a once well-performing turbine. "If it doesn't produce enough power to warrant being installed on a suitable tower, then it's not a serious wind generator. Why bother?"

Manufacturers prefer taller towers because they want their products to perform well and want to minimize turbulence-induced service and warranty claims. Taller towers also allow more flexibility in siting. If buildings and trees are present—and they usually are—a tall tower can redeem an otherwise unusable site. For household-size turbines, a minimum tower height by today's standards is 80 feet (25 m). And when trees are nearby, 100 to 120 feet (30 to 35 m) is the norm.

The height requirements for micro wind turbines are somewhat different. Micro turbines are often used in low-power applications, such as weekend cabins, where maximum generation isn't necessary (see figure 7-1, Micro turbine tower). Because of their relatively low cost, they're often used with inexpensive towers. These towers are generally not suited for heights above 60 feet (18 m). It doesn't make a lot of sense to install them on taller towers that cost three or four times as much as the turbine itself, unless you plan to eventually install a bigger turbine.

Southwest Windpower's Andy Kruse contends that what height tower people use with Southwest's Air series depends on what they want from the micro turbine and what they're willing to accept. Kruse argues that the Air turbines are so inexpensive that installing one on a taller and more expensive tower is seldom justified. The average tower height for the Air is only 25 feet (7.6 m), Kruse says.

"People feel comfortable with this tower height. Admittedly, this isn't the optimum, but users shouldn't be so concerned with tower height if they're willing to accept less-than-optimum performance."

There are other times where a nonoptimal tower height may be acceptable. At sites in Scotland, for example, Proven turbines are typically installed on freestanding towers only 6.5 meters (20 ft) tall. Scoraig Wind Electric's Hugh Piggott finds that in the open terrain of the Scottish Highlands, these towers work acceptably. Most importantly, these short towers pass muster with strict local planning codes.

Figure 7-1. Micro turbine tower. This relatively short tower is adequate for the Marlec 910F in this application, powering a few DC lights in the shed. For anything more demanding, a taller tower is necessary.

Buckling Strength

Next to the height, the most important criterion for a tower is its ability to withstand the forces acting on it in high winds. Towers are rated by the thrust load they can endure without buckling. Standards in the United States call on manufacturers to design their wind systems to withstand 120 mph (54 m/s) winds without damage. The thrust on the tower at this wind speed depends on the rotor diameter of the wind turbine and its mode of operation under such conditions—whether it furls the rotor or changes the pitch of the blades, for instance.

Two wind turbines of the same rotor area may require entirely different towers because of differing approaches to protecting the rotor in high winds. Small wind turbines that furl the rotor reduce thrust loads on the tower substantially compared to those that feather the blades. For wind turbines that furl the rotor, thrust reaches a maximum at the furling speed, and remains fairly constant thereafter. Thrust continues to increase with increasing wind speed on small turbines with mechanical governors.

For small wind turbines on tall towers, the drag on the tower in high winds adds significantly to the thrust loads the tower must withstand. In contrast, the rotor on a medium-size wind machine presents far more frontal area to the wind, proportionally, than it does on a small turbine; thus thrust on the rotor dominates.

All towers flex and sway with the wind—even the massive tubular towers that support commercial wind turbines. In a strong wind, the swaying at the top of a tall tower may be enough to give you the impression of being aboard a storm-tossed ship. One wind turbine dealer discovered this the hard way.

After the dealer finished wiring his newly installed wind turbine to the service panel, he

Drag Force and Thrust

The drag on an object in the wind—whether it's a tall building, tower, or wind turbine—is a function of air density, the area intercepting the wind, the speed of the wind, and a dimensionless coefficient that represents the object's shape and its angle to the wind. The drag force in the direction of the wind is the thrust: the force trying to push a tower over or blow a wind turbine off the top of its tower.

$$\text{Drag} = 1/2\ \rho A V^2 C_D$$

where ρ is air density in kg/m^3, A is the area in m^2 intercepting the wind, V is wind speed in m/s, and C_D is the coefficient of drag. A flat plate at right angles to the wind has a C_D of 1.1 while that of a closed cylinder is 0.6. It's important to know the maximum amount of thrust you can expect on the wind turbine and tower to determine the needed buckling strength of the tower and the size and depth of the anchors. For small wind turbines and their towers, it's best to assume the coefficient of drag for a flat plate. Some manufacturers include the thrust they expect their turbine to see under worst-case conditions in their list of specifications. Others will provide the data upon request. Tower manufacturers, such as NRG Systems, design their towers to withstand thrust loads at 120 mph (54 m/s) on the top of the tower, with a minimum safety factor of two.

was eager to see it in operation (an affliction that wind pioneer Jack Park diagnoses as "fire-'em-up-itis"). The wind was strong, blowing near the rated speed of his Jacobs wind turbine. To ensure that all was well and to get a bird's-eye view of his new investment, he unwisely climbed up the 100-foot (33 m), heavy-duty truss tower. Stopping just below the rotor, he decided to check the operation of the feathering governor by unloading the generator and letting the rotor speed increase. When an assistant disconnected the wind turbine, he suddenly found himself hanging on with all his might as the blades feathered and the tower sprang several feet back into the wind like a giant whip. He was lucky. If he hadn't been strapped to the tower and kept his wits, he could easily have been killed. (This example violates one of the fundamental safety rules of working around wind turbines: Never climb the tower when the rotor is spinning. For more on safety, see chapter 16).

Slender tubular towers are far more flexible than truss towers and visibly deflect in strong winds. Deflection isn't a problem unless the turbine and tower are mismatched. If the tower or the blades flex too much and at the wrong time, the blades could strike the tower. This dynamic interaction between the wind machine and tower is of major concern to manufacturers.

As the rotor and tower deflect in the wind, they begin to oscillate like the swaying spans of a rickety suspension bridge. Should the turbine and tower begin to sway in harmony, the oscillations could gradually increase in magnitude until they destroy the wind machine, tower, or both.

Though guyed towers may appear less secure, the reported accounts of tower failures in wind system applications involve nearly equal numbers of truss towers and guyed towers. In one widely discussed case, a truss tower failed when the bolts holding two 20-foot (6 m) sections together sheared. The tower manufacturer asserted that the tower was overloaded by the dynamic interaction of the turbine and the tower. Witnesses noted that the tower was vibrating wildly prior to the accident. The turbine manufacturer, Jacobs Wind Electric, countered that the failure was due to "bad steel" in the bolts. Bad steel or not, a wind system vibrating in resonance can exert tremendous force on the tower, creating loads well beyond its design limits.

It's this dynamic interaction between the wind machine and the tower that leads some manufacturers to restrict the types of towers for their wind machines. The pairing becomes increasingly important as size increases. Unlike small wind turbines, where even today there's a tendency to mix and match turbines

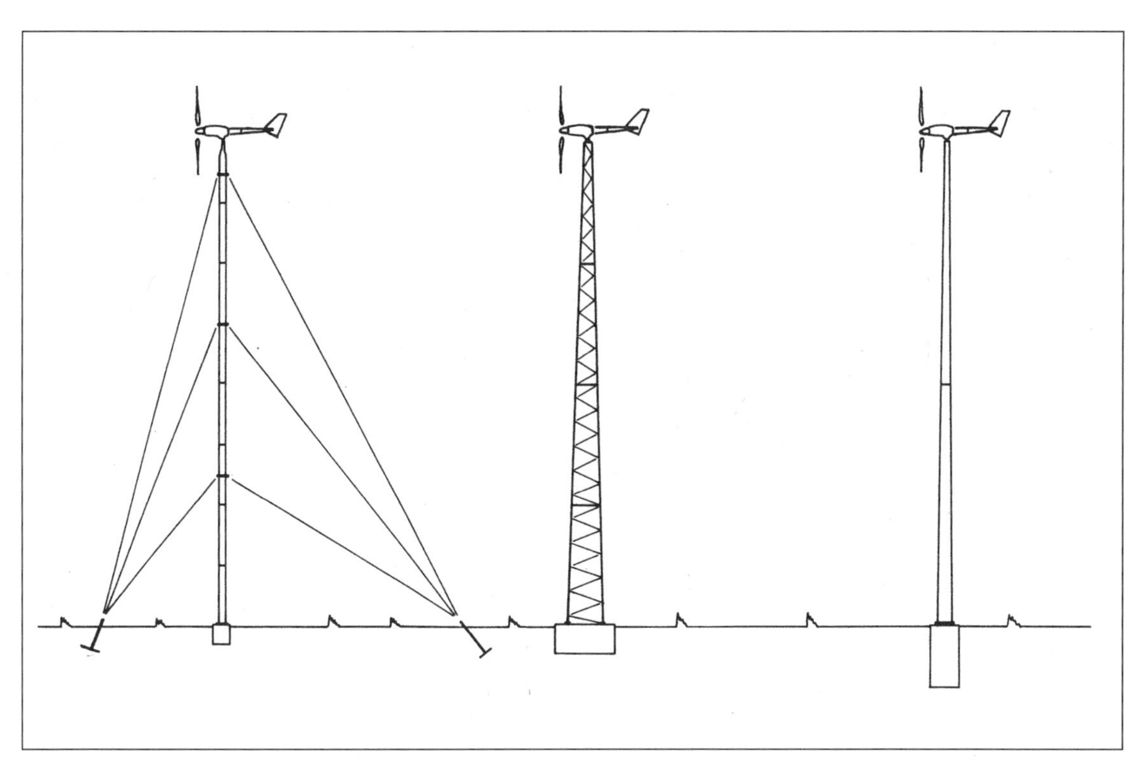

Figure 7-2. Tower types. For small turbines, guyed towers (left) are the most economical. Freestanding lattice (center) and cantilevered tubular towers (right) are more costly but also options. (Bergey Windpower)

and towers, commercial wind turbines are matched to a limited number of towers that meet the manufacturers' requirements for strength and dynamic response.

Tower Types

Towers fall into two categories: freestanding and guyed (see figure 7-2, Tower types). Freestanding towers, also known as self-supporting towers, are just that—freestanding. They depend on a deep or massive foundation to prevent the tower from toppling over in high winds, and they must be strong enough internally to withstand the forces trying to bend the tower to the ground. Guyed towers, in contrast, employ several far-flung anchors and connecting cables to achieve the same ends. Freestanding towers are more expensive than guyed towers, but they take up less space.

Freestanding Towers

There are two types of freestanding towers. The most common is the truss or lattice tower, so called because it resembles the latticework of a trellis. The Eiffel Tower is the best-known example of a truss tower (see figure 7-3, Truss tower). The tubular tower is another form of freestanding tower. Truss towers are typically more rigid than tubular towers.

Towers can be designed to withstand any load. But as the size of the wind machine increases, so do the weight and cost of the tower supporting it. The same is true as the tower increases in height. The components become heavier, harder to move, and more costly to ship.

In North America truss towers for small wind turbines are assembled from a series of 20-foot (6 m) sections. For small wind machines, the sections may be preassembled and welded together prior to delivery. For household-size turbines, the tower is shipped

Figure 7-3. Truss tower. Gustave Eiffel first used the compound taper seen on these 55-meter (180 ft) tall truss towers on the Cabazon site in California's San Gorgonio Pass. The rotor on the Z750 is 48 meters (157 ft) in diameter. Note the power line and buildings for scale. The Z750 was probably the last commercial wind turbine of this size to use a truss tower in North America.

"knocked down" or in parts, and must be assembled on the site.

Installation of truss towers usually requires a crane. The tower is assembled on the ground, then hoisted into place and bolted to the foundation.

Another method is to hinge the tower at its base. The tower is bolted together on the ground, the wind turbine attached, and the whole assembly tipped into place with a gin pole and winch, or a small crane.

Tubular Towers

Nearly all medium-size wind turbines are installed on tubular towers, though there are some notable exceptions. This wasn't always the case. During the great California wind rush of the 1980s, an equal number of turbines were installed on truss and tubular towers. The enormous size of today's turbines, however, as well as aesthetic demands, have led to the almost exclusive use of gently tapered tubular towers (see figure 7-4, Tubular tower).

Small turbines have also been installed on tubular towers, though infrequently. These slender towers resemble light standards (see figure 7-5, Pole tower).

Pole towers are made from tapered steel tube, steel pipe, wood, concrete, or even fiberglass. Though most are made of steel, pole towers of prestressed concrete have been used by some manufacturers. For small turbines, pole towers are available only in limited sizes and strengths. For example, the selection of wood and concrete poles suitable for wind machines is limited by the length of pole that can be shipped conveniently. Like truss towers, pole towers are difficult to handle without heavy equipment. It is primariliy for this reason that they have seldom been used in North America.

Figure 7-4. Tubular tower. A 44-meter (144 ft) diameter NEG-Micon turbine atop a 50-meter (160 ft) tower on the Whitewater Wash near Palm Springs, California. Access to the turbine's nacelle is via a ladder on the inside of the tower.

Figure 7-5. Pole tower. An old Enertech 1800 on a slender tubular tower near Palm Springs, California. Foot pegs or climbing rungs welded to the tower provide access to the turbine.

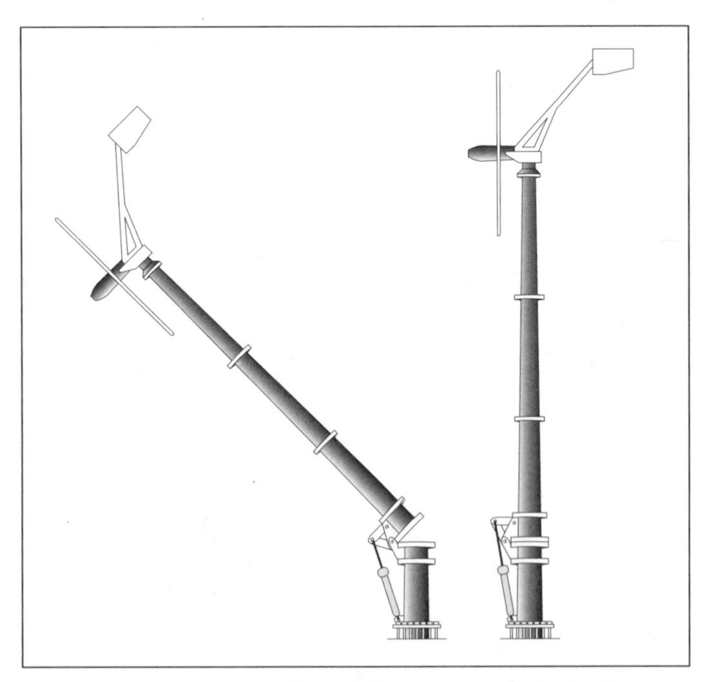

Figure 7-6. Hinged, freestanding tubular tower. This hydraulically operated hinged tower was developed for use at inaccessible Norwegian telecom sites. (Roheico A/S)

Installation of pole towers usually requires a crane. A pole tower, though, can be hinged at the base and tipped into place with a gin pole. When upright, the tower is then bolted to the foundation. Some European manufacturers erect hinged tubular towers with powerful hydraulic jacks (see figure 7-6, Hinged, freestanding tubular tower).

Many observers consider freestanding tubular towers more aesthetically pleasing than truss towers. This is certainly true in foreground views, but it isn't always the case. Surprisingly, tubular towers can be more visible at a distance than lattice towers, especially in silhouette. In arid regions lattice towers tend to blend into the landscape (see figure 7-3, Truss tower).

For small wind turbines, pole or tubular towers are significantly more expensive than guyed towers, but cost only modestly more than truss towers. Still, pole towers require a more substantial foundation than truss towers, which spread the overturning force over a wider base. If you're sensitive to cost, consider a guyed tower.

Guyed Towers

Guyed towers are, by far, the most common choice for small wind machines (see figure 7-7, Guyed lattice towers). They offer a good compromise among strength, cost, ease of installation, and appearance. Unfortunately they take up more space than freestanding towers, and they suffer from the potential danger that a guy cable will fail and the tower will come crashing down.

Guyed towers include a mast, guy cables, and earth anchors. The mast itself may be made from steel lattice, heavy-walled pipe, or thin-walled tube. In North America most guyed towers for wind machines up to 7 meters (23 ft) in diameter use masts of welded lattice made from steel tube and rod. These masts are popular because they're mass-produced for the telecommunications industry and thus are relatively inexpensive.

Figure 7-7. Guyed lattice towers. Bergey 1500 (foreground) and Bergey Excel (background) on guyed lattice masts at the U.S. Department of Agriculture's test station near Bushland, Texas.

Figure 7-8. Guyed pipe tower. Though a novel configuation for North Americans, the guy cable arrangement used on this Bornay turbine in Spain is common in Europe. Note the work platform. (J. Bornay Aerogeneradores)

They're also produced in a convenient range of sizes, from lightweight sections designed for radio antennas to heavy sections for mountaintop microwave dishes. Tower height is practically unlimited. A guyed tower using a lattice mast can be assembled by bolting sections together vertically, a section at a time, with a tower-mounted gin pole; or the entire mast can be assembled horizontally, on the ground, and raised upright with a crane.

Masts of steel pipe and tube are also popular (see figure 7-8, Guyed pipe tower). Masts of the desired height are assembled from several sections bolted or slipped together. Guyed towers using masts of steel pipe or steel tube are usually assembled on the ground and tipped into place with a crane or gin pole.

The strength of guyed pipe towers is a function of both wall thickness and the fourth power of diameter, says NRG Systems' Dave Blittersdorf. Thus large-diameter, thin-walled tube towers, such as those made by NRG, can be as strong as smaller-diameter, thick-walled pipe.

Towers may use three to four guy cables at each level, and often require two or more levels. Four guys are used where the site or method of erection requires them. Tilt-up towers, by necessity, use four guys at each level.

Under special circumstances, more guy cables and anchors may be necessary. During the early 1980s Fayette Manufacturing used a novel guying layout on its turbines in the soft soils of California's Altamont Pass. Fayette employed guyed towers of steel pipe and anchored the guy cables by driving large screws into the ground. To lessen the risk that

Table 7-1
Guyed Masts for Two Mini Wind Turbines

Manufacturer	Model	Rotor (ft)	Rotor (m)	Mast	Nominal Dia. (in)	Nominal Dia. (mm)	Tower Height (ft)	Tower Height (m)	Guy Levels	Radius (ft)	Radius (m)	Screw Anchor* Dia. (ft)	Screw Anchor* Dia. (m)	Screw Anchor* Length (ft)	Screw Anchor* Length (m)
Southwest Windpower	H40	7	2.1	pipe	2.5	0.064	65	20	4	33	10	0.8	0.25	5	1.5
							80	24	5	40	12	0.8	0.25	5	1.5
Bergey Windpower	XL1	8.2	2.5	tube	4.5	0.114	64	20	3	35	11	0.5	0.15	5.5	1.7
							84	26	4	50	15	0.5	0.15	5.5	1.7

Note: * For normally cohesive soils.

a screw anchor would pull out of the ground, it used two anchors and accompanying guy cables at each of three guy points. (The tower was guyed at only one level.) This lessened the loads on each anchor, reducing the chance that any one anchor or cable would fail.

The loads on a guyed tower that might snap a cable or pull an anchor out of the ground are determined by the thrust on the tower and by the guy radius—the distance from the tower to the anchors. The guy radius is a critical aspect of guyed towers, and is dependent on the site, the loads imposed on the tower by the wind turbine, and the stiffness of the mast. Because guyed towers need a lot of space, there's a tendency to set the anchors too close to the tower. This can affect the dynamics of the tower and the loads it can endure. Consequently some manufacturers precisely specify the guy radius (see table 7-1, Guyed Masts for Two Mini Wind Turbines).

Guy radius is limited by the compressive loads the mast can withstand before it buckles, and by anchor construction. When the anchors are too close to the tower, for example, the mast may buckle in high winds or the anchors may fail. As a rule of thumb, the guy radius shouldn't be less than one-half the height of the tower. The guy radius can be as great as you like. The tension in the guy cables and the compression of the tower continue to decrease as distance from the tower increases. Usually there's no reason for going beyond three-fourths of the tower's height. However, Unarco Rohn, manufacturer of lattice masts for guyed towers, recommends a guy radius of 80 percent of the tower's height.

When the mast is stiff and the tower short, only one guy level may be needed. There are usually two or more guy levels on most guyed towers. One anchor is used to guy all levels. The topmost guy prevents the tower from overturning, and the lower guys prevent the tower from buckling.

Unarco Rohn's 25G lattice mast is suitable for self-furling wind turbines up to 3 meters (10 ft) in diameter. For masts 80 to 110 feet (24 to 33 m) tall, Unarco Rohn suggests three guy levels spaced equally apart, beginning one rotor radius below the top of the tower. This will allow sufficient clearance between the topmost guys and the rotor, while giving the tower the necessary strength and stiffness.

Novel Towers

If you can think of it, it's been tried—trees, silos, and even rooftops—typically with little lasting success.

Rooftop Mounting

Many unfortunate souls have discovered to their dismay that installing an Air 403 on their roof was a good way to destroy a night's sleep. It seems like a fine idea at first glance:

The building gets the turbine above the ground and eliminates the need for a tall tower.

Yes, the old Zenith and Wincos windchargers of the 1930s were sometimes mounted on rooftops. But times and expectations were different then. Ranchers were delighted just to be able to listen to a nightly radio program. They could tolerate the noise of the shiny new turbine on the roof. Today we expect our machines to work quietly in the background. We also know much more about the wind.

Much has been written about the small Jacobs windcharger that was installed in the late 1970s on a tenement in the Bronx. True, it was done once, and it can be done again. But what's the point? The Bronx project was intended as a challenge to Consolidated Edison Company, New York City's utility. It succeeded, and it proved that electricity could be fed back into the utility's network without destroying the city. Later the turbine was removed.

Rooftop mounting of small wind turbines remains controversial. Few topics can stir more heated debate among the small wind community as can rooftop mounting, and notably Southwest Windpower's aggressive marketing of the concept. The Arizona manufacturer of the popular Air series suggests installing its micro turbine on rooftops as a way to compete with the simplicity of mounting photovoltaic panels. Its advertising generates howls of protest from critics such as Wisconsin's Mick Sagrillo.

Southwest Windpower eventually tempered its advertising by suggesting that buyers install the Air turbine on "as tall a tower as possible." Though it never abandoned rooftop mounting, the company now stresses that the building should be unoccupied. Southwest Windpower's founder, Dave Calley, proudly says he lives with four of the machines on his roof.

All wind turbines vibrate, and they transmit this vibration to the structure on which they're mounted. All rooftops create turbulence that interferes with the wind turbine's operation. Even if Southwest Windpower engineers were able to design a sophisticated dampening system that isolated the wind turbine from the structure, they couldn't eliminate the power-robbing and damaging turbulence created by the building.

Worse, a rooftop-mounted turbine can provide a nasty surprise, as an owner in upstate New York learned. One stormy night his Air turbine destroyed itself—and then plunged through his roof. That was the end of his experimentation with rooftop mounting.

As they might say on Manhattan's Lower East Side, "Rooftop mounting? You gotta be kidding me. Fahgeddaboudit."

To avoid rooftop turbulence, the wind turbine must be raised well above the roofline. This often negates any potential savings on the tower and increases the complexity of mounting the wind turbine and installing it safely.

Few who consider rooftop mounting ask whether the building can support the loads created by both the wind turbine and tower. The wooden roofs of homes in North America can't support more than a micro turbine at best. A reinforced concrete roof on a commercial or industrial building might be able to withstand a slightly larger turbine. Can the roof, then, handle the dynamic loads—the vibrations—that the tower will transmit to the structure? If the building is an unoccupied warehouse, the vibrations won't bother anyone, but if it's an office building, they may prove annoying.

Rooftop mounting has been tried and, with the exception of the founder of the company that encourages its use, the technique has been found wanting. It's simply not worth the trouble. As they might say on Manhattan's Lower East Side, "Rooftop mounting? You gotta be kidding me. Fahgeddaboudit."

Silos

Farm silos also seem ideal for a low-cost tower. They're already in place,. usually stand well above surrounding farm structures, and are relatively close to where the power will be used. In the early 1980s Alcoa envisioned strapping its Darrieus turbine to thousands of silos across the vast heartland of North America. The company first found that not all silos were structurally suited to the task. Installation and service, it then learned, would also be difficult. Alcoa abandoned the market, and silo mounting hasn't been used since.

Trees

If you have to mount your turbine on a tree because your site is wooded and you can't afford a tall enough tower, then wind isn't for you. Trees seldom occur right where you would like your tower. Nor is there usually one lone tree that reaches well above all others. The turbine will also be difficult to install and service. For a tree to be of long-term use, it must remain alive. That's unlikely unless you're a skilled arborist like Ian Woofenden.

Woofenden operates two tree-mounted wind turbines on Guemes Island in Washington State: one a Whisper 1000, the other an African Wind Power 3.6. As a professional, he knows what he's doing. Most won't. Woofenden installed climbing rungs, fall protection systems, and guy cables, as well as ensuring that he didn't gird either tree's living cambium. Even then Woofenden found the 3.6-meter (12 ft) sweep of the AWP was more than he felt comfortable with on a treetop looming over his house. (After watching his turbines in a fierce squall, I'd agree.)

For these reasons, "trees don't make good towers," says Mick Sagrillo. "They're hard to climb safely. They sway too much. Dead trees rot and fall over. Enough said."

Farm Windmill Towers

Though certainly not novel, farm windmill towers are ubiquitous in North America. Because of their abundance, there's always a temptation to buy a used water-pumping windmill tower and adapt it for a small wind–electric system. Their utility is limited, however, unless you plan to use the tower as it was intended.

American farm windmill towers are a special case of the freestanding truss tower. Most farm windmills have been installed on light-duty towers. Farm windmill towers typically have a greater taper than the truss towers used for small wind turbines. (The height of a water-pumping windmill tower is proportionally about five times the width of the base. In contrast, the height of a small wind turbine tower is nine times the base width.) This design enables the water-pumping windmill tower to use less steel in the legs and braces and a less substantial foundation than required for a tower supporting a similar-size modern wind turbine.

Farm windmill towers are also short. Most are no more than 40 to 50 feet (15 m) tall, particularly in the western United States. Ken O'Brock, who distributes water-pumping windmills along America's East Coast, says farm windmill towers up to 80 feet (24 m) tall can be found in Amish settlements in Pennsylvania and Ohio. These are exceptions to the rule, and the Amish are not about to part with their towers.

The most commonly used farm windmill is only 8 feet (2.4 m) in diameter. Towers used with these machines are not suited for larger wind turbines. A lightweight tower designed for an 8-foot farm windmill may be adapted to a 7-foot (2.1 m) diameter Whisper H40, but nothing larger. The AWP 3.6, for example, which sweeps a diameter of about 12 feet (3.7 m), presents three times the frontal area of the H40 and needs a much more substantial tower.

Wooden Poles

In North America wooden poles are as commonplace as farm windmill towers. And, like farm windmill towers, they're frequently considered a choice for a cheap tower. They're strong, rigid, and cheap when bought in quantity. Wooden poles can be installed by a crane or utility truck with a special boom.

Wooden poles are classified according to their circumference 6 feet (2 m) from the butt end. Poles of a given class and length are rated to carry approximately the same load. They can handle even greater loads when guyed. Pole lengths suitable for wind systems are found only in Class 4 or better. A Class 4 pole is strong enough for small, self-furling wind turbines up to 5 meters (16 ft) in diameter.

The inexpensive wooden poles used by utilities in North America are too short for most wind turbine applications. Utilities often use poles only 40 or 50 feet (15 m) long. For the minimum 60-foot (18 m) tower height needed by a small wind turbine, a 70-foot (21 m) pole is necessary. Longer poles are available, but the cost rises rapidly with lengths beyond the standard sizes used by utilities. Longer poles are also more difficult to transport.

At one time Northern Power Systems' Jito Coleman promoted the use of wooden utility poles with NPS's HR3 turbine. The idea never caught on, and though wooden poles are abundant, they've never been widely used for wind machine towers. They're difficult to climb, and the heights available are limited to the length that can be conveniently shipped. They're also unsightly, and even if you don't mind their looks, others might—especially the local zoning officer.

Steel Pipe

Well casing or water pipe is a frequent choice for guyed pipe towers supporting micro wind turbines. The strength of water pipe or well casing comes from its thick walls. Steel pipe is readily available and inexpensive. Do-it-yourselfers often choose steel pipe for these reasons. Unfortunately most micro turbine tower kits using steel pipe use it in 20-foot (6 m) lengths that are extremely difficult to handle safely.

Concrete

Yes, concrete. Unlike the wooden poles used in North America, utility poles in continental Europe are often constructed of reinforced concrete. It was a natural step to use concrete for electricity-generating wind turbines. Since the late 1970s wind turbines from the giant multimegawatt models in Europe to household-size versions in the United States have been mounted on concrete towers. Grumman Aerospace installed several of its experimental wind turbines on concrete towers during the 1970s. And German manufacturer Enercon has built a reputation around the sound-deadening qualities of the spun concrete towers it uses. Enercon has installed wind turbines up to 66 meters (220 ft) in diameter on concrete towers in Germany.

Access to the nacelle on the slender concrete towers used by Enercon requires an external ladder. Transport of the massive sections is also difficult, but not much more so than the large-diameter tubular towers that would otherwise be required.

Other Considerations

Sometimes the site or local regulations may limit your tower options. For example, a small lot could preclude a guyed tower. Your choice would then be restricted to a freestanding truss or pole tower. The final choice will be determined by an evaluation of the difference in relative cost, appearance, and ease of installation.

Space

Guyed towers occupy more space than freestanding towers. Normally this isn't a drawback; where it is, guyed towers can be adapted to small lots by placing the anchors closer to the tower. If you must do this, ask the manufacturer or contact a structural engineer to run through the numbers for you. They may find that the tower has an ample safety

Figure 7-9. Guyed, hinged tower. Vergnet's 2 kW prototype at a test field in southern France. The hinged tower is lowered with the attached gin pole and an electrically powered griphoist.

margin with the shorter guy radius. On the other hand, if you're approaching the limits of the mast, you may be forced to use a free-standing tower.

Sites with a lot of traffic, either vehicular or pedestrian, will also limit your choice to free-standing towers. Many a guyed Wincharger tower was felled by a rancher on a tractor. Not only is there the danger of losing your wind system, but the guy cables themselves may also present a hazard to errant snowmobiles, for instance. To avoid tragedy, guy cables should be kept out of traveled ways. And they should be well marked.

Maintenance

Another factor to consider in tower selection is the maintenance of the tower and the wind machine. Will the tower have to be painted, for example? The steel used in most small wind turbine towers is galvanized. To provide corrosion protection, individual steel members or welded subassemblies are dipped into molten zinc. When parts are shipped from the plant, galvanizing presents a shiny silver finish. It soon oxidizes and takes on a dull gray luster. Galvanized surfaces do not require painting or other treatments.

Access to the turbine shouldn't be overlooked, either. The turbine will require at least occasional inspection, if not maintenance. Some provision must be made for getting to it without hiring a crane. Truss towers can be ordered with climbing lugs or rungs. Guyed towers using lattice masts can be climbed using the cross-girts. Tube towers, whether freestanding or guyed, require the addition of climbing lugs if these are not part of the package.

For small wind turbines, hinged towers are advantageous (see figure 7-9, Guyed, hinged tower). Instead of climbing up to the turbine or working from a boom lift, you bring the machine down to ground level. Hinged, tilt-down towers may be safer to work around once they're on the ground, but safely lowering them can be hair-raising if you're inexperienced. More than once installers have found their trucks being dragged across the ground as tower and turbine came down with a thump. To work properly and safely, tilt-down towers should be designed by professionals and used with due care. Griphoist hand winches are ideal for raising and lowering micro and mini wind turbines. For larger turbines, power-assisted griphoists or electric winches may be necessary.

Ease of Installation

Towers should also be easy to install. You may have found an inexpensive tower that will do the job you want, but if it takes a lot of effort to install, the savings may be offset by greater labor costs. If the tower sections are so large that you need a crane or other heavy equipment just to get them off the truck, you'll need an experienced crew with equipment. You pay less for handling and shipping when the sections can be moved around by hand.

> Griphoist hand winches are ideal for raising and lowering micro and mini wind turbines.

When considering how a tower for a small wind turbine is installed, also look at how the wind machine will be mounted on top of the tower. A tower adapter, sometimes called a stub tower, is needed between the tower and the wind turbine. This isn't a concern with manufacturer-designed tower-and-turbine kits, but it is when you want to use a nonstandard tower. Also consider how to ensure that your tower is plumb once installed—that is, that the tower is vertical. If it isn't, the wind turbine may not yaw freely.

Tower Selection

Start by choosing as tall a tower as practical. Mick Sagrillo tells his workshop students that a tower two to three times the height of the nearest obstruction is optimal for a cost-effective tower. Don't skimp. For micro turbines, such as Southwest Windpower's Air models, a suitable tower can easily cost as much as, if not more than, the turbine itself.

Because towers add so much to the total cost of small wind turbine installations, do-it-yourselfers often try to cut corners by building their own tower or using secondhand materials. Unless it's for a micro turbine, where the forces are limited, avoid economizing on the tower. It usually leads to trouble. Dave Blittersdorf, the designer of NRG's tilt-down tower systems, argues that in the long run it's cheaper and more reliable to buy the wind turbine and tower as an integrated package.

If you plan to use a tower not recommended by the manufacturer, here are a few questions to consider. Has this type of tower been used for wind machines by anyone else? If so, how has it performed? What, if any, were the problems encountered? What does the manufacturer think? Will it honor its warranty? If not, why not?

If you will raise the tower yourself, plan ahead. Carefully considering how to safely raise the tower with the least fuss will avoid big headaches later.

Now let's put it all together. Let's assume you want to install a mini wind turbine such as either Southwest Wind Power's Whisper H40 or Bergey Windpower's XL1. Unlike the past—when manufacturers offered a bewildering variety of tower choices—your options today are far more limited. Both companies offer tilt-down tower kits for ease of erection (again, see table 7-1, Guyed Masts for Two Mini Wind Turbines). Bergey uses NRG's nesting sections of thin-walled tube. Southwest Windpower requires the user to buy thick-walled pipe locally. Both companies suggest screw anchors. If your soil conditions are too sandy or swampy for screw anchors, both provide details on how to bury the screw anchors in a block of concrete.

Though both manufacturers offer shorter towers, wind turbines of this size should be installed only on towers 60 feet (18 m) tall and greater. Mick Sagrillo recommends towers at least 60 feet tall for micro turbines, at least 80 feet (24 m) tall for turbines up to 1,500 watts, at least 100 feet (30 m) tall for household-size turbines up to 10 kW, and 120 feet (37 m) tall for 20 kW units and larger.

A final caveat applies to homeowners and professional wind developers alike. If at all possible, use a tower with a pleasing appearance. It isn't enough to look around and point an accusing finger at other obtrusive objects on the horizon that have found acceptance, such as utility poles and transmission towers. Wind systems shouldn't be an embarrassment to the community. You'll be happier in the long run—and there will be fewer objections from your neighbors—if the tower you select is aesthetically pleasing.

8

Cutting Costs, Not Corners

> I have not learned the doctrine that cheapness is the only thing in the world we are to go for. I do not believe that the great object of life is to make everything cheap.
>
> **—Senator Teller, during the debate on the Sherman Antitrust Act, *U.S. Congressional Record* 2561 (1890)**

"You'll never get there from here," said the rancher in disbelief. "Not without a four by four."

"We'll give it a try, anyway," said the easterners.

After an hour of fruitlessly searching the banks of the Powder River for a rumored windcharger, they were about to give up when they impulsively decided to follow a hunch. "If you were a homesteader, and you were settling this land in southeastern Montana, where would you plant yourself?" they asked themselves.

They pointed their small truck toward the horizon beyond the breaks and set off. Heading across the dry rangeland, they bounced over pungent sagebrush and crashed down steep arroyos, wheels churning in the sand. Soon they could see an old farm windmill in the distance, and then a sod house came into view. Finally the object of their search appeared, the pot of gold at the end of the rainbow—an old Jacobs windcharger. There it sat on a rusting tower, a "home light plant" with a shed full of glass batteries at its base and a shoulder-high pile of antlers stacked nearby. It was absolutely still, and then a slight breeze caused the old mill to creak. A coyote howled in the distance.

That's how some of us in today's wind industry started our careers in the 1970s, leading expeditions to remote parts of Montana in search of once abundant windchargers (see figure 8-1, End of the hunt). To bypass the high cost and poor reliability of the wind machines then being produced, we scoured the Great Plains, tracking down, buying, and rebuilding the windchargers of another era. It was a colorful and exciting period in America's reacquaintance with wind energy. Those days are gone forever. Cost-conscious farmers in North America are more likely to buy a used wind turbine from a California wind farm and have it delivered to their door than to go off in search of elusive windchargers of the past. Fortunately today there are other sources for used machines, as well as other ways to cut the cost of buying and installing a wind power system.

The wind turbine itself comprises one-third to one-half the total cost of a micro wind system, and about two-thirds of the cost for household-size machines and

Figure 8-1. End of the hunt. The author sitting atop an old Wind King in July 1977 just prior to lowering the turbine to the ground. (Don't try this at home!)

> If you must build your own wind turbine . . . keep it small and keep it simple.

larger. Towers become a less significant percentage of the total cost as wind machines increase in size. Installation can account for anywhere from as little as a tenth the total cost of a big turbine to as much as a third of the total cost for a small machine. Because the wind turbine accounts for so much of the cost, homeowners frequently try to save money by building their own.

Building a Small Wind Turbine Yourself

If you're thinking about building your own wind turbine, ask yourself some hard questions. First, why? Because you like to work with your hands? Or is your intent to save money? Tinkering is a valid reason. You'll learn a lot about natural forces, mechanics, and Murphy's Law: If anything can go wrong it will. If you want to build your own wind machine as a economical alternative to buying a commercially available wind system, think again.

It can be done. You need only visit Denmark's Jutland Peninsula for proof. But very few home-builts outside Denmark work reliably. And often home-builts are more expensive than expected. They can also be dangerous. Manufacturers with teams of competent engineers have a hard enough time keeping their turbines operating. You must be exceptionally talented to do better with fewer facilities and no technical support. Building a wind machine that will work reliably and safely is beyond the skills of most homeowners. There are already too many home-built contraptions standing as derelict monuments to the mistaken belief that anyone can build a wind turbine.

This warning applies to towers, as well. Many try to reduce costs by building their own towers or using whatever just happens to be lying around. Water-pumping windmill towers, for example, are unsuited for all but micro turbines. By building your own tower or using one unsuited to your wind generator, you may not only be shortening the life of your turbine, but also endangering yourself and your neighbors. As Murphy himself would cynically say, "There's never enough money to do it right, but there's always enough to do it over."

The problem for home builders is dynamics. It's not too difficult to calculate the static loads operating on the wind turbine and tower. Anyone with a little background in math and physics can master the equations. Figuring out how to deal with the dynamic loads is altogether more involved. There are several rotating or moving components that make up a wind generator and tower: rotor, transmission (where one is used), generator, yaw mechanism, and tower. The interaction of all these moving components in varying winds is almost unpredictable. Dangerous harmonics can develop among these components, causing dynamic loads to exceed the static loads for which they were sized. Design teams try to predict when these harmonics will occur and how to prevent them from doing damage. You, however, will be doing it solely by the seat of your pants.

If you must build your own wind turbine, choose a design where the forces involved are manageable, such as designs featuring sail wings. Keep it small (less than 3 meters, or 10 feet, in diameter) and keep it simple (use furling only; leave variable pitch to the pros). Blades using sails are inexpensive, can be made in any size, and are easy to work with.

DIY Brake-Drum Windmill

Scoraig Wind Electric's Hugh Piggott offers plans for turning a used brake drum into a 7-foot (2.1 m) diameter wind turbine that can churn out 300 to 400 watts. The brake-drum windmill is a proven design that Piggott has operated at his remote, windswept headland in northwest Scotland since 1993.

The beauty of these plans can be found in Piggott's use of conveniently handy scrap-yard parts. The design is based on the rear brake drum used by Ford trucks widely available in both Europe (transit van) and North America (F-250).

Another plus is Piggott's elimination of slip rings and yaw bearings. Slip rings bedevil the design of many small wind turbines as well as home-builts. They're not necessary, and are seldom found on many wind turbines built in Europe. Piggott wisely avoided them, instead substituting a simple pendant cable. This greatly simplifies the design, as does his use of pipe on pipe for a simple and hardy yaw system that allows the turbine to respond to changes in wind direction.

The brake-drum windmill also incorporates the durable "inside-out" alternator design found in several popular—and successful—small wind turbine designs. With this alternator configuration there's no need to build a complicated hub that attaches the blades awkwardly to a small-diameter shaft. Instead a simple plywood sandwich holds the blades tightly to the rotor and this assembly is mounted directly to the generator housing: the brake drum. In small wind turbines it doesn't get more straightforward than this.

And like all reliable small wind turbines today, these plans use self-furling to protect the product of your labor in high winds. Piggott's an expert on this technique to limit the speed of the wind turbine's rotor. The simple design found in these plans would be helpful to not a few commercial wind turbine companies that haven't quite mastered the art.

The plans include instructions for building the entire wind turbine, from carving the blades to building a permanent-magnet, direct-drive alternator. As such, Piggott's brake-drum windmill is for hard-core aficionados who know their way around the shop.

Brake-drum windmill. Plans for this do-it-yourself wind turbine can be found on the Internet. See the appendix for details. (R.A.M. Design)

Figure 8-2. Home built, but well built. This locally built 22 kW Smedemestermølle has operated more than 20 years on the west coast of Denmark near the Folkecenter for Renewable Energy. Note the work platform and railing. The rotor is 10 meters (33 ft) in diameter.

They are also unlikely to fly apart and cause damage or injury. Under severe loads (high winds), the sail cloth simply tears away, leaving the rotor intact.

Plans for a Cretan sail windmill can be obtained from the National Centre for Alternative Technology in Wales. Two other sources for competent do-it-yourself plans are those by Scoraig Wind Electric and Kragten Design (see the appendix for details.)

Smedemestermølle

While do-it-yourselfers in Great Britain and North America have typically built small wind turbines of only a few kilowatts or less, backyard inventors and grassroots activists in Denmark at one time experimented with turbines considerably larger: 5 to 6 meters (16 to 20 ft) in diameter capable of 5 to 10 kW.

Danes already knew that wind energy worked, and they already knew what approach worked best for them. They shared a common cultural heritage of how their nation had used wind-electric turbines in the past. Moreover, their famous Gedser mill was still atop its tower south of Copenhagen, exhibiting the traits that would later become known as the "Danish" design; three blades; upwind, active yaw; and pitching blade tips for overspeed control. There was also another cultural advantage. Danes still valued craft skills. There were many small metalworking shops with highly skilled craftsmen throughout the country. These master mechanics, *smedemester,* knew how to build machines.

In the mid-1970s to early 1980s experimenters, hobbyists, and small metalworking shops began building wind turbines designed to supplement electricity from the local utility by driving induction generators. The wind turbines gradually grew in size: first from 7 to 15 kW, then to 22 kW, and from 22 to 30 kW. By 1982 they reached the then incredible size of 55 kW.

During this period, organizations such as the Folkecenter for Renewable Energy circulated plans that relied on readily available industrial components, such as generators and gearboxes, that made local assembly possible. Suppliers arose for generic wind turbine blades and other components suitable for these turbines and the machines then being built by the burgeoning commercial manufacturers. (Blades are one of the critical components that often torment backyard tinkerers.) With a set of plans, blades, and a skilled *smedemester*, wind turbines could be built anywhere in the country. Many of these early wind turbines are still operating today (see figure 8-2, Home built, but well built).

Buying a Used Wind Turbine

Though building your own wind turbine isn't a realistic way for most people to cut costs, buying a used wind machine could be beneficial. It can be fraught with risk, as well, but buying used wind turbines gives you a better starting place, assuming you buy a reliable machine. Whether it's a used contemporary wind turbine or a used wind-

Secondhand Wind

It takes one tough and determined hombre to make secondhand wind machines work. Ask Bill Young of Medicine Bow, Wyoming. He knows from experience. Young bought Hamilton Standard's multimillion-dollar WTS4 from the U.S. Bureau of Reclamation in 1989 for only $20,000. He then invested another $80,000 to rewind the 4 MW generator and get the turbine back into operation, for a total cost of $40/kW. Young then ran the machine single-handedly for 13 months beginning in November 1992. During this period the two-blade turbine generated 7.4 million kilowatt-hours—as much as it had produced during four years of operation in the 1980s—more than paying off his investment. Unfortunately in January 1994 the turbine destroyed itself when the blades struck the tower. It took Young nine months to fully bring the giant beast under control. He finally had to hire ironworkers from Casper to corral the dangerously wayward rotor.

Bill Young, one tough hombre.

Like a rejected lover on the rebound, Young sought another machine to keep his wind power dreams alive. In the spring of 1995 he bought a used—very used—Nordtank 65 kW in Palm Springs for $15,000. That was just the beginning of another saga. While a tractor-trailer carried the nacelle and tower to Wyoming, Young hauled the three fiberglass blades back with an underpowered truck and trailer, breaking down so often along the way he felt like the Joad family in *The Grapes of Wrath*. He then invested another $39,000 to rebuild the Flender gearbox and install the turbine. But he eventually put the old Nordtank back online in the fall of 1996. It stands today as a testament to Young's determination.

Bill Young's secondhand Nordtank. In the background is the abandoned WTS4, a multimegawatt monstrosity from a failed research program in the1980s.

Figure 8-3. Pot of gold at the end of the rainbow. A pre-REA Jacobs windcharger on an abandoned homestead in southeastern Montana, 1977. (A sheepherder's wagon is visible in the foreground.) The Jacobs windcharger is the most sought-after pre-REA wind machine.

charger, you can find out how well it worked in the past and what problems you can expect, rather than starting from scratch with a home-built.

Today there are several sources of used machines in North America: 1930s-era windchargers, small household-size turbines that were built during the late 1970s and early 1980s, and the much larger turbines used in California. Unfortunately there was a lot of junk built for all three markets. There are also suppliers of used wind turbines in Denmark and Germany.

Used Windchargers

During the 1930s, when only 10 percent of U.S. farms were served by central-station power, literally hundreds of thousands of small wind turbines were in use on the Great Plains. These "home light plants" provided the only source of electricity to homesteaders in the days before the Rural Electrification Administration brought "high-line" electricity to all.

The market for small wind turbines blossomed in the 1930s as crude "crystal" radio sets were rapidly replaced with more powerful—and power consuming—radios using vacuum tubes. Batteries initially met the need, but batteries needed frequent charging. The solution was the "windcharger." Through skillful promotion by Zenith Radio and wind turbine manufacturer Wincharger, small radio chargers began sprouting from rooftops across the plains states.

Radio advertisements soon whetted the homesteader's appetite for electrical appliances and power tools. In response, windchargers grew steadily in size; by the end of the era Wincharger was building turbines 14 feet (4 m) in diameter. Jacobs Wind Electric, another grand old name of the era, at its peak employed 260 people building what Marcellus Jacobs, in a bit of marketing hyperbole, called the "Cadillac of windchargers."

In response to the oil embargoes of the 1970s, many starry-eyed idealists roamed the American plains buying these long abandoned windchargers. (Some of the budding windsmiths who once reconditioned these junk windmills were tending wind power plants in the 1990s.) Through their efforts, hundreds of the old machines were put back into operation. These scavengers left behind only a few of the pre-REA windchargers, but at least one was visible off a remote Montana highway in 1999. Although the probability of finding an old windcharger hidden in a shed on the outskirts of some dusty cow town is now remote, hunting for one can still be a rewarding experience.

Many areas of the United States were not served by utility power until President Franklin Roosevelt's REA brought federally subsidized power to the hinterlands. Some regions didn't receive power until well into the 1950s. As the rural electric cooperatives extended service to more and more remote locations, the home light plants previously used were no longer needed. Often they were taken down and sold for scrap. Some were sold to neighbors who had not yet been "electrified." Others fell to the ground as rust, disrepair, and violent storms took their toll.

In the regions where they were once used, windchargers can be found almost anywhere: packed away in the back room of an old store, buried beneath a farmer's junk pile, or hanging in the barn. Some have been found still in their original crates. All of the easily accessible generators have been bought.

Not all windchargers were created equal (see figure 8-3, Pot of gold at the end of the rainbow). By far the most desirable—and the most valuable—of the old windchargers is the Jacobs home light plant. It was the most reliable and one of the largest generators built during the period. The most common brand, on the other hand, was Wincharger, the "Chevrolet" of pre-REA wind turbines. The Wincharger, though much maligned, can be profitably rebuilt and used in remote power systems, according to Mick Sagrillo, who has rebuilt his share of windchargers. More importantly, there may be some Winchargers still available (see figure 8-4, Rebuilt Wincharger).

During the heyday of salvage operations in the mid-1970s, this brand was often passed over in deference to the more valuable Jacobs generator. Wincharger produced a wide variety of models. The early ones used a two-blade wooden rotor counterbalanced by dual air brakes or buckets. These models ranged from 6 feet (1.8 m) to 12 feet (3.7 m) in diameter, with power ratings from 200 to 1,250 watts. The smaller Wincharger models were direct drive: The blade was bolted

Figure 8-4. Rebuilt Wincharger. The Wincharger was a popular brand of home light plant. This 14-foot (4.3 m) diameter, 1.5 kW version was produced into the 1950s.

directly to the shaft of the generator. Later models were more sophisticated and sported four extruded aluminum blades 14 feet (4.3 m) in diameter. Rather than air brakes, this model varied the pitch of two blades (not all four) to control rotor speed via two heavy weights in the governor. The rotor drove a 1- to 1.5-kilowatt DC generator through a single-stage transmission. Most models were painted yellow and used a triangular tail vane that distinguished them from water-pumping windmills and the Jacobs windcharger.

Jacobs took a different approach to building a home light plant, and the appearance of its turbines reflects this. Jacobs's generators used a

Figure 8-5. Rebuilt Parris-Dunn. The vertical furling of the Parris-Dunn was a radical concept among 1930s windchargers. Northern Power Systems borrowed the concept for its HR series of modern windchargers. Note the spring used to regulate furling sensitivity.

rotor with three wooden blades that spanned 13.5 feet (4 m) in diameter. Rotor speed in high winds was controlled by varying the pitch of all three blades simultaneously. Early models used what Jacobs labeled the flyball governor. Later versions used a blade-actuated governor that took advantage of the centrifugal force acting on the blades themselves to change blade pitch. The tail vane is also distinctive.

Unlike Wincharger, Jacobs didn't use a transmission. Its direct-drive generator is a massive affair of copper and iron. The turbine with blades, governor, generator, and tail vane weighs 500 to 600 pounds (225 to 275 kg), depending on the model. The generator came in several versions. Most were 32 volt, but a late model introduced to compete with REA generated 110 volts DC. Early models were rated 1,200 to 1,500 watts, with the last version reaching 3,000 watts. The Jacobs generator is the most sought after because it was built to survive the elements—and has. Some Jacobs generators have been running for more than 30 years. Sagrillo found one on a farm in southern Minnesota that had been in operation for 60 years. Its reputation is well earned, and it is the best windcharger of the pre-REA era you can find. In fact, these old Jacobs are superior to many modern small wind turbines.

Another windcharger worth seeking is the Parris-Dunn (see figure 8-5, Rebuilt Parris-Dunn). It used a two-blade, wooden rotor 6 to 12 feet (1.8 to 3.7 m) in diameter. Very few of the larger models still exist, but you may stumble across one of the smaller ones. The Parris-Dunn was unique because of its approach to limiting rotor speed in high winds. Like the Jacobs, it too was direct drive. But the blades were bolted rigidly to the shaft of the generator. They didn't change pitch. In high winds the hinged generator and attached rotor gradually tilted vertically out of the wind. When the winds subsided, the rotor–generator combination fell back toward the horizontal running position. Though extremely simple, this system worked reliably. They're also easy to rebuild—a good reason to pick one up if you have the chance.

Ranchers, like everyone else, don't like to be rushed. Resist the urge to stuff a wad of bills into the owner's hand and haul away your windcharger. That machine you covet has probably been on the family homestead for 40 years or more, and the family might like to mull over the deal. The rancher's father may have installed that windcharger, and he might want to "leave it right there where Dad put it." Bargaining is half the fun of buying, and good bargaining requires consideration of more than your own interests.

Even after buying the generator, you are still a long way from erecting it in your backyard. First, you have to get it off the tower. To do that you'll need pulleys, rope and cables, a gin pole, a safety harness, hard hats, utility belts, and assorted tools. Add to that list an adventurous spirit tempered with caution.

Lowering a wind generator from a tower is dangerous work. There's no chance to test the equipment under full load until the generator swings free. Everything must work then, or else. An excellent source of information on how to both remove and rebuild windchargers is Michael Hackleman's *The Homebuilt Wind Generated Electricity Handbook.* Hackleman devotes a whole chapter to the mechanics of raising and lowering windchargers.

Lake Michigan Wind & Sun usually has a few rebuilt windchargers in inventory, and advertisements for these and later wind turbines can be found in *Home Power* magazine (see the appendixes for details).

Used 1970s Machines

The market for small wind turbines in the United States collapsed in the mid-1980s and with it the fortunes of many small wind turbine companies. Only one, Bergey Windpower, survived the tumult.

Altogether there are 5,000 to 6,000 turbines from this era still operating in the United States. Most, about 80 percent, are interconnected with local utilities. The remainder charge batteries. In small turbines, the names to look for are Bergey and Enertech—in that order. Bergey Windpower built wind machines suited for both battery charging and utility power. Enertech turbines used induction generators and are unsuited for battery charging.

Bergey turbines use three flexible fiberglass blades to drive a purpose-built, permanent-magnet alternator. The Bergey turbines orient themselves upwind of the tower with a tail vane, and their yellow-and-white paint scheme is easily distinguishable from other turbines of the period. Any of the Bergey turbines are worth salvaging. Bergey's principal competitor was Enertech.

Enertech's design approach, though extremely popular in its day, was far less successful over the long term than the more prosaic Bergey. Enertech used a three-blade, downwind rotor to drive an off-the-shelf induction generator through a small gearbox. The downwind rotor characterized the high-tech, aerospace designs then prevalent. The company's most popular models were the 1500 and 1800, which used a rotor 13 feet (4 m) in diameter. While never a particularly robust machine, the turbine is fairly easy to rebuild. Even the larger 4 kW and 5 kW models, says Sagrillo, can be profitably rebuilt. Be aware that the tip brakes on the end of each blade are an essential component. It's dangerous to operate the turbine without them.

There were a lot of knock-offs of Enertech's popular models as new manufacturers tried to cash in on the once booming market for small turbines in the early 1980s. Nature's Energy Technology and Fight Bills were two. These turbines were even less reliable than the unreliable Enertechs they imitated. Avoid them. Hummingbird, Whirlwind, Aeropower, Swiss Elektro, and a host of other machines from the 1970s are also best left on the scrap heap.

Jim Sencenbaugh built two models of his small battery-charging turbine that have won high regard. Few are available, and most of those are found on the U.S. West Coast. The Sencenbaughs would be a good find.

Marcellus Jacobs came out of his Florida retirement in the 1970s and with his son Paul launched a new Jacobs Wind Electric Company. They based the business, like its predecessor, in Minneapolis, Minnesota. But the turbines they produced were a far cry from those of Jacobs's heyday. The new Jacobs used a right-angle gear drive—a "hypoid" in Jacobs terminology—alternator, and synchronous inverter. About the only parts of the turbine that resembled the earlier machines were

Figure 8-6. Modern Jacobs. These are modern Jacobs wind turbines on a wind farm in Hawaii in 1987. Many Jacobs wind tubines in Hawaii and California have been removed and are now circulating on the used turbine market.

the ungainly tail boom and articulating blade governor. The company was first sold to Control Data Corporation and then eventually liquidated. Still, several thousand units were produced. Nearly 1,000 were installed on wind farms in Hawaii and California (see figure 8-6, Modern Jacobs).

Used California Turbines

Just as salvaged windchargers from the 1930s once fed a thriving market for wind systems in the 1970s, used turbines from California could become the backbone of low-cost wind power for farmers and ranchers on North America's Great Plains. Already, used wind turbines from California have found their way to small wind projects in Minnesota, Iowa, North Dakota, and the Canadian province of Alberta.

There are a lot of machines to choose from. There were more than 13,000 first-generation wind turbines installed in California during the early-1980s wind rush, many of which are still standing. Altogether there's 1,600 MW of aging wind turbines in California. These machines have seen continuous heavy-duty use, most for nearly 20 years. Some are unsalvageable and suited only for the scrap heap. Those that are reusable will need a a lot of work before they're put back into service. Some of these may also need design modifications to improve their reliability and safety. Even so, they could be worth the trouble.

Danish and American Designs

As a rule, many first-generation American designs from the 1980s can't be salvaged, while most Danish designs can. The reasons can be seen in the physical differences between them and in the way they operate. Danes placed the rotor upwind of the tower (see figure 8-7, First-generation Danish wind turbines). Americans loved downwind configurations. Danish wind turbines appeared far more massive than American machines: Their blades were thicker and their nacelles bulkier than those on U.S.-designed turbines. Danes operated their turbines at relatively slow speeds. American turbines operated at high speeds. In short, Danish designs projected a solidity and durability missing from the frail, frantically flailing U.S. turbines of the period.

The difference in appearance betrays an underlying difference in design. First-generation European wind turbines—the nacelle and rotor—weigh up to three times more than American designs relative to the area swept by their rotors. For example, Vestas's V17 and U.S. Windpower's model 56-100 intercept about the same area of the wind

Figure 8-7. First-generation Danish wind turbines. Shown here are several 65 kW Bonus (foreground) and several Nordtank (background) turbines near Tehachapi, California, in early 1986. For scale, note the truck. Both turbines use 7.5-meter (25 ft) Aerostar blades. In the pantheon of wind turbines in California, Bonus, Vestas, and Windmatic turbines are preferred, followed by Nordtank and Micon. Almost any Danish wind turbine is superior to an American-built wind machine from the same era.

California Wind Turbines

There are a large number of salvageable wind turbines in California. Nearly any Danish turbine could be a good find. There are literally thousands of Danish turbines in the 15- to 16-meter (50 to 53 ft) size class that will eventually be replaced by larger turbines. These Danish turbines, while not indestructible, could see productive second or even third lives when rebuilt and operated with proper care. There are also 300 German turbines (Aeroman) that would be suitable for relocation.

Jacobs, Enertech, and the Carter 25-kilowatt model are also salvageable. Nearly 1,500 units built by these American manufacturers were at one time operating in California. U.S. Windpower's 56-100s may also be usable, but they are more maintenance-intensive than the Danish machines.

Unless you're in the junk business, stay away from Fayette, Storm Master, Century, Windtech, ESI, Windshark, Wenco, Polenko, Bouma, and the Carter 300. These wind turbines are more trouble than they're worth and have consistently and substantially underperformed the Danish turbines. Many have already been scrapped.

Aeroman. These first-generation German wind turbines have performed reliably in California since 1985. Unlike their Danish counterparts, which use three fixed-pitch blades, these 40 kW turbines use two variable-pitch blades.

First-Generation Wind Turbines Originally Installed in California—Suitablility for Reuse (Sorted by Rotor Diameter)

Model	Country of Origin	Orientation	Number of blades	Nominal Rotor Diameter (m)	Nominal Rotor Diameter (ft)	Capacity (kW)	Units
Preferable							
Aeroman	D	u,a,m	2	12.5	41	40	320
Bonus	DK	u,a,m	3	15–16	49–52	65–100	640
Micon	DK	u,a,m	3	15–16	49–52	65–75	720
Nordtank	DK	u,a,m	3	15–16	49–52	60–75	1,130
Vestas	DK	u,a,m	3	15–17	49–52	65–90	1,700
Wincon	DK	u,a,m	3	15–16	49–52	65	100
Windmatic	DK	u,a,m	3	14–17	46–56	65–95	340
							4,950
Salvageable							
Jacobs	US	u,a,tv	3	7–8	23–26	18–20	630
Carter	US	d,p	2	10	32	25	350
Enertech	US	d,p	3	13.5	44	40–60	550
USW 56-100	US	d,p	3	17.6	58	100	4,000
							5,530

Notes: Orientation: u, upwind; d, downwind; p, passive; a, active; m, mechanical; tv, tailvane
First generation: defined as 50 kW and derivatives less than 100 kW, less than 19 m in diameter.

Figure 8-8. U.S. Windpower 56-100. This 100 kW American-built wind turbine is potentially salvageable but also notoriously more maintenance-intensive than a comparable Danish machine. For example, note that the blades have been removed from one of the turbines.

Figure 8-9. Carter 25 kW. Because of its relatively small size—10 meters (32 ft) in diameter—and because it uses a hinged tower, this early American wind turbine is servicable by anyone familiar with farm machinery.

stream, yet the specific tower head mass of the V17 contrasts sharply with that of USW's 56-100: 27 kg/m^2 versus 10 kg/m^2 (see figure 8-8, U.S. Windpower 56-100). At one time there were some 4,000 of USW's 56-100 operating in California. Carter and Storm Master turbines were even flimsier than USW's 56-100, but there were fewer of them.

Lightweight U.S. Designs

Few wind turbines epitomize American lightweight design philosophy better than those designed by the Carter family (see figure 8-9, Carter 25 kW). They built two models, a 25 kW and a 300 kW, both of which used a flexible downwind rotor. As in other American wind turbines of the period, the drive train was highly loaded. For example, the 300 kW model was driven by a 23-meter (80-ft) rotor (see figure 8-10, Carter 300). Most other wind turbines of comparable size would limit such a turbine to 200 to 250 kW.

Figure 8-10. Carter 300. The epitome of American wind turbine design of the early 1980s: a downwind, flexible two-blade rotor. Because of its unwieldy size, poor reliability, and limited supply of spare parts, the Carter 300 is unsalvageable. Avoid it.

Figure 8-11. Enertech E44. Slightly larger and less reliable than the Aeroman, this 40 kW American-built wind turbine can be operated successfully when well maintained. Some E44s have been in continuous use for nearly 20 years.

Figure 8-12. Storm Master. Another first-generation American design to avoid. Numerous attempts to salvage these 40 kW turbines have all failed. This is a Storm Master with laminated wood blades near Tehachapi.

Figure 8-13. Windtech (early 1985). One of the most significant failures of the U.S. Department of Energy's research program in the 1980s. Avoid this 52-foot (16 m) diameter turbine. Suitable only for scrap. Note the use of aerial lift.

Oak Creek Energy Systems' Hal Romanowitz says, tongue in cheek, that the Carter turbines "are not a low-maintenance machine." Never one to shy away from giving his opinion, Romanowitz declares point-blank, "They were not a durable machine." The Carter turbines, like those of Enertech, were pushed to their technological limits. "All the parts are highly stressed," Romanowitz adds (see figure 8-11, Enertech E44).

Other turbines characterizing lightweight designs of the early 1980s include Storm Master, Windtech, and ESI. *Storm Master* was a misnomer. Some Storm Masters failed within a few hours of their installation in the early 1980s. This 12-meter (40 ft), 40 kW turbine had only slightly more specific mass (9 kg/m^2) than that of the Carter turbines (4 to 7 kg/m^2). (See figure 8-12, Storm Master.) Nearly 400 of the machines, with their whip-like pultruded fiberglass blades, were at one time installed in California.

One of the most spectacularly unsuccessful designs to evolve from the federal wind program in the United States was that of United Technologies Research Center. This first-generation American design featured a downwind rotor incorporating passive yaw and passive pitch control of its slender pultruded fiberglass blades. UTRC used a torsionally flexible spar that allowed the blade to change pitch toward stall via an ungainly system of pitch weights and metal straps. Though technically elegant, the flexibility of the blades induced complex and ultimately uncontrollable dynamics between the wind and the rotor. Windtech commercialized a 15.8-meter (52 ft) version, and more than 200 of the problem-prone machines were installed in California (see figure 8-13, Windtech [early 1985]).

Figure 8-14. Biting the dust. Scrap ESI turbine buried in the sand of the Whitewater Wash near Palm Springs. None of the ESI turbines is worth salvaging. Note the characteristics of a flawed design: teetered two-blade hub, planetary gearbox, brake on the high-speed shaft (back end of generator).

ESI was another commercial spin-off from the U.S. wind program. The ESI turbines were distinctive for their fairly large diameter, the high speed of their rotors, and the tip brake at the end of each blade. The tip brakes, or flaps, were intended to protect the turbine during high-wind emergencies. The design used two fixed-pitch, wood-epoxy blades downwind of its hinged, lattice tower. Two versions were introduced. Nearly 700 of the ESI-54 (16.4-meter) and 50 of the ESI-80 (24-meter) were installed in California (see figure 8-14, Biting the dust).

Mark Haller, a longtime wind hand who survived California's tumultuous wind rush, warns against using Storm Master, ESI, Enertech, Kenetech (U.S. Windpower), or "anything that isn't Danish." Romanowitz, another old-timer, says that any Danish turbine is salvageable.

What Are They Worth?

California wind turbines have been beaten to death by those who are pros at squeezing every last cent out of a piece of machinery. Some of the turbines have been poorly maintained to boot. To buy a used machine from California, "you shouldn't pay more than scrap metal prices, and that's probably what it's worth," advises wind energy researcher Eric Eggleston. On top of that, "expect to completely rebuild it."

As Eggleston suggests, one way to gauge the value of a used wind turbine is by determining its worth as scrap metal. The scrap value of old wind turbines in California is subject to the whims of the market. Some turbines are worth no more than the scrap value of the metal used to make them. Others contain marketable components such as generators and gearboxes—or towers.

In the late 1990s Oak Creek's Romanowitz was selling used Carter 25 towers. Because the towers were hinged, they could be lowered to the ground with a winch. Several enterprising homeowners in the Tehachapi area used them with the 7-meter (23 ft) diameter Bergey Excel.

American-built ESI 54s and 54Ss were

available in late 1999 "as is" for not much more than their scrap value. The much heavier Danish machines were worth considerably more.

Old Danish turbines also have higher scrap value because the generators and gearboxes are salvageable. There are so many Danish wind turbines still operating in California that there's a ready market for the spare parts they contain.

In 2002 Tehachapi companies were selling used Nordtank turbines "on the ground"—that is, they have already been removed from the tower—for about $5,000. Eggleston says an Enertech E44, in good condition, with tower may be worth as much $5,000 to $8,000, whereas a more sought-after Danish 65 kW turbine with tower could fetch up to $20,000, possibly more. Energy Maintenance Services in Gary, South Dakota, has created a niche for itself by reconditioning Danish 65 kW turbines and adding new controllers. They have been installing the rebuilt turbines on the original 80-foot towers in the upper midwest for about $65,000, including warranty.

Jacobs Designs

The Jacobs turbines of the 1980s don't fit neatly into either the European or the modern American category. They were a throwback to the 1950s. Their only nod to modernity was their abandonment of the direct-drive DC generator and the use of a much larger rotor than their design of the 1930s.

The 1980s Jacobs used a rotor 7 to 8 meters (23 to 26 ft) in diameter to drive an alternator mounted vertically in the tower. Unlike other California turbines, which all used induction generators and could be directly coupled to the grid, the Jacobs alternator required a synchronous inverter to produce utility-compatible power. Mike Edds, who repairs these inverters, says that many of the turbines and towers are still available for the 400 to 600 Jacobs turbines that were once operating.

Dutch Valley Produce

Dutch Valley Produce Ltd. operates a feedlot with 2,000 head of cattle in southwestern Alberta. The Hutterite colony of 20 to 30 residents also operates three used wind turbines and sells the electricity to Trans Alta, the local utility. It bought the three Windmatic 15S turbines in the early 1990s from Wind Power Inc. in nearby Pincher Creek. Eli Walter, who manages the farm, says the group installed the turbines under the Southwest Alberta Renewable Energy program (SWARE), which guaranteed premium payments and made the project feasible. Wind Power Inc. bought the Danish turbines from a wind farm in California.

The Anabaptist sect does minor service itself, but contracts major repairs to Wind Power. "They [the turbines] have their downtime," says Walter, and "they need an aspirin here and there."

The turbines cause no loss to grazing, he notes. He sometimes looks out his window at the machines, wishing the wind would blow. "It's the nicest industry around. In the future you're going to see these things pay well, but it's the unfair playing field that has depressed their use here [in Canada]." If there were a new SWARE program today, he would do it again, only this time he'd opt for the bigger machines that are now available.

Dutch Valley Produce. Two of three used California turbines providing a cash crop for a Hutterite colony in the Canadian province of Alberta.

Like others in the used turbine business, Edds maintains that it's important to know where the machines were used and how they were treated. The "machines at REV [Renewable Energy Ventures] were well maintained," says Edds. "They at least used new parts"—unlike some other operators, who cut corners by using parts salvaged from other turbines.

Yet to Mick Sagrillo they don't justify the freight to Wisconsin. "Why buy a rebuilt Jacobs from California," he asks, when you could get a new one from Wind Turbine Industries in Minnesota for about the same price?

Steve Turek of Wind Turbine Industries warns against using any Jacobs turbines from wind farms in California or Hawaii. "They've worked pretty hard," says Turek in classic midwestern understatement. "It's like buying a used lawn mower from a golf course."

Windmatic 15S—A Workhorse

The total number of Danish machines once in California having found a second life on America's Great Plains is modest by wind industry standards. Most have been installed by independent developers, such as Wind Power Inc.'s Dale Johnson in Alberta, EMS in South Dakota, and Winway's Ty McNeal in Iowa. For many the turbine of choice has been the Windmatic 15S, a 15-meter (50 ft) diameter turbine with a lattice tower.

McNeal calls the Windmatic 15S the "DC-3 of [used California] wind turbines."

McNeal installed his first used Windmatic 15S near Sioux City, Iowa, in 1992. He followed with three of the 65 kW machines for a farm near Britt in 1994 and several more machines in the following years.

Like McNeal, Dale Johnson is still sold on the old Windmatic 15S. If someone wanted to use 1980s-era turbines, "Windmatic would be my choice—without a doubt," he says. They are "a real workhorse." Johnson should know: Altogether he installed about a dozen used machines in Canada, including old Nordtanks. One of Johnson's customers, rancher Erni Sinnott, went through two E44s before he finally gave up on the Enertechs and switched to a used Windmatic.

One reason Johnson and McNeal prefer Windmatics was the Danish turbines' use of an oversize gearbox when some other Danish manufacturers didn't. The gearbox was rated at 150 kW on the 65 kW machine. Another reason was Windmatic's placement of the brake on the drive train's low-speed shaft. This arrangement works much more reliably than placing the brake on the high-speed shaft, such as on the Nordtanks and Micons, two other well-known Danish brands.

Johnson advises that the Stork (Dutch) and Alternegy (Danish) blades used on Nordtank turbines were also not nearly as reliable as the LM blades used on the Windmatic turbines. "The LM [Glasfiber] blades have worked well," requiring few repairs, he adds. The spoiler on the LM blades was less troublesome than the pitchable tip on the Stork and Alternegy blades.

McNeal calls the Windmatic 15S the "DC-3 of [used California] wind turbines"—a reference to the rugged and reliable aircraft of the 1930s. "They're substantial machines," he notes. The Atlantic Wind Test Site's Carl Brothers agrees. The used Windmatic 15S that Brothers installed in 1992 is "one of our more reliable turbines," he says, though they've since added their own controller. Periodically, Brothers adds, they replace the brake pads on the disk brakes. "In general, Danish turbines are serviceable, though there's a lot more hours on the turbines in California now" than when they bought their machine.

Limitations of California Turbines

There are stumbling blocks to using California turbines. One is their generators. Most farms and homes in the Americas have only single-phase electrical service, while the induction generators used on most California turbines produce three-phase. If this limitation could be overcome, the rural market in the Midwest could absorb a large number of these machines, says Oak Creek's Romanowitz.

Another concern is the turbines' operating speed. "California turbines all run too fast for long life in rural applications," says Romanowitz. Danish and German machines were originally designed to run on Europe's 50 hertz (Hz) electrical networks. When used in North America where the electrical system is 60 Hz, the turbines operate at least 20 percent faster than originally intended. In those rural markets where California turbines could be put to use, the wind resources are typically less energetic than those in California's windy passes. For these markets, the turbines could be de-tuned by reducing rotor speed without sacrificing performance. According to Romanowitz, "Nordtanks and Micons are well suited to running day in and day out" if they are de-rated.

Buying a used Danish wind turbine is "like buying a used truck," says Jason Edworthy of Nor'Wester Energy in Alberta. "You can get a dandy one, or you can get a lemon." And he cautions, "Though [Danish] machines are built like tractors, they've got to be maintained." Likewise, Eggleston warns that "nothing lasts forever." California turbines "have so many hours on them that to replace the gear sets alone may cost more than they're worth." And that assumes the parts are available.

Service companies such as Enxco, Haller says, should have access to parts and technicians familiar with Micon and Bonus turbines. Any well-maintained turbine, he adds with the voice of experience, should have a long paper trail tracing its repair and retrofit history. "These [machines] are not cheap, even when used. They should have documentation. If the information isn't there, don't buy it. Where there was no continuity of service, the records will be poor, as will have been the maintenance. Then it's buyer beware!"

Romanowitz argues that these turbines can operate for another 20 years provided they are adapted to more conservative loads. The rural market is a "great place to expand wind power." Yet the salvage market has never taken off, even though there are ample machines available. The reasons are simple and the same ones that stymie wind development everywhere outside a few giant wind farms in the Americas: The price paid for the electricity generated is too low, and there are still too many regulatory obstacles inhibiting interconnection with local utilities.

Used Wind Turbines in Europe

California isn't the only source for used wind turbines. There are several companies selling used turbines in Denmark and Germany. Dansk Vindmølleformidling's Leif Pinholt will ship used Danish turbines from Denmark worldwide. He's delivered several machines to Sweden when their Danish owners upgraded to bigger, more modern turbines. Advertisements for used wind turbines available in Europe can be found in European trade magazines.

Caveat Emptor (Buyer Beware)

There's no escaping it: Buying a used wind machine is risky. There's the risk of accident from working on bulky machinery in awkward places (atop a tower), and the risk of buying a wind machine that may not work reliably. There's no simple way to tell how well a used turbine will perform or what's wrong with it until you've taken it apart. If you're averse to risk, buy a new turbine from a reputable manufacturer.

When buying a used wind machine, whether or not it's from a dealer, the rule is caveat emptor, buyer beware, "and be very careful who you buy it from," says Sagrillo. Still, he adds, "There's nothing wrong with used equipment. I built a business [Lake Michigan Wind & Sun] on it." Sagrillo specifically warns against what he calls "Rustoleum rebuilts" peddled by some less reputable dealers. These are machines that look rebuilt because they have a fresh coat of paint, but are not.

After years of shopping in supermarkets and department stores, few of us have well-developed bargaining skills. Our trust in products offered for sale has grown through extensive advertising and standardization. But not long ago, everyone was haggling with the vendors at the local market. There you examined the goods carefully, decided how much they were worth to you, and began the exchange. It's still that way with used cars. Much remains hidden beneath the hood, so a great deal of faith is placed on the truthfulness of the seller with regard to its inner workings. The same is true with used wind machines. You must make not only a careful examination of the goods, but also a careful examination of the seller. Are the claims reasonable and verifiable?

Another important question to ask yourself is whether the wind turbine you're considering will meet your needs. Will the wind machine be used to heat water, produce line-quality power, or charge batteries? The answer will determine the degree of reliability and the type of generator required. If the wind machine will be used in a remote power system, then it must use an alternator or DC generator, and it had better work reliably. If it's line-quality power you want, then a wind machine with an induction generator may be suitable.

Related to how you plan to use the wind system is why you want it in the first place. If you want a wind machine principally to tinker with (sure, you'd like it to be a paying proposition as well), then you're free to take more risks than if your primary need is a wind turbine that will consistently generate usable energy.

If you buy a used machine, use it as it was intended. Avoid mating a synchronous inverter with a battery-charging wind generator, for instance. It's better to use it for charging batteries or heating domestic hot water through resistance heaters.

The first question to ask the seller of any used wind machine is: Why is it for sale? Is it because the previous owner traded up, died, moved, or simply got fed up? If the latter, what's wrong with the turbine? What caused the problems? What were the headaches? Did the previous owner, for example, tire of climbing the tower every month to change the brushes or tighten some bolt? If so, how do you plan to deal with this problem?

Are parts readily available? Where can parts easily be found? How much do they cost? If parts are unavailable because the company went bankrupt, how difficult will it be to make them? There may be some folks out there making parts for their own units who could easily make a few more, if need be. How will you find them?

The seller should be familiar with these questions and have answers for you or at least tell you where to look for the answers. The classified advertisements in trade magazines or other energy publications are a good place to check for spare parts. If you can't find them there, you'll have a hard time finding parts, period. Does the seller offer a warranty? If so, what kind: the manufacturer's or the seller's own? How long does it last? A used wind machine may be bought directly from the previous owner or through a dealer. By buying the machine directly, you save by cutting out the middleman's markup. At the same time, buying direct forces you to install the turbine and tower yourself. By buying from a dealer you may pay more, but you gain some assurance that the turbine will work and will be installed properly. The dealer can also be more easily held accountable. In a direct sale, the previous owners may want to dump the wind machine and wash their hands of the whole affair as quickly as possible.

Take a hard look at the overall economics. Are you saving enough by buying a used machine to justify the risk that it will work or work as well as it should? Check with local dealers. You may find that it doesn't cost much more to buy a new wind system instead.

The Spirit of Adventure

Imbued with the Holy Spirit and a sense of adventure, a small group of progressive nuns have become a beacon on the windswept plains near Richardton, North Dakota. Though the Benedictine Sisters live a contemplative life, they are also pragmatic.

The 24 permanent residents of the Sacred Heart Monastery wanted to control their utility bill. They also wanted to give something back to the rural community of which they are a part. After learning about the potential of wind energy from their contacts at the North Dakota Resource Center, the sisters thought those seemingly ever-present and often annoying prairie winds could be the answer.

Sacred winds. One of two rebuilt Silver Eagle turbines now powering the Sacred Heart Monastery near Richardton, North Dakota. (Benedictine Sisters)

They then bought two used turbines in neighboring Montana. But as others have found, you can be disappointed unless you know what you're buying and who you're buying it from. The sisters were. Still, with the help of Tad Miller—regional representative for wind developer Enxco—they were able to put both turbines online in the summer of 1997 for $120,000.

"I'd do it differently today," says Sister Paula Larson, the monastery's prioress. "I'd probably go to a company such as Enxco and get one sized to our specific needs." They would also prefer someone who could do the whole job from start to finish. The nuns found themselves acting as a general contractor, a role that was unfamiliar to them.

The Silver Eagle–brand turbines they bought have a checkered history, and the problems the sisters encountered were not surprising. The turbines are an American copy of a Danish Nordtank and were built during the tax-credit-driven wind rush of 1985. Like the Nordtanks, the Silver Eagles use old Aerostar 7.5-meter blades, which give the machines a 15-meter (50 ft) diameter.

The most difficult tasks, says Enxco's Miller, were negotiating a parallel generation contract with a timid local utility, West Plains Rural Electric Cooperative, and installing a new controller that would meet the utility's unnecessarily stringent requirements. "They [the utility] were terrified" of what could happen, he says, and they quite literally "threw the book at us." Ironically, the "book" was interconnection standards for the fossil-fuel plants they understood.

The sisters hoped that the turbines would offset one-quarter of the monastery's consumption. If they did, the machines would pay for themselves in 10 years. The monastery saved nearly 40 percent on its electricity bill in the first year. It did even better the second year: The 240,000 kWh produced by the turbines cut the bill almost in half. "We're way ahead of our projections," says Sister Paula.

Some three-fourths of the electricity generated was used on site, offsetting purchases from the utility at $0.092 per kWh. The remainder was sold back to the co-op for only $0.014 per kWh. "We should say *gave back,*" at that price, says Sister Paula. "It's appalling, from a spiritual viewpoint, that conservation is penalized."

All told, the turbines operate about one-half the time, spinning out electricity for the monastery. The Silver Eagles have performed well for such early machines: capacity factors of 16 to 22 percent and annual specific yields of from 500 kWh/m^2 to 700 kWh/m^2 during the first two years of operation. This approaches the performance of some of the best midwestern sites where modern turbines are producing 700 to 800 kWh/m^2.

"They've been a good investment," says Sister Paula. But there was another equally important reason for the project. The monastery has made a spiritual commitment to help preserve the endangered rural life of the Dakotas, as well as to protect the environment. The wind turbines fulfill both goals.

The twin turbines are visible from I-94, the principal east–west corridor across North Dakota, and have become a major landmark. They've had so much traffic from the highway that they plan to erect a kiosk explaining how they are using the wind turbines, and spread the word that increased use of wind energy would benefit North Dakota's economy.

The sisters' success has dispelled several myths prevalent in a state long known for coal mining and smoke-belching power plants. Critics said that no one in North Dakota could maintain the complicated machines. "We showed that wasn't true," says Sister Paula. The monastery's own staff operate and service the turbines. Others doubted that the turbines would work in the state's extreme cold. Doubting Thomases said they would "freeze up." "They don't," answers Sister Paula bluntly. Some warned that the turbines would be too noisy for the contemplative life. But Sister Paula found that "when the turbines are running, we only hear the howl of the wind. There's been absolutely no disturbance to our life."

The sisters proudly note that using the winds that sweep over the prairie is part of their heritage. A photo from 1916 shows a group of nuns standing in front of their monastery with an old water-pumping mill in the background. Their two turbines are now part of their faith in a rural economic revival based on stewardship, not exploitation.

Assembling a Kit

Kits are one way to cut the cost of micro and mini wind turbines. Kits for the serious wind enthusiast or do-it-yourselfer should contain—at a minimum—a professionally designed turbine. Because the design and construction of the wind turbine is the most demanding of all the components in a wind system, it's best left to those who know what they're doing. All you should be required to do is bolt on the blades and hang the tail (if one is used).

A complete kit should also contain a tower designed for the wind machine (not a tower of lightweight TV antenna masts, as some have advertised), all tower hardware, wiring, conduit, and electrical connectors needed, and a detailed assembly and installation manual. That's a tall order. No manufacturer or distributor presently provides such complete kits. (You're essentially asking for the same service and packaging a wind machine dealer receives from the manufacturer.)

Many of the smaller items, such as the wiring and the conduit, can be purchased locally. But unless you're familiar with the ins and outs of wiring and your local electrical code, you could run into problems. This is where a good installation manual becomes important. It not only tells you exactly what you need to do the job right, but also warns you about problems that may develop and how to deal with them.

Buying a wind system kit is similar to buying a computer through the mail. You can save a significant amount of money, but you don't have ready access to someone who can help you correct a problem. When you buy a wind machine from a dealer, you always have someone to turn to for repairs.

With a professional kit, you're not so much building a wind turbine as you are providing final assembly and installation. For example, Bergey Windpower offers an installation kit, including a guyed tower, for its mini wind turbine, the XL1. The components can be handled easily and installed with a minimum of risk by following the instructions in the installation manual. The wind machine comes assembled, except for the addition of the blades and the tail vane. Because installation accounts for 25 to 35 percent of the total cost for this size of wind machine, assembling and installing it yourself can produce considerable savings. Kits such as these are limited to wind machines less than 5 meters (16 ft) in diameter. The larger wind machines are more difficult and dangerous to install, and manufacturers seldom offer them for owner installation.

Working Together

There's another way for individuals to reduce the cost of owning a wind system: working together. It's an approach common in northern Europe, particularly in Denmark. Collective action is a way for small investors to combine their financial clout. Joining together enables Danes to acquire the most cost-effective equipment possible, whether it's to process cheese, bake cookies, or generate electricity.

Cooperatives—The Danish Approach

Cooperative wind development is a natural outgrowth of Danish cultural and agricultural interests, according to Asbjørn Bjerre of Danmarks Vindmølleforening (Association of Danish Windmill Owners). The objective of the Folketing, or Danish parliament, in providing incentives for wind cooperatives, says Bjerre, was to encourage individual action toward meeting Danish energy and environmental policy. Through this program nearly any Danish household can effectively generate all its own electricity with wind energy. It worked, and the concept has also caught on in Germany and the Netherlands.

Danish law encouraged mutual ownership of wind turbines *(fællesmølle)* by exempting owners from taxes on the portion of the wind generation that offset a household's domestic electricity consumption. A wind co-op would then buy a wind turbine, site it to greatest advantage, sell the electricity to the utility, and share the revenue among its members. This enabled a group to buy the most cost-effective turbine available, even though it may have generated more electricity than any individual member needed.

Kennemerwind

Denmark has no monopoly on cooperatives. Many of the first wind turbines installed in the Netherlands were installed by co-ops, though conditions were far less favorable than in Denmark. One of the largest is Coöperatieve Windenergie Vereniging Kennemerwind. The group operates 10 turbines in all, 9 of which are part of a group of 15 turbines along a canal in Noord Holland.

Kennemerwind has no bank loans or debt. It raises all of its capital from members, most of whom invest as little as 50 euros (about $50), though some have invested as much as 10,000 to 15,000 euros. The co-op views its members' investment as a 15-year loan to be repaid. While the driving force behind the co-op's 650 members is their desire to produce clean energy, Kennemerwind consistently pays an annual dividend of 7 percent to shareholders. Members can reinvest their dividends, and nearly all do.

Kennemerwind has a contract with Nuon, a private utility. This includes a guaranteed tariff of about 0.075 euro per kWh plus a 0.015-euro-per-kilowatt-hour tariff that depends on the price of fossil fuel.

From 1989 through 2002, the co-op generated more than 15 million kilowatt-hours. With the addition of several new Lagerwey turbines in the mid-1990s, Kennemerwind produces from 1.5 million to nearly 2 million kilowatt-hours yearly, enough electricity to meet the needs of 500 to 650 Dutch households. During the life of the co-op, the wind turbines have earned more than a million euros.

Kennemerwind. The four Lagerwey turbines (right) along the Noord Holland Kanaal are part of the Kennemerwind co-op. Each turbine generates 150,000 to 200,000 kWh per year.

Figure 8-15. Tændpibe *fællesmølle* The 35 turbines at Tændpibe near Ringkøbing on the west coast of Denmark's Jutland Peninsula are cooperatively owned by 508 local families. Nearby are 65 similar turbines owned by the local utility.

In the Danish context, many cooperative ventures were assembled as limited liability companies in response to the vagaries of Danish law and tax policy. These associations funded installation of single turbines, clusters of turbines, and sometimes small wind farms. One example of the latter is the Lynetten cooperative that owns four out of seven 600-kilowatt wind turbines on a breakwater within the port of Copenhagen. (The other three turbines are owned by Copenhagen's municipal utility.)

Danish wind turbine cooperatives have had a profound effect on the development of wind energy. Prior to the late 1980s, nearly all wind turbines in Denmark were installed individually or by cooperatives. Nearly all the early wind plants, such as that at Tændpibe near Ringkøbing, are owned cooperatively (see figure 8-15, Tændpibe *fællesmølle*). The 2,100 wind cooperatives in Denmark accounted for about 50 percent of the nation's total installed wind power in the mid-1990s. Some 100,000 Danish households, or nearly 5 percent of the population, own a stake in a *fællesmølle,* or cooperatively owned wind turbine.

> Ringkøbing's Landbobank (farmers' bank) [has] financed so many wind turbines that [it's] been dubbed the "wind farmers' bank."

Until recent utility restructuring, cooperatives were paid 85 percent of the retail price of electricity for their generation. They also received payment for the carbon dioxide tax, and a portion of the electricity tax paid by all consumers on domestic consumption.

One often-overlooked aspect of Danish success with wind co-ops and single-turbine installations by farmers is the backing of Danish banks. As in buying a house or other costly capital investment, buyers seldom pay cash for the entire amount. They finance a portion of the cost. To the envy of many commercial wind plant developers elsewhere in the world, Danish banks and finance societies willingly provided 10-year loans for 60 to 80 percent of the installed cost. Some, such as Ringkøbing's Landbobank (farmers' bank), have financed so many wind turbines that they've been dubbed the "wind farmers' bank."

Danish wind co-ops vary in size from a single turbine to small wind farms. Turbines today are so big that they can produce more than one million kWh per year, requiring more than 100 members per machine. The annual shareholder meeting is often a time for celebration as members gather for a communal picnic.

The success of Danish co-ops and risk-taking farmers can best be seen in the township of Sydthy in northwest Jutland, where winds sweeping across the great Limfjord from the North Sea produce 130 percent of the electricity consumed by the township's 12,000 inhabitants. The area's wind turbines, all cooperatively or individually owned by local residents, are net exporters of wind energy. During blustery spring months, the turbines produce three to four times the electricity consumed locally.

The foundation upon which community wind was built in Denmark was the assurance of a fixed minimum price for the electricity generated. With it, farmers and local residents and—just as importantly—their banks could reliably project an income stream that would justify the investment.

Despite changes in Danish policy toward renewable energy, cooperative action remains an important avenue toward local ownership. Half of the Middelgrunden wind farm's twenty 2-megawatt turbines installed just offshore from Copenhagen were developed cooperatively. (The remaining 10 turbines are owned by the municipal utility.) Organizers believe the project would not have been built without the public support engendered by local ownership. Despite numerous obstacles, the cooperative sold 40,500 shares for 570 euros each, a price set as low as possible to enable broad participation. Altogether, 8,500 investors bought shares.

Sydthy Kabellaug

Cooperatives are just one example of Danish communal action. When Thy Højspændingsværk, the local high-voltage distribution company, refused to connect a group of new wind turbines proposed by farmers in windy northwest Jutland, the farmers banded together and formed Sydthy Kabellaug to build their own high-voltage collection system. The joint-ownership company then laid 16 kilometers (10 miles) of cable to 26 different wind turbines in the township of Sydthy and connected directly to the high-voltage system, bypassing the local distribution company. Most, like pig farmer Niels Mogens Sloth, installed Vestas V27s. In Sydthy this 225 kW wind turbine can generate 600,000 to 900,000 kWh per year, depending on the site. Farmers typically paid about 300,000 euros each for the turbine, installation, and their share of the cable company. The farmers were so successful that the distribution utility has since followed their example and offered its services to other farmers in the area.

Sydthy Kabellaug. Some of the 26 privately owned wind turbines in the jointly owned Sydthy cable company. Most of the 225 kW turbines are owned by the area's pig farmers.

Cooperative Net Billing

Innovative Canadians and Vermonters have attempted other approaches. The Toronto Renewable Energy Co-op had hoped to install two medium-size turbines on the city's lakefront using what board member Jim Salmon calls cooperative net billing. The co-op would deliver electricity from the turbines to Toronto Hydro, the local utility. The turbines' production would then be used to offset the retail rate on co-op members' electricity bills in proportion to the amount of investment each had in the turbine. Instead of co-op members installing small turbines on their own property, says Salmon, the co-op would have bought large turbines and installed them at good sites where they would be most productive. In this way, members gained economies-of-scale by using more cost-effective turbines at more productive sites.

For a host of regulatory reasons, however, the Toronto co-op had to abandon the concept in its infancy. It adroitly changed direction and turned toward an investment co-op as in Denmark, selling shares to its 500 members for the construction of two turbines (see figure 8-16, Toronto's windshare). The first machine, a 750 kW Lagerwey turbine, was installed in a prominent location on the city's harborfront at the end of 2002. The second turbine will follow in a similarly prominent location.

In Vermont renewable energy activists proposed what they called "group net metering" as a way for home and condominium owners to share in the cost of installing a wind turbine sited to best advantage. The Vermont proposal would have permitted condominium owners to install a wind turbine and connect it to the grid. Electricity from the turbine would then equally offset electricity consumed by each condominium. Likewise, in rural areas where homes are more widely dispersed, a group of neighbors could band together and install a turbine, offsetting their individual consumption. As in Toronto, implementation of the concept was thwarted, but the dream remains alive.

Buying Clubs

Until the Toronto Renewable Energy Co-op's WindShare program was launched, the Danish model of wind development had seen little application in North America. Co-ops are not foreign to the continent. There are many agricultural cooperatives in midwestern dairy states and in Canada. And a few rural electric co-ops have installed wind turbines to sell "green power" to their customers.

Cooperatively owned wind turbines that sell their generation to the utility work economically only when the utility pays a sufficient price for the electricity. In much of the United States and Canada, Danish-style cooperatives are hindered by the low price paid by the utility—often only 25 percent of the retail rate—for wind-generated electricity. These low rates force individual consumers to install wind turbines sufficient only for their own consumption, and discourage cooperative ownership.

Buying clubs are a concept related to co-ops that may work in the American context of low utility prices. Buying clubs are more adaptable to the small wind turbines that make the most sense when the rate paid by the utility is so low. A buying club pools money from individuals to buy a quantity of a particular product, in this case small wind turbines. The bulk purchase gives the club more leverage when negotiating price with the manufacturer than that of an individual. A club may be able to save 10 percent, possibly much more, for example, by qualifying for a dealer's discount.

The USDA's Dr. Nolan Clark has long thought such consumer consortiums that buy wind and solar hybrids collectively from manufacturers offered promise. In one government aid program to the Third World, U.S. manufacturers sold their wind turbines at bargain-basement prices. The government agency didn't accept the initial quotes from the manufacturers and used its buying clout to force prices down. If homeowners and other individual buyers could organize multiple-turbine purchases, as the government agencies did in the aid program, they could possibly negotiate the same deals. As

Figure 8-16. Toronto's windshare. The Toronto Renewable Energy co-op installed this 750 kW Lagerwey at a prominent location on the city's harborfront. Dubbed "Windshare," thee turbine is co-owned by 500 individual investors and Toronto Hydro, the municipal utility.

with many innovative strategies, group purchases have yet to materialize.

Wind Plant Co-ops

There are even greater economies possible when installing multiple machines in one location. This is the principle behind building wind power plants. Financiers pool money from multiple sources, negotiate a quantity discount from the wind turbine manufacturer, gain economies-of-scale through multiple identical installations, and negotiate with the utility for a higher purchase rate from a position of greater strength than an individual can muster.

Once again Canadians are trying adapt this technique for use by cooperatives in North America. The Ontario Sustainable Energy Association hopes to expand the Toronto co-op's WindShare model into a consortium of community co-ops to build one 10 to 20 MW wind plant on the shores of Lake Ontario. It may jointly develop this LakeWind project with provincial municipal utilities, like the Danes have done with wind farms in and offshore of Copenhagen.

While unique in North America, German community wind activists have built numerous projects similar to Ontario's proposed LakeWind. At Paderborn in the state of North-Rhine-Westphalia, 91 investors built a *Bürgerwindkraftwaerke* or citizen-owned wind plant containing 11 Vestas 1.65 MW V66s. Most of the shareholders in the 18 MW Paderborn wind plant live nearby. Nearly one-third of all wind capacity operating in Germany is locally owned by similar investment co-ops. In the northern state of Schleswig-Holstein, as many as two-thirds of the wind turbines are owned jointly *(Bürgerbeteiligung)* by those living in the area.

9

Buying a Wind System

He that will not be counseled cannot be helped.

—John Clarke, *Paroemiologia*

Selecting a wind system entails gathering information (as you're doing now), sorting through it (weeding out the hype), and determining which combination of product, manufacturer, and dealer best meets your needs, with the least risk and the best chance of success. Selecting the combination that's right for you is much like buying a car. You don't buy a car solely on what's under the hood or because of the transmission it uses. You look at the complete vehicle. The same is true with a wind machine. You weigh the pluses and minuses of each component, then add the reputation of the manufacturer and, finally, the cost. And with wind turbines, as with most things in life, you get what you pay for.

Most of the following information applies to small wind turbines, but the process described is useful when buying any wind turbine, regardless of size.

Choosing a Product

First, determine the size range that meets your energy needs. Never forget that conserving energy is always cheaper than trying to produce it. Mick Sagrillo advises his workshop students that spending $1 on improving energy efficiency saves $3 to $5 in wind system costs.

In general the cost-effectiveness of small wind turbines increases with increasing size; that is, their specific price in dollars per square meter of swept area decreases as the turbines become larger (see figure 9-1, Small wind turbine specific price).

Keep in mind that it's energy you're after, not power. Avoid the false promise of high power ratings. In strong winds "power is liberally available," says Scoraig Wind Electric's Hugh Piggott, who wants high performance at low wind speeds for his battery-charging systems. Piggott is willing to sacrifice some performance in high winds to win usable amounts of electricity at lower wind speeds, which occur more frequently. Once again, the best indicator of how much energy a wind turbine will capture is the area swept by the rotor—or its surrogate, rotor diameter—and not its power rating.

Next, find the product that offers the most for your money. A complete wind system includes more than just a wind turbine. It includes a tower, as well as the conductors or cables to carry electricity from the wind turbine to your home and all the electrical fittings required to do so safely. On a small wind turbine these fittings can add significantly to total cost.

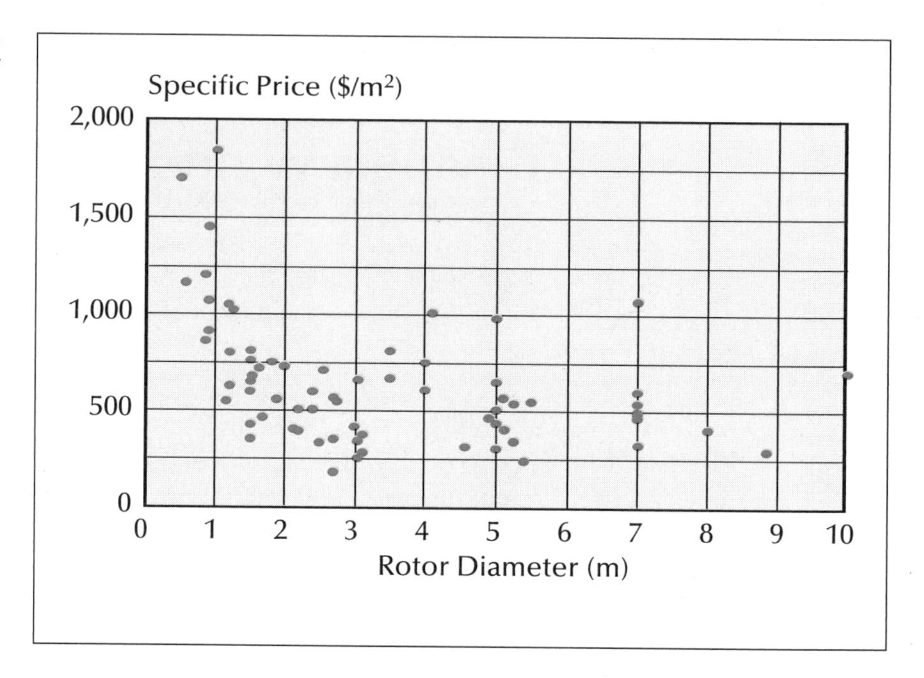

Figure 9-1. Small wind turbine specific price. The relative cost of small wind turbines decreases with increasing size. Though the total cost increases, the cost relative to swept area or per kWh decreases as wind turbines become larger.

Don't limit your product evaluation to energy output. Reliability, maintenance, and the soundness of the firm manufacturing it are each equally important.

To gauge reliability, "ask the man who owns one," says an adage. Track down owners of the same wind machine or those owning models made by the same manufacturer. Ask them how well it has performed. What kinds of problems, if any, developed? Are they satisfied? If they had to do it over, what would they do differently?

Standard Power Rating

For those who insist on using rated power to evaluate small wind turbines—despite frequent admonitions to use rotor swept area instead—Wind Power introduces standard power ratings in the appendix. These ratings assume each wind turbine is capable of generating 200 watts per square meter of rotor area. Thus these ratings will overstate the performance of some turbines and understate the performance of others, especially those with alternators using neodymium magnets. Use these standard power ratings to compare one small turbine with another. Remember, the standard power ratings in the appendixes are based upon swept area and nothing more. "Your performance may vary."

Call the local utility for information, but consider the source when assessing the response. Many utilities view wind turbines as a nuisances—unless they own them. It's unlikely you'll hear a glowing endorsement of independently owned wind turbines from your local utility company.

Above all else, remember, "If it sounds too good to be true, it probably is."

Find test reports on the product you're considering. Some universities, such as West Texas A&M's Alternative Energy Institute, have tested specific wind turbines. Individuals have also tested small wind turbines and published their findings in trade magazines (see figure 9-2, Performance testing). Most testing has been performed by government or private laboratories: the Atlantic Wind Test Site in Canada, the National Renewable Energy Laboratory in the United States, ECN in the Netherlands, Risø in Denmark, Windtest Kaiser-Wilhelm-Koog and the Deutsches Windenergie Institut in Germany.

The Atlantic Wind Test Site on Prince Edward Island is one of the few places that tests North American and European products side by side. The U.S. Department of Agriculture's Bushland, Texas, experiment station conducts

> "If it sounds too good to be true,
> it probably is."

tests of turbines in water-pumping applications. (For a list of test stations, see Government-Sponsored Laboratories in the appendix.)

Standardized Tests

There is international consensus that the only way to properly test small wind turbines is "in the field." Truck or wind tunnel testing is useful to designers for observing and fine-tuning the furling or regulating behavior of their turbines, but not for measuring performance.

Truck testing, in which a small wind turbine is mounted on top of a pickup truck and the truck then driven down an abandoned airport runway, has sometimes been used in the United States. Truck testing is "one tool in the [designer's] toolbox, but only one," says Mick Sagrillo. Both the American Wind Energy Association's performance testing standard and the proposed European standard for small wind turbines exclude performance measurements of wind turbines in wind tunnels or by truck tests. Neither truly replicates how a wind turbine will perform under real-world conditions.

Wind tunnel tests usually overstate performance. This phenomenon was seen several times in the 1970s during development of government-sponsored wind turbines in the United States, notably with McDonnell Aircraft's giromill. Consumers never see the performance measured in a wind tunnel.

AWEA adopted a performance testing standard in 1988. It was specifically compiled for small wind turbines, but the method is similar to that used internationally today to measure the performance of wind turbines of all sizes.

AWEA's standard was ". . . intended to provide consumers . . . with an equitable basis for comparing the energy production performance and operating characteristics" of different wind turbines. It avoided any discussion of reliability or durability, except as it affected testing of the power curve. (You can't measure a power curve if the turbine's not operating.)

At the heart of the standard was a detailed description of how power curves should be measured. Further, the standard required all manufacturers to prepare a "test report" describing the techniques used to measure

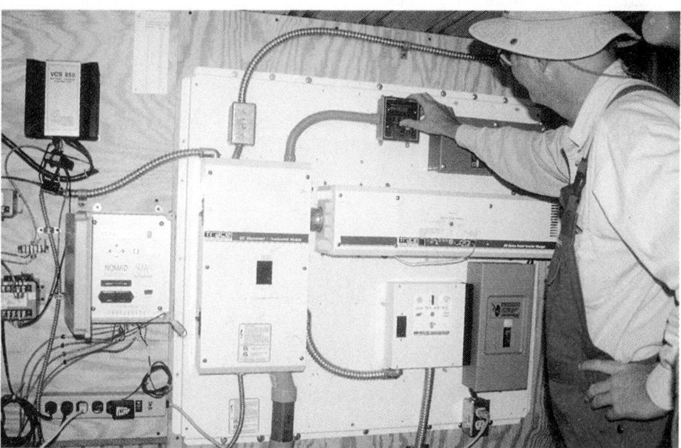

Figure 9-2. Performance testing. *Top:* A boom-mounted anemometer is used to measure the performance of an Air 303 at the Wulf Test Field in the Tehachapi Pass. *Bottom:* A power center and electronic instruments are used to measure the performance of micro and mini wind turbines at the Wulf Test Field: power and current transducers (far left), Second Wind Nomad data logger (left), and diversion controller (center bottom) for dump load.

Much More Than a Turbine

Buying a wind system is much more than just buying a wind turbine. For micro and mini wind turbines, the "balance of system" costs for the tower, cables, switches, and connectors can amount to a substantial part of the total cost. Take the installation of Southwest Windpower's popular Air series, for example. Properly installing the turbine on a relatively short tower close to the load it will power can easily cost more than twice the cost of the wind turbine alone. You can certainly do it for less, but why should you? If you want to do it right, you'll need these components to minimize problems in the long term.

Equipment Cost for Adding Micro Wind Turbine to Existing Off-Grid System

SWP Air model on 45-foot (14 m) tower with 100 foot (33 m) cable run	
AirX	$700
NRG 45-ft tilt-up tower	$700
3 NSI Connector blocks	$30
Wire mesh strain relief	$30
10 ft flexible, liquid-tight conduit	$50
Post for disconnect switch	$15
Fused, rain-tight disconnect/stop switch	$100
50 ft Sch 40 PVC conduit	$30
Fittings	$30
2x100 ft # 8 Insulated Cu conductor	$70
100 ft #8 Bare Cu ground	$30
Ground rod	$15
30 A DC Circuit breaker (for load center)	$15
	$1,815
Turbine/Total	39%

Notes: Assumes tower is tower length from battery location.
Assumes existing approved load center and batteries.

their power curves and the results of their measurements. This report would then be available to the public upon request.

The first draft of AWEA's performance standard appeared in 1979. After a decade of bickering, AWEA's standards committee finally approved it. More than two decades after the standard was first proposed, there were still no small wind turbines on the U.S. market that complied with its provisions or with European testing conventions.

American manufacturers' reluctance to embrace any standards whatsoever goes back to the rebirth of the wind industry in the United States in the mid-1970s. This recalcitrance results from a uniquely American fear that standards stifle engineering creativity.

Of course, standards haven't hurt manufacturers of medium-size wind turbines. Very few, if any, large wind turbines sold on the international market can obtain financing without a performance test conducted by an independent testing laboratory.

According to measurements made at the Wulf Test Field, many small wind turbines won't deliver the performance promised in their product literature (again, see figure 9-2, Performance testing). Sagrillo warns that manufacturers of small wind turbines can be "wildly enthusiastic" about their products. He chuckles when he adds, "Your performance may vary."

To find a compendium of measured power curves on both large and small wind turbines, buy a copy of the Bundesverband Windenergie's Windenergie Marktübersicht (Market Survey). BWE's annual survey lists product specifications for a host of wind turbines, including power curves from certified testing laboratories worldwide, noise measurements, and electrical characteristics such as power factor and voltage flicker. Windenergie Marktübersicht is a treasure trove of information about wind turbines on the international market (see Wind Turbine Market Surveys in the appendix for details).

Where test reports are not available, talk to the manufacturer or dealer. Ask why no reports are available, or why the product isn't listed in the BWE or other market surveys.

Operational History

Determine what kinds of tests have been conducted, for how long, and the highest winds experienced. Not all wind machines are built to the same standard. In one case during the early 1980s a wind turbine designer sized his machine to withstand a maximum wind speed of no higher than 90 mph (40 m/s) in a parked condition. He asserted that no winds above that speed had ever been measured near his site in western Pennsylvania. This claim was false, and consequently the design of the wind machine was suspect, more so when considering that all other U.S. manufacturers at the time were designing their products to withstand a maximum wind speed of 120 mph (54 m/s).

Knowing how long the tests were run or how long a particular model has been in service is especially important. Unscrupulous manufacturers have frequently resorted to touting their products as "extensively tested" when they haven't been. In one case the new product had been in operation during only a mild summer in Ohio, an area of moderate winds. The first time this "extensively tested" product was installed at AEI's windy West Texas test field, it suffered severe blade flutter, experienced a brake failure, and the tower fell over—all within the first hour.

In another notorious case, the "extensive tests" were conducted on a bench-scale model! The manufacturer hadn't even bothered to test the turbine outdoors before it began selling the turbine to wind farm developers in California. Eventually this company was prosecuted for fraud, but not before walking off with investors' money.

Reputable, well-tested products, in contrast, have been in unattended operation for months, if not years, at numerous sites in widely different wind regimes. Well-tested products have endured hurricane-force winds without damage. Manufacturers with well-tested products can provide documentation on the performance of their machines over time, under harsh conditions.

When evaluating operational history, don't be alarmed by occasional reports of defects. You're looking for trends. If every wind turbine of a particular model has thrown a blade and is still throwing them, then there's a good chance the one you're looking at will, too.

Wind turbines shouldn't be held to any higher standards than we hold other machines. After more than 80 years of development, automobiles are still being recalled by the thousands for manufacturing and design defects, yet we continue to buy and use them. We try to minimize the risk of buying a lemon by trying to select a model with the least potential for problems. Reputable manufacturers of wind machines make mistakes like everyone else. Your challenge is to find one that makes fewer mistakes than the rest.

Design defects usually appear within the first year of operation. Like automobiles, new products must undergo a period of debugging. Unexpected problems will undoubtedly arise and must be corrected. These problems are greatest when the product is first introduced, and decline thereafter. Ideally you'd like a wind machine that has been on the market for several years, and one that operates successfully in a range of environments. This isn't always possible. You have to rely on your own judgment.

Product Specifications

Examine the promotional literature describing the wind machine. Are the estimates of energy output reasonable? Do they stress generator size while ignoring energy output altogether? Most manufacturers will present a list of parameters that succinctly describe how the wind machine performs, how it functions, and what it can be used for. This will include the annual energy production, power curve, and power form.

Estimates of annual energy production could ultimately replace the rated power at rated wind speed currently used to describe the size of a wind turbine. But it hasn't happened yet. More often than not, the annual energy output is presented as a graph of estimated generation at various wind speeds at hub height. Occasionally, the AEO may also be presented as a table of annual generation at various average wind speeds.

Power form indicates how the power will be used. For small wind systems interconnected with the utility, this will be given as the nominal voltage and frequency. For battery-charging wind systems, power form should indicate the DC voltage. For interconnected wind turbines, power form should also include the number of phases. In North America, for example, most homes and many small farms use single-phase service. Large farms and businesses use three-phase service.

Most product specifications should also include the cut-in and cut-out wind speeds, maximum power, average noise level in dBA, maximum design wind speeds, rotor speed, and overspeed control.

Product literature should always specify the type of noise data presented. Preferably, noise data will be presented as emission source strength or sound power levels. If noise data is presented in sound pressure levels, however, the distance from the wind turbine must be specified; otherwise the information is meaningless. Specifications for most European wind turbines include sound power levels.

The maximum wind speed is the speed the turbine was designed to endure unattended without suffering damage. Notations may also indicate the maximum measured or tested wind speed the turbine has survived without damage.

Rotor speed is the number of revolutions per minute of the wind turbine's rotor. For a wind turbine driving a two-speed induction generator, there are two rotor speeds: the rpm when operating on the low-power generator or windings, and the rpm when operating on the primary generator or full-power windings. In a variable-speed wind machine, rotor speed is given as a range of values.

Overspeed control is a concise description of the method used to protect the wind turbine in high winds or during a loss of load.

The following is a summary of product specifications for a hypothetical wind turbine. The power curve and estimates of annual energy output are often presented separately.

- Power form: 3 phase, 440 volts AC, 60 hertz (50 hertz in Europe).
- Sound power level: 98 L_W dBA at 8 m/s wind speed.
- Cut-in speed: 10 mph (4.5 m/s).
- Cut-out speed: 50 mph (22 m/s).
- Maximum wind speed: 120 mph (54 m/s) design, 112 mph (50 m/s) tested.
- Rotor speed: 60 rpm.
- Overspeed control: blade stall, brake, pitchable blade tips.

Manuals

As you consider your wind system purchase, also ask to see a copy of the owner's manual. These are often available for free download from manufacturers' Web sites.

Is the manual well written and sufficiently detailed to tell you what's needed to install, operate, and maintain the wind system safely? For example, are there instructions for starting and stopping the wind machine? Is a parts list provided?

Engineers, when ordering expensive equipment, often ask for a copy of the operator's manual from each company competing for a contract. They then compare them. This offers the engineers a better understanding of the equipment, and also gives them a feel for the manufacturer's approach to problems that may be encountered by the user.

Even good manuals can let you down. Bob Gobeille learned this when he installed an 850-watt wind turbine and it immediately blew the recommended fuses in a storm, leaving his turbine operating unloaded. Later he learned that the manual had specified the wrong fuses, yet the manufacturer never corrected the manual.

Maintenance

Now examine maintenance. What maintenance is required, for example? How often must maintenance be performed? How difficult is it to perform? Must someone climb the tower, or is a cursory examination from the ground adequate? If parts must be replaced, are they readily available or must they be specially ordered? What kind of materials are used in the blades? Are they made of wood, wood composite, or fiberglass? What kind of corrosion protection is provided? Are exposed surfaces painted or galvanized? These are a few more questions to be answered when evaluating the difficulty and hazards of maintaining a wind turbine.

When can maintenance be performed? For medium-size turbines, the answer is simple: anytime you choose. The turbines are all designed so that you can safely turn off the turbine and stop the rotor. This is not true of many small wind turbines.

Controls

There are times when you need to stop the rotor of a small wind turbine or otherwise bring it under control. Imagine that something goes wrong at three o'clock in the morning in the middle of a gale, and you're standing at the base of the tower in your underwear wondering what to do. It's times like this when you wish all small turbines included a reliable means of stopping or slowing the rotor to a safe speed.

Many small turbines use dynamic braking of the generator to slow the rotor when desired—with varying degrees of success. Often dynamic braking is most effective in moderate winds, and least reliable when you really need it.

"If a wind turbine's more than 1 kW," says North Dakota wind enthusiast Mike Klemen, it should have "a brake that will work under any wind conditions." For Klemen, if the generator isn't sufficiently robust to stop or significantly slow the rotor through dynamic braking, then there needs to be some other means, mechanical or aerodynamic, of controlling the rotor under emergency conditions. For a small wind turbine, this could be as simple as a furling winch.

If the turbine will be serviced on the tower, it should also include a parking brake or locking pin on the rotor shaft to keep the rotor from turning while the turbine is being serviced. Locking pins to prevent the main shaft from rotating are now standard on all commercial-size wind turbines. These simple devices would have prevented the deaths of three men in two separate accidents if they had been more widely used. Similarly, there should also be a means of preventing the turbine from yawing unexpectedly whenever someone is on the tower.

Robustness

With the revival of interest in small wind turbines during the early 1990s, some products built only for light-duty service have been introduced. They're simple and inexpensive. For these reasons they've proven popular. Unfortunately these wind turbines are not particularly rugged. They may work fine—most of the time. But their longevity is site-dependent. One severe storm could wreck them beyond repair.

How do you judge robustness? First, if the product literature indicates that the turbine was designed for light or moderate wind regimes, take the manufacturer's word for it. Assume that the turbine is not rugged enough to withstand a windy site.

In general, heavier small wind turbines have proven more rugged and dependable than lightweight machines. Wisconsin's Mick Sagrillo is a proponent of what he calls the "heavy metal school" of small wind turbine design. Heavier, more massive turbines, he says, typically run longer (see figure 9-3, Small wind turbine specific mass). *Heavier* in this sense refers to the weight or mass of the turbine relative to the area swept by the rotor.

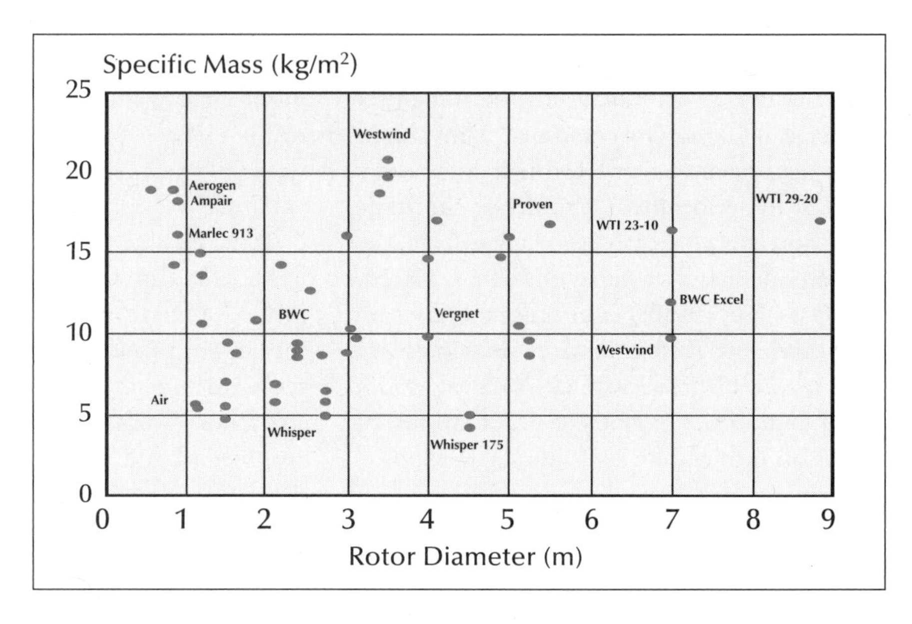

Figure 9-3. Small wind turbine specific mass. The mass or weight of a small wind turbine relative to the area swept by its rotor is a good indication of ruggedness. Wind turbines with a higher specific mass are generally more rugged than those with a lower specific mass. Typically turbines with a higher specific mass cost more than lightweight machines, but they last longer and operate more reliably.

By this criterion, a turbine that has a relative mass of 10 kg/m2 may be more robust than one whose specific mass is 5 kg/m2.

Greater specific mass doesn't guarantee that a certain wind turbine will operate more reliably over a longer period of time than a lighter-weight turbine, but it's a good indicator. Among medium-size turbines this question was settled long ago: More massive, solidly built turbines swept the marketplace, assigning to the scrap heap the flimsy, lightweight turbines once produced in the United States.

Another indicator of longevity is rotor speed. Higher rotor speeds, while possibly providing greater aerodynamic performance, impose more wear and tear on small wind turbines than comparable turbines with more languid operation. Sagrillo likes to compare the legendary longevity of the pre-REA Jacobs to contemporary designs. The 3 kW Jacobs was designed to operate at 200 rpm. The modern Whisper 175, also 3 kW, operates at 600 to 800 rpm. "Which would you prefer?" he asks.

Intangibles

Take a good close look at the wind turbine itself. How is it assembled? Does it look like it was welded together in a backyard shop? Do the parts fit snugly, or have they been made to "press fit" with a hammer? Now step back and look at the wind machine as a whole. How do you feel about it? Is it something you would be proud of, or will you have to put it "out back" so no one will see it? Wind machines are highly visible objects. Their appearance is important, both for your satisfaction and for acceptance by the community.

Evaluating Vendors

After you have dissected the technology, you must evaluate less tangible factors such as the longevity of the manufacturer and the reputation of the dealer.

Manufacturers

How long has the company been in the wind business? What's its track record? Does it have sufficient financial resources to honor its warranty commitments? There's no easy way to find answers to these questions. In most cases you'll be dependent on the dealer for information. Even when you do get the answers, it's hard to determine what's important and what isn't. For example, a well-established company that has been in business for several years is a better risk than one just starting out.

Likewise, partial or full ownership by a major national corporation usually indicates that a company has ample financial reserves to survive a major warranty recall. Nevertheless, a corporate executive who has no personal stake in the company can much more quickly decide to cut his losses and cease production during hard times than can the owner-entrepreneur who has put his own sweat and blood into the business. Only you can decide which business you want to place your bets on.

Ventilators and Squirrels in a Cage

Like ducted turbines (see chapter 6, Evaluating the Technology) a perennial favorite of hucksters and charlatans is, for lack of a better word, squirrel cage rotors. Many are nothing more than rooftop ventilators repackaged as "wind turbines." As ventilators, they work fine. It's when you try to couple them to a generator that you quickly learn why wind turbines use two or three slender, airfoil-shaped blades. Most hucksters, however, never progress that far. They never build actual wind turbines—or if perchance they do, they never measure the "wind turbine's" performance. Of course, they wildly exaggerate the potential of these breathtaking new inventions.

Doug Selsam, himself an inventor, has tried to understand why consumers—and the news media—are so gullible. His explanation: Ventilators and squirrel cage rotors are easy to understand; modern wind turbines are much less so. After all, a rooftop ventilator with its entire swept area covered with blades looks like it will capture more wind than a modern wind turbine with only a few blades, some with—unbelievably—only one.

In a 2002 Internet scam, a company peddling ventilators as "wind turbines" claimed that its product would produce nearly five times more electricity than a conventional wind turbine of the same size. Naturally, for this "superior" performance it would charge two to three times more than for a real wind turbine. The company asserted that it was "thinking outside the box," a catchphrase of 1990s management gurus. It certainly was; it wasn't even close to the box. It was on another planet where the laws of physics don't apply.

Dealers

Your choice of dealer—if you use one—is determined primarily by the wind machine you want and where you live. Most dealers, to round out their product line, represent more than one company. Even so, within a certain locale there will be only one dealer for each brand. (Manufacturers want to ensure a healthy dealer network, so they limit the number of dealers selling their product.) Proximity is important. If repairs or service are needed, particularly during an emergency, you don't want a dealer who lives on the other side of the continent.

Determine if dealers are reputable by checking with their previous clients. Have they been prompt in making repairs, or have they taken their time while hustling new sales? Dealers should have references available for such an inquiry. If not, are they willing to provide them?

Also call the manufacturer and check whether the "dealer" is authorized to sell its product. In one instance, a so-called dealer was selling a popular brand without the manufacturer's authority to do so. This dealer had declared bankruptcy previously, leaving a number of clients high and dry without spare parts or service for their ailing wind machines. In this case he was selling a used wind machine—that is, until the authorized dealer blew the whistle. The whole sad affair could have been avoided by a single phone call to the manufacturer.

Don't be misled by membership in various organizations as a claim of legitimacy. Some dealers and manufacturers use membership in a trade association, such as the American Wind Energy Association, as a promotional tool. It's one of the oldest marketing tools in the book, and the one most often used when no other credentials exist. Anyone can join an association, and most associations don't police their ranks. Suspect anyone trying to cash in on checkbook credibility.

What to Expect

If you buy through a dealer or use a contractor to install the turbine, you have a right to expect that all work will be performed according to standard practices and local building and electrical codes. The work should also be performed in a timely manner, and all construction debris removed from the site before the job is considered finished.

Don't expect overnight miracles. If delivery of a component has been delayed due to circumstances beyond the dealer's control, you shouldn't hold the dealer accountable. The dealer should make a reasonable effort to expedite the installation of your wind system or its repair, but don't expect the dealer to jump at your every request. Keep in mind that dealers operate a business and that they may have other commitments. At the same time, the dealer should fulfill those obligations stated in the contract or implied during negotiations.

Contracts and Warranties

To ensure that you get what you pay for, put it in writing. Demand a written contract and warranty, and consider having an attorney look them over. Installing a wind system is a major investment, akin to buying a car—or a house. It's worth the added cost of getting good legal advice. You may need an accountant's advice, as well.

You need to know specifically what is included in the price you've been quoted. If you plan to do any of the work yourself, the contract must spell out exactly where the dealer's responsibilities end and yours begin. The contract should also describe exactly what you must do to meet the terms of the warranty. Who has the final say, for example, as to how the work should be done? How will disputes be resolved, should they arise? What is covered by the warranty? What isn't? How long does it last?

Most small wind machines come with a three-year warranty. Extended warranties are sometimes available—for a price. Because of the differences in warranties among manufacturers, it's wise to read the fine print. Check whether the warranty is transferable or assumable by the manufacturer if the dealer goes bankrupt. Ask who pays for shipping or for the fieldwork on warranty repairs, and whether damaged or defective parts must be returned before replacement parts are shipped.

When Southwest Windpower introduced its Air series, the early turbines were unreliable; owners frequently had to send the machines back for repairs. Though the company paid for the repairs, it did not pay shipping or the cost of turbine removal and reinstallation.

Another aspect of the contract is the terms of sale. The contract should state the amount of payments, as well as how and when they should be made. In general, you will pay most of the cost for the wind machine and its installation in advance. You don't drive off the lot with a new car until you have handed over your check. Similarly, you shouldn't expect the dealer to install the wind system without your first paying a hefty deposit.

Terms vary from one dealer to the next. Usually a down payment is made to secure your order. Then full payment is required for the turbine and tower when they are ready to be shipped. Some dealers require payment for the turbine, tower, and installation in advance. In most cases 5 to 10 percent of the total contract is held by the buyer until the wind machine has been installed and operates properly.

In multiple-machine purchases such as for a wind farm, the buyer has more leverage with the manufacturer and can obtain written assurance that the wind machine will generate power as advertised and that it will be available to generate power a minimum percentage of the time. In Denmark, where there's a strong association of wind turbine owners, manufacturers have been held responsible for performance claims on turbines sold to farmers and individuals. Unfortunately such assurances are often not offered to purchasers of small wind machines in North America.

Case Studies

In a classic example of how not to go about it, the rural cooperative in Du Bois, Pennsylvania, bought a wind system from a manufacturer in nearby Clearfield, the forerunner of Fayette Manufacturing. The co-op members didn't contact anyone about the company or its product. Nor did they investigate the company's claims. If they had, they would have found that the wind machine was a prototype (not a well-tested wind machine ready for commercial sale), that the manufacturer couldn't possibly build and install the wind machine for the contracted price (one of those "it's too good to be true" deals), and that the wind machine couldn't do what the manufacturer said it would. In short, neither the rural co-op nor the manufacturer knew what it was doing. The machine was installed, and, as you would expect, it never worked. It stood for years along Interstate 80 as a testament to ignorance.

Cappy Reece

Some buyers take a more studied approach. Capitola Reece wanted a wind turbine that would pay for itself and work reliably. One more thing was certain: At 74, the retired teacher wasn't about to climb the tower and fix it herself. She also acknowledged she didn't know the first thing about wind turbines or even where to buy them. So the first thing she did was visit her local library.

There she read about a fellow who was promoting wind energy, and wrote to him. After reviewing his promotional literature, Cappy, as she's called, arranged for him to visit her site. She had her questions ready. If it was going to cost her money to get him to visit her site, Cappy was going to squeeze every penny's worth out of him. She had the site picked out, copies of all her utility bills, and a notebook full of questions. "How much wind do you think I have? How much does it cost for an anemometer? For a wind turbine? What tax credits are available? Is it noisy? What maintenance is required? How many have been installed? How many have you installed? How well have they worked? How much energy could I produce here? What does the utility think about all this? How will it affect my taxes and insurance rates? When the utility lines go dead, will it still work?"

After the inquisition, she took the dealer to her proposed site. Bad news. He would be glad to install an anemometer, but he would just be taking her money, because there were too many trees nearby. Though not tall, they were tall enough to block the flow at the anemometer. The results would be less than the speed at the nearest airport. Would that data do? he wondered. He gave her his estimates based on the airport, and left. *Well,* Cappy thought, *we'll just check this guy out.* She called the state energy office, the manufacturer, and a previous client with a similar wind turbine. He seemed all right.

Still, she wanted her attorney to look over the dealer's contract and offer of warranty. She also wanted to talk to her township supervisor about the need for a zoning variance and to the utility about the interconnection. No variance was required for her rural site. The utility didn't know a thing about the particular wind machine or the dealer, but did warn her that the few wind machines installed in their area had not worked well. That didn't deter Cappy. She'd figured the company would be less than thrilled with the idea.

The attorney had some objections; so did her bank. The contract called for a sizable amount of money for a rather novel purchase, and it called for most of it up front. "It won't do," said her attorney. He demanded changes in the contract and terms of payment. He wanted to pay after installation. The dealer balked—too great a risk for his small business.

Figure 9-4. Art and Maxine Cook's HR3. This is a 1982 photo; after more than 20 years, the turbine is still in service.

But a compromise was reached. A portion of the payment would be held in escrow by the attorney until the wind machine was installed and operating. The dealer agreed. His needs were met by knowing that the money was earmarked for him and was in safekeeping.

The three parties met and signed the contract. The dealer then ordered the equipment. The contract stipulated that the dealer had 90 days to install the wind machine and get it running. Within two months it was operating, but a problem developed just before the final payment was made. Because the escrow account hadn't been released, the dealer hustled to make the needed repairs. Cappy was satisfied, and enjoyed years of nearly trouble-free service.

Some Do's and Don'ts

Do plenty of research. It can save a lot of trouble—and expense—later.
Do visit the library. Books remain amazing repositories of information.
Do talk to others who use wind energy. They've been there. You can learn from them what they did right and what they'd never do again.
Do read and, equally important, follow directions.
Do ask for help when you're not sure about something.
Do build to code. In the end it makes for a tidier, safer, and easier-to-service system.
Do take your time. Remember, there's no rush. The wind will always be there.
Do be careful. Wind turbines may look harmless, but they're not.
Don't skimp, and don't cut corners. Taking shortcuts is a surefire way to ruin an otherwise good installation.
Don't design your own tower—unless you're a licensed mechanical engineer.
Don't install your turbine on the roof, despite what some manufacturers may say!
And, of course, don't believe everything you read in sales brochures, or on the Internet.

Art and Maxine Cook

Art and Maxine Cook believe in renewables. Wind and solar energy have been part of their lives since 1978, when they installed a farm windmill to pump water from their well. In their quest to live their dream, they didn't skimp. They were willing to pay for products that worked.

In 1981 the Cooks installed a top-of-the-line 3 kW wind turbine on their farm in western Pennsylvania's Somerset County (see figure 9-4, Art and Maxine Cook's HR3). For more than two decades Northern Power Systems' HR3 has sat atop a 60-foot (18 m) truss tower on the Cooks' lawn, reliably producing wind-generated electricity through winter snows and summer sun.

The wind turbine is part of a hybrid power system that includes 4.5 kW of Kyocera photovoltaic panels and an 8-foot (2.44 m) Aermotor water-pumping windmill. The farm windmill is also still going strong.

The HR3 was designed for charging batteries at harsh telecommunications sites. Northern Power Systems gave the turbine the *HR* moniker for "High Reliability." The small New England company built it to last. And it has. "It just works," says Art Cook.

What's the HR3's secret for longevity? "It's massively built," says Art. "I could have used something lighter for half the price, but I would've gone through three of them by now." The specific mass of the turbine, its weight relative to the area swept by the rotor, is one and a half times greater than a Bergey turbine, and three times that of a Whisper 175.

Art also gives the HR3 high marks for serviceability. He follows an old Pennsylvania adage: "The best thing you can put on your fields is your footsteps." As he explains it, "You've got to be out there, looking and listening. You've got to climb up and check it out."

This attitude may also explain why the Cooks' HR3 is still operating on its original set of blades. Not that repairs haven't been needed.

Occasionally the turbine is hit by lightning, necessitating a climb up the tower to replace blown rectifier diodes. Art also replaces the leading edge tape on the blades every few years. And when he replaced the Lundel alternator's core after it failed, he cut off the rotted tips of the wooden blades, cutting the HR3's rotor diameter down from 16.4 feet to 15.5 feet (from 5 to 4.7 m). "It's a better machine now than when we bought it," says Art. The turbine, after more than 20 years in use, still produces 4.2 kW at 27 mph (12 m/s).

The Cooks have made other changes, as well. After 18 years Art eliminated his battery backup and opted for conventional net billing with SMA's Sunny Boy inverter.

In the early 1980s Art told anyone who would listen that someday there would be windmills all along the ridges of Somerset County. "They said I was crazy," he grins. They don't say that now. Near the Cooks' home are Pennsylvania's first two wind farms. Each of the giant turbines is 300 times the size of the little HR3. "I am glad I've lived to see it," says Art.

"People ask me what return I've gotten from renewables," he adds. "For me, the pleasure and enjoyment I've gotten from it are priceless. That's how I measure its worth." To Maxine Cook, the answer is even simpler: "It keeps Art off the streets."

If only the co-op in Du Bois, Pennsylvania, had had as much common sense as Cappy Reece, or the dedication of Art and Maxine Cook, it could have avoided public embarrassment and installed a working and productive wind turbine, instead of a monument to hubris.

In general, doing it right the first time may take longer and cost slightly more, but you'll be a lot happier in the long run.

10

Interconnection with the Utility

Night has fallen and the sky is clear. December's chill winds whip the Lake Erie shoreline. Drifting snow swirls about the fence posts and outbuildings of George McClain's small farm. The whistling wind rises in crescendo and then dies away in an unpredictable ebb and flow. A faint whirring, rhythmic and ever present, can be heard. Dark, saberlike shapes sweep the starry sky.

"Looks like it's going to be a cold one tonight," George predicts.

His two kids, scampering around in their flannel pajamas, flee their mother as she readies them for bed. Darlene, both mother and partner in the McClains' dairy, responds, "George, don't you think we ought to turn the electric heat up? I feel like I'm coming down with something."

"Yeah, Daddy," the kids chime in, "just turn it up to 70 like we used to."

"Now, you kids know better than that," he says. "Christmas will be here soon, and we want to buy that new car we've been waiting for so long, don't we?" He winks at Darlene. "We only get one more check from Pennelec before the new year, and I want to sell them just as much power as we can. On a night like this everybody's going to be switching on their electric heaters. We need to save every kilowatt we can. The more we save, the more we can feed to Pennelec. I'll bet we can make $50 by morning more if this weather holds. Those turbines will really be turning out the juice in winds this high. Just listen to 'em hum."

When this passage was first written in the early 1980s, it sounded far-fetched. A family that awaits winter's winds, and looks to the local electric utility as a source of income? What was far-fetched then has become commonplace. Farmers such as the fictional McClains have been replaced by pig farmer Niels Mogens Sloth in northwest Denmark, grain farmer Peter Ahmels in northern Germany, or rancher Eli Walter in Alberta, Canada; all are raising a new cash crop: electricity.

It's true. In most countries you can interconnect your wind system with the local utility. And yes, you can even sell power to it. Despite the important role small wind turbines play in providing power to remote sites, more wind machines are now interconnected with electric utilities than are used in battery-charging applications. In 2003 there were more than 60,000 wind turbines generating utility-compatible electricity worldwide, the majority in northern Europe.

As you can imagine, though, it's not as simple as it first sounds, especially in North America, where there are many regulatory roadblocks. You may jump all the other hurdles only to find that the utility is slower than molasses in January when it comes to granting your request to interconnect with its lines. For most utilities in the United States, interconnection is a low priority, and your request may become buried under a stack of paperwork.

This was the case in the Pittsburgh suburb of Fox Chapel. The dealer, Bill Hopwood at Springhouse Energy Systems, had contacted the utility about the same time he began the zoning variance application. Yet even though the zoning board took months to make its decision, and it took another couple of months to install the wind machine, the utility still hadn't made up its mind what to do. Eventually it decided to let the interconnection proceed as a goodwill gesture toward the client, the Western Pennsylvania Conservancy. Since then the wind turbine has been in service for nearly two decades—without mishap—as part of the Conservancy's renewable energy program.

Farmers, homeowners, and businesses in the United States can generate electricity in parallel with electric utilities as a result of the Public Utility Regulatory Policies Act (PURPA for short), a part of the 1978 National Energy Act.

In Europe a system far more consumer-friendly—and far simpler—was adopted. Even today interconnecting wind turbines with the grid in Europe is a much more common experience than in North America. In Denmark, Germany, and the Netherlands, the interconnection of individual wind turbines or small clusters of machines is such a common occurrence that it's a regular part of utility operations.

In Denmark, Germany, and the Netherlands, the interconnection of individual wind turbines or small clusters of machines is such a common occurrence that it's a regular part of utility operations.

PURPA in the USA

For the United States, PURPA was downright revolutionary. Though multifaceted, PURPA is most widely known for Section 210, which requires utilities to buy power from, and sell power to, cogenerators and small power producers. And they must do so at reasonable rates. Unfortunately the rate, or tariff, was never specified.

Congress entrusted the Federal Energy Regulatory Commission (FERC) with the responsibility for drafting the regulations resulting from the act. The regulations were then implemented by state regulatory authorities. State public utility or public service commissions have jurisdiction over individual utilities and their compliance with the law.

PURPA affects nearly all utilities in the United States, both regulated (those that come under the purview of state utility commissions) and unregulated (those utilities exempt from state jurisdiction). Investor-owned utilities (IOUs, as they're known), public power corporations such as the Tennessee Valley Authority, and small rural cooperatives all have to comply with PURPA's provisions. There are, however, important exceptions. Some state utility commissions don't have direct authority over rural electric cooperatives and municipal utilities. Co-ops and municipal utilities fall under FERC's jurisdiction.

In one step PURPA removed two major barriers to more widespread use of wind turbines and other alternative sources of electricity in the United States. First, it exempted small power producers from restrictions of the Federal Power Act. Previously a household-size wind system could have been considered a utility and regulated as such by the state public utility commission. The paperwork burden alone would have buried many small power producers. Second, PURPA assured wind system users of backup power, and it stipulated how utilities were to charge for this standby supply.

Prior to PURPA, some utilities charged discriminatory rates for backup service. In effect they said, Sure, we'll sell you power when you need it, but boy, are you going to pay for it. PURPA put a stop to such practices. Utilities now can't penalize small power producers—overtly—by charging unreasonable rates for standby service.

Through PURPA, Congress sought to encourage development of renewable energy by removing barriers to its use. PURPA also created a powerful financial incentive that was absent before. Not only must a utility allow small power producers to generate electricity in parallel with the utility's own system, but it must also pay for it.

Under FERC regulations the price paid by the utility must reflect the costs the utility avoids by not having to generate the power itself. This avoided cost can be more or less than the retail rate. During the early 1980s, the avoided cost in parts of the United States was near, or exceeded, the retail price for electricity. But more often, the avoided cost was far below the retail price.

The regulations went even farther. They allowed state utility commissions, when ruling on the price paid by a utility, to take into account escalating costs over the life of the contract with a small power producer. If the state utility commission chose to encourage alternative energy, it could establish a levelized rate. In the case of a wind system designed to run for 20 to 30 years, the levelized rate would be much higher than today's avoided cost, assuming utility rates escalate over time. Levelized rates offer greater revenue in the early years than nonlevelized tariffs, accelerating payback and increasing the return on investment.

PURPA fundamentally changed the way Americans look at power generation, conservation, and supplemental power sources. As seen in the opening scenario, it altered our view of energy conservation from one of conserving to save money to one of conserving to earn money. It encouraged decentralization and offered decentralized energy investment opportunities, as well. Anyone who could afford a wind machine and had a windy site could become a small power producer.

PURPA, coupled with high buyback rates, also affected the size of the wind systems homeowners or farmers could choose. As a supplemental power source, wind systems were originally looked on as a means of reducing utility bills, and they were sized to produce only a portion of the user's consumption.

Because the cost-effectiveness of wind turbines increases with increasing size, sales of surplus wind-generated electricity to the utility at or near the retail rate encourages the consumer to seek the most economically sized wind system on the market. Thus homeowners or farmers are more likely to install a larger turbine than they would otherwise need, simply to offset their own consumption.

It is only a short conceptual step from buying a turbine larger than that needed to offset domestic consumption to buying two, three, or even more turbines. Space, the level of risk you're willing to take, and the availability of capital are the only limits. Like the McClains, farmers who began looking at the wind as a way of reducing their utility bills soon recognized it as another resource that could be tilled. They could harvest the wind that swept over their fields, producing a new crop—a cash crop on contract, at that. This simple realization resulted in clusters of wind turbines—and from clusters, wind farms eventually grew (see figure 10-1, Interconnection with the grid).

Direct sales to the utility also gave potential users more flexibility in siting. Under PURPA, users can simultaneously purchase power from the utility and sell power to it. Let's say you own some rolling farmland. Your house and barn rest snugly at the base of a tall hill with trees all around. It's a beautiful setting, but a lousy place for a wind machine. It just so happens, however, that a utility line crosses the top of the hill, which has been cleared for a pasture. The top of the hill is ideal for a wind turbine, but a good distance from the house.

What to do? Install the wind machine on the hill and sell all your power directly to the utility. At the same time you will continue to buy power from it for your house and barn.

The revenue from the wind turbine will offset your consumption much as they would have if you had installed the turbine near the house (see figure 10-2, Medium-size interconnected turbine).

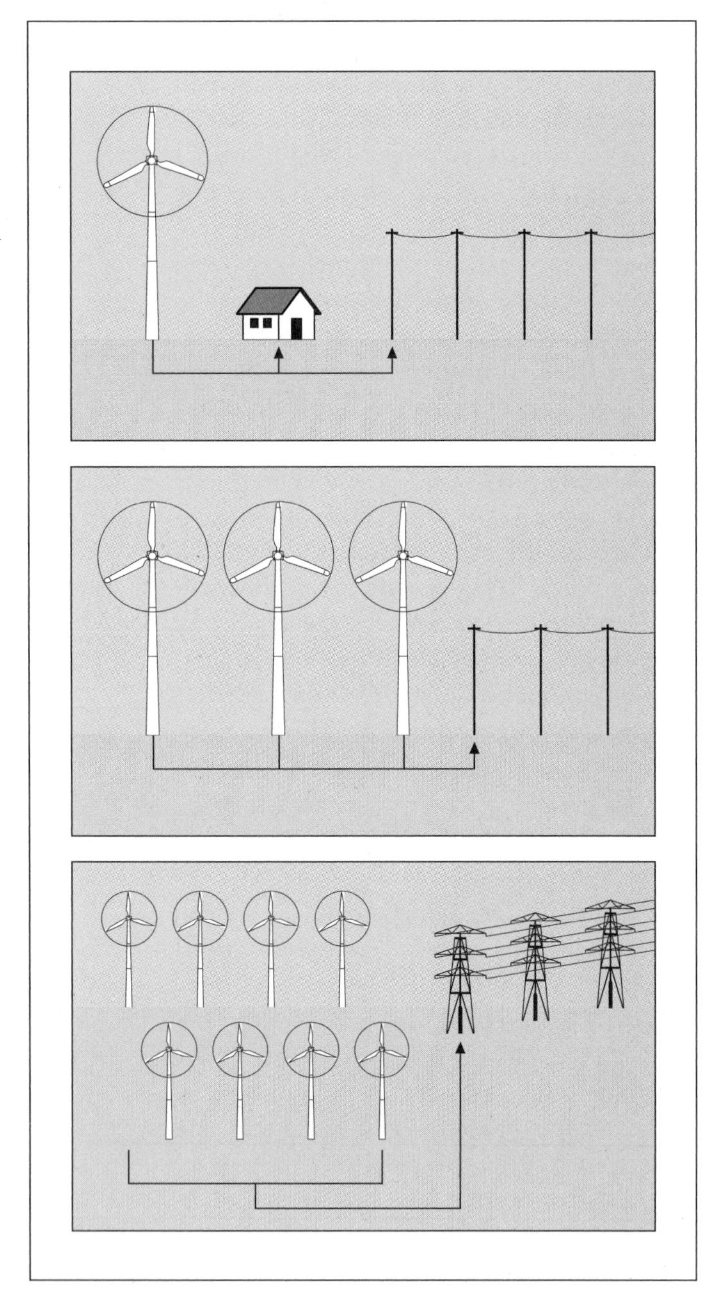

Figure 10-1. Interconnection with the grid. Household-size turbines *Top:* can be connected on the customer's side of the kWh meter. *Middle:* Medium-size turbines, either singly or in clusters, are connected at the utility's distribution voltage. *Bottom:* Large arrays of wind turbines, or wind farms, are connected at transmission voltages.

Practical Application of PURPA

That's the good news. The bad news is that during the past two decades utilities across the United States have convinced regulators to use the average cost of fuel to determine the avoided cost. The average cost often includes cheap natural gas, financially subsidized hydroelectricity, and environmentally damaging coal. Originally FERC had intended that utilities pay small power producers their incremental cost of energy as well as capacity. By the early 1990s few utilities were paying for avoided generating capacity because, they argued, there was no need for new power plants. That situation is bound to change as electricity demand in North America continues to grow.

Worse, the plague of utility deregulation that swept the continent in the 1990s fundamentally altered the relationships among consumers, their utilities, and PURPA. No one knows how this will be resolved; the situation will surely remain fluid well into the first decades of the new millennium.

Ever the pioneer, California rushed into a disastrous experiment with deregulation that brought a virtual halt to wind development in the Golden State for nearly a decade. During the early 1990s California utilities were paying only 30 percent of the retail rate for wind-generated electricity not under existing contracts. As a result there were few wind turbines installed during this period in a state once world renowned for its development of renewable energy. Only after the state passed legislation permitting net metering did sales of household-size turbines begin once again. Meanwhile commercial wind developers were still struggling to recover from the financial and political fallout of California's 2000 to 2001 power crisis.

Though the specifics of PURPA's implementation vary from state to state and from utility to utility, one provision of PURPA is common to all: the requirement that the utility charge for any interconnection costs it incurs. This is normally in the form of a one-time fee for the installation of additional equipment that the utility deems necessary for the safe operation of the small power system. These charges vary, and some utilities use these fees to discourage distributed generation by their customers.

All in all, PURPA gives the small power producer a little bargaining power where before there was none. If you want to install a wind system and it meets certain safety standards, then the utility must permit you to interconnect with its lines, and it must offer you standby or supplemental power at reasonable rates. These rates cannot discriminate against you because you are using a wind generator. The utility must also buy any excess power you produce. The exact price or tariff paid is subject to interpretation, however—and therein lies the problem. PURPA doesn't specify a price, only how to derive the price.

Figure 10-2. Medium-size interconnected turbine. This 250 kW Nordex turbine stands in a northern Iowa farm field and feeds the local utility's distribution network under the state's net-metering system. Note the three-phase conductors on the cross-arm. Medium-size turbines using three-phase generators are connected at distribution voltage via a transformer (not visible).

Net Metering

Output from a wind turbine fluctuates over time as the wind gusts and subsides. When you superimpose a wind machine's varying output onto a utility customer's varying demand for electricity, there will be times when excess wind-generated electricity is produced. For example, at night during a winter storm when the wind is howling and there's little electricity being consumed, the excess electricity generated will flow to the utility. Ideally you would like to bank this excess with the utility until it's needed, when you're using more electricity than you're producing. Net metering, or net energy billing as it's also called, makes this possible.

More than half the states in the country have instituted net metering. Specifics vary from state to state. Many allow you to run your kilowatt-hour meter backward, selling any excess energy at the retail rate until net sales equal net consumption. The utility balances your account—sometimes monthly, sometimes yearly. In effect, you're storing excess generation with the utility instead of storing it in batteries. Because of the great seasonal variation in the wind, it's to the small power producer's benefit to balance net-metering accounts annually, and an increasing number of states, such as California, are doing so.

Annual averaging balances seasonal cycles of high winds in springtime and low winds in summer. When the wind system closely matches your needs, you won't owe the utility any money and it won't owe you. This allows

Wind without Limits

Of the states with net metering, only a few permit use of any size wind turbine the customer chooses. In this age of customer choice, the consumers should indeed have a choice about the size of the wind turbine they want to use under net-metering programs. Most programs, however, limit wind turbines to 10 kW or less. Some permit up to 50 kW. A few go as far as permitting 100 kW. An even smaller number permit wind turbines up to 1 megawatt. The reason for setting limits? Politics. Caps on wind turbine size protect utility markets. Utilities accept net-metering programs with low caps because the programs pose no serious threat.

Still, utilities have little to fear. Net metering is self-limiting. Only those whose consumption can absorb all the production from a large turbine will choose to net meter. Customers such as Iowa's Schafer Systems and Spirit Lake School will opt for larger turbines, because it makes economic sense to offset as much of their load as possible. Those for whom a 10 kW turbine is a closer match to their needs will choose a 10 kW machine. Artificial limits are unnecessary.

the utility to settle the account without issuing checks for surplus electricity, saving it substantial administrative costs. It also gives you more money by minimizing sales to the utility at less than the retail rate.

Utilities often purchase generation in excess of the balance on a net-metering account at the wholesale price, the "avoided cost," or worse. In some states that have deregulated their electricity markets, excess generation is literally given to the utility company. There's no requirement that the utility even pay for the surplus electricity!

Net-metering programs, and especially those permitting annual averaging, encourage consumers to install a wind turbine that will produce as much electricity as they consume. Unfortunately net metering is often limited to small or household-size wind turbines. Attorney Tom Starrs argues that utility customers should be able to offset their consumption by generating their own electricity regardless of turbine size, just as they are entitled to reduce their consumption by conserving energy when they switch off an unneeded appliance. But that isn't the case in many states.

Under net metering, only one kilowatt-hour meter is needed: the same one you already have. Most kilowatt-hour meters will work in either direction. The meters accurately measure what you use and what you deliver. The National Renewable Energy Laboratory found in a study of kilowatt-hour meters that what inaccuracies did exist occurred in the utility's favor. Despite this evidence some utilities fear running their meters backward and insist on installing two unidirectional meters: one to measure consumption and the second to measure deliveries to the utility.

Unlike PURPA, net metering in effect sets a price, the retail rate, albeit for a limited amount of wind-generated electricity. For advocates of renewable energy, though, net metering merely nibbles at the potential. Electricity feed laws deliver.

Electricity Feed Laws and Quotas

The classic difficulty for proponents of renewable energy is to make provisions for both interconnection with the grid and payment for the electricity delivered. In the United States, PURPA paved the way by permitting interconnection with the grid. However, PURPA fell short because Congress avoided the question of price, leaving that determination initially to FERC and ultimately to state regulatory commissions.

Clearly the price paid is critical to the financial success of any venture. Renewable sources of energy are no exception. Two approaches have emerged for determining the price paid per kilowatt-hour for wind-generated electricity. One is a bidding or tendering system; the second uses a fixed price. In the former, a quantity of capacity, or quota, is determined politically, and the price derived from bidding among would-be wind developers. In the latter, the price is determined politically, and the amount of capacity that results is a function of an open market.

Britain's Non Fossil-Fuel Obligation, or NFFO, is an early example of what Aalborg University's Frede Hvelplund calls a political quantity-quota system. The Renewable Portfolio Standard (RPS) is a contemporary version being used in parts of North America. In

Feed Laws Deliver Renewables

The concept of electricity feed laws is simple: They permit the interconnection of renewables with the grid, and they specify the price paid. Via a public-policy debate, society (a parliament, congress, or state assembly) determines a rate to be paid for every kilowatt-hour generated with renewables. That's all there is to it. No cumbersome bureaucracy. No secret bidding. No sweetheart deals. A level playing field for all: farmers, homeowners, small businesses, municipal goverments—everyone.

Feed laws pay only for actual wind generation; they don't pay for crackpot inventions that don't work or—equally as damaging—the tax scams once so common in the United States. Wind turbines supply about 20 percent of Denmark's electricity and more than 3 percent of that used in Germany, a country with one of the world's largest industrial economies. In the northern state of Schleswig-Holstein, wind turbines provide more than 20 percent of electricity consumption. All of this is the result of feed laws.

Feed laws also create dynamic markets, spurring innovation and a host of supporting enterprises. All but one of the major wind turbine manufacturers on the global market are found in countries with electricity feed laws. The Danish feed law, for example, created a stable demand for wind energy that over time built the world's most successful wind turbine industry. As a result Danish wind turbine companies, or their affiliates, are among market leaders in every country where wind turbines are sold.

More than a decade into the wind boom fueled in Germany by the *Stromeinspeisungsgesetz,* or electricity feed law, nearly a dozen manufacturers were serving the needs of the German wind turbine industry. Since 1999 German farmers, businesses, and other landowners have been installing more than 2 billion euros' worth of wind turbines per year. Such success surpasses any previous attempt to develop renewable energy sustainably. At the turn of the millennium, Germany operated twice the wind-generating capacity as did the entire United States, a country with more than three times the population and more than 26 times the land area. Germany's feed law has made it the world's wind energy powerhouse.

Spain, another country following the feed law path, may someday rival Germany. With half a dozen domestic manufacturers, Spanish wind developers are installing more than 1 billion euros in wind turbines per year. By 2000 Spain rivaled the United States in total installed wind capacity.

No other program has delivered more renewable energy than electricity feed laws. None. Neither net metering, nor renewable portfolio standards, nor tax credits, nor even PURPA has produced more wind-generated electricity than the feed laws used in Europe. To put it simply, feed laws work!

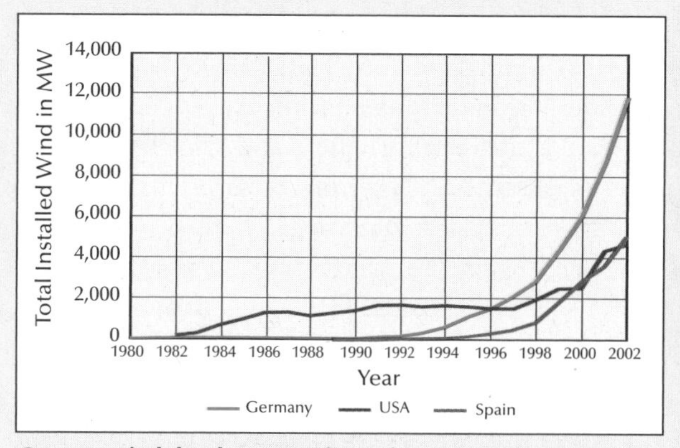

German wind development. The rocketing growth of wind energy in Germany during the 1990s was a direct result of its groundbreaking electricity feed law. With the introduction of its own feed law in the mid-1990s, Spain's wind development soon rivaled that of the United States.

the NFFO tendering system, a regulatory agency, under a mandate from a legislature or parliament, issues a call for tender of a specified amount of generating capacity. Companies then propose projects and submit sealed bids to provide that capacity at a certain price. The agency typically then selects the lowest bidders. Wind developers compete against each other to build projects at the lowest price. This forces wind developers to seek the windiest sites available, which, in the case of Great Britain, have often been on mountaintops.

Hvelplund calls the second approach the political price system. It's more commonly known as Renewable Energy Feed-In Tariffs (REFITs), or simpler still, as electricity feed laws. In the American context, feed laws are equivalent to PURPA with a fixed price. Bernard Chabot, an economist with France's ADEME (the Agency for Environment and Energy Management), estimates that quota systems resulted in the installation of some 2,000 MW of wind turbines by early 2003; fixed-price systems, more than 16,000 MW

Feed laws are extremely simple—as they need to be. (Part of the problem with utility regulation—and deregulation—in North America has been its mind-numbing complexity.) Every kilowatt-hour generated by a wind turbine or a PV panel is paid a fixed price. There are no negotiations, no dickering, and no bickering with the utility. All transactions are transparent. Everyone knows the ground rules, and the price paid is the same for everyone, whether for a pig farmer or a wind company. There are no backroom deals, no secret bids. Danes pioneered the concept, and Germans refined it.

Danish Feed Law

The objective of the Danish parliament, or Folketing, in creating the Danish feed law was to encourage individual action toward meeting Danish energy and environmental policy. For two decades Danish energy policy enabled farmers, businesses, and homeowners to install wind turbines that they owned outright or in which they owned a share.

Danish law encouraged individual as well as cooperative purchase of wind turbines by exempting the turbines from taxes on the portion of the wind generation that went to offset a household's domestic electricity consumption. Cooperatives would buy a wind turbine, site it to best advantage, sell the electricity to the utility, and share the revenue. This enabled city dwellers who couldn't install a wind turbine on their own property to buy the most cost-effective turbine available—even though it would generate more electricity than any one individual household used—and install it in the countryside.

Through this program nearly any Danish household could effectively offset all its electricity consumption with wind energy. Many did. Wind turbines dot the landscape from the capital, Copenhagen, to The Skaw, at the very tip of the Jutland Peninsula. Under the Danish feed law, utilities bought electricity from individually and cooperatively owned wind turbines at 85 percent of the utility's retail rate. The tariff also included a rebate of the carbon dioxide tax, as well as a rebate for 50 percent of the electricity tax paid by consumers on their domestic consumption. The total buyback rate averaged about $0.10 per kilowatt-hour.

Through the mid-1990s Denmark's 2,100 wind cooperatives accounted for one-half of the nation's total installed wind capacity. Some 100,000 Danish households, or nearly 5 percent of the population, own a stake in a *fællesmølle*, or cooperatively owned wind turbine. Many of the remainder are owned by farmers, such as Niels Mogens Sloth, whose Vestas V27 stands at his pig farm outside Hurup in northwestern Jutland.

Unfortunately the deregulation whirlwind cutting a swath through renewable energy programs around the world finally reached even environmentally conscious Denmark, bringing the Danish feed law to an end in the early 2000s.

German Feed Law

Meanwhile, Germans could see the success of their northern neighbor and were determined to follow suit. In 1991 Germany's conservative government introduced the *Stromeinspeisungsgesetz* (literally, the "electricity in-feeding law"), requiring utilities across the country to pay 90 percent of their annual average retail rate for purchases from clean energy sources such as wind turbines—indefinitely. The law, encompassing only a few paragraphs, resulted in an explosion of wind-generated electricity.

If, say, the average price in Germany was 10 cents per kilowatt-hour, then farmer Peter Ahmels could earn 9 cents per kilowatt-hour for every unit of electricity he sold to his utility. Ahmels quickly seized the opportunity and installed the first of two wind turbines on his 100-hectare (240-acre) farm in the northern state of Niedersachsen—wind turbines that he, not the utility, owns.

And because of the feed law, Ahmels could install more cost-effective turbines than the small turbines permitted under net-metering programs in North America. His first turbine was 300 kW; the second 500 kW. Ahmels's neighbors, along with farmers all across northern Germany, soon began following his example.

It was German farmers, says wind program manager Martin Hoppe-Kilpper of the Institute for Solar Energy Research, and not utilities or wind farm developers, who launched one of renewable energy's most visible success stories. By the early part of the 21st century wind turbines could be seen in every part of Germany, from the dikes at Kaiser-Wilhelm-Koog on the North Sea, to the hilltops of the Eifel Mountains, to the new states of the former East Germany.

Early in 2000 the German parliament, the Bundestag, replaced the old *Stromeinspeisungsgesetz* with a new feed law intended to double renewables' contribution to Germany's electricity supply, from 5 percent to 10 percent, by 2010. Under the new program, German utilities will pay 0.9 euro per kilowatt-hour for the first five years of operation of wind turbines installed in 2002. From the 6th year through the 20th year, the price paid varies with the performance of the wind turbine relative to that at a "reference site" (see table 10-1, Electricity Feed Laws for Wind and Solar Energy in Europe in 2002).

The price was chosen after parliamentary debate influenced by technical reports from the German Wind Energy Institute (DEWI) and the Institute for Solar Energy Research (ISET) on various renewable energy technologies. The reference site was chosen to represent a typical inland wind resource with an average wind speed of 5.5 m/s (0.12 mph) at 30 meters (100 ft) above ground level. Turbines that generate more than what would be produced at the reference site are paid less; those that generate less are paid more. At the best coastal sites, for instance, production could exceed the reference site by 150 percent, in which case payment would fall to 0.06 euro per kilowatt-hour under the new program after year five.

It was German farmers, says wind program manager Martin Hoppe-Kilpper of the Institute for Solar Energy Research, and not utilities or wind farm developers, who launched one of renewable energy's most visible success stories.

The tariffs are revisited biannually to monitor how the program is meeting the Bundestag's objectives. The price will then be adjusted accordingly.

German parliamentarians chose higher tariffs for less windy sites to encourage development in low-wind areas of central Germany. Their intent was to disperse wind turbines

Table 10-1
Electricity Feed Laws for Wind and Solar Energy in Europe in 2002

Country	(Years)	Full Load (Hours)	Capacity Factor	Tariff (€/kWh)	Conditions & Limitations
Wind Energy					
Germany	1–5			0.09	0.089 in 2003
Interior	6–20[a]			0.084	At reference site. Actual value varies by site.
Coastal	6–20[a]			0.06	Approx., actual value varies by site.
Spain					
< 50 MW				0.0628	
or				0.0290	Premium above market price.
France					
Continental < 12 MW[b]	1–5			0.0838	
Low wind	6–15	2,000	0.23	0.0838	
Moderate wind	6–15	2,600	0.30	0.0595	
High wind	6–15	3,600	0.41	0.0305	
Overseas Territories	1–5			0.0915	Includes Corsica.
Low wind	6–15	2,050	0.23	0.0915	
Moderate wind	6–15	2,400	0.27	0.0747	
High wind	6–15	3,300	0.38	0.0457	
Solar Photovoltaic[c]					
Germany					
< 5 kW				0.5000	For first 1,000 MWp.
Spain					
< 5 kW				0.4000	
< 25 MW				0.2000	
France					
Continental				0.1525	
Overseas Territories				0.3050	

Notes: 1€~$1 US

[a]Sites with 150% of "reference yield" are paid 0.06 €/kWh, those with less the 0.09 €/kWh tariff is extended two months for every 0.75% that the yield is under 150% of the reference yield.

[b]First 1,500 MW. After 1,500 MW reference full-load hours is reduced to low wind, 1900 hours; medium wind, 2,400 hours; high wind, 3,300 hours.

[c]Other technologies, such as biomass and geothermal, are also included.

across the landscape, rather than concentrating them in the windiest locales. This eases siting, as well as integration of the turbines into the electricity network.

The revised feed law also spells out, for the first time, how to calculate the costs for grid connection and for any necessary reinforcement of the distribution system. The new law then equitably apportions these costs to both parties: the wind turbine owner and the grid operator. The law also increases the transparency of how these costs are determined by allowing the wind turbine owner to use third-party consultants and contractors rather than relying on the good faith of the utility or grid operator.

French Feed Law

Just as Denmark began dismantling its successful electricity feed law, France began instituting its own version. In 2000 France became the latest in a growing roster of countries adopting electricity feed laws. Like the revised German feed law, the French feed law employs a tiered system, what ADEME's Chabot calls an "advanced tariff."

The French feed law adapts the German model to French political and cultural conditions. The French approach limits the size of wind projects that qualify for the fixed-price payment to 12 MW. Developers of projects greater than 12 MW must negotiate a contract with Electricité de France (EdF) outside the fixed-tariff system, negating all but the biggest projects.

As with the German feed law, payments are the same for all projects for the first five years. And as in the revised German program, the tariffs for the following 10 years are derived from the performance of the turbines relative to reference values.

Projects were limited to 12 MW as part of a grand compromise between powerful political forces and the then ruling red–green coalition in the national assembly. Following World War II, any power plant greater than 8 MW was nationalized and became part of EdF, the state-owned monopoly. During debate on liberalizing France's electricity market, EdF wanted to continue limiting independent generators to 8 MW. The compromise reached was 12 MW.

The French feed law pays one price for wind turbines in windy areas and higher prices for turbines located at less windy sites. This is intended to avoid the concentration of wind turbines in only the windiest locales—a practice that has brought bitter opposition to wind farms in Britain's highlands.

Using the continental designation of "full-load hours," the French program pays one price for low-wind sites, a lower price for moderately windy sites, and an even lower price for windy sites. Full-load hours are the amount of time that a wind turbine would take, operating at peak capacity or "full load," to generate the electricity produced annually. Thus a wind turbine on the windy coast of Brittany would be paid less than a turbine near Carcassonne, in the south of France.

Turbines operating in French overseas territories, such as the island of Guadeloupe in the French West Indies, are governed by a separate multitiered scheme. Electricity in the territories is generated mostly with fuel oil and thus is more expensive than in metropolitan France. For this reason the feed law tariffs in the territories are higher than those in continental France.

As in Germany, all projects—whether one turbine or twelve 1 MW turbines—require the approval of local or regional planning authorities. After the first 1,500 MW of wind is reached, the payment is reduced by lowering the amount of full-load hours that qualify for the higher payments. At a low-wind site, for example, the amount of full-load hours that qualify for the highest payment is reduced from 2,000 to 1,900 hours.

The French feed law has been so successful, reports ADEME's Chabot, that in only two years 700 MW in building permits have been applied for, compared to the 30 MW operating in all of France when the program was launched. Today wind turbines are spinning amid the vineyards of southern France, and more are to come. Chabot adds that more than 19,000 MW of applications for interconnection with EdF's grid had been filed by the end of 2002.

Feed Laws Elsewhere

Each country that has adopted feed laws has adapted them to its own needs. Spain is one example. As in France, the Spanish feed law grew out of efforts to liberalize the electricity market. In late 1997 Spain opened its electricity market to nonutility renewable-power projects. The following year a royal decree established tariffs for projects less than 50 MW and guaranteed them access to the grid. Under the Spanish feed law, wind turbine owners can opt for either a fixed price or a "market" price, plus a premium. The total tariff is designed to approximate 80 to 90 percent of the average price paid for electricity. Nearly all wind turbine owners have opted for the fixed-price tariff, about 0.066 euro per kilowatt-hour.

The feed law resulted in a veritable explosion of wind capacity, creating a bustling wind industry in less than a decade. Patterns of settlement and landownership differ in Spain from those in Germany and Denmark. In contrast to those in northern Europe, wind projects in Spain are massive, more like those in North America. To qualify for regional incentive programs, however, wind projects must use domestically produced wind turbines, not turbines imported from Denmark or Germany. Spanish wind projects use Spanish wind turbines built by Spaniards.

Seeing the dramatic success in Spain and the creation of new manufacturing jobs, Portugal and Greece have adopted similar feed laws. South Korea and Brazil also have comparable programs.

Wind turbines have operated for more than 5,000 million hours interconnected with local utilities worldwide without serious incident.

Quotas

The British bidding or "tendering" system has concentrated wind development and ownership in the hands of a few large firms, mostly subsidiaries of engineering conglomerates or divested electric utilities. The system has also engendered a caustic and well-organized backlash against wind energy that nearly halted wind development in the British Isles. Moreover, the cumbersome British system has required a host of solicitors to read through the tea leaves of the various regulatory directives. Despite the professed objective of market liberalization, the NFFO quotas were reminiscent of the command-and-control policies of previous eras.

In contrast, German wind development has been geographically dispersed and open to both cooperative and individual ownership. The German market was liberalized in the true meaning of the world. According to one survey, more than 200,000 people were members of wind cooperatives in Germany in 2001.

Feed laws encourage competition among manufacturers, and as a consequence spur technological development. Quotas encourage developers to build as big a project as feasible to gain economies-of-scale in site infrastructure and financing instruments. Each is the result of a political decision.

Power Quality and Safety

Utilities on both sides of the Atlantic were initially reluctant to interconnect their lines with wind turbines because of concerns about safety and power quality. To the chagrin of wind energy's critics, though, there have been few problems with safety, voltage flicker, harmonics, or other technical issues.

Collectively wind turbines have operated for more than 5,000 million hours interconnected with local utilities worldwide without serious incident. Most of this experience has been with wind turbines driving induction generators, a technology utilities understand well.

Wind systems, whether using inverters or induction generators, now have produced line-quality power for almost three decades without endangering utility equipment or personnel (see figure 10-3, Medium-size utility-compatible wind systems). Despite this record, some utilities may still have questions that need to be addressed.

The utility will require a signed agreement or contract stating the responsibilities of both parties, how costs are shared, and how payment will be determined. This agreement often demands that you address the utility's concerns about power quality, power factor, and safety.

In most countries, utilities have a franchise from the state to supply electrical power within a restricted territory. The company is required by this franchise to provide reasonably reliable service to its customers. It is

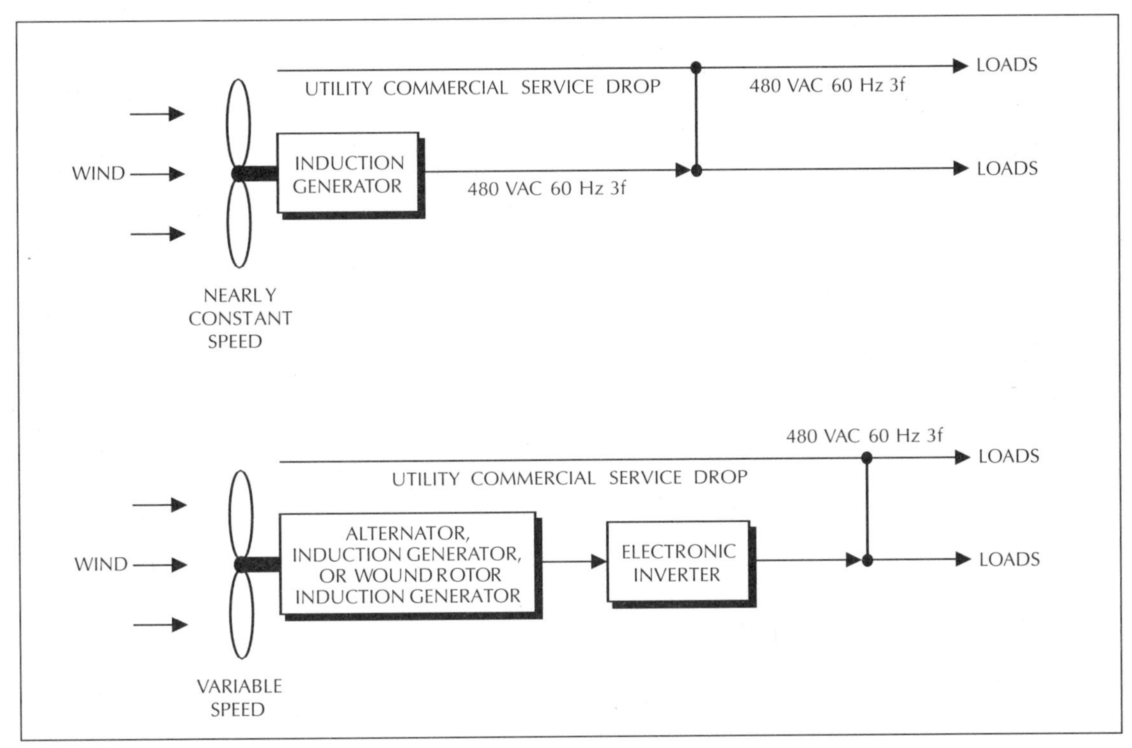

Figure 10-3. Medium-size utility-compatible wind systems. A schematic of the interconnection of wind turbines using induction (asynchronous) wind generators or alternators and power electronics. The voltage is typical of farms, ranches, and small businesses in North America.

responsible only to the point of delivery known as the service drop—the spot where the utility's lines reach the building or premises. From this point on the customer is responsible for the installation, operation, and maintenance of all other wiring—including that of a wind turbine.

In the United States, the tariff under which a utility operates allows it to refuse service when the safety of its equipment or linemen is threatened or when it believes service to customers may be compromised. The utility's only means of ensuring a safe interconnection with a wind machine is to refuse service when it has doubts.

The safety of its employees is the utility's principal concern, as it should be. Managers may fear that an interconnected wind system could energize, or deliver power to, a downed line during a power outage and electrocute a lineman. This fear is not entirely unfounded. Linemen have been injured by improperly wired emergency generators, and from the utility's perspective a wind turbine is little different.

Most inverters and all induction generators are line-synchronized; without the presence of the utility's line, they can't generate power. Nevertheless, it's possible for induction generators, in rare circumstances, to "self-excite," or provide their own excitation. "Islanding," as it's called, requires a rare match among capacitance, generation, and load. While the confluence of such conditions is fleeting, it's also possible and potentially damaging.

Utilities require that all wind machines designed for interconnection with their lines must disconnect themselves from the line during an outage and must not be able to self-excite. To preclude this, relays or electrical switches are placed on the utility side of the wind turbine's inverter or control panel (see figure 10-4, Schafer Systems' interconnection). When utility power is present, the AC

Figure 10-4. Schafer Systems' interconnection. Phil Littler displays the three-phase interconnection of Schafer Systems' 225 kW Vestas turbine in Adair, Iowa. That's all there is to it.

relay, or contactor, is energized, completing the electrical circuit. If utility power is lost for any reason, the spring-loaded relay is de-energized (turned off), opening the circuit and disconnecting the wind system from the utility line.

Some inverters, such as the Trace (now Xantrex) SW series, are designed for standby or uninterruptible power systems. Coupled with batteries, these inverters are able to provide power to a load when the utility line is down (see figure 10-5, Interconnection with battery backup). Because of the risk that the batteries could power the inverter and energize the downed utility line, these inverters incorporate electronic switches that prevent the inverter from sending power into the grid.

No matter what protection system is used, utilities require a disconnect switch to be visible and accessible to their personnel. When servicing a line nearby, linemen can throw the disconnect switch, severing the connection between the wind turbine and the grid.

Power Factor

Self-excitation becomes a problem when capacitors are used on the wind turbine side of the contactors. Wind turbine manufacturers and utilities frequently use capacitors to correct "power factor." Capacitors on the utility side of the AC contactors are disconnected from the wind machine whenever the contactor opens. Any capacitors used with a wind system for power factor correction should be placed on the utility side of the interconnection or should be designed to bleed down (discharge) after power is removed. This prevents self-excitation as well as precluding any electrocution hazard to those servicing the wind turbine.

Why does a utility or wind generator need capacitors to correct power factor? To understand the answer you need to know the difference between true and apparent power. (Yes, they both exist, and there is an important difference.) Power in watts, as you recall, is the product of voltage and current. This is generally true, but when we're dealing with alternating current we need to add another parameter to the equation: phase angle.

Don't let phase angle scare you. It merely describes the degree to which rising and falling voltage is in phase with the rising and falling current. If current rises from zero at the same time voltage rises from zero, the two waveforms are said to be in phase.

When this occurs, the cosine of the phase angle (0 degrees) equals unity (1), and true power is the product of voltage and current. In this case, true power and apparent power are equal, and the power factor—the ratio of

Figure 10-5. Interconnection with battery backup. Greg Nelson logs the number of kilowatt-hours sold back to San Diego Gas & Electric Company with his sine wave inverter. This inverter and a large battery bank (not visible) enable Nelson's wind and solar hybrid system to power his rural home in the event of a power failure.

true power to apparent power—is 1. This is the ideal. Unfortunately the real world seldom looks like this.

Loads on the utility's lines cause the current waveform to shift slightly. Current either leads (starts rising earlier) or lags (starts rising later) voltage. In rural areas, where power must be transmitted over long distances, the length of the line itself is sufficient to cause current to lag behind voltage.

At this point you may cry out, "Who cares?" The utility cares. And it cares a lot. When current and voltage are out of phase, true power decreases (the cosine of the phase angle becomes less than 1), yet apparent power remains the same. To deliver the same amount or real or billable power as apparent power increases, the utility must deliver more current. Beyond a certain point, to avoid these resistance losses and to maintain proper voltage, the distribution system must be expanded at a cost to the utility and its ratepayers. Apparent power, as Bruce Hammett at Wind Energy Conversion Systems in Palm Springs explains, is the power to excite or, as he elegantly puts it, "wake up the iron" in the windings of a motor or generator. The induction, or asynchronous, generators used in many wind turbines consume apparent power to energize or "excite" their fields.

The utility's trusty watt-hour meter measures only true power. Therein lies the dilemma. The utility must generate apparent power while it's able to charge only for true power. And true power is always less than apparent power (the power factor is normally either 1 or less than 1). It gets shortchanged. The utility tries to correct this—to keep the power factor as close to unity as possible—by adding banks of capacitors to its lines.

The foregoing applies to wind systems as well as to electric utilities. It is of importance to you because the power factor of your wind generator may cause the utility to take corrective

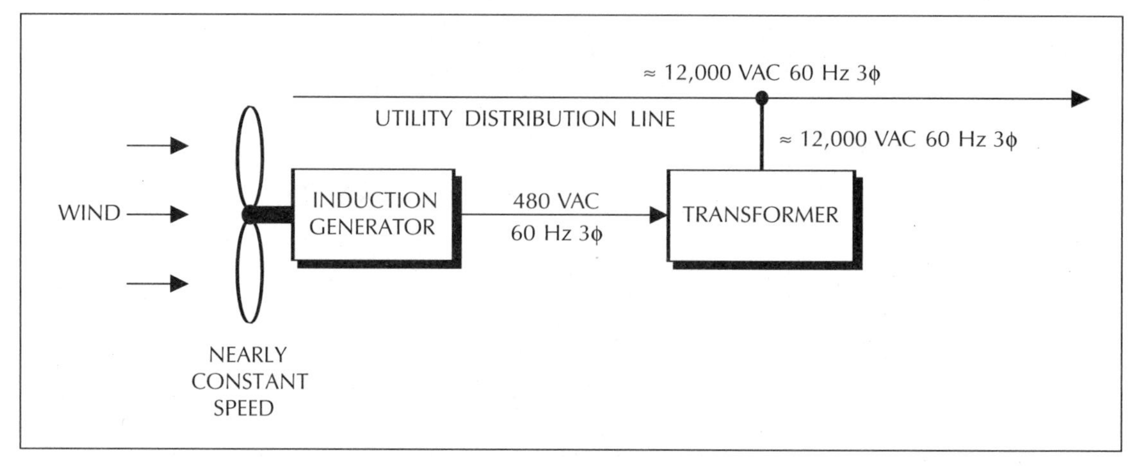

Figure 10-6. Typical medium-size wind system. A schematic of the interconnection of a medium-size wind turbine using an induction (asynchronous) generator via a transformer with a North American utility's distribution line.

action by installing capacitors, and then charging you for them. Power factor is also a favorite whipping boy for opponents of wind power within electric utilities.

If problems arise, you need to remember that the utility pays only for true power when you sell to it. (You are now in the utility's shoes; that is, you're trying to maximize the production of true power by getting your power factor as close to unity as you can. Values above 90 percent are desirable.) Don't think you're selling them horsemeat while charging them for prime beef, as some utilities have implied.

Moreover, your wind system appears to the utility as just another load (albeit a negative load) calling for power factor correction. It's no different than if you added a couple of freezers or new power tools to your home, as far as your impact on the utility is concerned. Utilities don't charge you for power factor correction when you install a new freezer, do they? No, of course not. And they shouldn't charge you for power factor correction when you install a wind generator, either.

Large wind turbines, clusters of turbines, and wind farms do require power factor correction. This is normally part of the wind turbines' electrical system, or it can be added separately. Large wind turbines using variable-speed rotors and power conditioning—industrial-scale inverters—not only can compensate for the reactive power needed to excite their generator windings, but can also deliver reactive power to the grid.

Voltage Flicker

Another problem arises with certain wind systems using induction generators. On some early models, the rotor was parked until the cut-in wind speed was reached. The rotor was then motored up to synchronous speed by drawing power from the utility. The effect on the utility was similar to the start-up of a compressor motor in a refrigerator or freezer. There was a momentary surge of current until the rotor came up to speed and the wind began driving the rotor. Because the power rating of wind generators is usually larger than that of most household appliances, the magnitude of the in-rush current can be large enough to cause a slight voltage drop in the line. The result may be voltage flicker. Your lights may dim briefly whenever the wind generator starts. This isn't a serious problem, but it can be annoying. The utility, though, may claim that an induction wind machine will detract from the level of service it offers other customers, and may require a dedicated transformer to mitigate the problem (see figure 10-6, Typical medium-size wind system). The transformer isolates the voltage drop to the customer using the wind machine and often eliminates the problem entirely.

In sparsely populated rural areas, most homes or farms already have a dedicated transformer; that is, only one customer is served by the transformer. In the suburbs, on the other hand, one transformer may be used for a number of customers. You can tell if you have a dedicated transformer by taking a look out your window. The utility primaries, or high-voltage lines, are carried at the top of the utility pole. Leads from the primaries are attached to the top of the transformer. The low-voltage lines, or secondaries, are attached to terminals on the side of the transformer. The service drop to your home is always from the secondary (low-voltage) side of the transformer. If the secondaries are strung directly to your house and to no other, the transformer is dedicated to your service drop. But if the transformer is located several poles away and the low-voltage lines serve several other customers, the transformer is communal and your wind turbine could affect your neighbors.

The degree of in-rush current or voltage flicker varies from one wind turbine model to the next. Early U.S. designs were the most notorious because they motored the rotor up to operating speed. Wind turbines today use free-wheeling rotors instead. When there is sufficient wind to turn the rotor at synchronous speed, the AC line contactors are energized and the wind system is brought online. There is still some in-rush current, but it's minor in comparison to what it was with the turbines of the early 1980s. Most medium-size turbines and all megawatt-size turbines use electronic controllers that connect to the grid "softly," minimizing any voltage flicker.

Harmonics

Inverters are not without their power-quality faults. They can produce current harmonics in various degrees. This does not affect most electrical appliances but, theoretically, can cause electromagnetic interference with television, radio, and telephones. The degree of interference depends on the inverter, its size and location, and the level of electrical noise on the utility's lines with which it is connected. (Power from the utility is itself never free from harmonics.)

This interference may be noticeable, but not necessarily objectionable. For example, one wind machine owner thought his machine was performing normally when his wife walked into the kitchen and wanted to know why he had turned the wind turbine off. Amazed, he asked her what she meant. She answered that she had grown accustomed to the faint humming it made on her radio, but she couldn't hear it anymore. He proceeded to check the control panel, and found that the inverter was down with a blown fuse.

Current harmonics haven't presented any insurmountable problem to utilities in either North America or Europe. Again, a transformer can be useful in attenuating any disturbance to the customer's immediate vicinity. Utilities in North America set very high standards, however—some say too high—for harmonic distortion from electronic inverters. In 2002 Underwriters Laboratories shocked the renewable energy community in the United States when it decertified Trace's SW series of inverters for failing to meet the harmonic standards. Trace (Xantrex) quickly corrected the problem, but not before every utility in the country had been alerted.

Technical specifications on the power factor, voltage flicker, and harmonic distortion of wind turbines on the international market can be found in Bundesverband Windenergie's annual Marktübersicht (Market Survey). See the appendix for details.

Dealing with the Utility

Electric utilities are businesses, and their employees are charged with making them

Aussie Built—Aussie Operated

In the late 1980s one of Australia's few wind turbine manufacturers, WestWind, adapted the Danish design approach to conditions down under. One of the few turbines that resulted from this venture was installed in the state of Victoria in late 1987. Through early 2001 the 16-meter (52 ft) diameter Breamlea turbine cumulatively generated more than 1 million kWh. During this period the turbine operated 18,000 hours on the generator's low-speed windings and 46,000 hours on the high-speed windings. In 2000 the 60 kW turbine produced 104,000 kWh at an average wind speed of 5.9 m/s (13 mph). The turbine had been operated for much of this period by a dedicated group of volunteers. Despite its success and its supporters, the turbine was taken out of service in 2001 because of the low price then being paid by the local utility.

profitable. Utilities are often large institutions. And as in any big bureaucracy, the right hand frequently doesn't know what the left is doing.

Corporate policy, as expounded by company managers, may be adamantly opposed to the interconnection of wind turbines of any size. Yet the word may not have drifted down to the lower levels where the work gets done. You may call up and be surprised to find the staff friendly and curious about your project. They may go out of their way to be helpful. This was the case with Metropolitan Edison, the once infamous Pennsylvania utility with nuclear reactors at Three Mile Island. Despite management's public opposition to renewables, Met-Ed's staff was extremely cooperative with the interconnection of a small turbine near Harrisburg.

Of course, there's the other type: utilities that make supportive pronouncements about clean energy sources and brag about all they're doing to improve the environment, but when you contact them, tell a different story. Mike Bergey of Bergey Windpower reports that in some cases it has taken more time to reach agreement with the utility than to build and install the wind system. In a particularly flagrant example, the Los Angeles Department of Water and Power simply told one wind turbine dealer it didn't want small wind turbines connected to its lines. The dealer chose to build a stand-alone system instead of challenging the powerful publicly owned utility.

Because utilities are bureaucracies, nothing happens quickly. You should notify them of your plans months in advance. Now that wind turbines have become much more commonplace, they may have a clearly defined policy and may be able to promptly give you the answers you seek. More often than not the particular employees handling your account won't have dealt with a small power producer before and will cover themselves every step of the way with time-consuming approvals.

Utility deregulation has complicated matters in North America. Many utilities are merging, changing their names, and moving offices. Employees often don't know who is responsible for small power producers, or even if they will have a job next week. Philosophy professor Corky Brittan learned this to his regret with Montana Power. After 18 months of fruitless negotiations, he still didn't have a contract!

You need to appreciate the utility's point of view, because you're dependent on its cooperation. The difference today is that we have almost three decades of actual experience with wind turbines operating on utility networks around the world. Utilities are far less likely to balk at interconnection today than they were 10 years ago. Utility executives dread nothing more than a self-righteous customer strutting into the office and demanding that the utility kneel down and graciously grant a trouble-free interconnection. If you want a fight, there's no better way to find one.

Wind turbine manufacturers have experience dealing with utilities, and they can be helpful in obtaining a fair contract. They can also assist with any technical issues that the utility might raise. Often the manufacturer will talk directly with the utility engineer handling your application. This can facilitate a quick agreement, because the two speak the same technical language.

After you have run the technical gauntlet there may be a few more obstacles to overcome. Liability insurance is one. Some utilities, principally the rural electric cooperatives, may insist that the small power producer purchase liability insurance in case of an accident. Here's their rationale: If a lineman is injured by a wind machine, the utility's workers' compensation insurance pays the lineman's claim. The insurance company that covered the utility's compensation sues the small power producer for recovery of its money. The lineman may also sue for damages. The small power producer loses in court, can't pay, declares bankruptcy, and the insurer has to absorb the loss. The utility's insurance rates rise, and you know who pays the bill in the end: the ratepayers. This scenario hasn't taken place with a wind system, but electric utilities have been sued for a lot less when they were not at fault—and have lost. The rural co-ops, because they are small, can't afford mistakes. They're gun-shy when it comes to liability.

The cost for this insurance isn't prohibitive, but it can cut deeply into the total revenues of a small wind system. Before you acquiesce, determine how the utility treats customers with emergency or standby generators. Are they required to have the same insurance? If not, why not? If they are not required to have insurance, the utility may be discriminating against you. The line contactors on your wind system respond in much the same manner as automatic transfer switches on standby generators. (Technically anyone with a standby generator would fall into this category, whether using a manual or an automatic transfer switch.) After decades of experience with interconnected operation from wind turbines worldwide, and no reported cases of off-site electrocution, there's simply no longer any reason for utilities to impose liability insurance.

Another contract provision that might raise your ire is the "hold harmless" clause. Read the contract's fine print. If you don't understand what it says, get an attorney to look it over. Pennsylvania Power & Light once proffered a contract that held the company harmless against all claims arising from damage or injury due to the interconnection, whether the fault of the small power producer or the fault of the utility. One dealer inserted the following to make the contract more equitable: ". . . so long as the act(s) giving rise to the claim were not due to the negligent conduct of [the utility], its employees and agents." In this way, the utility becomes responsible for any damage it causes. It's only fair.

Should a dispute develop, consult the responsible authorities. In the United States contact the state's public utility commission. The commission can mediate the dispute or bring action against the utility through the regulatory process—if it has jurisdiction. Many states also have a separate agency that acts on behalf of the consumer. These often act in the role of ombudsman and facilitate conflict resolution without the expense and enormous delays of the regulatory process. Above all, be patient. The regulatory bureaucracy is painfully slow. But sometimes all it takes is a phone call from the right person for the utility to suddenly see the light and mend its ways.

Cooperation is growing between utilities and the wind industry as more utilities gain operating experience with modern wind turbines. Bergey Windpower reports that most utilities it deals with require only one phone call from the manufacturer to grant permission for the interconnection. Some utilities have also allowed net energy billing, even when they were not required to do so.

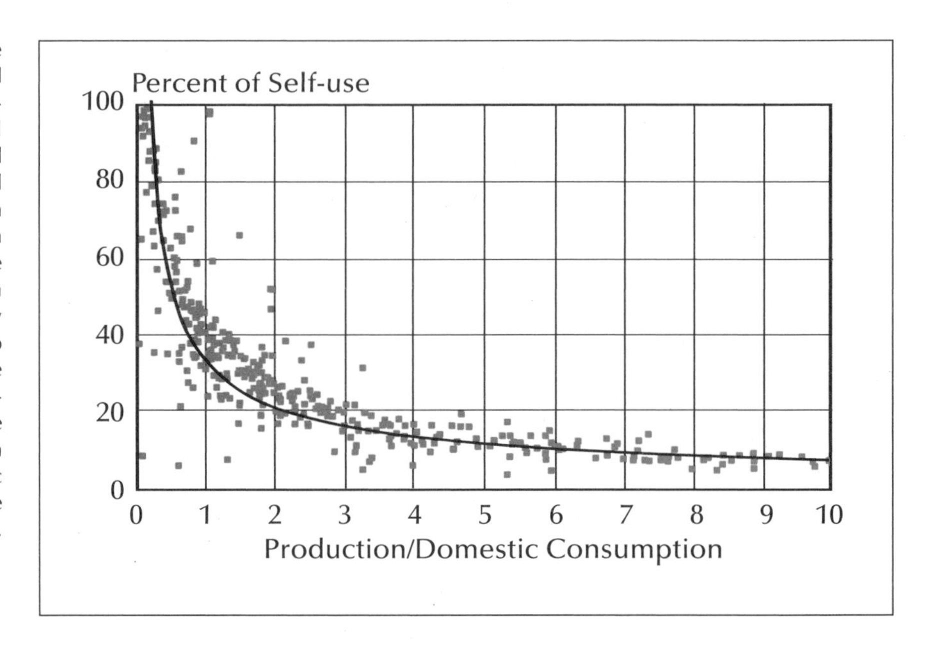

Figure 10-7. Degree of self-use. The amount of household energy demand supplied by a wind system as a function of the wind turbine's annual energy output and total household consumption. The most detailed records on this relationship come from Germany. This chart is derived from data on more than 400 household-size wind turbines. There is some scatter in the data because some owners wisely shift discretionary consumption to periods when wind is available. To use 100 percent of a wind turbine's generation on site, the turbine must be sized to produce only one-fourth of the household's domestic consumption. (Institut für Solare Energieversorgungstechnik, ISET).

Sizing—What's Right for You

From the previous discussion of PURPA, it's apparent that finding the right size turbine for an interconnected application in the United States isn't difficult. In most cases, a wind machine that produces less than your total consumption is preferred. Where the utility buys wind generation at approximately the retail rate, install the biggest wind machine you can afford. But before you can estimate what size wind turbine is best, you must first determine how much energy you need.

How much electricity do you consume? Do you know? If you're like most people, you know how much you pay each month but you don't have the foggiest idea of how much electricity you use. To find out, you don't have to do anything more complicated than pull out your old electric bills. There are at least two items of importance on them. One is your consumption in kilowatt-hours. The other is the total cost. From these you can calculate how much electricity you use each year and what you're paying for each kilowatt-hour of electricity.

To encourage energy conservation, most utility bills clearly tell you how much electricity you consume each month and how that amount compares with past consumption. Some utility bills show your consumption over the past 12 months. If your bill doesn't have this feature (an electric use profile), you'll have to tally up monthly consumption for the entire year yourself.

The average North American household without electric heat uses approximately 20 kilowatt-hours per day or 600 kilowatt-hours per month. The typical household consumes about 7,200 kilowatt-hours per year. Californians and people living in the Southwest consume less, about 6,000 kilowatt-hours per year. European households use half the amount that the typical North American family does. Dutch homes use about 3,200 kilowatt-hours per year.

Homeowners with electric heat use considerably more than those without. Depending on the climate, the size of the building, and how well it's insulated, an all-electric home will use from 15,000 to 80,000 kilowatt-hours per year.

If you plan to use the wind turbine to reduce your heating bill from other fuels, apply the above technique to bills for heating oil or natural gas.

Use your utility bills to learn more about your pattern of consumption. In the process,

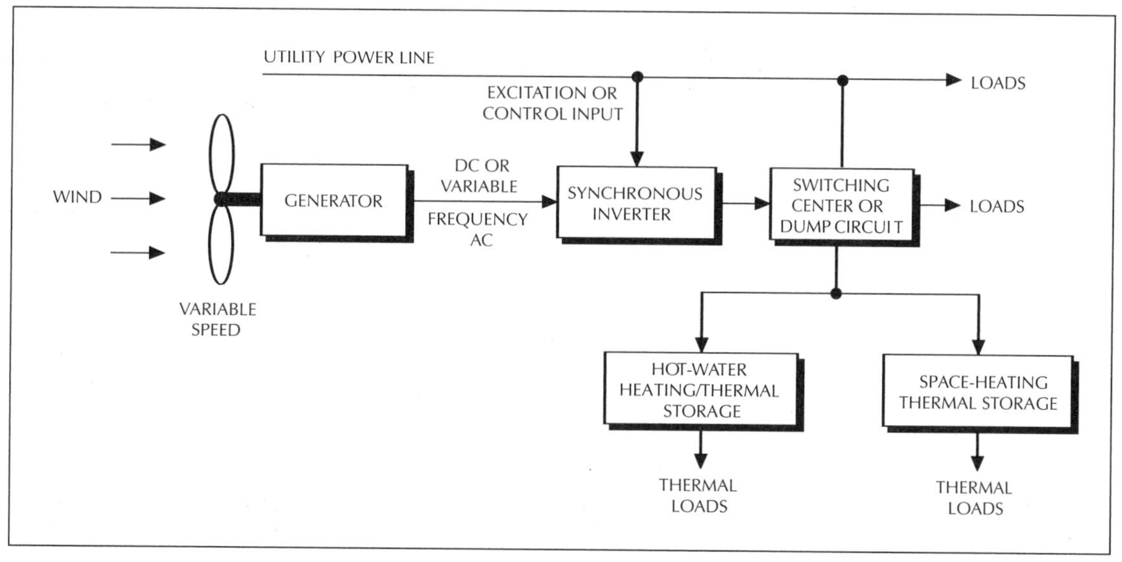

Figure 10-8. Dump circuit. Excess electricity is dumped or diverted to alternative loads rather than being sold back to the utility at less than the retail rate.

you will become more attuned to how you use electricity and you'll begin to find ways to use less. It's cheaper to conserve energy than to generate it.

Consumption varies not only from month to month, but also by time of day. For a typical home without electric heat, use peaks with peak activity around the house: early morning as family members ready themselves for work or school, again at lunch, and then in the evening around dinnertime. Because wind speed also fluctuates throughout the day, output from the wind machine may not match consumption. The wind may howl at night when consumption is lowest and be calm during the day when use peaks. In an independent power system this mismatch is tempered by the storing of excess energy in batteries. In an interconnected application excess energy flows to the utility. During light winds the utility makes up any shortfall from the wind turbine.

Ideally you would like to run your kilowatt-hour meter backward, selling any excess energy at the retail rate and buying what you need when you need it. In this way you use the utility as a battery, storing excess electricity until you need to use it. Net metering makes this possible.

Yet many states and provinces in North America don't permit net metering, and many utilities pay significantly less for the energy they buy back than for the energy they sell. Often the utility will replace the kilowatt-hour meter with two ratcheted meters: one to register purchases the other to register deliveries to the utility.

If you determine that the sale of surplus generation is not in your interest, then it's often necessary to size the wind turbine so that it rarely produces excess electricity. Studies of household-size turbines in Germany indicate that a wind turbine meeting only one-fourth to one-third of your demand may be as large as practicable under such circumstances (see figure 10-7, Degree of self-use).

Dump Loads

Instead of limiting yourself to a small wind turbine to avoid selling electricity to the utility at a loss, you have the option of using a larger, more cost-effective turbine and diverting any surplus to another load. Use a diversion controller to direct excess energy to a standby load or to space heating (see figure 10-8, Dump circuit). For example, you could install another tank for your domestic hot water. As in the wind furnace concept, wind-generated electricity is used to heat water in the tank, which is then fed to your hot-water heater. The preheated water reduces your consumption of conventional energy, whether you heat water with electricity or with natural gas. The extra tank allows you to dump excess electricity into thermal storage whenever you're not using all the electricity produced. Your water heater can then draw from this tank whenever it needs makeup water. In this way you have stored electricity as low-grade heat.

You may wonder why you need the dump circuit and an extra tank if you already have an electric hot-water heater. When the tank's thermostat calls for electricity to heat the water to a certain temperature, it draws automatically from the house's circuits, which the wind machine is supplying. That's the problem. It draws power only when needed. It's the boss. We want to use the wind-generated power when it's available, and that's not always when there's a demand for it.

Once you've mastered the concept of using excess electricity to preheat domestic water, you can turn your attention to other loads. If you heat your house with hot water you can preheat the water in your furnace: Just put in a bigger storage tank. Voilà, we're right back where we started, with the wind furnace concept. You have the same advantages as before, plus the wind machine supplies all your regular electrical loads.

While the vast majority of wind-generated electricity today is delivered to the utility network from wind turbines "tied" to the grid, there are many places where there is no utility service. Wind turbines that supply standalone power systems off the grid are the subject of the next chapter.

11

Off-the-Grid Power Systems

Prior to the development of interconnected wind turbines, wind generators had historically been used for powering remote sites where utility power was nonexistent (see figure 11-1, Off-the-grid wind systems). These home light plants used wind machines and banks of batteries sized to carry the household through winter winds and summer calms. Occasionally the dealer would throw a backup generator into the mix to charge the batteries during extended calms. The high cost, poor reliability, and maintenance requirements of these early systems discouraged all but the hardiest souls from living beyond the end of the utility's lines.

That's no longer true. Home power systems have become so mainstream, says Wes Edwards, that homes using them now qualify for mortgages. Edwards, a licensed electrician who has lived off the grid in northern California since 1974,

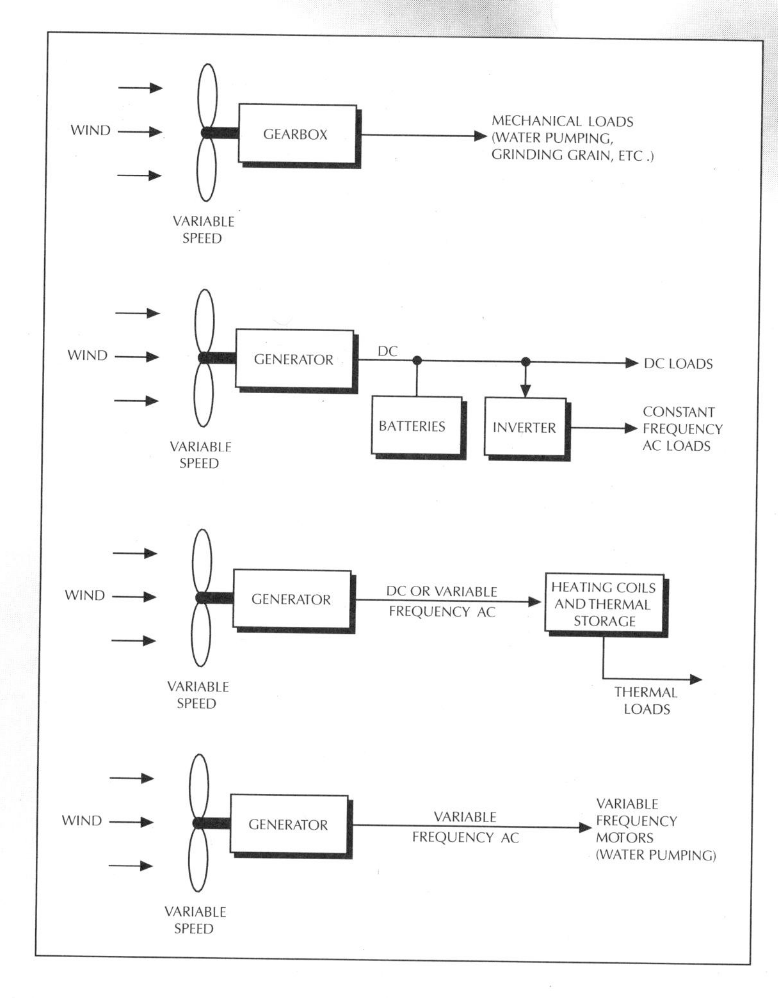

Figure 11-1. Off-the-grid wind systems. Several techniques exist for using wind machines in standalone applications. Historically, wind turbines have been used to charge batteries (windchargers) in remote power systems, to pump water mechanically (farm windmills), or to grind grain (European windmills). Today wind turbines can be used to drive AC motors directly in specialized pumping applications. Wind turbines can also be used to generate heat at remote sites

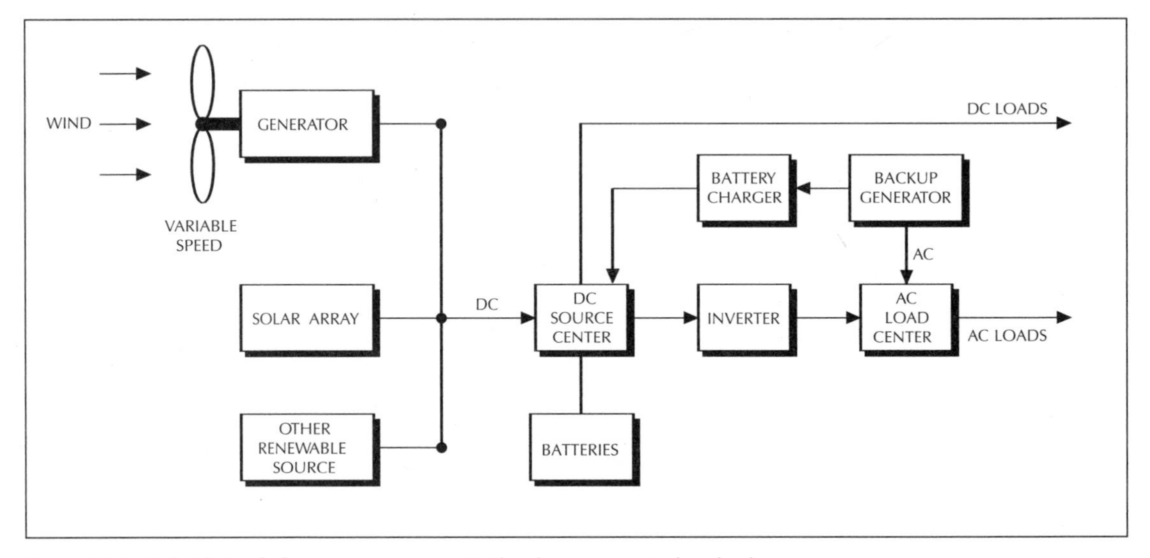

Figure 11-2. Hybrid stand-alone power system. With advances in wind and solar energy, remote power systems are no longer dependent on any single technology. The addition of a backup generator provides even greater flexibility in sizing the system's components.

says that today's improved inverters, low-power appliances, and widespread availability of photovoltaic panels have revolutionized stand-alone power systems.

Data compiled by Pacific Gas & Electric Co. (PG&E) confirms Edwards's observation. In an early-1990 study of its service area, PG&E found the number of stand-alone power systems mushrooming at a rate of 29 percent per year. The firm expects the market to continue expanding as urbanites increasingly move to rural areas not currently served by the California utility. The business prospects looked so enticing that PG&E even toyed with the idea of providing the stand-alone home power systems itself, instead of building additional power lines.

It just doesn't make sense today to design an off-the-grid system to use only wind or only solar.

New technology has made this possible. Until photovoltaic's entry into the market, remote power systems were solely dependent on wind machines (in some special cases, small hydro systems) and backup generators. The modularity of photovoltaics (PVs) has transformed the remote power system market by enabling homeowners to more closely tailor power systems to their needs—and their budgets.

Despite wind's advantages, wind turbines are less modular than PVs. Although the output of some micro turbines is no more than that of most PV modules (50 to 100 watts), each additional turbine requires a separate tower and controls. When scaling up the output from a remote power system, it's easier to add more modules to a PV array than it is to add more wind turbines.

Wind is also far more site-specific than solar. It's safe to assume that nearly everywhere on earth the sun will rise and set every day. Not so with wind. Wind follows daily and seasonal patterns that are less predictable. In the midlatitudes of the Northern Hemisphere, the winds are strongest in winter and spring and weakest during summer. Fortunately this pattern happens to coincide with the attributes of solar energy. Winds are generally strongest when the sun's rays are weakest, and winds are weakest when solar radiation reaches its peak. For this reason wind and solar are ideally suited for hybrid systems that capitalize on the advantages offered by each technology.

Hybrids

With advances in solar and wind technology, it just doesn't make sense today to design an off-the-grid system using only wind or only solar. Hybrids offer greater reliability than either technology alone, because the remote power system isn't dependent on any one source (see figure 11-2, Hybrid stand-alone power system). For example, on an overcast winter day outside Pittsburgh, Pennsylvania, when PV generation is low, there's more than likely sufficient wind to make up for the loss of solar-generated electricity.

Wind and solar hybrids also permit the use of smaller, less costly components than would be needed if the system depended on only one power source. This can substantially lower the cost of a remote power system. In a hybrid system, the designer need not size the components for worst-case conditions by specifying a larger wind turbine and battery bank than necessary.

Hybrids often include a fossil-fuel backup generator for similar reasons. In effect, stand-

Tale of Two Cities

At many midlatitude sites, wind and solar are complementary resources. Wind generation peaks during winter when solar insolation is at a minimum. Similarly, in summer a solar array will shine while a wind turbine may sit in the doldrums. Amarillo, Texas, and Pittsburgh, Pennsylvania, illustrate this relationship. No two cities could be more different: Pittsburgh's an old steel town; Amarillo, an old cow town. Their wind resources are equally diverse. Amarillo's windy; Pittsburgh's not. Even so, a wind and solar hybrid power system can provide usable amounts of energy in either location.

Consider a hypothetical hybrid sized so that both the wind and solar systems have approximately the same collector area. You wouldn't size a hybrid system this way; it's simply a means to compare the potential contribution of each resource. In this example, six panels of 75 W photovoltaic modules are equivalent to one Whisper H40.

While the solar resource is 40 percent greater on the high plains of Texas than in the humid eastern United States, the wind resource is 180 percent more productive. Sizing an off-the-grid power system must account for such large differences. The Amarillo data illustrates that at a windy site, a wind turbine can contribute far more energy to a hybrid system than a similar-size PV array. At a low-wind site, such as Pittsburgh, wind and PV are equally productive.

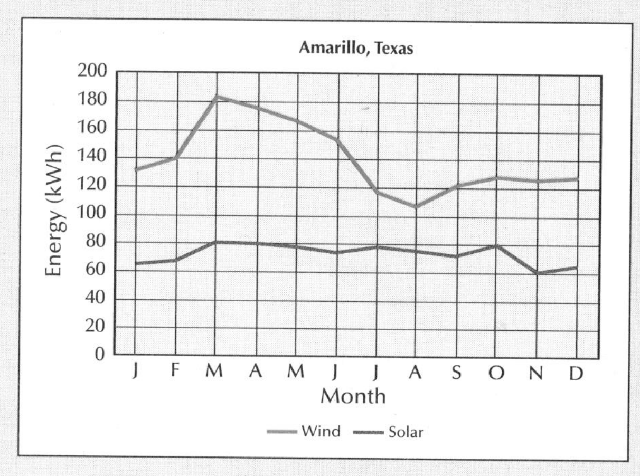

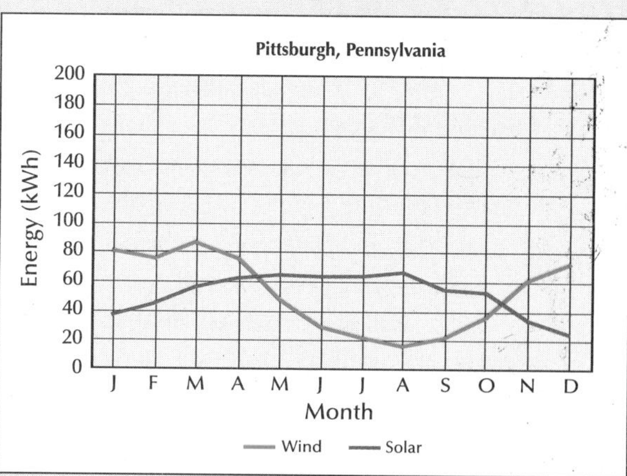

Hybrid wind and solar generation. Wind and solar resource for Amarillo, Texas (top); Pittsburgh, Pennsylvania (bottom).

alone systems substitute the fuel and the maintenance of the backup generator for a larger wind turbine or more solar panels. Depending on the size of the backup generator and the power consumption at the time, the generator can top up discharged batteries and meet loads not being met by the combined wind and solar generation.

Despite the improving cost-effectiveness of PVs and small wind turbines, the initial cost of a hybrid system remains high. To keep costs down, it behooves users to reduce demand as much as possible. Fortunately the development of compact fluorescent lights and energy-efficient appliances now makes this possible with little sacrifice. Today's energy-efficient appliances permit homeowners to meet their energy needs with smaller, less expensive power systems than were once necessary.

Table 11-1
Residential Energy Consumption

	(Therms/hour)	(kWh/hour)
Heating		
Small gas furnace	0.6	
Large gas furnace	1	
Space heater		1.5
Baseboard heater		3
Electric furnace		10
Heat pump		3–5
Air Conditioning		
110 volt window unit		1.5
220 volt window unit		2.6
Central		4.5
Portable fan		0.2
Water Heating		
Electric		300–400
Gas	20–30	
Heat pump		175–225
Refrigeration		
16 cu. ft. frost-free refrigerator		100–150
20 cu. ft. frost-free refrigerator		115–180
10 cu. ft. manual defrost refrigerator		35–60
15 cu. ft. frost-free freezer		70–150
Lighting		
General		50–200
Laundry		
Electric clothes dryer		5/load
Gas clothes dryer	22/load	.5/load
Washing machine		
Cold		.25/load
Warm wash, cold rinse	.11/load	.25/load
Hot wash, warm rinse	.33/load	.25/load
Appliances		
Stereo		0.03
Color TV		0.23
B&W TV		0.07
Vacuum cleaner		0.75
Microwave/5 minutes		0.1
Toaster/use		0.08
Toaster oven		0.5
Electric range		
Oven		1.33
Surface		1.25
Cleaning/use		6
Gas Range		
Oven	0.09	
Surface	0.07	
Cleaning/use	0.5	

Source: Pacific Gas & Electric Co., Market Research Dept.

Reducing Demand

The first place to start is by reducing demand. Most North Americans can easily halve their electricity consumption. If The Company Store's advertisement for down comforters "to take the chill off an air-conditioned room" doesn't strike you as absurd, then it's time to ask yourself the question of how much you're willing to pay to generate your own electricity. Reducing your consumption by conserving and increasing efficiency improves the services that a renewable power system can provide by stretching each kilowatt-hour to do as much work as possible.

Most North Americans can easily halve their electricity consumption.

To reduce demand, perform an energy audit of your lifestyle. Knowing how, where, and when you use energy is even more important for an off-the-grid system than for an interconnected wind turbine. Determine what appliances you plan to use at your remote site, and estimate how much electricity they will consume (see table 11-1, Residential Energy Consumption).

Conserve as much as possible. It's always cheaper to save energy than to generate it with a hybrid power system. In other words, the return on investment for conserving energy is higher than that for producing it in an off-the-grid power system (see table 11-2, Return on Investment of Conservation Measures in an Off-the-Grid Power System). To maximize the value of your renewable power system and minimize its cost, carefully pare your electricity consumption to the minimum needed for the services you require.

Table 11-2
Return on Investment of Conservation Measures in an Off-the-Grid Power System

	ROI (%)
Compact flourescent lighting	100–140
New refrigerator	10–30
Replace PC with laptop	10–20
Solar photovoltaics	3–12

Source: Rahus Institute

Turn off all unneeded loads. This should be obvious, but like the example of grabbing a quilt because the air-conditioning has made the room too cold, it isn't. The National Renewable Energy Laboratory found that at one hybrid system in Chile, villagers left lights on 24 hours per day—despite NREL's plea to turn them off!

Decide if there are any electric appliances, such as electric hot-water heaters or electric stoves, that can be switched to gas or other fuels. It makes no sense to squander your

Cutting Consumption

When California's power crisis struck in 2001 my wife, Nancy Nies, and I carefully examined our domestic consumption. We found that we were already at the baseline for our climatic zone: the sunny San Joaquin Valley. *Baseline* is California's term for what energy planners calculate is the average electricity consumption in a particular region. At the height of the crisis, many wealthy Californians charged that no one could actually live on the baseline. People can. We did. But we found we could do even more, without hardship.

We looked at lighting, refrigeration, laundry, and cooling. We had been using fluorescent task lighting and notebook computers for several years—notebook computers use one-tenth as much electricity as desktop models.

We replaced the few remaining incandescent lamps with compact fluorescents. Then we replaced our refrigerator. This alone cut our consumption 600 kWh per year. We also dusted off our "solar clothes dryer" and began using the clothesline instead of the electric dryer.

Our peak consumption occurs in summer, due to air-conditioning. During the cooling season, we turned up our thermostat 2°F and found we could live comfortably with the addition of a few strategically placed fans. We also opened our windows at night to cool the interior, and closed them as soon as outside temperatures began rising.

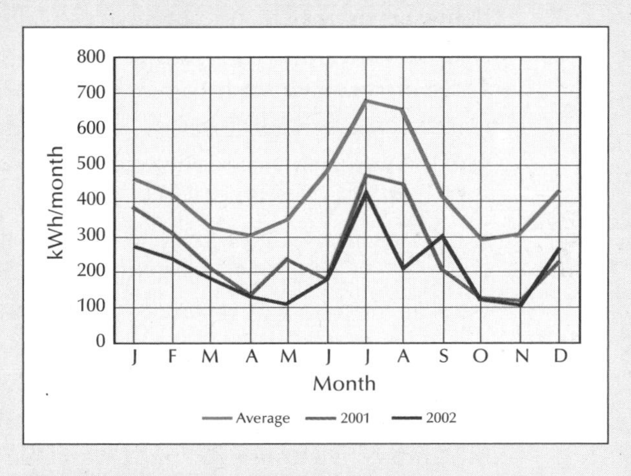

Nies-Gipe electricity consumption. The Nies-Gipe household was able to cut its consumption by 40 to 50 percent in 2001 and 2002 with simple conservation measures.

By living more consciously we cut our consumption 40 to 50 percent. Our consumption of 2,500 to 3,000 kWh per year is about half that of the typical California household.

hard-earned electricity on inefficient appliances or on uses to which electricity isn't well suited. Heating is one of them. Heating with gas, oil, propane, or wood is far more economical at a remote site than heating with electricity.

Though cooking consumes little energy overall, electric stoves have high peak power demands that will affect the size of inverters and other hybrid components. Cook with gas or propane, or use a microwave oven instead.

PG&E found, in its study of off-the-grid systems, that most remote generation is used for lighting, refrigeration, and water pumping. (Remote sites are seldom served by municipal water sources.) Lighting is the easiest to tackle. Compact fluorescent lamps can reduce lighting demand significantly. Task lighting—lighting only those areas where light is needed—daylighting, and simply turning off lights when they're not needed can all cut lighting consumption by two-thirds.

Similar savings can be achieved with refrigeration. Modern refrigerators use as little as 300 kilowatt-hours per year, a fraction of what they used in the 1970s. Sunfrost refrigerators, the efficiency champions, use even less. Generally, if your refrigerator is more than 10 years old, it should be replaced—and whatever you do, don't put that old refrigerator in the garage. Drive a stake through its heart.

Depending on the size of the house and on the climate, air-conditioning can double the consumption of an otherwise energy-efficient home. If you must have air-conditioning, ask yourself whether an evaporative cooler will suffice. Swamp coolers use far less electricity and work well in arid climates, such as the southwestern United States.

For energy-efficiency ratings of appliances from washing machines to refrigerators, and for tips on reducing your energy consumption without reducing your comfort, read Consumer Guide to Home Energy Savings by the editors of *Home Energy* magazine (see the appendix for details).

The key is to remain flexible. Sacrificing the lifestyle you desire isn't necessary, but some modification of behavior often proves beneficial. For example, cutting back on energy-intensive discretionary loads on days when the power supply is reduced extends battery life and leaves a little extra in storage available, should you need it for more important loads such as pumping water. Not unlike our ancestors, learn to synchronize your behavior with the weather. Do the laundry when it's windy or on a bright sunny day. In this way, you take full advantage of the fuel when it's available.

Turning off unneeded appliances isn't much of a burden for those who are energy-conscious; it's already become second nature. But for those going "cold turkey" from a highly consumptive lifestyle where energy's undervalued, it can be a rude awakening. In such cases it might be wise to gradually reduce your consumption until you're ready to make the transition to producing your own power.

The average North American household should be able to reduce its consumption to about 3,600 kilowatt-hours per year, or about 10 kilowatt-hours per day. This isn't spartan living. Most Europeans live comfortably on this amount or less. How much you're able to reduce your consumption will determine not only what size system you need, but also whether you should wire for DC or AC, and at what voltage you should operate your power system.

AC and DC Systems

All stand-alone power systems produce and store direct current (DC). Photovoltaic (PV) arrays produce DC directly. Most wind machines produce alternating current (AC), which must then be rectified to DC, as in an automotive alternator. Direct current is then stored in batteries until needed. The exception to this scenario is the backup generator, which can provide constant-frequency AC for AC loads while also recharging the batteries.

There are three sides to the stand-alone power system: generation, storage, and loads. Since most of the loads will be supplied by the batteries most of the time, the choice becomes whether to feed the loads with DC directly or with AC through an inverter.

For any major off-the-grid application, such as a full-time residence, wire for AC and wire to local building codes. Wiring to code simplifies construction, because most electrical fittings readily available are designed only for AC. Wiring to code also makes mortgage financing possible. While electronic inverters add complexity and reduce the power system's overall efficiency, modern energy-efficient appliances make up for any losses in the inverter.

Don't discount DC entirely. A small vacation cabin or motor home can function quite satisfactorily with DC appliances (see figure 11-3, On the road and off the grid). In even larger systems there may be some major loads, like refrigeration or water pumping, where DC wiring is justified to limit inverter losses. Sunfrost refrigerators are built in both DC and AC versions for just such situations.

Figure 11-3. On the road and off the grid. Low-voltage DC systems are ideal for motor homes and small vacation cabins. Here two Air modules charge a motor home's batteries outside Tehachapi, California. (Nancy Nies)

An equally important decision is charging-system or DC voltage. The voltage of the charging system determines the size of the cables needed to conduct current from the solar array and wind turbine to the batteries, as well as the number of batteries. For cabin-size systems that use DC directly, 12 volts is preferred because of the large number of appliances available for the recreational vehicle market.

As demand increases, so too does the optimal system voltage. Richard Perez, editor of *Home Power* magazine (not to be confused with *Home Energy* magazine), says that systems requiring less than 2 kilowatt-hours per day, such as vacation cabins, should stick with 12 volts; those using up to 6 kilowatt-hours per day should go with 24 volts; and those using more than 6 kilowatt-hours per day should opt for 48 volts. Systems that use 24 and 48 volts typically use inverters for powering conventional AC appliances.

Except for the smallest applications, wind turbines, PV arrays, and batteries should be sized for 24 to 48 volts. Long cable runs or heavy consumption may require a 120-volt system. But the large number of batteries needed for 120 volts is generally prohibitive except for those systems designed to meet consumption approaching 20 kilowatt-hours per day. High-voltage systems do permit greater flexibility in siting your renewable power sources than do either 24-volt or 48-volt systems. If you need to move your solar array into a more exposed position or move your wind turbine to a hilltop, the 120-volt system may be your only choice to keep resistance losses in the cable, and the cost of the conductors, to a minimum.

Sizing

At extremely small loads—say, a few lights and a radio—a 12-volt PV system comprising one or two 50-watt panels that power DC appliances directly makes economic sense. For such applications PVs are easier to work with and less expensive than wind. But as loads increase above a few hundred watts, wind and solar hybrids become increasingly more attractive. At the loads typically found in most households wind is more cost-effective than solar alone. The generation from small wind systems at good sites costs about half that of solar.

As with the wind, the output from a solar array isn't constant over time, over seasons, or from one region to another. PV systems in the southwestern United States and plains states produce at their rated output about five hours per day. For example, a 200-watt system will generate 1 kilowatt-hour per day, or 365 kilowatt-hours per year.

Often after installing a stand-alone power system, a homeowner finds that the noisy backup generator runs more than anticipated. Because PV modules are more modular and far easier to add incrementally than wind turbines, most people simply add more solar panels as needed. Many PV arrays are mounted on the roof and, for much of North America, are simply tilted at an angle equal to the latitude. Some users adjust their panels seasonally to optimize performance. To maximize winter production, tilt the panels at the latitude plus 15 degrees. To maximize summer generation, set the tilt angle to the latitude minus 15 degrees.

Just as wind–electric generation can be increased by installing turbines on taller towers, so PV generation can be improved by mounting the array on a tracker. In winter the tracker boosts performance a modest 10 to 15 percent. But during summer, when the winds are most likely to be light, trackers really shine, producing 40 to 50 percent more energy than a fixed array. For more information on estimating the performance of a PV array, see Joel Davidson's book *The New Solar Electric Home,* or *The Real Goods Solar Living Sourcebook.*

How many PV panels and how big a wind turbine will you need? For all systems greater than a few hundred watts, design for both wind and solar. "It's senseless to install a PV-only system," says Mick Sagrillo. If you do, and you undersize the PV array because of the expense, the result is a merely a "PV-assisted generator set."

In PV-only systems, designers size the arrays to compensate for minimum winter insolation. By adding wind, says Sagrillo, you can size costly PVs for maximum summer insolation, reducing the total number of PV modules needed. In a temperate climate, this takes best advantage of each resource by emphasizing solar-electric generation when wind-electric generation is often at a minimum. Sagrillo argues that "in a PV-hybrid system a tracker is critical" because you want to start generating as soon as the sun rises in

U.S. Solar and Wind Data

Solar radiation data and some wind data for major U.S. cities, suitable for downloading into a personal computer spreadsheet, can be found the National Renewable Energy Laboratory Web site at:

http://rredc.nrel.gov/solar

NREL maps of wind resources in power density can be found at a different address:

http://rredc.nrel.gov/wind

Climatic summaries for major cities can be ordered from the National Climatic Data Center Web site at:

http://lwf.ncdc.noaa.gov/oa/ncdc.html

Micro Hybrid Power System

Remote power systems using a micro turbine and a few solar panels have proven popular for those who want to use renewable energy in a limited application. The low cost of their components also makes them perfect for hobbyists and backyard experimenters. These low-power hybrids can be packaged with small DC–AC inverters to power consumer electronics, but many are used with DC appliances obtainable through specialty houses serving the recreational vehicle market.

Micro turbines are the wind energy equivalent of solar walklights: inexpensive and easy to install. They're light enough that you can pick one up and carry it home in your arms. Because most micro wind turbines are used for vacation cabins where there's less stringent demand on performance, they're often installed on shorter towers than their big brothers. Aerogen, Ampair, Marlec, and Air models are often installed on guyed masts using readily available galvanized steel pipe. With due care, micro turbines can be installed by do-it-yourselfers.

Micro Hybrid Power System

		Collector Area			AEO	
		(m^2)	(ft^2)	Total	(kWh/yr)	($/kWh/yr)
Wind	SWP Air	1.1	12	$2,000	429	4.7
Solar	4 panels, 300 W	2.4	27	$2,700	548	4.9
Battery storage	4 kWh			$400		
Inverter				$500		
Total				$5,600	1,000	5.6

Assumptions: 6 m/s (13 mph) average wind speed at hub height; 5 sun hours per day; owner-installed, 45-foot (13 m) hinged tower.

the morning. A hybrid wind–solar system requires sizing and siting the PV array for summer conditions.

Sagrillo's rule of thumb is to spend two-thirds of your budget on generating sources for wind and one-third for the PV array. Batteries should be sized for six times total capacity. If you follow Sagrillo's advice, your system will overproduce in spring and fall when the two resources typically overlap.

Determining the number of PV panels you'd need to meet Sagrillo's criteria is relatively straightforward. Finding the right size wind turbine is less so. In part this is due to the limited choice of wind turbines available for battery-charging applications. It's also partly due to the confusing "rated power at rated wind speed" nomenclature most novices rely on.

In a study of the economics of hybrid power systems at low-wind-speed sites—where most such systems are installed—NREL's Peter Lilienthal found that small wind turbines with relatively large-diameter rotors relative to their generator capacity offered better returns than small turbines with small rotors and high power ratings. Large-diameter, low-rated power turbines, says Lilienthal, reduce the need for both battery and PV capacity in a hybrid system compared to using a more aggressively rated wind turbine. Lilienthal's study once again confirms that what counts is rotor swept area and not generator size.

Scoraig Wind Electric's Hugh Piggott agrees. "For practical purposes," he says, "it's much more useful to have a large rotor and a small generator." A big rotor, says Piggott, "gives a gentle charge all the time, which is what batteries like best." In a storm, turbines with high power ratings but small rotors can generate "lots of watts in the middle of the night, usually far more than you need."

Cabin-Size Hybrid Power System

These entry-level systems are suitable for vacation cabins and also well adapted for rural second homes that could someday be upgraded into permanent residences. The sample cabin-size hybrid system detailed here uses six 75-watt solar panels, a modified-square wave inverter, and sufficient batteries to provide 8 kilowatt-hours of storage on a 24-volt system.

Though easily installed with rudimentary tools, this hybrid system will generate nearly 8 kilowatt-hours per day on average—more on windy spring days, less during summer and fall. The wind turbine provides about two-thirds of the system's total generation. The turbine in this example is Bergey's XL1; it and the tower were designed for owner installation.

Cabin-Size Hybrid Power System

		Collector Area			AEO	
		(m^2)	(ft^2)	Total	(kWh/yr)	($/kWh/yr)
Wind	BWC XL1	4.9	53	$3,500	1,963	1.8
Solar	6 panels, 450 W	3.6	41	$3,800	821	4.6
Battery storage	8 kWh			$800		
Inverter & load center				$2,000		
Total				$10,100	2,800	3.6

Assumptions: 6 m/s (13 mph) average wind speed at hub height; 5 sun hours per day; owner-installed, 65-foot (19 m) hinged tower.

Inverters

Whether you're using a PV array, a wind machine, or a hybrid stand-alone system, you will require an inverter to operate conventional AC appliances. Most inverters produce a modified AC sine wave that can serve a wide variety of AC loads—from sensitive electronics, such as computers and stereos, to washing machines.

To determine the size of the inverter needed, add up the demand from all appliances that are likely to operate at the same time. The inverter should be sized to handle both the surge requirements of the induction motors in refrigerators and washing machines and their continuous demand when operating for extended periods. Small appliances often demand one and a half to two times their rated current when they first start. Large appliances such as washing machines and refrigerators can draw three to four times their rated current when first switched on. For example, an electric motor using 500 watts could require as much as 1,500 to 2,000 watts when starting.

Inverter ratings vary, but all manufacturers list both continuous and surge capacity. Refer to the continuous output rating. This is what the inverter can actually supply over a long period without failure, keeping in mind that few loads operate continuously, and those that do draw little current. Give yourself some room. Sandia National Laboratories recommends sizing the inverter to 125 percent of the expected load.

A 2-kilowatt inverter should run most minor loads and some major loads, like a washing machine or a microwave, when operated singly. Two 2-kilowatt inverters or one 4-kilowatt inverter may be necessary if there's any chance the refrigerator, washing machine, well pump, and microwave may operate simultaneously. Microwaves, hair dryers, and similar loads draw a lot of power (1 to 1.5 kilowatts) but are only operated for short periods. They may influence the size of the

Household-Size Hybrid Power System

The sample household-size hybrid system detailed here should be ample for most off-the-grid homes with a good wind and solar resource. At 3.6 meters (12 ft) in diameter, the African Wind Power turbine is a fairly hefty machine for owners to install themselves. This system includes almost 1 kilowatt of PVs and the more expensive sine wave inverters that some find necessary, as well as a large battery bank. While such a system is capable of producing an average of nearly 16 kilowatt-hours per day on Texas's high plains, for example, most sites will be less productive. North Americans will find it necessary to closely monitor their consumption or provide a backup generator to supplement periods with little wind or sun.

Household-Size Hybrid Power System

		Collector Area		AEO		
		(m^2)	(ft^2)	Total	(kWh/yr)	($/kWh/yr)
Wind	AWP 3.6	10.2	109.5	$4,700	4,072	1.2
Solar	12 panels, 900 W	7.2	41	$3,800	1,643	2.3
Battery storage	16 kWh			$1,600		
Inverter & load center				$5,00		
Total				$15,100	5,700	2.6

Assumptions: 6 m/s (13 mph) average wind speed at hub height; 5 sun hours per day; owner-installed, 65-foot (19 m) hinged tower.

inverter needed, but they contribute little to total energy consumption.

The inverter should also provide fused protection from the various sources of generation, and should offer power factor correction for inductive loads. Ideally the inverter should also have disconnect switches on both the AC and the DC sides, and load management switches to limit certain loads from exceeding the inverter's capacity. Never try to operate an electric dryer, electric hot-water heater, or electric stove on a stand-alone power system unless you're using it as a dump load for excess generation. Use bottled gas (propane) instead. These electric appliances consume inordinate amounts of electricity and place an unreasonably high current demand on the inverter.

Some inverters also offer optional battery-charging functions. If you don't use one packaged with the inverter, you'll need to add to your system a battery charger that's capable of handling multiple inputs: those from the renewable sources and the backup generator. Without a battery charger there's no way to ensure that the batteries stay properly charged.

Batteries

No electrical generator works 100 percent of the time, even those of the utility. Batteries permit a renewable power system to coast from one spurt of power to the next, from windless night to sunny day. They're integral to a successful home power system.

To illustrate why batteries are important, consider the operation of an Air 403, a high-performance micro turbine, one June at the Wulf Test Field (see figure 11-4, Daily generation from Air 403). June is one of the windiest months of the year in the windy Tehachapi Pass. Even in a windy month, however, there are days when the small turbine produces only a fraction of a kilowatt-hour. During June 9, a particularly windy day with an average wind speed of 16 mph (7.1 m/s), the Air module generated 1.8 kWh. For the

Figure 11-4. Daily generation from Air 403. Batteries are used to store energy produced during windy days, such as during June 8–12 in this example, for windless days, June 13–14.

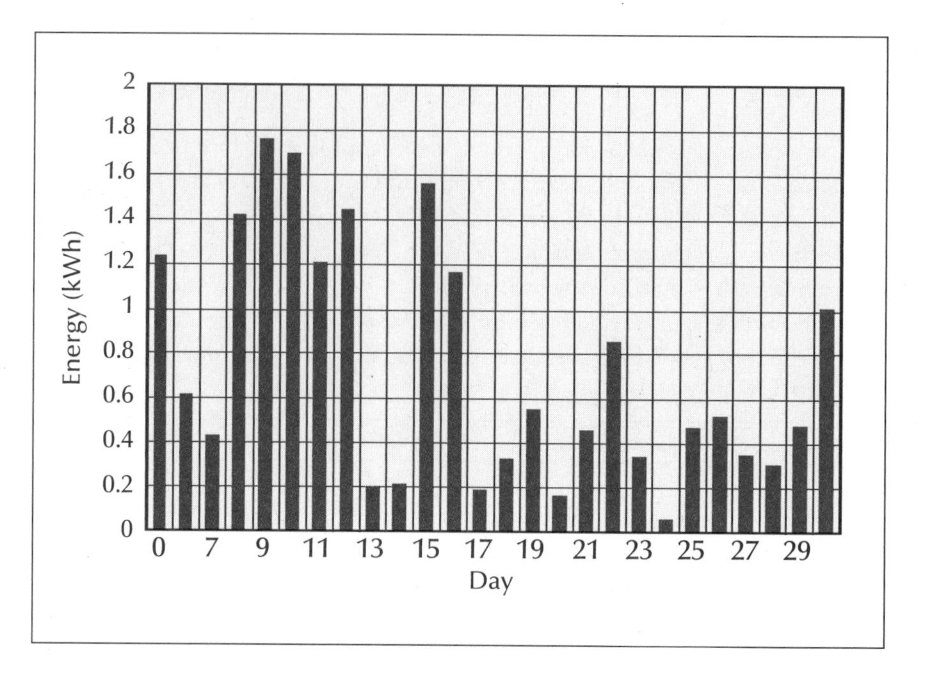

month, the Air 403 generated nearly 17 kWh at an average wind speed of about 11 mph (5 m/s). Batteries enable such a fluctuating source of generation to store production on windy days for use later, when the winds are less productive.

Batteries do add significantly to the cost and complexity of a hybrid system. Conventional lead-acid batteries, the type most commonly used, cost about $100 per stored kilowatt-hour.

There's a bewildering array of batteries to choose from. Avoid making your selection purely on price. Batteries for stand-alone power systems must be capable of numerous deep discharges. Cheap automotive or truck batteries ($50 per kilowatt-hour) might be suitable for a tinkerer, but not for a remote power system.

Golf-cart batteries, while less than ideal, are popular with do-it-yourselfers and offer good value for entry-level systems. Above all else, use batteries that are designed for the frequent charge–discharge cycles found in remote power systems.

When considering the cost of batteries, don't overlook shipping. Lead and cadmium are among the densest materials known, and shipping batteries made of them any distance at all incurs a hefty freight charge.

Batteries can be finicky. They don't like to be over- or undercharged. To prevent permanent damage, lead-acid batteries shouldn't be discharged to more than 80 percent of their capacity. Note also that lead-acid batteries don't tolerate temperature extremes well.

The capacity of lead-acid batteries decreases markedly with temperature. They're least effective in the dead of winter. They can even freeze, particularly when discharged and the electrolyte becomes more water than acid. One way to avoid damage in cold climates is to store batteries in a cellar below the frost line. But don't bring them indoors without proper ventilation.

Batteries should always be isolated from living areas and from sensitive electronics (see figure 11-5, Suggested battery and inverter placement). The gases given off by lead-acid batteries are not only highly corrosive but also highly explosive.

It's also a good policy to separate batteries from inverters, switches, and service panels. This prolongs the life of the electrical components while guarding against a spark igniting the hydrogen given off during high charge rates.

Size batteries and other fixed hardware to the size of system you eventually want. Avoid undersizing batteries, inverters, and wiring. With a lead-acid battery bank, you're locked in if you find that storage is insufficient after a few years of operation. Unlike solar panels, batteries can't be simply added to the system in small increments. A battery bank is a fixed entity, so additions must include a complete new set of batteries of the proper voltage, properly wired.

Undersize the PVs, if anything. Make up the difference in lost production with the backup generator. You can incrementally add PV modules or a larger wind turbine as you can afford them. It's much more difficult to go back and rewire your house, replace your battery bank, and switch out your inverter than it is to add a few more solar panels.

From field experience with hybrid systems in Chile, NREL found that batteries should not be discharged below 20 percent of their normal state-of-charge, or 1.75 volts per cell. Run the backup generator for at least 30 minutes every two weeks, says NREL, and equalize the batteries every three months. Don't use the backup generator to finish charging the batteries. The latter is a task well suited to a battery-charging wind turbine in a period of good wind. NREL suggests stopping the backup generator when the batteries reach 80 percent of their state-of-charge, then finishing the charging with the wind turbine.

Battery-charging wind turbines should always include their own charge controller or regulator. The charge controller will limit battery voltage by either disconnecting the turbine from the load or activating a diversion load when the desired battery voltage is reached. For example, in a 24-volt system a controller might respond when battery voltage reaches to 28 VDC (2.35 volts per cell), and effectively stop charging for a moment until battery voltage falls below 25.2 VDC (2.1 volts per cell), when the controller signals that charging can resume. Like the voltage regulator in an automotive alternator, today's charge controllers use solid-state electronics that perform these tasks many times every second. Some wind turbine charge controllers include a manual switch for using the wind turbine to equalize the batteries during a period of high winds.

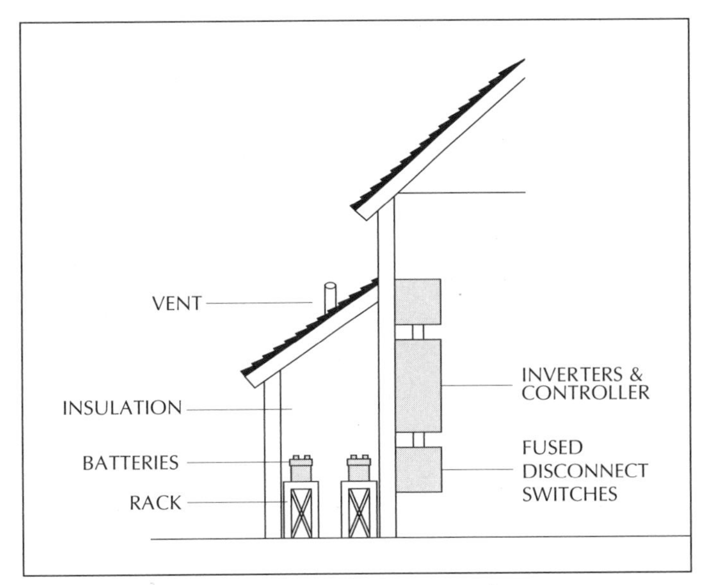

Figure 11-5. Suggested battery and inverter placement. Where possible, batteries should be housed in a well-ventilated room separate from the living quarters.

Backup Generators

In a properly sized power system with ample battery storage, the backup generator may not be used at all, particularly if you're willing to adapt your usage to the resource. Operate those discretionary loads when the wind is blowing or the sun shining. Do your laundry, for example, only when there's a surplus of power.

Because most remote sites need propane for heating, cooking, and domestic hot water, it's relatively simple to use the same propane to power a backup generator. Propane is modular, portable, and offers good utility.

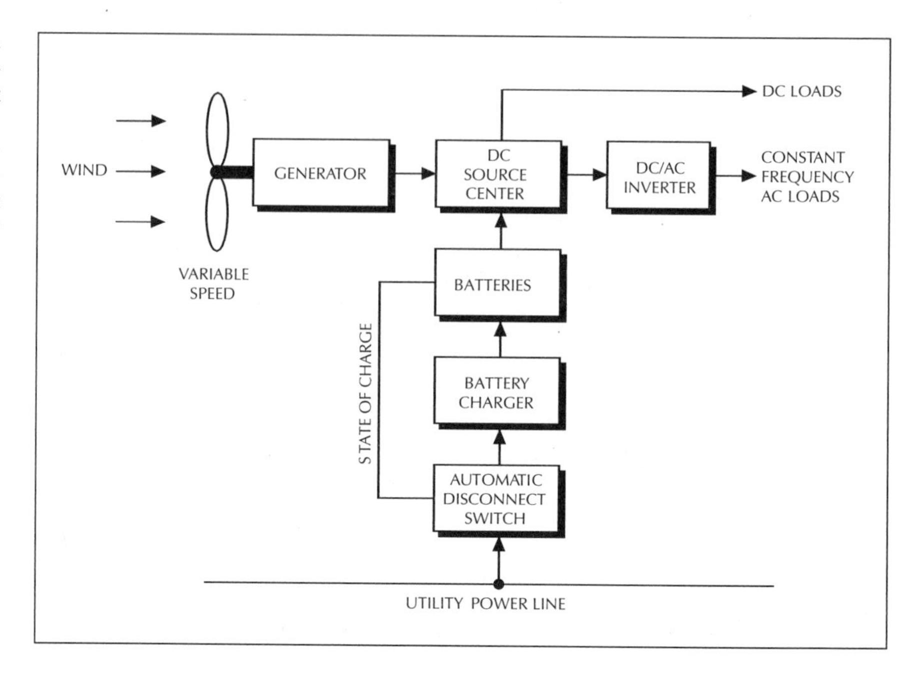

Figure 11-6. Utility backup. This is a common configuration for a semi-autonomous home or business that remains connected to the utility grid. Electronic controls monitor the batteries. If the batteries become discharged, the automatic disconnect switch closes, directing utility power to the battery charger. In practice, inverters designed for this application perform their task silently and seamlessly. Standby losses, however, can be significant.

A backup generator, or gen-set, allows you to design a system with less battery storage than otherwise needed. You can substitute fuel costs for extra batteries. Pacific Gas & Electric found in its survey of California stand-alone systems that 80 percent of the respondents used a backup generator. But the gen-sets provided only 2 percent of the system's total energy, suggesting that most off-the-grid power systems had enough storage for almost all conditions. The backup generators, even if seldom used, provided a valuable service: peace of mind.

Because generators operate most efficiently near full load, it's best to run them only after lead-acid batteries have fallen to about 20 percent of their full charge. The generator can then run at full output until the batteries are brought back up to 80 percent of full charge. The free fuel sources, wind and solar, can then top up the batteries with long-duration charging cycles. This limits the overall running time of the generator, extending its life. It also keeps fuel consumption down. Automatic controls are available for monitoring the batteries' state-of-charge; these will start and stop the generator as needed.

For long life, look for generators that operate at 1,800 rpm, not 3,600, and that have self-starting capabilities. Avoid the inexpensive portable generators used at construction sites. They're not well suited for remote power systems.

Utility Backup

An alternative is to use utility power, where available, to make up for discharged batteries (see figure 11-6, Utility backup). An automatic transfer switch senses the load and switches between inverter-supplied power and the utility's. This system permits users to gradually wean themselves from utility power by adding components incrementally.

If the renewable power sources consistently don't provide enough power, and utility service is nearby, you can bring in utility service to charge the batteries. You substitute utility power for the backup generator. On the surface this may seem to defeat the purpose of a stand-alone system. But the independent power system still serves all the loads, not the utility. The stand-alone system will still provide service if utility power is interrupted. This technique and the inverters that perform this function were designed for areas of the world where utility power is unreliable.

Utility backup is not unlike the uninterruptible power systems widely used to protect computers from power outages. Many critical computer systems operate continuously from batteries. The utility is used only to charge the batteries. The loads draw current from an inverter that's constantly powered by the batteries.

Be advised that inverter standby losses in these systems can add up. Bob Gobeille in Fort Collins, Colorado, found that one popular inverter for this application "burns," as he says, 50 watts continuously in the "sell" mode—that is, when the batteries are fully charged and there's excess generation that can be sold back to the utility. *Home Power* magazine's Joe Schwartz has seen these inverters consume 0.75 to 1 kilowatt-hour per day.

Stand-Alone Economics

In general users of remote power systems don't expect their wind and solar generation to compete with the cost of utility-generated electricity. They typically install a remote power system because the cost of extending utility power to their site is even more costly. California utilities charge new customers about $10 per foot ($33,000 per km) for overhead line extensions (50 percent more for buried lines). The situation's no different in Europe. Electricité de France charges rural residents in France 20 euros per meter ($32,000 per mile) to extend utility service.

Under these conditions a stand-alone system can pay for itself in the first year, if it's more than 0.5 mile or 1 kilometer from the utility's lines. If you're considering a stand-alone power system in North America on purely economic grounds, in general it's cheaper to bring in utility power if your home is less than 1,000 feet (300 m) from existing utility service.

Extending the line may not be the only cost you incur. Many utilities require a minimum purchase of electricity to justify extension of the line. In Pennsylvania, West Penn Power Company at one time required a minimum monthly payment of $100 to $200 for a period of five years from customers requesting line extensions.

Other Stand-Alone Power Systems

The use of wind turbines in hybrid power systems for telecommunications and village electrification are essentially variations on remote power systems for residential use, but each has unique requirements that distinguish it from home light plants.

Telecommunications

Telecommunications demand reliability. Wind machines used in telecommunications encounter more extreme weather, operate more often (sometimes in excess of 7,500 hours per year), and must function unattended for much longer periods of time than home power systems or even commercial wind power plants typically do. Only robust wind machines using fully integrated direct-drive designs perform satisfactorily in the rugged environments characterized by telecom sites (see figure 11-7, Telecom hybrid power system).

A site in McMurdo Sound, Antarctica, illustrates the severe conditions that wind turbines must endure to serve telecom applications. Shortly after installation a Northern Power Systems' model HR3 operated for 12 hours in the furled position during a fierce Antarctic storm. The radio station eventually went off the air when the exhaust stack for the backup generator blew away. After the worst of the storm had passed, the HR3 dropped back into its running position, recharged the system's batteries, and brought the station back to life. Twice during the first two years of operation anemometers at the site blew away, once after recording a wind speed of 126 mph (56 m/s). Since then the site has endured even

Figure 11-7. Telecom hybrid power system. Several of Southwest Windpower's Air Industrials, and a photovoltaics array, power a mountaintop telecommunications site. (Southwest Windpower)

stronger winds. The project has been so successful that it has been expanded with two more HR3 turbines. Northern Power Systems designed the HR3 for just such demanding applications.

Hybrid systems using small wind turbines have been able to substantially reduce fuel consumption at telecom sites in Canada and the United States. At a 6.3 m/s (14 mph) site on Calvert Island off the coast of British Columbia, two Northern Power Systems HR3 turbines, in conjunction with a 1.2-kilowatt solar array and 84-kilowatt-hour battery bank, were able to cut the diesel generator's operating time substantially. Overall the hybrid system reduced fuel use nearly 90 percent, at half the maintenance cost of a conventional diesel system. At Norway's Hamnjefell telecom station above the Arctic Circle, an HR3 turbine has met 70 percent of the site's loads since 1985.

At a similar installation atop Duncan Mountain in Idaho, OnSite Energy (a subsidiary of Pacific Power & Light) has operated a Bergey Excel since the mid-1980s. In the late 1990s OnSite replaced the rotor, rebuilt the alternator, and returned the turbine to service.

Village Electrification

Nearly two billion people live without electricity. Extending utility service from the cities to remote villages in developing countries—where most people live—is costly, can be difficult to finance, and takes years of struggle. To surmount these problems, some villagers have turned to small diesel genera-

> The benefits of providing even small amounts of power to remote villages are magnified because so little electricity is needed to raise the quality of life.

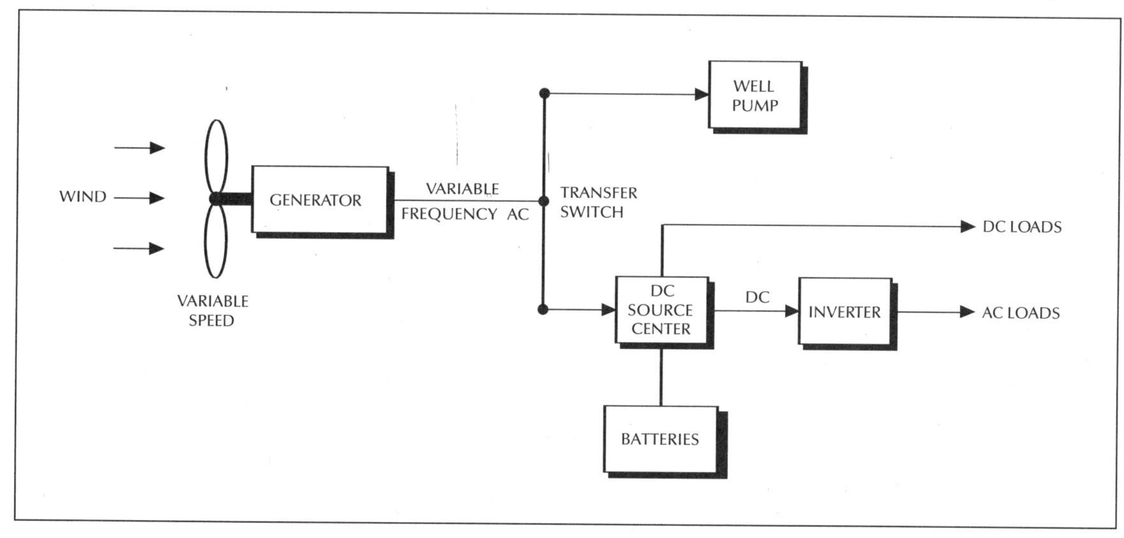

Figure 11-8. Village power. One scheme for using small wind turbines to serve a variety of loads in a village distant from central station power.

tors. But the generators are expensive to operate and often unreliable. Fortunately more and more developing countries are finding that renewable energy is a less expensive, more reliable, and quicker way to meet the electrical needs of their rural citizens.

Village power systems must meet standards for ruggedness, simplicity, and reliability similar to those demanded by mountaintop telecommunications sites. Though the weather may not be as demanding, Third World villages are distant in both time and space from the technical support and spare parts found in the developed world.

The benefits of providing even small amounts of power to remote villages are magnified because so little electricity is needed to raise the quality of life. A Bergey Excel may supply only one home with electric heat in North America, but it can pump safe drinking water for a village of 4,000 in Morocco.

The typical village system might use two or more wind turbines, batteries, an inverter, and a backup generator (see figure 11-8, Village power). And like hybrid home light plants, village power systems could also include a solar array. The key is to use as much power as possible directly, instead of

Village Self-Reliance

Village power systems can meet not only domestic demands but also those of light industrial and commercial uses. Synergy Power's Eric Kalkhurst suggests giving priority to light industrial uses to encourage economic development. As with increasingly popular microcredit programs, villagers can use revenue from the light industrial loads to expand the power system themselves instead of relying on often distant and sometimes indifferent central governments for ongoing aid. For example, villagers could use electricity from a hybrid power system to refrigerate freshly caught fish for sale at a higher price in nearby markets. From Kalkhurst's perspective in Southeast Asia, such a system would provide power for five to six hours of light industrial loads and four to five hours of domestic demand per day.

Kalkhurst's strategy also deals with the mischievous genie unleashed in the past by village power systems: seemingly limitless demand. Soon after installation, everyone wants a personal lamp, television, and other appliances, and the system quickly collapses. Kalkhurst's pay-as-you-go approach provides a mechanism for villagers themselves to finance expansion of the power system.

Figure 11-9. Island power. Using *les alizés,* or "the trade winds," a cluster of household-size wind turbines not only power La Désirade, an island in the French West Indies, but also export electricity to the nearby main island of Guadeloupe. (Vergnet)

storing it in batteries and running it through an inverter. This reduces both initial cost and complexity, while delivering more of the power system's energy to do useful work.

If power is used directly to pump water, grind grain, or run other loads not dependent on utility-grade electricity, the need for batteries is diminished. The batteries and inverter then need to be sized only for those loads that must use constant-frequency AC. The output from the wind turbine can be manually switched from the direct loads, such as water pumping, to the batteries and inverter as needed. For example, the operator monitors the water level in a storage tank and the batteries' state of charge to determine where the power should be directed. The operator is also responsible for starting the backup generator when the power system can't meet demand. Eliminating automatic switches decreases the likelihood that a minor component could fail and imperil the entire system. It also ensures that one person is always responsible for operation of the power system, a lesson learned from past demonstration projects.

Inhabitated islands share many characteristics with isolated villages. Since many islands are also windy, integrating wind energy with existing island power systems has long been an attractive option for reducing the high cost of imported diesel fuel.

Power engineers once thought that wind turbines could off-load only a small portion of an island's diesel network without endangering the stability of supply. Because it's also damaging to run the diesel engines at low loads, wind could never be used to meet all or even most of an island's electrical demand. Electronic controls developed in the 1990s, however, have driven significant advances in marrying the fluctuating wind resource with an island power system's fluctuating load.

Northern Power Systems' Lawrence Mott points to St. Paul Island as an example of a hybrid wind–diesel system where wind energy contributes a large portion of electrical demand. NPS installed one Vestas V27 on the island in Alaska's Pribiloff Islands group. The 225 kW turbine joined two 150 kW diesel generators. No batteries were provided for storage. As in many household-size hybrid power systems, electronic controls in the St. Paul system dump excess wind generation into resistive heaters producing hot water, a commodity always in demand on the windswept island in the Bering Sea. The electronic controls, coupled with a synchronous condenser, enable the system to operate with the conventional wind turbine—alone. The controls keep voltage and frequency constant without operating the diesel generators. When the wind turbine is insufficient to meet demand, the system starts the diesel.

In the early 1990s, Electricité de France contracted with French wind turbine manufacturer Vergnet to operate a wind plant on La Désirade, a small island east of Guadeloupe in the French West Indies. The goal: to cut in half the cost of serving the island. Vergnet successfully demonstrated that a cluster of household-size turbines could deliver a significant percentage of the islands' electricity.

Since then the project has been steadily expanded, harnessing *les alizés,* or the Caribbean's trade winds. Subsequently an undersea cable was laid to Guadeloupe, and now La Désirade not only is self-sufficient, but also exports electricity to the main island (see figure 11-9, Island power). All in all Vergnet's turbines generate 165 percent of the island's electrical consumption.

In the next chapter we'll look at another way to use wind turbines in a stand-alone application: for pumping water.

12

Pumping Water

For those in industrialized countries served by community water systems, where water—like electricity—is available on demand, it's hard to imagine the importance of a water pump to the world's rural population. Lifting or carrying water accounts for much of the energy expended in Third World villages. It's also a major load for North American homes beyond the reach of utility lines. In fact, settlement on America's Great Plains wasn't feasible until experimenters developed a reliable means of pumping water from deep wells. The technology that made settlement possible was the American water-pumping windmill, and it forever changed the face of the landscape.

Windmills That Won the West

Three technological innovations made settlement on the Great Plains possible: the Colt .45, barbed wire, and the farm windmill. Yes, the windmill—that ubiquitous symbol of rural America—ranks right alongside the six-shooter, says Walter Prescott Webb, a noted historian.

Webb writes in his book *The Great Plains* that the water-pumping windmill was so essential for life in what was then known as the Great American Desert that settlers warned newcomers, "No women should live in this country who can't climb a windmill or shoot a gun." Promoters extolled the virtues of a land where "the wind pumps the water and the cows chop the wood."

Promoters extolled the virtues of a land where "the wind pumps the water and the cows chop the wood."

In the semiarid lands west of the Missouri River, the wind did indeed pump the water. Unlike in the eastern United States, few streams coursed across the surface, and seldom were aquifers in reach of simple hand-dug wells. Water was there but at depths that required pumping by machines—wind machines. (And those poor homesteaders who couldn't find wood for their hearth on the treeless landscape burned cow chips from their bovine lumberjacks.)

The nation's westward migration both caused and was aided by the growth of a great midwestern industry that built windmills. By the late 19th century 77 firms were assembling windmills in one form or another. Catalogs of the day bristled with choices. Sears Roebuck and Wards offered their own house brands.

In 1854 Daniel Halladay invented the first fully self-regulating windmill. Until then, millers had to manually turn the spinning rotor out of the wind or reef (roll up)

Figure 12-1. Traditional farm windmill. This Dempster No. 9 is found at the Windmill State Recreation Area near Gibbon, Nebraska. In this classic design the thrust on the pilot vane (small) furls the rotor toward the tail vane (large) . A crescent-moon weight regulates the wind speed at which the multiblade rotor furls toward the tail vane.

the sails during storms. Halladay changed all that by constructing a multiblade rotor made up of several movable segments that reefed automatically in strong winds (see figure 6-40, Halladay rosette or umbrella mill).

Early models of Halladay's rosette or umbrella mill used a tail vane to point the rotor (*wheel* to old-timers) windward. Later "vaneless" versions did away with the tail vane by orienting the wheel downwind of the tower.

Halladay's "patent" mills were immediately popular with farmers for watering livestock. Because his mills could be left unattended, they were ideal for remote pastures where water was scarce. But it wasn't until the post–Civil War boom in railroad construction that the fledgling windmill industry began to flower.

Water was as essential to running a steam locomotive as coal. As the transcontinental railroads pushed westward across the plains, the water-pumping windmill came into its own. Huge windmills (even by today's standards) with rotors up to 60 feet (18 m) in diameter pumped a steady stream into storage tanks at desolate way stations. Through skillful marketing, Halladay's chief rival, the Eclipse, created a name for itself as the "railroad" mill.

Invented in 1867 by the Reverend Leonard Wheeler, the Eclipse used fewer moving parts and was both cheaper to produce and easier to maintain than Halladay's design. Wheeler built a solid wheel instead of the sectional wheels Halladay popularized. Wheeler protected his windmill in high winds by furling the rotor toward the tail vane. The idea was so successful that even Halladay's U.S. Wind Engine and Pump Company began producing similar versions under the Standard trade name. Other manufacturers also followed suit (see figure 12-1, Traditional farm windmill).

The stage was set. The technology existed and an industry was in place. Demand grew rapidly as the railroads poured settlers onto the prairie. The industry blossomed like the

Dempster Still Pumping

For more than 125 years Dempster Industries has been manufacturing farm windmills in the small midwestern town of Beatrice, Nebraska. Manager David Suey takes pride in the fact that Dempster's traditional "Chicago-style" wind pumps sell themselves. "We don't spend a single nickel" on marketing, he says.

The Dempster design has remained virtually unchanged for the past 40 years, though the company no longer offers the bigger 16- and 18-foot windwheels. Dempster's mills are known for their tapered roller bearings. These are longer lasting, says Suey, than the Babbitt bearings used on other farm windmills.

Wind pump sales were brisk prior to the year 2000, due to Y2K fears. Normally most sales in Dempster's farm windmill line are in spare parts to the long-lasting turbines.

Before switching to the more familiar design in the 19th century, Dempster built "umbrella mills." The horse-shaped counterweights used on these early downwind machines are much sought after by antiques collectors, and can be found in dusty corners of antiques stores throughout the United States.

These days Dempster builds submersible pumps and specializes in farm equipment.

Dempster Annu-Oiled. A 14-foot (4.3 m) diameter Dempster mechanical wind pump at a park in Beatrice, Nebraska. The moniker refers to changing the oil only once per year, in contrast to more frequently for other brands.

Dempster Wind Pumps (American Farm Windmill)

Windwheel (rotor) Diameter		Mechanical Power			Tip Speed		
(ft)	(m)	(hp)	(kW)	(rpm[1])	(m/s)	Ratio	(strokes/min[2])
6	1.8	0.19	0.14	128	12	1.8	32
8	2.4	0.26	0.26	107	14	2.0	32
10	3.0	0.53	0.40	78	12	1.9	26
12	3.7			63	12	1.8	21
14	4.3			54	12	1.8	18

Notes: [1] at 15 mph (6.7 m/s)
[2] Furling spring at maximum tension, 15-18 mph (6-8 m/s) 6,8, and 10-foot; 18-20 mph (8-9 m/s for 12, and 14-foot

Source: Dempster Industries

homesteaders' gardens that mass-produced windmills soon made possible.

The farm windmill was on its way to becoming an American icon. It quenched many thirsts, provided an occasional bath, and offered salvation through baptism in its associated stock tank. The clanking windmill sang a lullaby to many a sleepless child. The gentry found it useful, as well. To their country villas and suburban estates the windmill brought running water and the convenience of Sir Thomas Crapper's flushable commode.

This was an age of invention and empire building. Steel and the factory system were driving the Industrial Revolution to new heights. Into this milieu stepped LaVerne Noyes, a Chicago industrialist.

In 1883 Noyes hired an engineer, Thomas

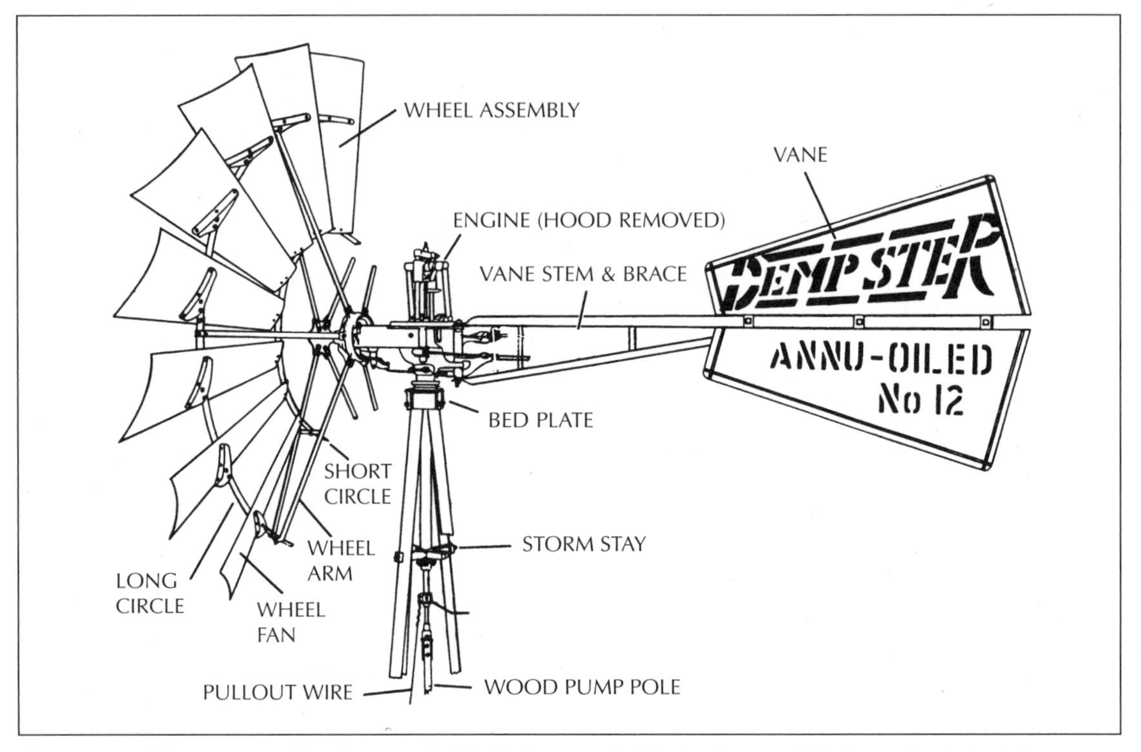

Figure 12-2. Chicago or American farm windmill. This is a standard design for a multiblade mechanical wind pump used worldwide. (Dempster Industries)

Perry, to develop a thresher. With his previous employer, U.S. Wind Engine and Pump Company, Perry had conducted more than 5,000 tests on different windmill designs. His had been the first scientific attempt to improve wind machines. By using a steam-driven test stand, Perry had designed a rotor nearly twice as efficient as those then in use. But Halladay's U.S. Wind Engine and Pump wasn't interested. Noyes was. He encouraged Perry's work.

Five years later when Perry's trade secrets were released, Noyes introduced the Aermotor. Derisively tagged the "mathematical" windmill by competitors, the Aermotor incorporated both Perry's design and Noyes's manufacturing sense.

Although it wasn't the first to use metal blades (Mast, Foos and Company's Iron Turbine used them in 1872), Aermotor's stamped sheet-metal "sails" revolutionized the farm windmill. Another innovation was Aermotor's method of furling the rotor. Rather than use a pilot vane as on the Eclipse, Perry offset the Aermotor's wheel from the center of the tower. High winds striking the rotor disk forced it to fold toward the tail, eliminating the need for the pilot vane. The furling force was counterbalanced by a spring that held the rotor into the wind. This arrangement worked so reliably that it's still used.

Noyes's ability to mass-produce windmills at Aermotor's Chicago plant reduced its cost to one-sixth that of its competition and led Aermotor to dominate the industry. Eventually Aermotor captured 80 percent of the market. Today if you see an abandoned windmill along a country lane, it's most likely an Aermotor.

The industry peaked in the early part of the 20th century, collapsing soon after with the extension of electricity to rural areas. Durability played a large role in the industry's demise in the days before planned obsolescence. On many homesteads, the family windmill has been in continuous use for generations.

After a century of refinement, the farm windmill performs its job well. It remains the signature of the Pennsylvania Dutch and Amish settlements in Ohio, Indiana, and Iowa, and of the tourist mecca on the Yucatán Peninsula, Mérida. The Amish, who use them for domestic water, and ranchers of the Southwest, who use them for stock watering, account for the few thousand sold in the United States each year (see figure 12-2, Chicago or American farm windmill).

Farm windmills, says Vaughn Nelson, founder of West Texas A&M University's Alternative Energy Institute, produce the equivalent of 130 million kilowatt-hours yearly, pumping water for livestock on the Great Plains. "They still make sense in a lot of places," he says.

Table 12-1
Wind Pumps Worldwide

	Units	Effective MW
Argentina	600,000	150
Australia	250,000	63
Brazil	2,000	1
China	1,700	0
Syria	5,000	1
Cuba	9,000	2
Colombia	8,000	2
Curaçao, N.A.	3,000	1
Nicaragua	1,000	0
South Africa	100,000	25
U.S. Southern Plains	60,000	15
	1,039,700	260

Data compiled from multiple sources.

Mechanical Wind Pumps

The Dutch, or European, windmill was well suited to shallow-surface sources, but water on the Great Plains was found deep below the surface. The mechanical wind pump, or farm windmill, enabled settlers on North America's Great Plains to pump from deep wells the relatively low volumes of water needed for domestic uses from deep wells. The farm windmill was a perfect match between the needs of settlers and the abundant winds found on the prairies.

The design proved so successful that it has been widely copied around the world. Today nearly a million remain in use, mostly in Argentina, the United States, Australia, and South Africa (see table 12-1, Wind Pumps Worldwide). Some have been in quasi-continuous operation for more than 80 years. Because it works, the farm windmill remains popular even today.

Farm windmills, like modern wind turbines, have their own arcane vocabulary and obscure methods for estimating performance. The following section first examines the factors affecting the pumping capacity of farm windmills, then compares these with those available from modern wind turbines. The description of the American, or classic, multiblade windmill technology uses English units. Tables for estimating the pumping capacity of modern wind turbines use both the English and the metric system, however, because the technique is applicable worldwide.

Pumping Head

The energy needed to pump water is a function of the volume and the height, or head, it must be lifted (see figure 12-3, Pumping head nomenclature). It takes as much energy to lift 10 gallons (38 l) of water 10 feet (3 m) as it does to lift 1 gallon (3.8 l) 100 feet (30 m). Sizing a wind machine to meet the need for electrical energy is simply a matter of estimating electrical consumption. But sizing a wind pumping system requires an estimate not only of how much water is used, but also of how far the water must be pumped from the well to its point of use.

The total dynamic head includes the distance the water must be lifted from the well as well as the height the water must be lifted to fill any aboveground storage tank (the discharge head). The dynamic head also includes any energy lost due to friction in the pipes (friction head). When estimating head, bear in mind that the static water level in any well

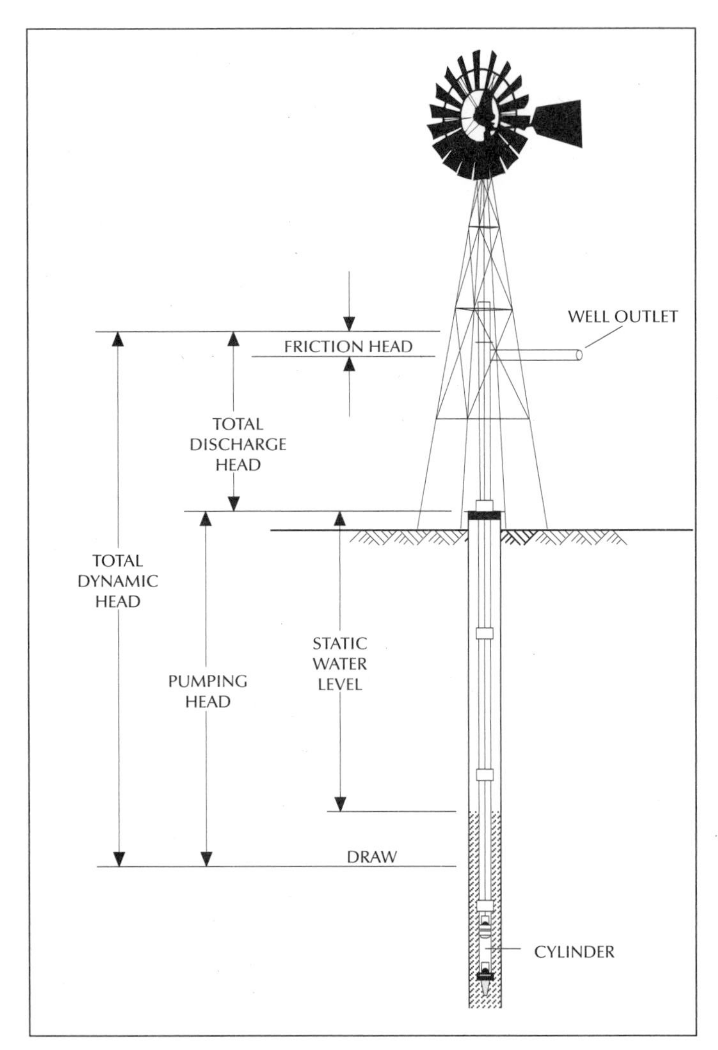

Figure 12-3. Pumping head nomenclature.

will be lowered, or drawn down, once pumping begins. The depth the water level falls in the well depends on the pumping rate. Thus the pumping head is always greater than the static head.

Friction in pipes is directly proportional to the rate of flow. You can transfer a given amount of water with large-diameter pipes and a low flow rate, or small-diameter pipes and a high flow rate (see table 12-2, Friction Head in Iron Pipe). For example, 1-inch pipe may be less costly than 2-inch pipe, but a pump must overcome more than 20 times more friction in the smaller pipe at a flow rate of 10 gallons per minute (38 liters per minute, l/min).

Farm windmill manufacturer Dempster Industries advises not to skimp on the size of water pipe. Using a pipe too small for the flow rate can significantly reduce the pumping capacity of the windmill because of the friction along the walls of the pipe. Dempster warns that you can add tens of feet to the dynamic head of a well with pipe too small for the job, effectively "making a deep well out of a shallow one."

For example, let's assume that you plan to install a farm windmill on a well 100 feet deep. You expect to pump about 10 gallons per minute. At this rate, the pump will draw down the static water level 5 feet (1.5 m). Let's say you plan to store water in a tank with an inlet 5 feet above the ground, up a small hill with a 5-foot gain in elevation. Now add the final details: a 100-foot run of 1-inch pipe that will need three 90-degree elbows to route the water into the storage tank. Even though the well is only 100 feet deep, the windmill must work against 145 feet (44 m) of dynamic head (see table 12-3, Sample Calculation of Total Dynamic Head).

Unless the tank is below the well outlet, you'll need to install a packer head to pump water into a storage tank. The packer head enables the farm windmill to pump water above the height of the well. In the past farm windmills used for providing domestic water sometimes had storage tanks built right into the towers; others delivered water to a nearby tank on a raised platform. In either case a packer head is needed when the outlet is above the well opening (see figure 12-4, Packing head).

Estimating Farm Windmill Pumping Capacity

Early farm windmills directly coupled the windwheel (rotor) via a crank to the sucker rod. The sucker rod lifted the column of water in the well. Thus the windmill would lift the sucker rod every revolution of the rotor. In light winds the weight of the water in the well would often stall the rotor,

Table 12-2
Friction Head in Iron Pipe

Water Flow Rate (gallons/minute)	(l/min)	Pipe Diameter (inches) 1	1.5	2	2.5	3
		Loss of Head in Feet per 100 Feet of Pipe				
5	19	3.25	0.4	-	-	-
10	38	11.7	1.43	0.5	0.17	0.07
15	57	25	3	1.08	0.36	0.15
20	76	42	5.2	1.82	0.61	0.25
		Loss of Head in Feet for 90-degree Elbows				
		6	7	8	11	15

Source: Adapted from Aermotor Co.

Table 12-3
Sample Calculation of Total Dynamic Head

	Head (ft)	(m)
Static head	100	30.5
Drawdown	5	1.5
Discharge head	10	3.0
Friction head		
Pipe	12	3.7
Elbows	18	5.5
Total dynamic head	145	44.2

bringing it to a halt. One of the great innovations introduced by Perry and others was back-gearing. Today most, if not all, farm windmills are back-geared and use a transmission to increase the rotor's mechanical advantage in light winds. This increases the farm windmill's complexity but also enables it to pump water more reliably in light winds. Most back-geared windmills lift the piston once every three revolutions. Back-gearing works for rotors up to 16 feet (5 m) in diameter. Farm windmills from 16 to about 30 feet (10 m) in diameter are crank-driven, or direct acting. Mechanical wind pumps greater than 30 feet in diameter are geared up to compensate for the slow rotor speeds and long stroke (see figure 12-5, Back-geared).

Farm windmills are available in a range of sizes. Australia's Southern Cross can be ordered with windwheels up 25 feet (8 m) in diameter.

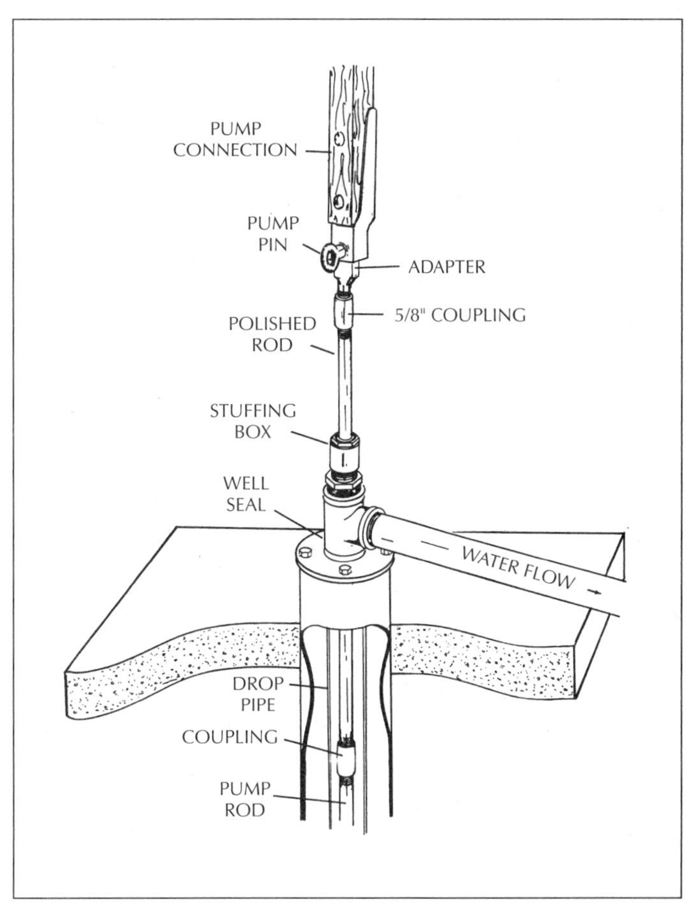

Figure 12-4. Packing head. A packing head is required for a wind pump to raise water above the height of the well outlet. (Dempster Industries)

The most common size in North America is the 8-foot mill, which is capable of pumping less than 10 gallons per minute (38 l/min) from depths of 100 feet (30 m). How much it can actually pump has often remained a mystery.

Figure 12-5. Back-geared. Most farm windmills, such as this Dempster model, are back-geared—that is, the rotor turns several revolutions per pump stroke. Rotor hub (far left), rotor drum brake, screw pump to oil rotor bearings, pitman arms, and gearing (right). Unlike many small wind turbines, farm windmills were often provided with a brake for stopping the rotor when service was needed.

Estimating the amount of water that a farm windmill might deliver is even more of a dark art than estimating the electrical output of a wind generator. Standard windmill pumping tables are based on instantaneous wind speed, pump cylinder diameter, and the depth of the well. These tables, which were probably derived during the late 19th century (no one knows for sure), give you little idea of how much water can be delivered within a given wind regime. They do illustrate, though, that the performance of the farm windmill is strongly influenced by the relationships among the windmill, the pump, and the pumping head. Too big a pump and the windmill will stall; too small a pump and the windmill will operate less efficiently than it might otherwise.

The common 8-foot (2.4 m) windmill, when matched with a well cylinder 2 inches in diameter, will pump about 3 gallons per minute (11 l/min) from a well about 140 feet (43 m) deep (see table 12-4, American Farm Windmill Pumping Capacity). According to Aermotor, the pumping capacity remains the same for the same size well cylinders among windwheels (rotors) from 8 to 16 feet (2.4 to 5 m) in diameter. By varying the pump's stroke—its vertical travel—Aermotor uses the increased power from the larger-diameter rotors to increase the height through which the water can be lifted. A 10-foot (3 m) farm windmill will pump water at the same flow rate as the 8-foot rotor, but will lift the water half again as much.

For any given rotor, the stroke can be adjusted to vary the proportions between the volume pumped and the head. Adjusting the windmill to the short stroke decreases the volume that can be pumped by one-fourth, but enables the windmill to work against a one-third greater head. For the same head the

Table 12-4
American Farm Windmill Pumping Capacity

Pump Cylinder Diameter (inches)	Flow Rate (gallons/minute)	l/min	Maximum Total Pumping Head in Feet Windwheel (Rotor) Diameter (feet) 8	10	12	14	16
2	3	11	140	215	320	460	750
2.5	5	19	95	140	210	300	490
3	8	30	70	100	155	220	360
3.5	11	42	50	75	115	160	265
4	14	53	40	60	85	125	200
4.5	18	68	30	45	70	100	160
5	22	83	25	40	55	80	130

Note: Assumes a 15–20 mph (7–9 m/s) wind speed, stroke set for maximum capacity.
Source: Adapted from Aermotor Co.

shorter stroke will permit the rotor to start pumping in lighter winds.

Nolan Clark and his staff at the U.S. Department of Agriculture's experiment station in Bushland, Texas, undertook a series of tests to determine just how much water the farm windmill could pump under standard conditions. He and his scientists found that on average, the typical 8-foot (2.4 m) diameter windmill can pump 1 to 2 gallons per minute (4 to 8 l/min) under the conditions found on the Great Plains.

Pumping tables are the farm windmill equivalent of a wind generator's power curve. Unfortunately the pumping tables apply for only one wind speed bin, not for the full range of wind speeds needed to estimate production anywhere the wind isn't blowing constantly at 15 to 20 mph. However, Alan Wyatt at the Center for International Development has devised a simple formula for calculating the potential output from mechanical wind pumps operating at various average wind speeds and pumping heads, when the windmill is properly matched to the pump.

The farm windmill is much less efficient at converting the energy in the wind to useful work than are modern wind turbines. Although farm windmills can deliver instantaneous efficiencies of up to 15 to 20 percent in low winds, the average operating efficiency is only 4 to 8 percent. Assuming an average conversion efficiency of 5 percent and a Rayleigh wind speed distribution, Wyatt calculates the daily or monthly volume in cubic meters:

$$m^3 = (0.4 \times D^2 \times V^3)/H$$

where D is the rotor diameter in meters, V is the average daily or monthly wind speed in m/s, and H is the total pumping head in meters.

Farm Windmill Conversion?

Because of the prevalence and apparent simplicity of water-pumping windmills, newcomers to wind energy often turn their sights toward adapting farm windmills for generating electricity. Eric Eggleston, a wind engineer, has experimented with farm windmill conversions. His advice: "Forget it." For its part, farm windmill manufacturer Dempster agrees and states flatly, "No practical mechanical conversion has been put forward to convert the water pumping windmill to the generation of electricity."

While Eggleston acknowledges that the farm windmill's multiblade rotor is superior to that of simple drag devices, it underperforms modern high-speed wind turbines in generating electricity. One measure of this is the tip-speed ratio. For farm windmills, the maximum tip-speed ratio is about 2. In contrast, the tip-speed ratios for high-performance wind turbines is between 4 and 10.

Table 12-5
Approximate Daily Pumping Volume of American Farm Windmill in m^3/day and gallons/day

		Average Annual Wind Speed									
	(m/s)	3		4		5		6		7	
	(mph)	6.7		9.0		11.2		13.4		15.7	
Pumping Head											
(m)	(ft)	(m^3/dy)	(gals/dy)	(m^3/dy)	(gals/dy)	(m^3/dy)	(gals/dy)	(m^3/dy)	(gals/dy)	(m^3/dy)	(gals/dy)
10	30	10	2,700	24	6,300	47	12,300	80	21,200	128	33,700
20	70	5	1,300	12	3,100	23	6,100	40	10,600	64	16,800
30	100	3	900	8	2,100	16	4,100	27	7,100	43	11,200
40	130	3	700	6	1,600	12	3,100	20	5,300	32	8,400

Notes: Rotor Diameter = 3.05 m (10 ft)
Source: Center for International Development, Research Triangle Institute

For a site with an average wind speed of 6 m/s (13.4 mph), a 10-foot (3 m) diameter farm windmill will pump about 27 cubic meters per day (7,100 gallons per day) from a well 30 meters (100 ft) deep (see table 12-5, Approximate Daily Pumping Volume of American Farm Windmill).

Like the estimates of annual energy output using swept area, these tables of the expected pumping capacities of farm windmills are only crude approximations of what may actually occur (see figure 12-6, Estimated pumping volume).

New Technology for Mechanical Wind Pumps

The farm windmill no longer has a monopoly on wind pumping. Today there are more options: from novel mechanical wind pumps, to air-lift pumps, to the improved wind–electric pumping systems now available.

Experimenters have developed several devices for improving the operation of the traditional farm windmill. The simplest is a counterbalance to the weight of the sucker rod. Farm windmill rotors tend to speed up when the rod begins its downward journey. On the up stroke, the rotor slows down as it lifts both the weight of the rod and the weight of the water in the well. (This is most noticeable in light winds.) The change in speed changes the tip-speed ratio of the rotor and its efficiency. To steady the speed of the rotor and maintain the optimal relationship between the rotor and wind speed, some designers have added weights or springs to counterbalance the weight of the sucker rod.

> The farm windmill no longer has a monopoly on wind pumping.

Another approach tackles a more fundamental problem with wind pumps. As noted before, the power in the wind increases with the cube of wind speed. But the pumping rate of mechanical windmills varies linearly. If the stroke of the windmill is adjusted for optimal production in high winds, the windmill will perform poorly, if at all, in low winds. Ideally the stroke should vary with wind speed to more closely match the pumping capacity with the power available. In one variable-stroke design, that of Don Avery of Hawaii, the modified wind pump produced twice the water pumped by the traditional farm windmill.

Though these innovations sound appealing, neither mechanism is widely used. Manufacturers haven't adopted the technology, and they continue building farm windmills the same way they have for the past 100 years. Dutch researchers, on the other hand, have successfully designed modern versions of the farm windmill. The modern wind pumps developed by CWD in the Netherlands use only 6 to 8 blades of true airfoils, in contrast to the 15 to 18 curved steel plates found on the American farm windmill. As a result, they

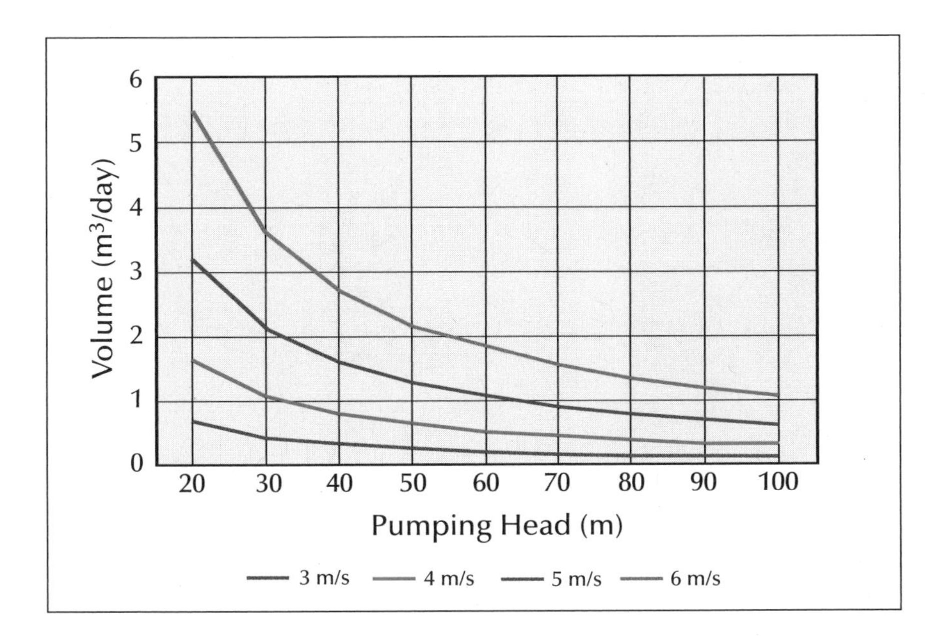

Figure 12-6. Estimated pumping volume. A graphical summary in metric units of a technique for estimating the pumping volume of traditional farm windmills, developed by Alan Wyatt at the Center for International Development.

use fewer materials to do the same job, and are simpler and less costly to manufacture than the American farm windmill. Both aspects are important for Third World countries such as India. Though nearly twice as efficient as the traditional design, modern wind pumps haven't proven as rugged. They may be best suited for regions with light winds.

One drawback to the farm windmill is the lack of flexibility in siting. The farm windmill must be located directly over the well. This usually isn't the best location for a wind turbine. In hilly terrain water is found at the bottom of swales, whereas the wind is often found on hill crests nearby. One solution for low-volume applications has been the introduction of air-lift, or bubble, pumps. The rotors on these wind pumps drive an air compressor, which pumps air through plastic tubing. The use of pliable plastic tubing allows flexibility in locating the wind pump to best advantage (see figure 12-7, Air-lift wind pumps).

Similarly, in a wind-electric pumping system the wind turbine can be sited where the wind is strongest and the turbine will perform best. The gain in performance more than offsets the cost of the electrical cable to the well.

Electrical Wind Pumps

"Many people are simply unaware that you can use a small wind turbine to pump water," says Brian Vick, an engineer at USDA's Bushland, Texas, station. For the same rotor diameter, wind-electric pumping systems can deliver two or more times as much water as the traditional farm windmill. For example, in the windy Texas Panhandle, a Bergey 850 will pump two and a half times the volume of

Wind Pumping Research

The field of wind–electric pumping is continuing to advance. You can find the status of current research on wind pumping in North America by visiting USDA's project page at:

www.ars.usda.gov/research/programs.htm

or by contacting West Texas A&M University's Alternative Energy Institute at:

www.wtamu.edu/research/aei

In Great Britain, IT Power has developed several new-generation wind pumps. Visit IT Power's Web site for details at:

www.itpower.co.uk

Figure 12-7 Air-lift wind pumps. Bubbles of air created by these mechanical wind pumps can be used to lift water in wells or aerate farm ponds during winter. *Left:* Bowjon air-lift pump near Warner Springs, California. Note the air compressor attached to the rotor. *Right:* Koenders wind pump in Alberta, Canada.

Table 12-6
Approximate Wind–Electric Measured Pumping Rates in Texas Panhandle

	Rotor Diameter (m)	(ft)	Swept Area (m^2)	Rated kW	Submersible Motor Rating (kW)	(hp)	Pumping Depth (m)	(ft)	Pumping rate (liters/min)	(gal/min)
Enertech E44[1]	13.4	44	141.3	40	22.3	30	85	279	946	250
Bergey Excel	7.0	23	38.5	10	5.6	7.5	60	197	227	60
Bergey 1500	3.05	10	7.3	1.5	1.1	1.5	30	98	87	23
SWP Whisper[2]	2.7	9	5.9	1	0.7	1	30	98	47	12.5
AC Solar Pump			18.6	1	0.6	0.75	30	98	42	11.2
Bergey 850[3]	2.4	8	4.7	0.85	0.4	0.5	30	98	38	10
Mechanical wind pump	2.44	8	4.7	0.5	0.4	0.5	30	98	15	4
DC Solar Pump			1.0	0.1	0.05	0.07	30	98	4	1

Notes: [1] No longer manufactured.
[2] Superceded by H80.
[3] Superceded by XL1.

Source: USDA ARS, Bushland, Texas

water as a much larger farm windmill (see table 12-6, Approximate Wind–Electric Measured Pumping Rates in Texas Panhandle). Still, determining when you want that water available can be critical in choosing a farm windmill or wind–electric pump.

Previously the only means of using the wind to pump water without utility power present was the installation of a multiblade farm windmill, or a small wind turbine com-

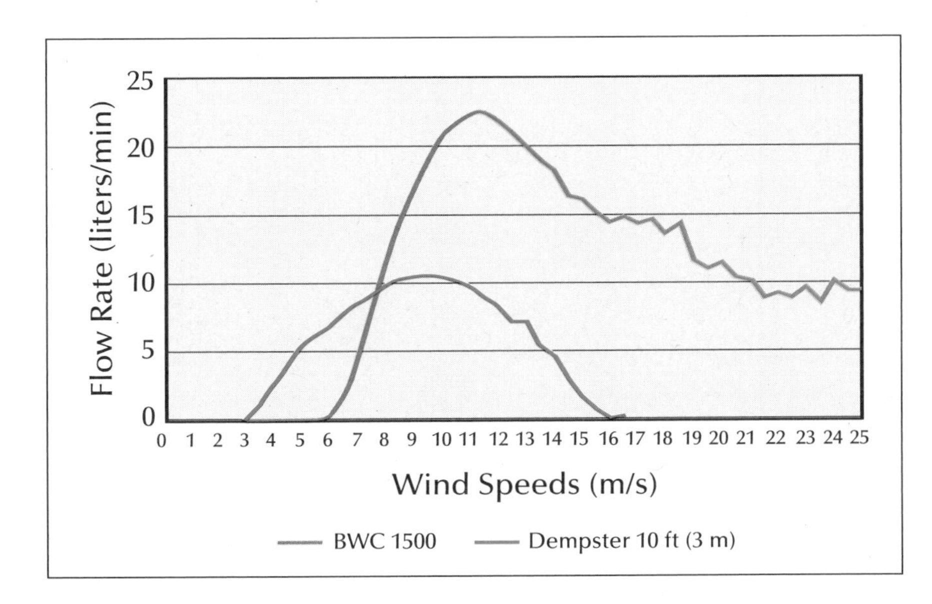

Figure 12-8. Wind pumping flow rate comparison. Pumping tests of a Dempster farm windmill and a Bergey 1500 wind turbine. The Dempster tested was 8 feet in diameter, but its performance was adjusted to represent that of a 10-foot (3 m) diameter farm windmill. Note that the farm windmill will pump water in light winds, when the wind-electric system has yet to begin operating. (Adapted from USDA ARS)

plete with costly batteries, inverter, and electric pump. Contemporary wind–electric pumping systems instead drive well motors directly. The key has been the development of electronic controls that match the pump motor load to the power available from the wind turbine at different wind speeds. This approach frees the wind turbine from the need for batteries and inverter. When wind is available, the wind turbine drives the pump at varying speeds, pumping more in high winds than in low winds. During a period of strong winds, these wind-electric pumping systems bank surplus energy as water in a storage tank, rather than as electricity in batteries.

Direct wind-electric water pumping simplifies matching the aerodynamic performance of the wind turbine to pumping by varying the load electrically, instead of by mechanically changing the stroke of the farm windmill. The control system is the limiting factor, says Nolan Clark, director of the USDA's Bushland, Texas, station, which pioneered work on the technology. With assistance from West Texas A&M's Alternative Energy Institute, Clark and his team of researchers developed electronic controls that made wind–electric pumping possible.

As an example of what wind–electric pumping can do, Clark notes that two Bergey Excels are used to irrigate cotton in West Texas. The turbines each pump 20 gallons per minute (76 l/min) from 300-foot (100 m) wells—when the wind is blowing. And that remains one of wind–electric pumping's weaknesses: performance in the low winds of late summer, when water demand is greatest. This can be partially compensated for by installing the wind turbine on a much taller tower than the farm windmill.

To test the performance of wind–electric pumping relative to that of mechanical wind pumps, Clark's team evaluated both types at the Agricultural Research Service's wind-pumping test facility. They compared the flow rate and the total volume of water delivered from a 240-foot (73 m) well by two wind-pumping systems. For the mechanical wind pump, USDA researchers used a Dempster farm windmill on a 35-foot (10 m) tower. For the wind–electric system, they used a 10-foot (3 m) diameter Bergey 1500 on a 60-foot (18.3 m) tower.

The Bergey, using modern, high-speed airfoils, produced a much higher flow rate than the Dempster (see figure 12-8, Wind pumping flow rate comparison). But this isn't the whole story. The Dempster outperforms the modern turbine at low wind speeds: from 3 to 8 m/s (7 to 18 mph). In the Texas Panhandle, which

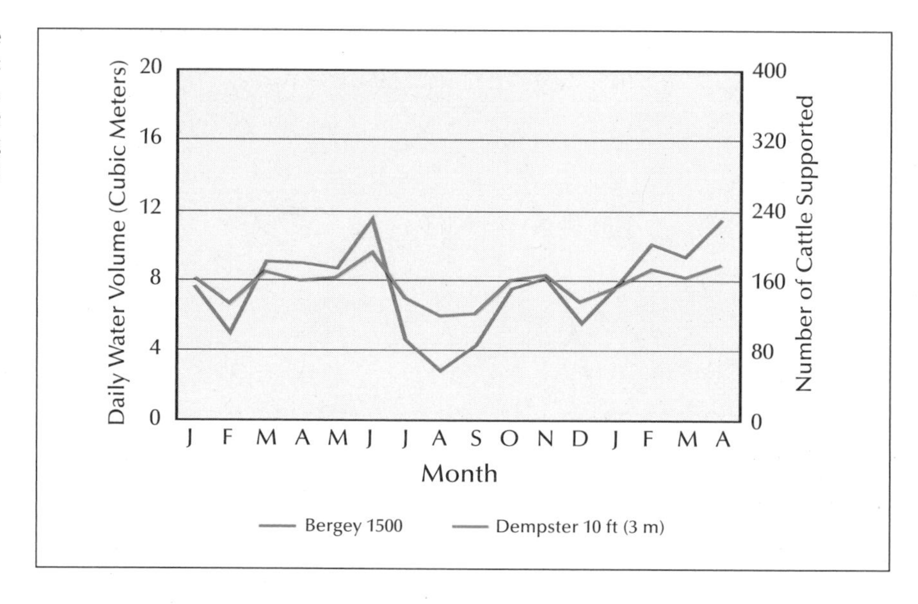

Figure 12-9 Wind pumping volume comparison. While the wind-electric system will pump more water overall, the farm windmill will pump more water when it's needed most: during summer. (Adapted from USDA ARS)

has a wind resource fairly typical of the American Great Plains, the Dempster will pump considerably more water during the hot summer months than the Bergey. During the critical month of August, the Dempster will pump twice as much water as the Bergey 1500 (see figure 12-9, Wind-pumping volume comparison).

Researchers at Bushland, however, estimated that increasing the Bergey's tower height from 60 feet to 100 feet (18 to 30 m) would allow the wind turbine to match the pumping performance of the farm windmill. Modestly increasing the tower height of a wind–electric turbine doesn't appreciably increase the total system cost, whereas tall towers for farm windmills are costly.

The tests confirmed an anecdotal report from the livestock manager of the site that for 15 years he'd never had to haul water. His 10-foot Dempster reliably met the needs for a herd that never exceeded 50 head. But when he switched to a wind–electric system of the same rotor diameter, the wind turbine couldn't meet the herd's need in the critical summer months, though the size of the herd had increased to 80 head during the test.

The farm windmill's high-solidity rotor and its direct conversion of the wind into mechanical power results in a very low cut-in wind speed, explains the USDA's Vick. For example, the cut-in speed for the Bergey 1500 is twice that of the Dempster. The Bergey outperforms the Dempster in spring, but falls short of the Dempster during summer months. "You don't get any thirstier at 18 mph than at 5 mph," says Wisconsin's Mick Sagrillo, suggesting that high performance is of little benefit if the water isn't available when you need it.

Despite this, there's still a role for wind–electric pumping. Vick himself has operated a Bergey 1500 since 1998 to pump water for his orchard and vegetable garden in the Texas Panhandle. In 2001 there was virtually no rain during the growing season, but the turbine was able to keep Vick's drip irrigation system supplied, saving his 130 fruit and nut trees from drought. Storage of the springtime surplus made the difference. Water storage for wind-powered irrigation systems is just as important as storage for livestock watering or domestic uses, says Vick.

As with so much of wind energy, the application determines which technology will work best. The classic farm windmill "is hard to beat in low-wind regimes," says Eric Eggleston, a wind engineer who has worked on both types of wind pumps. They do their job well, he adds.

Table 12-7
Approximate Cost for Traditional "American"("Chicago") Wind Pump

	Rotor Diameter				
	2.4 m 8 ft	3 m 10 ft	3.7 m 12 ft	4.3 m 14 ft	4.9 m 16 ft
Windwheel (rotor)	$2,500	$3,500	$6,500	$7,500	$10,000
Tower	$2,000	$2,500	$2,500	$3,000	$4,000
Sub total	$4,500	$6,000	$9,000	$10,500	$14,000
Pump (see Note)	$1,500	$1,500	$1,500	$1,500	$1,500
Total	$6,000	$7,500	$10,500	$12,000	$15,500

Notes: Tower: 10 meter (33 feet) lattice; drop pipe: 30 meter (100 feet); sucker rod: 30 meter (100 feet); pump rod: 10 meter (33 feet); drop pipe, $600; pump rod, $100; sucker rod, $400; stuffing box, $100; piston pump, $300.

Source: USDA ARS, Bushland, Texas

The costs of both farm windmills and small wind turbines of comparable size are similar. (See tables 12-7, Approximate Cost for Traditional "American" ["Chicago"] Wind Pump, and 12-8, Approximate Cost of Wind–Electric Water Pumping.) So the choice of which to use may be determined by factors other than cost and efficiency.

Like battery-charging home light plants, wind–electric pumping systems are capable of providing power to multiple loads. Specialists in village electrification are finding new uses for the direct coupling of wind–electric turbines to conventional motors. Small wind turbines in these applications can not only pump water, but also power motors for grinding (grains), cooling (vaccine refrigeration), and freezing (fish storage).

When large volumes of water are needed, wind–electric pumping can deliver. For example, Bergey Windpower's Excel can pump upwards of 800 gallons per minute (3,000 l/min) or pump against heads of 750 feet (225 m). It's well suited for the high-volume applications that might be found in Third World villages. If one turbine is insufficient to meet the need, several machines can be used at different points in the water distribution system or ganged together in a cluster.

The market for wind–electric pumping systems has not grown as rapidly as envisioned in the 1990s. Unlike the dozen manufacturers of farm windmills worldwide, only a handful of companies manufacture wind–electric pumping systems. The introduction of new progressive-cavity pumps designed specifically for wind and solar–electric applications could boost interest. The new pumps use high-efficiency motors that could make wind–electric pumping more competitive with mechanical wind pumps in light winds. While simplicity and versatility favor wind–electric pumping systems, only time will tell whether they will last as long as their mechanical counterparts.

Table 12-8
Approximate Cost of Wind–Electric Water Pumping

Manufacturer	SWP H80	Bergey	Bergey
Rated power	1 kW	1 kW	10 kW
Pipe diameter	1-inch	1-inch	4-inch
Wind turbine	$2,000	$1,700	$19,000
Tower	$1,200	$1,200	$5,700
Polypropylene pipe	$100	$100	$1,400
Motor	$400	$500	$1,300
Pump	$200	$400	$1,000
	$3,900	$3,900	$28,400

Notes: Tower: 20 m (60 ft); drop pipe: 30 m (100 ft)

Source: Adapted from USDA, ARS, Bushland, Texas

Estimating Water Use

Finding the right size wind machine for a water-pumping application is much like sizing a stand-alone power system to meet

Table 12-9
Total Average Daily Water Use

Humans	liters/person	gallons/person
Third World village	40	10
Third World urban	100	30
U.S. average	570	150
U.S. arid regions	950	250

your need for electricity. Sizing a wind-pumping system requires an estimate of both average and peak demand. Average demand corresponds to the total volume of water required and is similar to the overall electrical energy generated in a wind-driven remote power system. Peak demand corresponds to the maximum flow rate needed to meet simultaneous water uses, and it resembles the peak power demands that determine the size of inverters in a home light plant.

Water consumption is as much a reflection of lifestyle as electricity is, and each person's use varies with habit, culture, climate, and the availability of water. Third World villages consume as little as 10 gallons (40 l) per person per day, because water must be pumped by hand and often carried some distance (see table 12-9, Total Average Daily Water Use). Urban dwellers, because their water is often more plentiful, may use three times as much as Third World villagers. North Americans use considerably more.

As late as the mid-1970s, design manuals suggested that a rural homestead in the United States could manage with 50 to 100 gallons (200 to 400 l) per person per day. By the late 1980s, average consumption had risen to 150 gallons (570 l) per person per day. In arid regions of the United States, such as southern California, planners assume each household will use no less than 250 gallons (950 l) per person per day, not including the water needed for lawns, gardens, and ornamental plants. Watering a lawn or garden is little different from irrigated agriculture. Both require a lot of water. Watering with just one garden hose may use 300 gallons (1,100 l) per

Table 12-10
Daily Water Use for Farm and Home

Farm Animals	liters/animal	gallons/animal
Horses	50	13
Dairy cattle	70	19
Steers	60	16
Pigs	20	5.3
Sheep	10	2.6
Goats	10	2.6
Chickens	0.3	0.08
Household Uses	**liters/person**	**gallons/person**
Bath tub/filling	130	35
Shower/use	90–230	25–60
Flush toilet/use	10–30	2–7
Dish washing, hand/day	80	20
Dish washing, auto/day	40–80	10–20
Water softener	570	150
3/4-inch hose/hour	1,130	300
Laundry/use	110–190	30–50
Miscellaneous uses/day	90	25
Lawn/garden (1000 ft^2)	2,270	600

Source: Adapted from Sandia National Laboratories, Heller-Aller Co., and "Planning for an Individual Water System," AAVIM

hour. Livestock watering and other farm uses may also add dramatically to water requirements (see table 12-10, Daily Water Use for Farm and Home).

To estimate your average water requirements, sum the typical uses in table 12-10 and multiply by the number in your household, the number of livestock you expect to raise, and the size of any lawn or garden you plan to water frequently. The U.S. Department of Agriculture estimates that a typical farm may require up to 6,000 gallons (24,000 l) per day, including domestic uses.

Most homesteads without livestock or intensive gardening won't require nearly as much. Yet a family of three in an arid region of the Great Plains could quickly exhaust the daily supply provided by a standard farm windmill during the light winds common in late summer. If each person uses 250 gallons per day and the family tills a large garden, the household will need about 1,400 gallons

(5,300 l) per day. A 10-foot (3 m) diameter farm windmill (see table 12-5) at a 9 mph (4 m/s) average wind speed could pump about 1,600 gallons (6,000 l) daily from a well 130 feet (37 m) deep. There's little margin for error in such a system, and storage is necessary to assure an adequate supply.

Water use on remote homesteads becomes self-limiting. Families learn to reduce their water consumption, in part because they don't have unlimited power for pumping, or unlimited storage. Designers of rural water supplies have observed that people use far less water if they have to pump it by hand or carry it from a well. Those using a stand-alone power system become more sensitive to water conservation if they have to start a noisy backup generator to water their lawn. As with electricity, it pays to conserve. The less water used, the smaller the wind pumping system necessary to meet your needs, or the more surplus that can be stored for days of light winds. For wind–electric pumping and for conventional remote power systems where batteries power the well pump, more energy can be devoted to electrical appliances if less energy is devoted to water pumping. Many nonfarm families today live comfortably beyond utility lines on less than 100 gallons (400 l) per person per day. These modern-day pioneers have demonstrated that you can enjoy life in the country with only modest water and power requirements.

Yet they clearly need storage as well. Storing water in a large tank will provide a backup supply when winds are light and, just as importantly, will provide an emergency supply for fire protection.

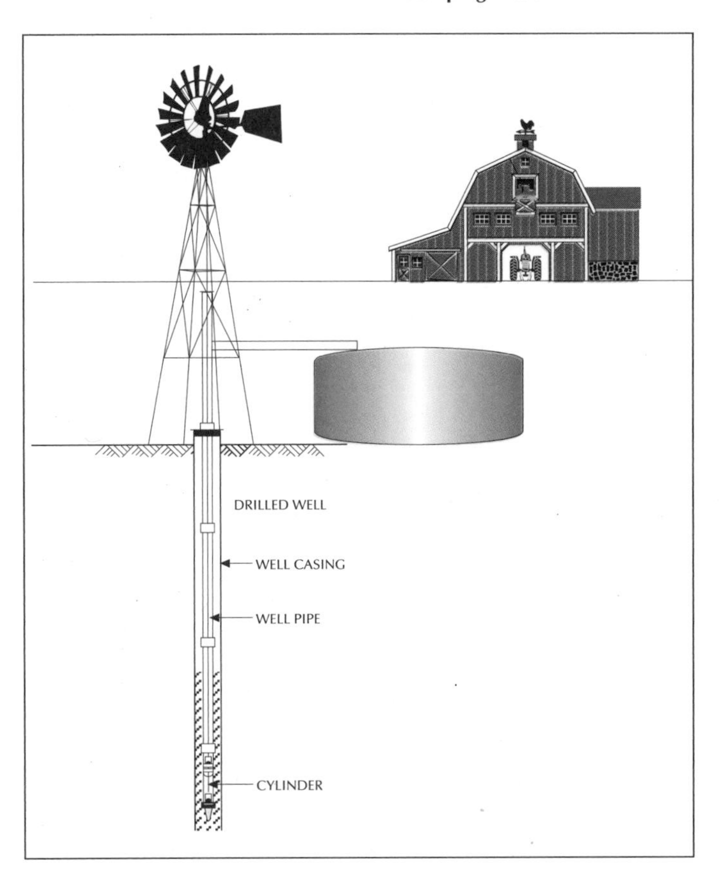

Figure 12-10. Farm windmill and storage tank. Storage is a necessary part of a wind-pumping system, whether it's for watering livestock or serving domestic needs.

Storage

Storing water in a tank is cheaper and more reliable than storing energy for pumping in batteries (see figure 12-10, Farm windmill and storage tank). A storage tank also provides the fire protection that batteries can't. For rural areas in western North America, fire protection is a critical requirement for any water system. Fire protection will often determine the storage needed and the maximum instantaneous demand placed on the water supply. Proper water system design requires determining both the average water usage and the water demand, or the flow rate at any one instant. The number of faucets that you expect to use at any one time plus other uses, such as a shower or washing machine, governs the maximum flow rate the water system must be capable of meeting. For the stand-alone power systems of the past, pumps were sized accordingly. The pump would then operate whenever water was needed, drawing power from the batteries.

In mechanical systems the farm windmill alone can't reliably provide the pressure to meet the flow requirement, because wind

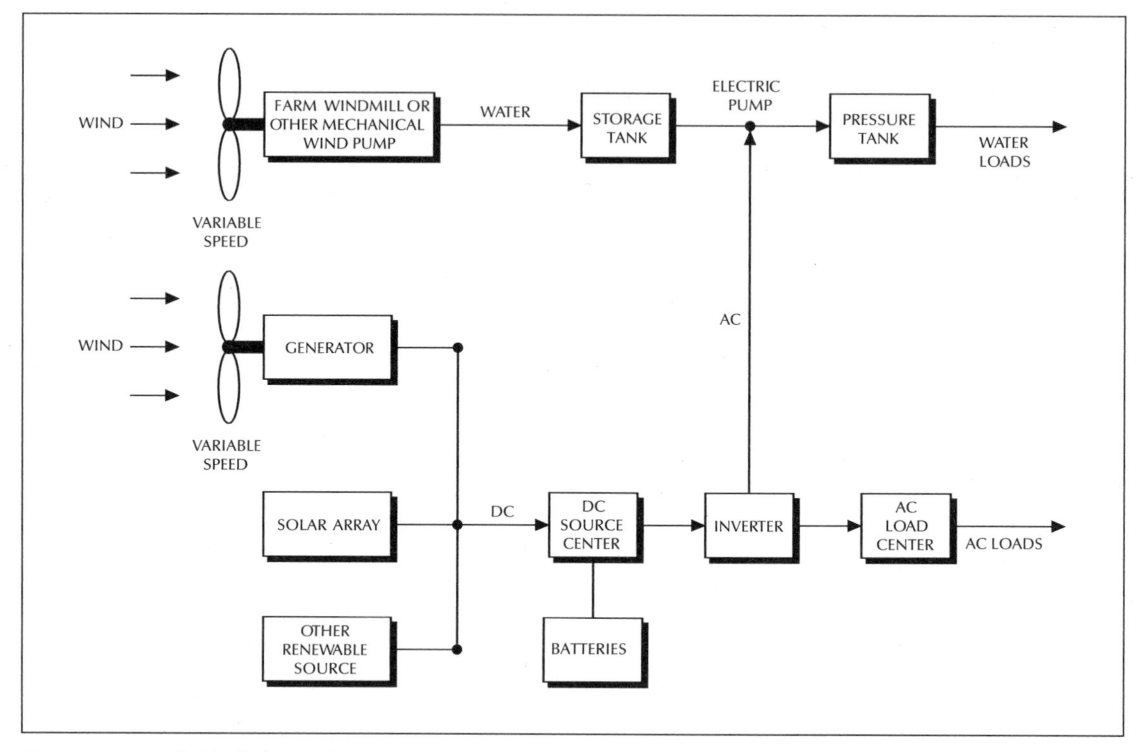

Figure 12-11. Hybrid wind pumping system. Most homes today require greater pressure than that provided by gravity-fed water systems.

speed and the pumping rate vary. In the past more or less constant pressure was provided by storing water in a tank on an elevated platform. The tank provided a modest amount of pressure by gravity.

Today nearly all rural water supplies in North America use accumulators, or pressure tanks, to provide the constant pressure demanded by contemporary consumers. Where fire protection isn't necessary, the storage provided by pressure tanks reduces the cycling of the well pump, allowing the pump to operate more efficiently than if it ran continuously whenever water was required.

Because pressure tanks must be charged by an electric well pump, they're not well suited for the true gravity-fed water supply typically used with mechanical windmills. But a farm windmill can be used to pump water, as available, into a storage tank. The rural water system can then draw water from storage and pump it into a pressure tank with an electric pump driven by a stand-alone power system (see figure 12-11, Hybrid wind-pumping system). Some have adapted electrically powered pump jacks to the reciprocating well pump used by the farm windmill. Thus if water needs to be drawn from the well and the winds are light, the remote power system switches on and drives the pump jack.

Ample storage shouldn't be avoided. Some building codes and insurance carriers may require it for fire protection. Richard Perez of *Home Power* reports that in northern California, fire codes may even require the addition of engine-driven pump jacks to ensure an adequate flow during emergencies.

Galvanized steel tanks are a common sight at rural homes in arid regions. Homesteads on the Mojave Desert of southern California store a minimum of 2,500 gallons (10,000 l) in unpressurized aboveground tanks, with many preferring 5,000-gallon (20,000 l or 20 cu m) tanks. In colder climates, cement cisterns can be buried below the frost line to prevent freezing. Ken O'Brock of O'Brock Windmill

Distributors, says this is a common practice among the Amish of Pennsylvania and Ohio.

For a comprehensive guide to designing a rural water system, consult *Planning for an Individual Water System* by the American Association of Vocational Instructional Materials (see the appendix for details).

Irrigation

Many applications for water pumping, notably irrigation, require considerably more capacity than the farm windmill alone can supply. Irrigated crops, intensive gardening, and large lawns use substantial volumes of water (see table 12-11, Daily Irrigation Use). High-discharge irrigation wells in the Texas Panhandle, for example, may pump upwards of 1,000 gallons (.40,000 l) per minute from depths approaching 400 feet (120 m). The American farm windmill isn't big enough to dent such a load.

Table 12-11
Daily Irrigation Use

Irrigated Crops	(m^3/ha)	(gal/acre)
village farms	60	6,400
rice	100	10,700
cereals	45	4,800
sugar cane	65	6,900
cotton	55	5,900
Lawn/Garden	240	26,100

Modern medium-size wind machines can be used for irrigation by either pumping water mechanically or producing electricity to run a large well motor. Because the wind is intermittent, and irrigation requires such large volumes of water that storage often isn't practical, the wind machine typically drives the well pump in conjunction with a conventional energy source. Coupling wind turbines with conventional sources is a fairly simple task with an electric well pump, but it isn't quite as easy with the engine-driven pumps commonly used on the southern Great Plains.

West Texas A&M University, in cooperation with the USDA, developed an ingenious device for mechanically coupling the varying mechanical output of a wind machine with an irrigation pump: the overrunning clutch. When the wind is strong and the wind machine is producing full output, it mechanically drives the well pump entirely on its own, via the overrunning clutch. When the winds are weaker the wind turbine assists in driving the pump. The conventional power source makes up the difference. During a calm spell the conventional power source operates the pump alone. The overrunning clutch enures that a constant volume of water is pumped regardless of wind speed. At the same time, the overrunning clutch enables using the wind to reduce the consumption of conventional fuels. Tests at the USDA's Bushland station in the early 1980s proved that the concept works. Unfortunately there are no wind turbines currently available that can mechanically drive the large irrigation pumps used on the Great Plains.

Using the wind to assist irrigation pumping is even simpler when the pump is driven by an electric motor. Wind machines producing utility-grade electricity can be connected directly to the well motor in a configuration identical to that for an interconnected wind turbine at a home or business. When wind is available, the wind machine offsets consumption of electricity from the utility. If the wind turbine produced a surplus of energy during periods when pumping loads were light, the excess is sold back to the utility.

The USDA has been operating an Enertech E44 at its water-pumping test laboratory in Bushland since the early 1980s (see figure 12-12, Long-lived Enertech). Nolan Clark estimates that in such an application on the Great Plains, wind turbines like the E44 will pump water 65 percent of the time in the Texas wind regime. With the addition of solar–electric pumping, the hybrid could provide water nearly 85 percent of the time.

Figure 12-12. Long-lived Enertech. This Enertech E44, used for wind–electric pumping experiments, has been operating at the USDA's Bushland, Texas, site since 1982. The turbine has logged more than 100,000 hours of operation and generates about 90,000 kWh per year.

Figure 12-13. High-volume wind pumping. One of a series of Lagerwey 18/80s at pumping stations in the polders of the Netherlands' Groningen province.

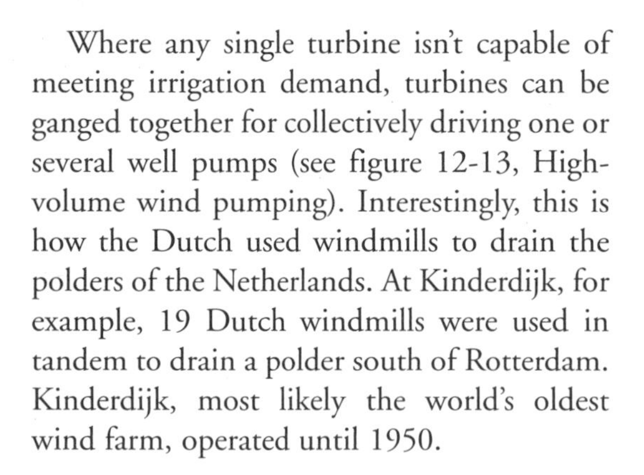

Where any single turbine isn't capable of meeting irrigation demand, turbines can be ganged together for collectively driving one or several well pumps (see figure 12-13, High-volume wind pumping). Interestingly, this is how the Dutch used windmills to drain the polders of the Netherlands. At Kinderdijk, for example, 19 Dutch windmills were used in tandem to drain a polder south of Rotterdam. Kinderdijk, most likely the world's oldest wind farm, operated until 1950.

Wind Pump Renaissance

American manufacturers of farm windmills no longer have a monopoly on the technology. Today the few remaining companies must compete head to head with other producers, such as Argentina's FIASA and Australia's Southern Cross, and with newly designed mechanical windmills developed by longtime wind pump designer Peter Fraenkel of IT Power.

Still, the windmill that "won the West" remains enshrined in the American psyche. Farm windmills and their place in rural life continue to intrigue Americans. Much later than the Dutch, who first called for preservation of their windmill heritage in 1923, Americans are slowly beginning to turn their attention toward preservation. In the mid-1990s four museums were under development in the United States for the display of rare farm windmill collections.

In explaining his fascination with farm windmills, Bryce Black calls them "aeliotropic." To him the farm windmill follows the wind like a sunflower follows the sun. Black, who has carved out a niche for himself as a farm windmill repairman in western Wisconsin, where some of the machines are still used, says "there's something timeless" about windmills. To Black the farm windmill was forged from fire, yet provides a living link between sky, earth, and water.

Beyond mere nostalgia for the traditional farm windmill, wind pumping remains an important use of wind energy. Spreading desertification and never-ending population pressure around the globe will, at a minimum, assure a continued demand for wind pumping and possibly create a renaissance among manufacturers of both traditional farm windmills and modern wind–electric pumping systems.

13

Siting

High winds blow on high hills.
—Thomas Fuller, Gnomologia

North American farmers, ranchers, and rural residents should encounter few barriers to erecting wind turbines of any size on their property. Siting in these circumstances is often simply finding the most exposed place for the wind turbine. Problems can arise in suburban and more densely populated areas, where some neighbors may not share your enthusiasm for wind energy. With care, consideration, and a good measure of patience, you should be able to allay any neighbor's concerns. Nevertheless, you should always make an honest appraisal of your site. You may find it unsuitable for wind energy because of physical constraints—too many trees and tall buildings, for example—or because of legal restrictions on how you can use your land.

In this chapter we look first at the physical restrictions on where you can place a wind turbine, then at the more complex topic of institutional restrictions that may limit or even prohibit the use of a wind turbine. The treatment here of the thorny issue of wind turbine siting is far from exhaustive. Entire books are devoted solely to aesthetics or noise. (See the bibliography for details on *Wind Power in View* and *Wind Turbine Noise.*)

Physical Restrictions

Is your site suited for a wind turbine? That's the first and foremost question. Do you have enough room? There must be not only sufficient space for the tower itself, but also enough space to install it safely. And as I've stated, putting the turbine on the roof is not a suitable option.

Wind turbines will not work for everyone, everywhere. But wind turbines, both large and small, are used in surprising places. Guyed towers for small turbines have been installed on city lots so small that the anchors have been placed in each corner of the backyard. Freestanding towers have been installed in equally cramped quarters; on occasion a crane has been required to lift the tower over the house and set it on the foundation.

Europeans, accustomed to greater population densities than are common in North America, are more tolerant of placing wind turbines in proximity to homes, businesses, and public places (see figure 13-1, Sidewalk siting). Commercial-scale wind turbines have been installed in parks, playgrounds, and parking lots, near soccer fields, and at busy truck stops and lock gates. They can also be found alongside canals, dikes, and breakwaters. In Germany it's common to see wind turbines lining the autobahn, while in Denmark rail passengers can watch wind turbines spinning in fields adjacent to the tracks.

Figure 13-1. Sidewalk siting. A Lagerwey 18/80 next to a fish processing plant along a frontage road in the port of Lauersoog, the Netherlands. The Dutch are accustomed to multiple use of their limited land area.

Yet there are limits, and it's wise to know what they are. Bergey Windpower, for example, recommends at least 1 acre (0.5 hectare) for its Excel model. Smaller turbines may need less.

Exposure and Turbulence

Wind turbines should always be located as far away from trees, buildings, and other obstructions as possible in order to minimize the effect of turbulence and maximize exposure to the wind.

Turbulence, rapid change in wind speed and direction, is caused by the wake from buildings and trees in the wind's path, and resembles the eddies swirling around a rock in a stream. Being buffeted by turbulence can be damaging to modern wind turbines because they use long slender blades traveling at high speeds. Turbulence can wreak havoc on a wind machine, rapidly shortening its life.

In Germany it's common to see wind turbines lining the autobahn.

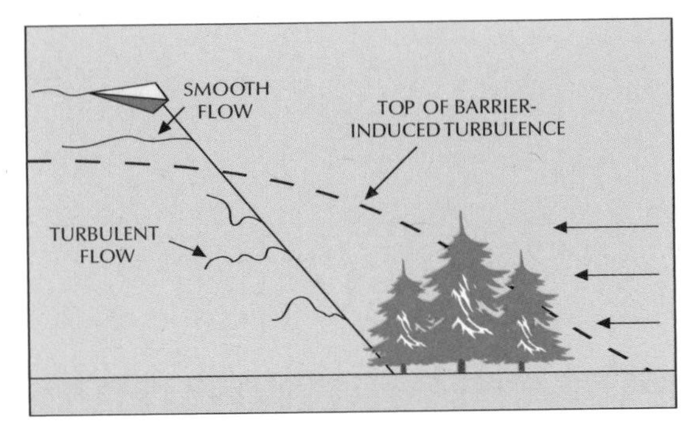

Figure 13-2. Go fly a kite. Trailing streamers from a kite is a simple yet effective way of detecting turbulence. (Battelle PNL)

Buildings and trees also drastically reduce the energy available to a wind turbine. One overriding lesson that has been gleaned from nearly three decades of working with modern wind turbines is that you can't overlook the effect of obstructions, whether buildings or vegetation. Though seemingly less a barrier to the wind than a building, trees, shrubs, and even low hedgerows can rob energy from the wind. It's for this reason that wind turbines are being installed on increasingly tall towers, some up to 100 meters (330 ft) in height.

When you're uncertain about the amount of turbulence over your site—go fly a kite. Tie streamers to the kite string and note how they flutter in the wind (see figure 13-2, Go fly a kite). It's a practical means of seeing the invisible—the swirls and eddies caused by obstructions—and a good way to learn firsthand about turbulence.

Locate the tower far enough either upwind or downwind to avoid the turbulent zone around nearby obstructions (see figure 13-3, Zone of disturbed flow). When this is impractical, use as tall a tower as possible to elevate the wind machine above the turbulence. If neither approach alone is sufficient, use some combination of siting and a taller tower.

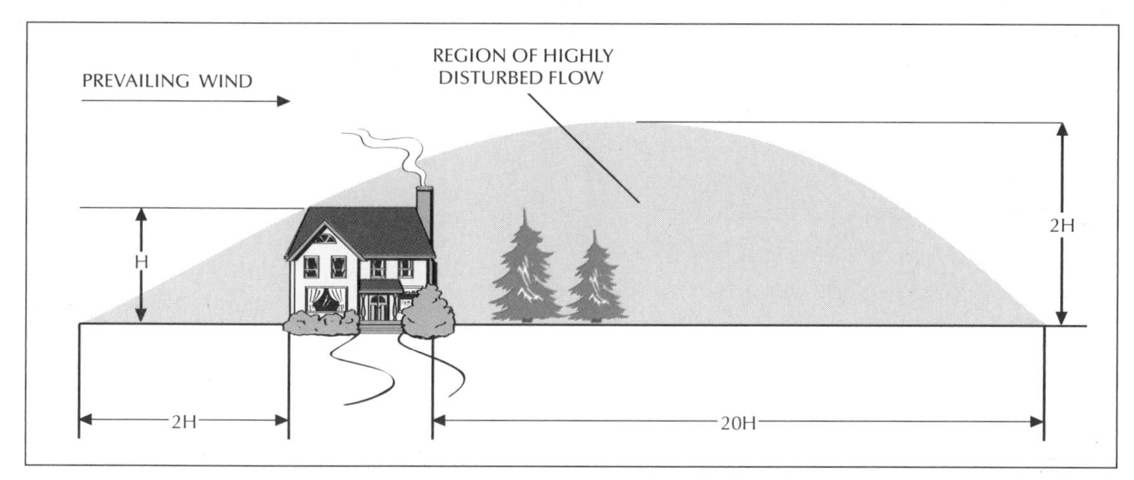

Figure 13-3. Zone of disturbed flow. Wind speeds decrease and turbulence increases in the vicinity of obstructions. The effects are most pronounced downwind but also occur upwind as the air piles up in front of the obstruction. The flow over a hedgerow or group of trees in a shelter belt is disturbed in a similar manner.

From years of experience, small wind turbine manufacturers, consultants, and users have derived a general rule of thumb: The entire rotor disk of the turbine should be least 30 feet (10 m) above any obstruction within 300 feet (100 m). If you've determined, for example, that a group of trees along a fencerow are 60 feet (18 m) tall, you'll need at least a 90-foot (27 m) tower (see figure 13-4, Clear of obstructions). To ensure the best performance, you should use an even taller tower.

The minimum tower height for medium-size turbines is equal to the turbine's rotor diameter. As mentioned previously, many are installed on taller towers. In forested areas of Germany it's not rare for the tower to exceed one and a half times the rotor diameter.

By all means avoid sites at the bottoms of creeks, draws, or ravines and at the bases of hills. If there's a hill on your property with a well-exposed summit, site the wind machine there instead of lower on the slope, even if the summit is some distance from where you plan to use the electricity.

Figure 13-4. Clear of obstructions. This Bergey 850 stands well above nearby willow trees at the Wulf Test Field in California's Tehachapi Pass. Raising small wind turbines above nearby obstructions is the single most effective way to increase performance.

Power-Cable Routing

Once you've selected the area where the tower will be erected, note how the power will be delivered to your load. At this stage you need to anticipate any problems that may develop

Property Values

Evidence that wind turbines affect property values one way or the other has been hard to come by. There was a cursory survey by a real estate agent of land sales near wind turbines in California's Tehachapi Pass. In this survey values actually increased in the vicinity of the wind farms. Why this happened is less clear. But property changed hands at higher values after the wind turbines went in than before. In one case property values increased surprisingly when agricultural land was developed for a mobile home park with a clear view of the wind turbines. In a Wisconsin survey of land sales near two operating wind farms, properties less than 1 mile (1.7 km) from the turbines traded at 141 percent of their assessed value, rising along with other land values in the area. The wind turbines didn't appear to have any negative effect.

The most extensive study in North America was conducted by the Renewable Energy Policy Project (REPP). This study examined property transactions near wind turbines at multiple sites in California, Iowa, and Minnesota, as well as sites in several other states. Altogether REPP evaluated more than 25,000 real estate transactions. After accounting for comparable sales outside the viewshed of the wind farms selected, defined as beyond 5 miles (8 km) from the turbines, the study found that viewshed property generally increased in value faster than property with no view of the wind turbines. REPP's report, "The Effect of Wind Development on Local Property Values," can be downloaded from the nonprofit group's Web site at www.repp.org.

later. They're easier to avoid than to solve. For example, a buried telephone line crossing your path may complicate digging a trench for laying the cable underground. Ideally the electric service from the wind machine will enter the building near where the utility's lines also enter.

In the past it was common for installers of small wind turbines to string the power cables on poles just like those of the electric utility. The consensus today is to bury all conductors, whether for a small or medium-size wind turbine. If the service entrance and meter are on the other side of the building from where you are planning to erect the tower, what is the best route for the laying the power cables to the service entrance? Are there any sidewalks, driveways, or roads in your path? How will you cross them? These are important questions, because the answers affect the cost of installing the wind system. They also determine how difficult it will be to meet certain institutional restrictions, such as the National Electrical Code in the United States.

Institutional Restrictions

Equally as important as finding the optimal site for the wind system is determining what legal requirements your local community places on structures such as wind turbines. In the United States land-use zoning, building codes, and protective covenants may all apply.

Planning Permission

Many who have installed small wind turbines in North America have had few problems, if any, with land-use restrictions. Either their property was not covered by regulations, or where it was permission was quickly and easily obtained. Many rural areas are not zoned at all, and where they are there are practically no restrictions on land that is zoned agricultural. The situation changes as you near cities, small towns, and residential neighborhoods. There the right to swing your fist ends where your neighbor's nose begins.

In some rural locales, wind turbines are specifically permitted unless there is an overriding reason to prohibit their use. In California's Kern County, for example, the use of wind energy is permitted in certain designated agricultural areas. Recently adopted state laws prohibit discriminating against the use of wind turbines relative to other similar land uses.

In most countries, planning approval (or

more broadly, the placing of restrictions on how land is used) is a responsibility entrusted to local governments by the public to protect general health and welfare. Officials will want you to show how your use conforms to the public's general agreement on what can and can't be done on land within a designated area.

Public officials have a moral and often legal obligation to treat you fairly. Above all, planning officials shouldn't discriminate against you because they're unfamiliar with wind turbines. Treat them cordially. One thing is certain: If you need a building permit, a zoning variance, or another form of planning approval, you want them on your side.

Building Permits

Where planning ordinances apply, you must conform to the law—period. Find out what the requirements are in your area by calling the local building inspector, board of supervisors, or planning office. You want to know how to obtain a building permit (where required) and who is responsible for issuing it (usually the building inspector). Get details. Whoever is responsible should provide a list of what you must do: the forms to fill out, the fees to pay, where and when to file, and any other information that you must supply. Then methodically deliver what's required.

The intent of this process is to determine conformance with the regulations governing your locale and to alert the public to your project. Whether you want to install 1 wind turbine or 100, take the initiative and contact anyone who might be affected, especially your neighbors. You have a responsibility to tell them what you're planning and why. Speak to them early in the project so that they feel consulted, rather than pressured into backing you. It's much better to talk with them informally over the back fence than in court or in a shouting match at a public hearing. If you get along well, there should be few problems, but if you've driven over your neighbor's prize rosebush for years, you'd better make amends. Objecting to your building permit is a great opportunity to even the score. You can head off conflict by respecting the needs of your neighbors. Treat them in the same way you would like them to treat you.

At a minimum you will be required to produce a plan or map showing the dimensions of your site and where the tower will be located. You can prepare this yourself. Drawings of the wind turbine, tower, and foundation with their specifications may also be required. The dealer or manufacturer can supply these.

Planning laws follow one of two approaches. One allows you to do whatever you want, unless specifically prohibited. The other approach prohibits you from erecting any structure unless it is specifically permitted. Where the latter approach is used, your application could be denied simply because no one has ever installed a wind turbine before.

In communities where this is the situation, you can sometimes get permission for a wind machine by bringing it under a permitted category such as radio or cell phone towers, TV antennas, or chimneys. Building officials may be empowered to make such a determination. If not, formal action before a public board is necessary. These officals must determine if your use conforms with the intent of the ordinance. Where it doesn't, or where the ordinance specifically excludes wind turbines or similar structures, you must obtain a variance from the regulation.

In the United States the zoning appeals board or board of adjustment is the final arbiter of permit approval disputes. This is a political body, and if there's a public outcry it'll respond accordingly within the limits of the law. Variances—variations from the law—give the zoning appeals board flexibility in meeting local planning objectives: the protection of the common good without undue restrictions. The board members will want to know whether your wind turbine detracts from your neighbors' use of their land, lowers the value of surrounding property, or endangers passersby. The burden of proof is on you, the petitioner.

Frequently the granting of a variance is little more than a formality. You may not even need to be present. But if the board has questions that you have not answered previously, or if the variance is contested, you'll need to be present and you'll need to be well prepared. On occasion unfamiliarity with wind turbines—even today—will fuel wild speculation about what they will do to the neighborhood. Often these fears can be quickly dispelled with the facts. Sometimes they can't. When contested, the public hearing can take on the appearance of an expensive courtroom battle, with opponents bringing in their own "expert witnesses" to counter your assertions. It can be rough—even humiliating—if you're unprepared, or if the hearing officers lose control of the meeting.

You have a right to a fair and impartial hearing. You also have a right to argue your case without intimidation—physical or verbal. Public meetings can quickly degenerate into mob rule if public officials and meeting organizers don't limit disruptive behavior. You have an obligation to stem rumors by immediately responding to wild or outlandish claims. Insist on proof or documentation of unsubstantiated charges. The list of real or potential problems wind turbines might cause can be endless, limited only by the human imagination. Hearing officers have an obligation to maintain civility. If they can't—or, worse, won't—hold them responsible.

In suburban housing developments or planned communities, deeds may contain restrictions, or covenants, on how the land can be used. These restrictions are intended to preserve the identity of the neighborhood. Take a look at your deed. Or call your attorney, realtor, or mortgage company for information. If there are any restrictions, these people will know how best to deal with them. For example, the restrictions may be unenforceable.

Also note the location of any easements on your property for utility rights-of-way. In the United States, easements transfer use of the land without transferring outright ownership. Easements are commonly used for a host of public purposes: power lines, underground telephone cables, pipelines, future roads or sidewalks, and so on. These could all limit your use of the land. You may be unable to encroach on these easements with your wind turbine even though you own the land, there are no restrictive covenants, and you obtained all the proper planning approvals.

Building officials are sometimes bewildered by a request to install a wind turbine. California's San Luis Obispo County officials demanded engineering calculations to assure them that Jim Davis's $1,000 wind system wouldn't pose a hazard to the public. Those calculations would have cost Davis a whopping $5,000 if the wind turbine manufacturer, Southwest Windpower, hadn't faxed him an 11-page document that satisfied authorities.

Bergey Windpower's Mike Bergey likens the permit approval process in some states to "medieval torture." Some projects have taken seven months to obtain a permit, says Bergey—far longer than the time needed to build, ship, and install the turbine.

Through experience, other building officials know what they need to ensure that wind turbines are installed properly. Jonathan Herr, for example, didn't have any problem winning approval to install an Air 403 in California's trendsetting Sonoma County. "I got the building permit over the counter," he says.

Height Restrictions

The most frequent limitation on the use of small wind turbines is a restriction on the height of the tower. In most residential areas of North America, there's a limit to the height of structures, usually 35 feet (11 m), a relic of the days when fire brigades had to pump water by hand. Variances to such ordinances can be obtained by pointing out other structures taller than the limit that have been allowed under the zoning ordinance: radio towers, chimneys, or utility poles. (Local officials seldom have control over utilities.)

When wind farm developers or users of medium-size wind turbines are hampered by such archaic restrictions, they can afford the legal assistance needed to change the law. Those wanting to use small wind turbines often are unwilling or financially unable to fight such height limitations.

In Great Britain some rural residents simply opt for a short tower to avoid the cost and the all-too-frequent controversy surrounding a request to install an appropriate-size tower. Similarly, North Americans also sometimes opt for the path of least resistance. NREL's Jim Green documented one case where the application for a permit to use a tower taller than the 35-foot limit cost more than the wind turbine. This may explain why Green found in a survey of six small wind installations in Colorado that "every wind turbine I saw could have benefited from being on a taller tower."

Figure 13-5. Obstruction marking. Italian authorities required obstruction marking of the towers (bands of red and white) on these Vestas V44 turbines above Montefalcone di Val Fortere, northeast of Naples, to alert pilots to the turbines' presence.

Obstruction Marking

In the United States when the height of the tower plus one blade length exceeds 200 feet (84 m), or you're within 1 mile (1.7 km) of an airport, you must register your plans with the Federal Aviation Administration (FAA). This allows the FAA to note an obstruction to aviation on maps and alert pilots to the hazard.

Small wind turbines normally operate well below this threshold, and no permit is needed. If there is any doubt about the need for obstruction marking, building officials may forward a notification to the FAA or advise you to do so as a precaution.

Still, medium-size wind turbines and megawatt-class turbines are being installed on increasingly tall towers that exceed this threshold. To grant a permit the FAA will require the use of a high-intensity flashing white light, or will require you to paint obstruction markings in red and white on the blades, tower, or both. Most of the very tall turbines in North America use flashing white lights. In daytime these lights are aesthetically preferable to the red-and-white banding seen on tall structures near airports, but they detract from the night sky, especially in western North America where there are few other light sources but the stars themselves.

Europe features a considerably higher

Figure 13-6. Stroll through the park. In this unstaged photo, mothers take their children for a stroll through a park above the village of Brooklyn within urban Wellington, New Zealand. The tower supports a Vestas V27, a 225 kW wind turbine. This particular turbine has consistently been one of the most productive in the world.

height threshold—100 meters (328 ft)—before action is required. Even so, regulations differ from one country to the next. Italy, reacting to the tragic collision of a low-flying U.S. Marine Corps fighter jet with a ski lift gondola full of skiers, has required obstruction marking of relatively short towers on ridgetops in the Apennines (see figure 13-5, Obstruction marking). Tall turbines in Germany are required to have obstruction markings only on the rotor blades. Early megawatt-size turbines in Germany incorporated a series of red-and-white-alternating bands on the blades. Later turbines use only one red band on the outer third of the blade.

No passerby has ever been injured or killed by a wind turbine.

Public Safety

The public has a legitimate interest in the safety of wind turbines and the hazards they may pose. There's no point in hiding the fact that several men have been killed while working on or around wind turbines. And one parachutist was killed on her first unassisted jump when she drifted into a wind turbine on the German island of Fehrmarn. But no passerby has ever been injured or killed by a wind turbine.

Wind turbines, like any large, rotating machinery, should be treated with respect, but there is no reason to fear them unduly. In many parts of the world, wind turbines are part of the community and found in public places (see figure 13-6, Stroll through the park).

In some communities in North America, towers must be set back from the property line a distance equal to their height. Officials reason that if the tower fell over it would not extend beyond the user's property and present a hazard to neighbors. If your lot is too small to permit this, you may want to reconsider wind power. Unfortunately this restriction

Figure 13-7. A Swiss watch it was not. Twenty years ago wind turbines were much less reliable than they are today. While it was uncommon even then for a wind turbine to destroy itself, some did. This Swiss Wenco operated only briefly before failing in California's Tehachapi Pass. All similar turbines were eventually removed after a complaint by the Sierra Club that they were an eyesore.

discriminates against wind turbines compared to other common structures.

We think nothing of other human-made and natural hazards that pose a risk similar to if not greater than that of a wind turbine. We've all seen homes sheltered beneath the branches of an old oak tree, where occasionally a storm-weakened limb crashes down onto the roof. We accept this hazard as the price we pay for the benefits the tree provides—shade and visual amenity).

The same is true for radio and television towers. In many ways they are similar to towers for wind machines. They are made of metal and extend visually above the roofline. The public has grown to accept them, and because their failure rate is so low, users often install them adjacent to occupied buildings.

Permitting authorities will be concerned that your tower could collapse. You must show them that the tower meets international standards for wind turbine design and applicable building codes, and that similar towers operate throughout your locale in a host of severe environments without incident. Though towers have failed, the occurrence is rare and far less frequent than that of falling trees or utility poles.

Falling Blades

Authorities will also be concerned that the wind turbine could throw a blade or, worse, fling itself off the tower. While infrequent, neither of these is unknown (see figure 13-7, A Swiss watch it was not). Once again, you must convince planning officials that the wind machine has been designed and built to accepted international standards and that there's little likelihood that it will throw a blade into the midst of a neighbor's lawn party.

Figure 13-8. Watch for ice. This sign at the Acqua Spruzza test site in Italy's Apennine Mountains warns, WIND TURBINE, WATCH FOR ICE. Europeans typically urge caution around wind turbines, but seldom exclude the public. In this case falling ice could be a hazard, but only during winter storms when few people are likely to visit the windswept site. Note that the roof and facade of the control building are constucted of native materials as used on similar structures elsewhere in this region.

You can best reassure officials that your wind turbine won't become airborne by citing the number of like turbines operating elsewhere and the number of years these turbines have operated without incident. Thus it behooves you to select a wind system with a proven track record: one where a host of units have operated reliably in a variety of applications for several years. If you plan to install a new, untested, or experimental wind turbine, expect authorities to demand more restrictive setbacks than for wind turbines in widespread use.

Falling Ice

A related question in cold climates is ice throw. Under certain conditions, often during cold winter nights when the wind turbine is becalmed, ice can build up on the blades. During the day sunlight warms the ice, loosening it so that the slightest motion sets the ice moving down the blade (see figure 13-8, Watch for ice). Occasionally, as at the 600 kW turbine outside the Bruce nuclear power station at Kincardine, Ontario, the turbine will throw the ice some distance. While no one has ever been injured by falling ice, it's prudent to discourage people from walking near the turbine during ice storms or shortly thereafter.

Ice is a common and accepted hazard in cities with severe winters, such as Montreal, Quebec. In such cities buildings have provisions for breaking rooftop ice sheets into pieces to minimize the hazards they pose when they eventually slide off. Similarly, some medium-size wind turbines destined for northern climes are constructed with heated blades that shed ice as it forms, so that it doesn't become a hazard.

Attractive Nuisance

The fear that a wind turbine could become an attractive nuisance—that is, attract the attention of vandals or children—is unique to North America. Generally a property owner is not liable for accidents to trespassers, but a different test is applied to the acts of children.

Swimming pools are thought to entice or attract children to trespass. Because children cannot discern the hazard presented by the pool, the community views it as a public nuisance, and if an accident occurs, a court can hold the owner liable. Permitting regulations allow attractive nuisances when they have met requirements designed to prevent accidents. Swimming pools must be fenced, for

Figure 13-9. Fencing unnecessary. Fencing of this transformer at a wind power plant in Colorado is unnecessary and detracts from the aesthetically pleasing array of wind turbines. Note that the tower is not fenced. The heavy door is locked.

example. The same ordinance may require that towers, such as wind turbine towers, be fenced as well.

Fencing isn't the only way to prevent someone from climbing a wind turbine tower. Electric utilities seldom use fencing. On their transmission towers they simply remove the climbing rungs to a level 10 feet (3 m) or more above the ground. You can do the same on a freestanding truss tower. Or you can wrap the base of a guyed lattice tower in sheet metal or wire mesh. These alternatives should be acceptable to planning officials because they accomplish the same goal as fencing while being less obtrusive. Utilities seldom erect fences around their utility poles or transmission towers. Imagine the outcry if every utility pole required a fence.

Medium-size wind turbines on tubular towers have no need of a fence to prevent unauthorized entry (see figure 13-9, Fencing unnecessary). The massive doors to these towers are securely locked. No child or common vandal could climb these towers. Of the thousands of wind turbines operating in Europe, nearly all are fence-free. Fencing of tall structures to thwart access by children and vandals is a peculiarly American phenomenon.

Avoid fencing wherever possible. Fencing increases the aesthetic impact of wind turbines by drawing unwarranted attention to the turbine with the message, *I am dangerous; stay away.* Or the equally offensive, *This is my wind turbine. Keep your hands off.* In the Tehachapi Pass, unfortunately, wind farm operators do both. They shield their wind turbines behind barbed wire and post signs that say KEEP OUT.

Aesthetics

Fences are just one facet of whether a wind turbine becomes a respected member of the community—or an unwelcome intruder. For some the appearance of a wind machine on the skyline is symbolic of responsible stewardship—a step toward a sustainable future. To others it's

Figure 13-10. Visual uniformity. This pleasing array of Ecotecnia turbines on Spain's Galician coast near Malpica is partly attributable to the visual uniformity of the turbines.

industrial blight and a call to arms. Concern about the visual effect wind machines may have on a landscape and the communities of which they are a part should not be dismissed lightly.

Much has been written about the place of wind turbines in the landscape and how to minimize their visual intrusion. For more on the topic, see *Wind Energy Comes of Age, Wind Power in View,* and *Wind Turbines and the Landscape.* What follows are some general guidelines. Most fall under the rubric of "Be a good neighbor."

Medium-Size Turbines

While medium-size and larger wind turbines are installed as single units, like small wind turbines, more often they're installed in clusters or large arrays—wind farms. When there are more than one or two turbines in visual proximity to each other, it is critical to provide visual uniformity of turbine, tower, color, and direction of rotation. This is the single most important step planners can take to successfully integrate wind turbines into the community. The turbines need not be identical, but they must appear similar (see figure 13-10, Visual uniformity).

As with any business, some wind projects succeed and some fail (see figure 13-11, Headless horsemen). The community has a right to demand that operators repair or replace any "headless horsemen"—towers without turbines on top. If the turbine is not returned to operation, then the turbine, tower, and support equipment should be promptly removed, and the site restored to its pre-project state.

Avoid visual clutter by designing arrays with open spacing. Don't place the turbines too close together. One Tehachapi wind farm operator placed his turbines so close together that their rotors tangled and the turbines had to be repaired, then moved.

There are already too many billboards littering the countryside. Wind turbines

Figure 13-11. Headless horsemen. Dead and dying Windmaster turbines on a wind farm in California's Altamont Pass. When wind turbines are no longer "used and useful," they and their supporting infrastructure should be promptly removed.

shouldn't contribute to this visual blight. Don't paint billboards or corporate logos on the tower or nacelle, and specify that the manufacturer provide a nacelle free of corporate advertising. The logo on the side of Vestas's turbines is the size of tractor trailer, but the company will provide the nacelles logo-free—when requested (see figure 13-12, Logo-free).

Bury all intra-project power lines and the transmission lines leading to the project site. Aboveground power and transmission lines at large wind projects detract from the otherwise rural character of the landscape, giving such projects an industrial feel (see figure 13-13, Bury power lines).

Always dress the turbine properly. Some manufacturers, such as Atlantic Orient, are so intent on cutting costs that they will sell and install wind turbines without nose cones (spinners) or nacelle covers. These wind turbines appear angular, mechanical, and, in a word, *industrial.* They say to neighbors, *We don't care what you think.* Similarly, some

Figure 13-12. Logo-free. Corporate logos on the sides of wind turbine nacelles are an unnecessary visual distraction from their clean lines. Riverside County prohibits logos on wind turbines, as here on Vestas's V27s near Palm Springs, California.

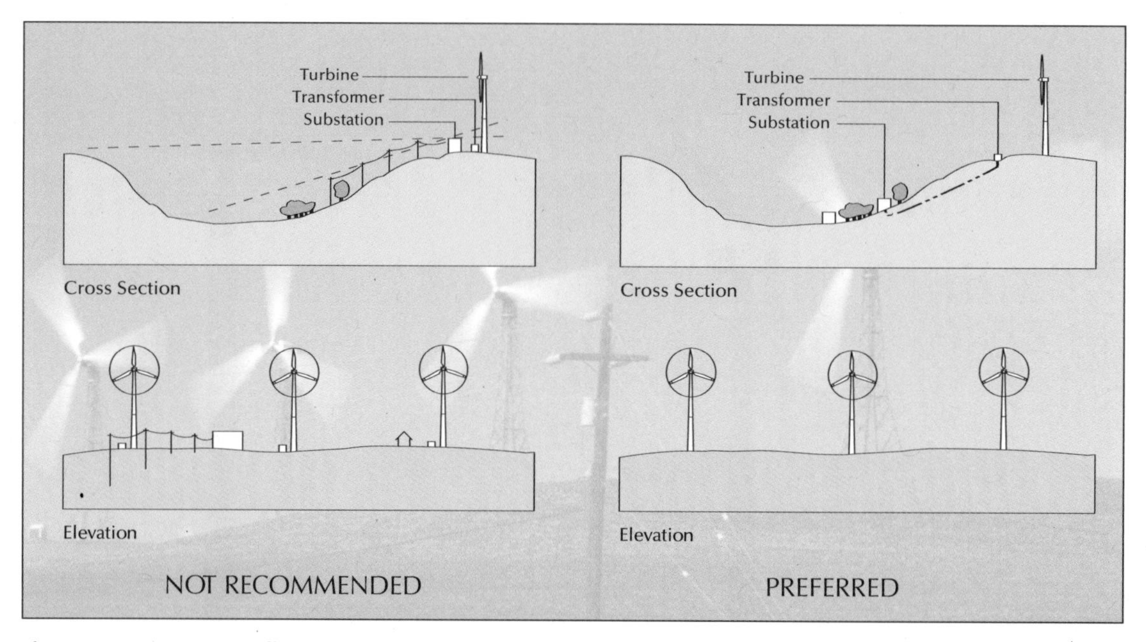

Figure 13-13. Bury power lines. Minimize visual intrusion by burying power lines and removing all ancillary structures from among the wind turbines. Where possible, place transformers inside the tubular towers often used on medium-size and larger turbines. (Chris Blandford Associates)

Figure 13-14. Dress turbines properly. A Kenetech KVS 33 operates in the San Gorgonio Pass without a portion of its nacelle cover. Kenetech (U.S. Windpower) turbines were notorious for losing their nacelle covers, which were seldom replaced.

California wind farm operators, in a misguided drive to squeeze every last cent out of aging turbines, remove the nose cones and nacelle covers or fail to replace them when they blow off (see figure 13-14, Dress turbines properly). Turbines in such projects become "junkyards in the sky," fueling wind energy's detractors.

Control erosion by minimizing or eliminating road construction, especially in steep or arid terrain. Too many unnecessarily wide roads can give an otherwise well-designed wind project the appearance of a mining site, instead of the pastoral scene wind advocates envision for the technology (see figure 13-15, Control erosion).

Harmonize ancillary structures with other structures on the landscape. Ancillary structures on a wind project should blend in with their surroundings. At the Italian test site of Acqua Spruzza, control and transformer buildings were built to resemble other rural buildings (see figure 13-8, Watch for ice). When Zilkha Renewable Energy needed an office building and maintenance shop for its Top of Iowa wind plant, it could have chosen a typical

Figure 13-15. Control erosion. Erosion gullies from excessive road construction in the steep, arid terrain of California's Tehachapi Pass give wind energy a black eye.

slab-sided metal building. Instead it took an abandoned barn and adapted it to company needs. The barn was more in keeping with other nearby farm buildings, Zilkha decided, than the industrial structure it would have used. Similarly, transformers should be placed inside the tower; where that's not feasible they should incorporate a facade that obscures their industrial features (see figure 13-16, Architectural transformer treatment).

Operators of wind projects should pick up any litter on their sites and eliminate on-site storage of spare parts, damaged wind turbines, oil drums, and other industrial detritus. Trash and litter quickly make a pastoral array of clean, modern wind turbines into an industrial site that just happens to use wind machines (see figure 13-17, Remove litter and boneyards).

Figure 13-16. Architectural transformer treatment. Where transformers cannot be placed inside the tower, transformers should be shrouded with an architectural treatment. The transformer on the right incorporates a facade to harmonize what otherwise would be an industrial structure with other nearby structures on the landscape. Open-cell concrete pavers harden the access track to the NEG-Micon tower on the left, allowing rain to percolate through to the groundwater table. This turbine and the Lagerwey 18/80 in the background are owned and operated by Noud de Schutter, a Dutch farmer in the Wieringemeer polder north of Amsterdam.

Small Turbines

Manufacturers of small wind turbines have paid far less attention to aesthetic design than

Figure 13-17. Remove litter and boneyards. Discarded wind turbines, blades, gearboxes, and other debris litter the former Zond site in the Tehachapi Pass. Some of the abandoned turbines in the background have since been removed.

manufacturers of larger turbines. Some small turbines are crude contraptions that a reasonable person may not want in the neighborhood. One, the 1980s Jacobs, is an ungainly design that looks like it came directly from a 1930s machine shop, which it did. Many manufacturers of small turbines could use a good industrial designer.

Try to incorporate the community's wishes when you're considering the type of tower to use. There's much less objection to the clean lines of a tapered tubular tower than to a wooden utility pole. Likewise, a tubular tower is more pleasing in foreground views than a truss tower. However, truss and lattice towers should be considered. In distant views, guyed lattice towers and truss towers become nearly invisible. From an aesthetic perspective, the type of tower that's most acceptable (whether truss, guyed lattice, or tubular) depends on the viewpoint and the distance between the observer and the tower. There are no definitive aesthetic guidelines for small wind turbines. (There are for wind power plants, but the two applications are quite different.)

If someone objects on aesthetic grounds, point out similar structures on the horizon that we've learned to tolerate, if not accept. You have as much right to erect a wind machine as the local radio station has to install a tower or the utility has to string a transmission line across town. While it's true that the utility's power lines and the radio's broadcast tower provide communitywide benefits, each person benefits individually. Your installation of a wind system differs little from the utility building a power line to your house. The appearance may differ, but the purpose remains the same.

Don't overlook some obvious ways to adapt the turbine to the community's or your own tastes. Patrick Campbell is a Kern County firefighter who has operated his Bergey Excel since 1998. Campbell is the type of customer who knows what he wants. And what he wanted was a wind turbine that matched the color of his home. Bergey Windpower obliged and painted the turbine to match. "Any customer can request it," says Campbell, but few do.

Avoid garishness. Don't string lights from your turbine for any reason. Be respectful of your neighbors and of the night sky. It's our common heritage to be enjoyed by all.

Anti-Wind Groups

There are organized anti-wind groups in most countries. These groups are distinct from and should not be confused with environmental organizations that may have legitimate concerns about the impact of large wind projects. Environmental organizations generally support the use of wind energy, though they may object to specific projects. Anti-wind groups oppose all wind energy for political or cultural reasons. Some of these groups are well funded, sophisticated, and utterly ruthless. While their ire is generally directed at wind farms, their broadsides don't make distinctions. They paint all wind turbines (large and small), all projects (big and little), and all wind turbine users (individual and corporate) with the same brush. These groups share information electronically. So don't be surprised if someone steps to the podium at a public hearing in Pipestone, Minnesota, and starts talking about wind turbines in Ryd-y-Groes, Wales, or the Causse du Larzac in France.

Noise

Like the appearance of a wind turbine and its placement in the landscape, noise is another frequent community concern. This concern is fueled in part by old reports of noisy wind turbines that were installed in California's San Gorgonio Pass during the early 1980s or by the giant General Electric turbine that operated briefly—very briefly—near Boone, North Carolina. The wind turbines that were the source of the problem are long gone, and manufacturers of medium-size wind turbines have made great strides in reducing noise. That's the good news.

The bad news is that manufacturers of small wind turbines began addressing the problem long after manufacturers of medium-size turbines, and only after some customers—and their customers' neighbors—complained. One model, Southwest Windpower's Air 403, was particularly notorious, though other brands were equally at fault. Fortunately manufacturers of small turbines are finally heeding customer demand for quieter products.

Noise is especially critical to siting small wind turbines because, as Carl Brothers, manager of Canada's Atlantic Wind Test Site, notes, "the smaller they are, the closer they are likely to be placed near someone's house." Mick Sagrillo, one of the founders of the Midwest Renewable Energy Fair, says, "Noise has a lot to do with acceptability." According to the outspoken Sagrillo, the public's occasional wariness toward small turbines could swiftly shift to outright prohibition if noise isn't addressed.

Despite all the technological progress, no operating wind machine is or will ever be silent. Wind turbines are audible to people nearby. Whether it's "noisy" or not is far more difficult to determine. Wind turbine noise is a field where the technical and the subjective meet head-on.

Noise, unlike visual intrusion, is measurable. And because noise is measurable, neighbors will "transfer" their concern about wind energy's aesthetic impact to the increase in background noise attributable to wind turbines. If wind turbines are unwanted for other reasons, such as their impact on the landscape, noise serves as the lightning rod for disaffection.

All wind turbines create unwanted sound, or noise. Some do so to a greater degree than others. And the sounds they produce—the swish of blades through the air, the whir of gears inside the transmission, and the hum of the generator—are typically foreign to the rural settings where wind turbines are most often used. These sounds are not physiologically unhealthful; they do not damage hearing, for example. Nor do they interfere with normal activities, such as quietly talking to your neighbor. But the sounds are new, and they are different.

Those who live in the rural settings where wind turbines are best suited do so because they prefer the peace and quiet of the country to the noise of the city. Longtime residents are accustomed to the relative quiet of rural life. They are familiar with the noises that exist, and have learned to live with them or even to find them desirable: the wind in the trees, the chirping of birds, the creaking of a nearby farm windmill, the hum of the neighbor's tractor. Rather than being nuisances, these sounds reinforce the bucolic sensation of living in the country.

The addition of new sounds, which most residents have had little or no part in creating and from which they receive no direct benefit, can be disturbing. No matter how insignificant they may be in a technical sense, these new sounds signify an outsider's intrusion. The effect is magnified when the source, such as a wind turbine, is also highly visible.

Table 13-1
Typical Sound Pressure Levels in dBA

Source	Distance from the Source (ft)	(m)	dBA
Threshold of pain			140
Ship siren	100	30	130
Jet engine	200	61	120
Jackhammer			100
Freight train	100	30	70
Vacuum cleaner	10	3	70
Freeway	100	30	70
Large transformer	200	61	55
Wind in trees	40	12	55
Light traffic	100	30	50
Average home			50
Quiet rural area at night			35
Soft whisper	5	2	30
Sound studio/quiet bedroom			20
Threshold of hearing			0

Decibels

First, some background. Noise is measured in decibels (dB). The decibel scale spans the range from the threshold of hearing to the threshold of pain (see table 13-1, Typical Sound Pressure Levels in dBA). Further, the scale is logarithmic, not linear. Doubling the power of the noise source—say, by installing two wind turbines instead of one—increases the noise level only 3 dB. This alone causes more confusion about noise than any other aspect, because a change of 3 dB is the smallest change most people can detect. Tripling the acoustic energy increases sound level 5 dB, an increase that is clearly noticeable. It takes 10 times the acoustic energy to raise the noise level 10 dB and double its intensity, or sound twice as loud.

For most discrete sources, such as wind machines, the distance to the listener is just as important as the noise level of the source. As in table 13-1, whenever noise is presented as sound pressure levels (SPL), the location is always specified, or implied, because sound levels decrease with increasing distance.

Weighting Scales

The perceived loudness varies not only with the sound level but also with the frequency, or pitch. Human hearing detects high-pitched sounds more readily than those low in pitch. The sound of a complex machine such as a wind turbine is composed of sounds from many sources, including the swoosh of the wind over the blades and the whir of the generator. Each source has a characteristic pitch, giving the composite sound a characteristic tonal quality. When measuring noise we try to take into account the way the human ear perceives pitch by using a scale weighted for those frequencies we hear best. The A scale is most commonly used. This scale ignores inaudible frequencies and emphasizes those that are most noticeable.

Impulsive sounds, those that rise sharply and fall just as quickly—like a sonic boom, for example—elicit a greater response than sounds at a constant level over time. Wind machines using two blades spinning downwind of the tower emit a characteristic *whop-whop* as the blades pass through the turbulent wake behind the tower. This impulsive sound and its effect on those nearby may be missed by standard A-weighted measurements. Many of the complaints about wind turbine noise near Palm Springs in the early 1980s were directed at the impulsive noise from two-blade, downwind turbines. Noise containing pure tones or impulsive sounds is perceived as louder than broadband noise. Broadband noise, such as the aerodynamic noise from the wind rushing over a turbine's blades, is composed of sounds across the spectrum of human frequency response. It is less intrusive than either impulsive noise or noise with distinct tonal components.

Exceedance Levels

Another component of noise is time. Noise ordinances specify a noise level that must not be exceeded during a certain percentage of the time. This complicates the task of estimating a wind turbine's noise impact. Unlike trains or airplanes, which emit high levels infrequently throughout the day, a wind turbine may emit far less noise, but do so continuously for days on end. Some find this trait of wind energy more annoying than any other. In windy regions the sound may appear incessant. The literature of life on the Great Plains is full of references to the ever-present sound of the wind. In the classic 1928 film *The Wind*, the sod-busting pioneer played by silent-screen star Lillian Gish is driven mad by the oppressive wind.

The time-weighting of noise is expressed as the noise exceedance level: the amount of time the noise exceeds a specified value. For example, L_{10} is the noise level exceeded 10 percent of the time; L_{90}, the noise level exceeded 90 percent of the time; and L_{eq}, the continuous sound pressure level, which gives the same energy as a varying sound level. A noise standard of 45 dBA L_{90} is stricter than a standard of L_{10}, because 90 percent of the time the noise must be below 45 dBA (see table 13-2, Selected Noise Limits, Sound Pressure Levels in dBA). Wind turbine noise emissions are measured in L_{eq} in order to calculate the sound power generated by the turbine.

Noise Propagation

Noise levels decrease with increasing distance as the sound propagates away from the source. Under ideal conditions sound radiates spherically from a point source, such as a helicopter, and for every doubling of distance the noise level decreases 6 dB. Wind turbines, however, seldom hover high above the ground like a balloon. They are earthbound, and their noise emissions spread outward hemispherically.

Over a flat reflective surface such as a lake, noise decays 3 to 6 dB per doubling of distance. The atmosphere and objects on the landscape absorb some of the noise energy, further attenuating the noise over distance. The International Energy Agency (IEA) assumes hemispherical spreading in its commonly used noise propagation model. This simple model also incorporates a modest amount of atmospheric absorption.

More complex noise propagation models account for ground cover and meteorological effects. Both can greatly influence noise levels. Temperature and wind shear, for example, refract or bend sound waves from those expected, and vegetation can attenuate or absorb more sound than the IEA model assumes.

The rate at which noise decays increases with increasing atmospheric absorption. Relatively close to the tower, within 100 to 200 meters (300 to 600 ft), atmospheric absorption has little effect. As distance increases—for example, from 200 to 400 meters (600 to 1,300 ft)—the decay rate with absorption increases to 7 dB with every doubling of distance. Thus the noise attenuated by atmospheric absorption can be important

Table 13-2
Selected Noise Limits, Sound Pressure Levels in dBA

		Commercial	Mixed	Residential	Rural
Germany					
Day		65	60	55	50
Night		50	45	40	35
Netherlands					
Day	L_{eq}		50	45	40
Night			40	35	30
Denmark[1]	L_{eq}			40	45
England[2]					
High speed	L_{50}				45
Low speed	L_{50}				40
Minnesota					
Day	L_{50}	75	65	60	60
Night	L_{50}	75	65	50	50
Minnesota					
Day	L_{10}	80	70	65	65
Night	L_{10}	80	70	55	55
Kern County, Calif.[3]	$L_{8.3}$			45	45
Riverside County, Calif.	L_{90}			45	
Palm Springs, Calif.[4]	L_{90}			50	60

Notes: [1] Not to exceed 45 dBA beyond 400 m from wind turbine.
[2] L_{50} approx. 350 m from the nearest turbine.
[3] $L_{8.3}$., not to exceed 50 dBA.
[4] 50 dBA if lot is actually used as residential.

in projecting noise levels surrounding a wind turbine.

Unfortunately meteorological effects vary with the season, weather patterns, and time of day. Vegetation may vary seasonally, as well. Row crops may be tilled in fall when deciduous trees also lose their leaves, removing much of the vegetation that dampens noise from nearby turbines. Moreover, nighttime temperature inversions refract sound waves, bending them back to earth, increasing the noise level over that estimated by simple models. Valley inversions during fall and winter produce a similar effect. Anyone living alongside a lake or river has experienced sound carrying great distances during wintertime inversions.

There is also little or no atmospheric absorption of extremely low-frequency sound. For these reasons, engineers are hesitant to incorporate greater atmospheric absorption into their noise propagation models. Thus the models remain conservative.

Multiple wind turbines complicate matters further. From relatively long distances, an array of turbines appears as a point source, and doubling the number of turbines simply doubles the acoustic power increasing noise levels 3 dB. As you near the turbines, they begin to act as a line source. The decay rate for line sources is 3 dB per doubling of distance, and not 6 dB for true spherical propagation.

Even the wind itself will influence how noise propagates. Noise levels are typically higher downwind of turbines, and even higher for downwind turbines.

Thus estimating the noise emitted by a single wind turbine or a large array is no simple matter and is fraught with uncertainty. Though noise, unlike aesthetic impact, is quantifiable, interpreting the results of field measurements and mathematical projections requires almost as much subjective judgment as it does objective analysis.

Ambient Noise

The total perceived noise is the logarithmic sum of the ambient or background noise and the projected wind turbine noise. Thus the noise generated by a wind turbine must always be placed within the context of other noises around it. Wind turbines near busy highways will hardly create a problem, no matter how noisy they are, though the noise from the wind turbine may still be identifiable above the background noise. Conversely, wind turbines, no matter how quiet, may be heard above ambient noise at great distances in the stillness of a sheltered mountain cove.

The wind itself often masks wind turbine noise by raising the ambient noise level. At exposed locations there will always be noise from the wind whenever the wind machine is operating, because the wind rustles the leaves in nearby trees or sets power lines whistling. Despite the masking effect of high winds, a wind turbine will still be audible to people nearby, particularly when they are sheltered from the wind.

The sounds emitted by wind turbines are easily distinguishable from those of the wind. The generator or transmission may produce a noticeable whine, for example, or the passage of the blades may generate more discrete sounds. The aerodynamic *swish-swish-swish* of three-blade rotors is a common wind turbine sound. These sounds may not be objectionable, but they are detectable. The whir of the compressor in a refrigerator is audible, for example, but few find the sound objectionable. Some have compared this situation to that of a leaky faucet. Once recognized, the noise is hard to ignore.

Where the background noise level is low, as in a deep valley sheltered from the wind, a new noise may be considered intrusive, particularly at night when few other human-made sounds are present or a nighttime temperature inversion has brought a deathly hush to the valley. Whether or not a noise is intrusive depends on the nature of the noise; that is, its tonal or impulse character, the perception of the noise source (whether the wind turbines are loved, despised, or merely tolerated), the distance from the source, and the activity (for example, whether you're sleeping inside with the windows closed or conversing with a neighbor in the yard). But no wind turbine, no matter how quiet, can do better than the ambient noise. It is the difference between ambient noise and wind turbine noise that determines how people react.

Will It Be Heard?

Yes. That's the short answer. If in the heat of a public debate on the noise from a wind turbine or that from a wind farm, resist the temptation to say *It won't be heard.* It will. Avoid the equally false statement *You won't hear it over the wind in the trees.* They will. The characteristic sounds from a wind turbine are distinguishable from the background noise of, for example, the wind in trees at great distances. While the noise may not be objectionable, it can be detectable to those who want to hear it.

Community Noise Standards

Local noise ordinances typically state the acceptable sound pressure levels in dBA at the property line or nearest receptor. Many noise ordinances differentiate between acceptable day and nighttime levels, and levels for sensitive land uses such as schools and hospitals. The noise levels that wind turbines must meet in Europe and the United States are surprisingly similar. Where they differ is in the exceedance levels.

California's Kern County, for example, limits wind turbine noise to 45 dBA at $L_{8.3}$ for sensitive receptors (see table 13-2, Selected Noise Limits, Sound Pressure Levels in dBA). $L_{8.3}$ is the noise level exceeded for five minutes out of every hour. Minnesota has two standards: 50 dBA at night in rural areas at L_{50}, the noise level exceeded half the time; and 55 dBA at L_{10}, the level is exceeded

10 percent of the time, or six minutes out of every hour.

All community noise standards incorporate a penalty for pure tones, typically 5 dB. If a wind turbine meets a 45 dB noise standard, for example, but produces an annoying whine, planning officers dock the offending turbine 5 dB. The operator must then lower the turbine's overall noise level 5 dB or eliminate the whine.

Despite compliance with community noise standards, operators of wind turbines still run the risk of annoying their neighbors. Whenever wind turbine noise exceeds the threshold of perception, there is the potential for complaints. Fluctuations in ambient noise and variations in the quality or tonal component complicate determining whether wind turbine noise will exceed the perception threshold and stimulate complaints (see table 13-3, Community Response to Noise from Sources Other Than Wind Turbines).

If there is a noise complaint public health officers will measure the sound pressure level using a sound level meter and will determine whether the wind turbine complies with the applicable ordinance. This was the situation New Zealand's Graham Chiu found himself in. He received a free noise test courtesy of the Wellington City Council after a neighbor complained about his Air 403. Chiu was found in violation and the noise control officer ordered the turbine shut off—permanently. Violation of the order could cost Chiu as much as NZ $200,000 in fines.

Table 13-3
Community Response to Noise from Sources Other Than Wind Turbines

Amount by which Noise Exceeds Background Level (dB)	Estimated Community Response	
	Category	Description
0	None	No observed reaction
5	Little	Sporadic complaints
10	Medium	Widespread complaints
15	Strong	Threats of action
20	Very strong	Vigorous action

Note: This table was derived for noise sources other than wind turbines, and neighbors could be either more or less sensitive to wind turbine noise than that indicated here.

Source: Harvey Hubbard, Kevin Shepherd, NASA, 1990.

As in Chiu's case, violating a noise ordinance can result in serious consequences, including removal of the wind turbine. Though not foolproof, there are mathematical models that can be used to project noise levels before a wind turbine is installed. These models use sound power to project noise levels surrounding a wind turbine.

Sound Power Levels

The International Energy Agency's model, for example, uses the acoustic energy generated by the wind turbine. Acousticians use field measurements of sound pressure levels (SPL), or L_p, to calculate the sound power levels, or L_w, emitted from the wind turbine. As if the similar-sounding names were not confusing enough, both measures use the same units, dBA. While sound pressure levels will always be specified at some distance from the turbine, the sound power level will always be presented at the source: the wind turbine itself.

The distinction is important. The sound power level of most wind turbines varies from 90 dBA to more than 100 dBA. For those familiar with sound pressure levels, this appears noisy. Yet a wind turbine emitting a sound power level of 100 dBA can meet a 45 dBA noise limit in sound pressure level, given sufficient distance from the wind turbine. The sound power level can be found by

$$(L_w) = (L_p - 6 \text{ dB}) + 10 \log (4\pi R^2)$$

where R is the slant distance from the turbine to the sound level meter, L_p is the sound pressure level measured by the meter, and –6dB is a correction to the meter reading to account for using a reflective soundboard (see figure 13-18, Noise measurement of a micro turbine).

Sound power data on many medium-size wind turbines is publicly available, for example in the German Wind Energy Association's annual *Windenergie: Marktübersicht (Market Overlook)*. There was little comparable data on small wind turbines outside Denmark prior to 2002, when data became available from NREL and the Wulf Test Field (see figure 13-19, Measured Air 403 plus ambient noise).

Noise measurements on wind turbines are recorded for two conditions. One condition is the turbine plus ambient; that is, the wind turbine operating as intended. (At the Wulf Test Field, for example, the micro turbines were charging batteries.) Another condition is ambient noise alone, or with the turbine parked. Once the difference between the turbine plus ambient and the ambient noise is determined, the sound power emitted by the wind turbine can be calculated.

For most small wind turbines there are only two conditions: operating and parked. The Air series of micro turbines, for instance, parks the rotor when the batteries are fully charged. Other turbines, such as Southwest Windpower's Whisper H40, divert charging to a dump load, keeping a load on the generator and limiting rotor speed. Bergey Windpower used a different approach on its

Figure 13-18. Noise measurement of a micro turbine. Beginning a sequence of noise measurements downwind from an Ampair 100 at the Wulf Test Field. The recording sound level meter is being inserted into the secondary windscreen mounted on the reflective soundboard. The sound pressure levels measured by the meter are used to calculate the strength of the noise emitted by the wind turbine.

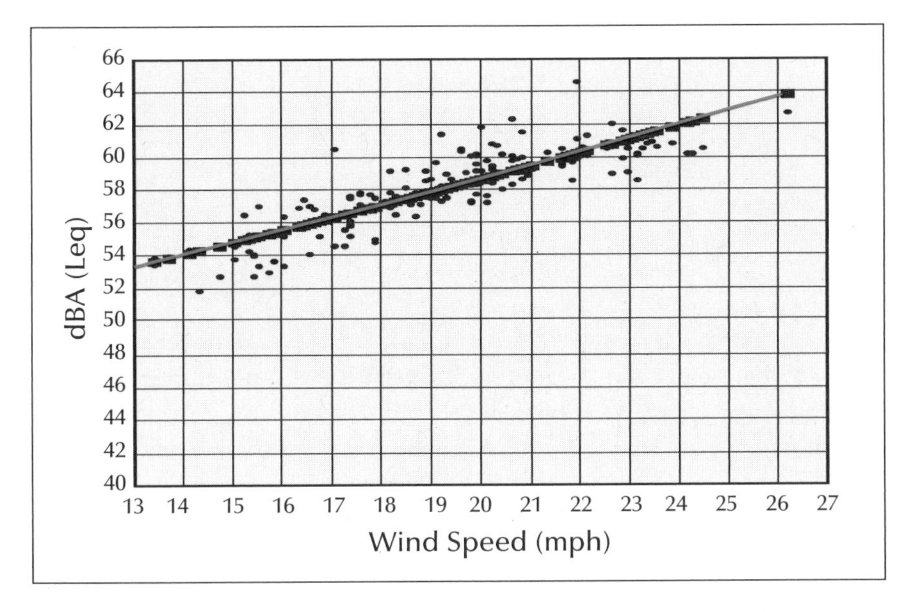

Figure 13-19. Measured Air 403 plus ambient noise. Sound pressure level measurements and linear regression for an Air 403 charging a constant load at the Wulf Test Field. These measurements reflect noise from the turbine plus the ambient noise, not turbine noise alone. This is the first of several steps in determining the noise emitted by a wind turbine. Measurements were made at a slant distance of 19.4 meters (63.6 ft).

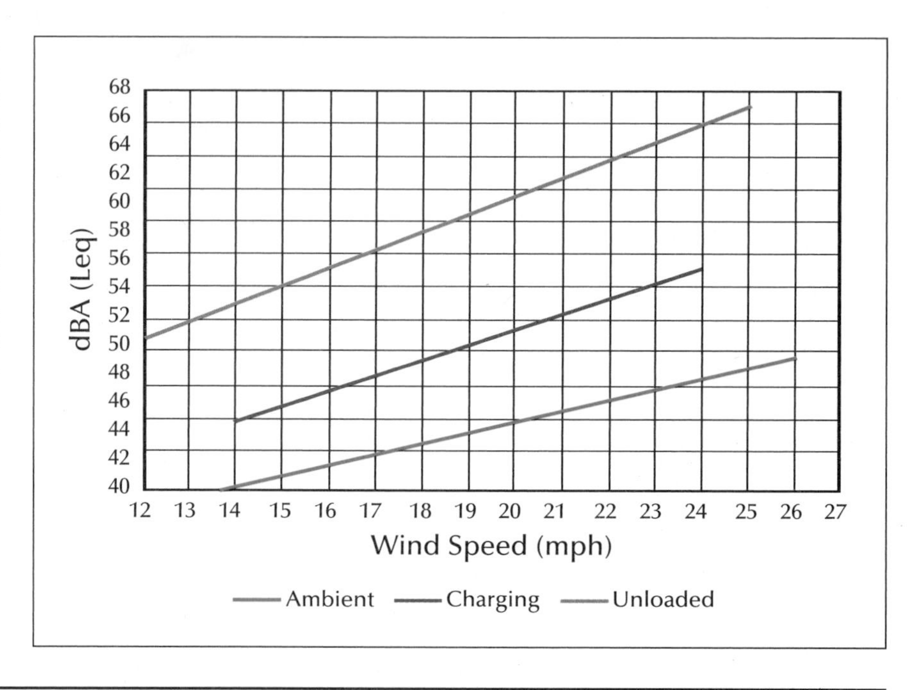

Figure 13-20. BWC 850 noise measurement summary. A linear regression of sound pressure level measurements of a Bergey 850 for ambient (turbine parked), charging, and with the turbine operating unloaded. The Bergey 850 unloads the generator when the batteries are fully charged, causing the rotor to spin faster, generating considerably more noise than when charging. Measurements were made at a slant distance of 28.04 meters (92 ft).

Table 13-4
Comparison of Noise from Small and Selected Medium-Size Wind Turbines

Turbine	Rotor Dia. (m)	Rotor Dia. (ft)	Swept Area (m²)	Rated Power (kW)	Sound Power @ 8 m/s ($L_{WA, ref}$)	Sound Power @ 10 m/s ($L_{WA, ref}$)	Data Source
Ampair 100	0.91	3	0.66	0.1	na[1]	na[1]	Gipe
Air 403	1.17	4	1.07	0.4	88	91	Gipe
AirX	1.17	4	1.07	0.2	80	na[1]	Gipe
Whisper H40	2.13	7	3.58	0.9	85		NREL
BWC 850	2.44	8	4.67	0.85			
Charging					82	87	Gipe
Unloaded					92	97	Gipe
Calorius	5	16	20	4.6	82		Risø
Gaia	7	23	38	6.5	88		Risø
Genvind	13	41	125	23.7	103		Risø
Furländer	13	43	133	30	93		TÜV
Gaia	13	43	133	11.6	89		Risø
Enercon E30	30	98	707	200	95	99	Wind-consult
Nordex N43	43	141	1452	600	101		Wind-consult
NEG-Micon	60	197	2827	1000	98	101	Windtest KWK
Enercon E66	66	216	3421	1800	101	103	Windtest KWK

Note: [1] Not applicable. Difference between turbine plus ambient and ambient was less than 5 dBA.

850, however. The Bergey 850 unloads the generator when the batteries are charged, releasing the rotor and allowing it to spin faster than when charging. For turbines such as the Bergey 850, then, measurements must reflect all three conditions. The Bergey 850 is noisiest when it operates unloaded (see figure 13-20, BWC 850 noise measurement summary).

To compare one wind turbine's noise to another's, you must derive the sound power level, L_{wa}, for a standard wind speed of 8 m/s (17.9 mph), and often 10 m/s (22.4 mph) as well (see table 13-4, Comparison of Noise

Sources of Small Turbine Noise

Noise from small wind turbines is largely a function of tip speed, blade shape—especially near the tip—and how the turbine regulates power in high winds. Unlike medium-size turbines, many of which operate at constant or near-constant tip speeds, nearly all small turbines operate at variable speeds. As wind speed increases, so does tip speed—and noise.

The Air 403, for example, would reach a tip speed of 90 m/s (200 mph) in winds of 10 m/s (22 mph), nearly twice that of medium-size commercial wind turbines. And the tip speed for the Air 403 would continue to increase until the blades begin to flutter. At that point the noise from the turbine has been described variously as like a hoarse shriek or the buzz of a chain saw. Similarly, when the BWC 850's controller unloaded the generator, the rotor would reach a tip speed of nearly 70 m/s (156 mph). While this may seem modest in comparison to the Air 403, the Bergey pultruded blade was quite different from the saberlike shape of the Air 403 blade and consequently was noisier.

Bergey turbines have used pultruded fiberglass blades since the late 1970s. These blades, while extremely durable, have a thick trailing edge. Jim Tangler, an aerodynamicist at NREL, attributes much of the noise from the older Bergey blades to this thick trailing edge. In contrast, the trailing edge of the Air 403 blade is so sharp, Southwest Windpower warns users to wear gloves when assembling the rotor.

Dave Blittersdorf of NRG Systems operates a Bergey Excel in the backyard of his home near Burlington, Vermont. A keen observer, Blittersdorf noted that the Excel was noisiest when the controller unloaded the rotor, leading to higher tip speeds. To keep the neighbors—and his wife, Jan—happy, he ensures that his turbine always operates under a load.

Wisconsin wind advocate Mick Sagrillo explains that aerodynamic noise can be especially noticeable in small turbines that furl the rotor to limit power in high winds. This behavior differs from one design to another, with a resulting difference in noise emissions.

Bergey turbines, Southwest Windpower's Whisper series, and African Wind Power's design all furl the rotor horizontally toward the tail vane. "There's less furling hysteresis in the AWP design and in the Whisper's angle governor" than in the Bergey line, says Sagrillo, and this is reflected in the noise characteristics of these turbines.

Small turbine manufacturers have heard the message that noise is a subject that won't go away. "Noise is a concern," says Bergey Windpower's Mike Bergey. In response the Oklahoma company has introduced new airfoils to replace the simple cambered blades that were once the hallmarks of the Bergey design. Dave Calley, Southwest Windpower's chief designer, acknowledges that noise was the "absolute number one complaint" about the Air 303 to 403 design. "It's a very important issue to us," says Calley, with the result that Southwest Windpower's AirX is significantly quieter than previous models in the Air series. Small turbines need not be noisy. The "Marlec is remarkably quiet," says Wisconsin's Sagrillo. Among household-size turbines, the 1930s-era Jacobs and the 1980s turbine of the same name were extremely quiet. And, Sagrillo adds, "the AWP and Proven 2500 are every bit as quiet as the Jacobs." What's quiet? To Sagrillo, a wind turbine's quiet "when you have to go outside to see if it's running." He says, "wind generators should be seen, not heard."

from Small and Selected Medium-Size Wind Turbines). The measurement and reporting techniques designed for medium-size wind turbines may not adequately describe the noise characteristics of small wind turbines. Small turbine noise may be most noticeable at wind speeds other than 8 or 10 m/s.

When the data is available, sound power levels can be calculated for a range of wind speeds (see figure 13-21, Calculated emission source strength). This enables comparisons that otherwise wouldn't be revealed using the standard reporting format. For example, measurements of the Ampair 100 at the Wulf

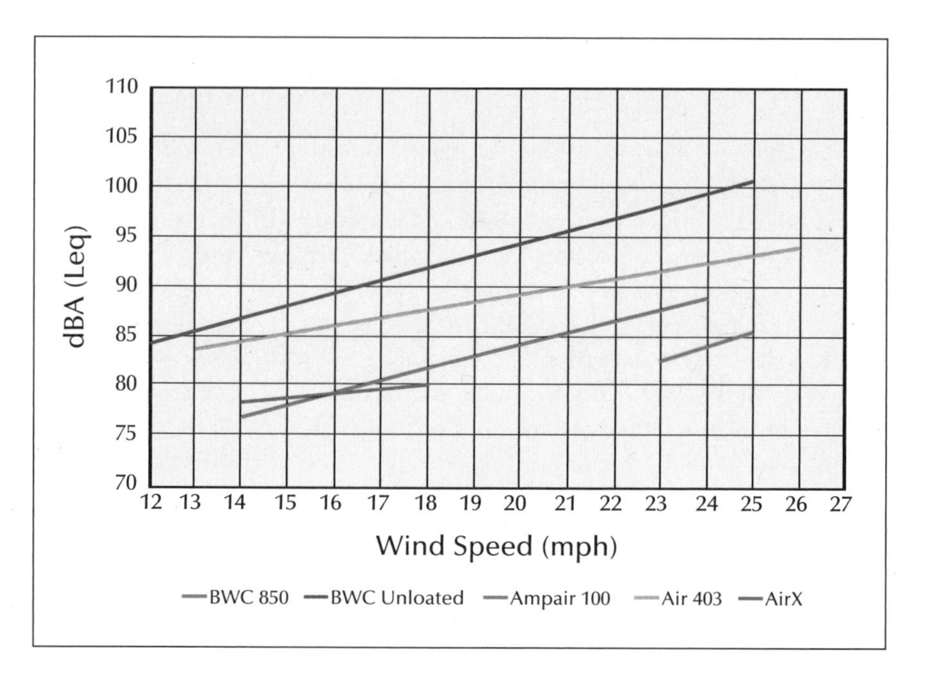

Figure 13-21. Calculated emission source strength. Measured sound pressure level data from the Wulf Test Field on the BWC 850, Air 403, AirX, and Ampair 100 was used to calculate the sound power level or emission source strength (L_{wa}). The sound power level was calculated for each turbine charging a load, as well as for the BWC 850 operating unloaded. As indicated by the Ampair 100, small wind turbines need not be noisy.

Test Field indicated that it was significantly quieter than most other turbines tested at wind speeds above 10 m/s (22.4 mph).

Wind Turbine Noise

There are two sources of wind turbine noise: aerodynamic and mechanical. Aerodynamic noise is produced by the flow of the wind over the blades. Mechanical noise results from the meshing of the gears in the transmission, where used, and the whir of the generator.

Unless there is a whistling effect from slots or holes in the blades, aerodynamic noise is principally a function of tip speed and shape. Aerodynamic noise is also influenced by trailing edge thickness and blade surface finish. The number of blades is also a factor. Neil Kelley, a researcher at the National Renewable Energy Laboratory, finds that the aerodynamic noise of two-blade wind turbines is greater than that of three-blade machines, all else being equal, because the two-blade turbines place higher loads on each blade for an equivalent output. Further, the type of rotor control, whether fixed or variable pitch, affects aerodynamic noise. On rotors with fixed-pitch blades, noise increases when the blades enter stall in high winds. But rotor diameter and speed are the primary determinants of aerodynamic noise. Many constant-speed, medium-size turbines operate at tip speeds around 40 m/s (90 mph) in low winds when their low-power windings are energized, and 50 to 60 m/s (110 to 130 mph) when the generator is fully energized. Some early experimental turbines, operating at variable speed, reached tip speeds of 100 m/s (224 mph).

Dutch researcher Nico van der Borg found that by using rotor diameter as a substitute for tip speed, he could approximate the noise emission of wind turbines. Larger-diameter wind turbines generate proportionally more acoustic energy than smaller machines. Van der Borg's model was derived from data on experimental wind turbines designed in the 1970s and early 1980s. Many of these early research turbines operated at very high tip speeds. Van der Borg compared them to commercial turbines available in the 1980s and estimated that the commercial turbines were as much 7 dB quieter than their predecessors (see figure 13-22, Calculated and measured noise emissions). Later turbines are even quieter.

Van der Borg's model can also be used to answer the question of whether small wind tur-

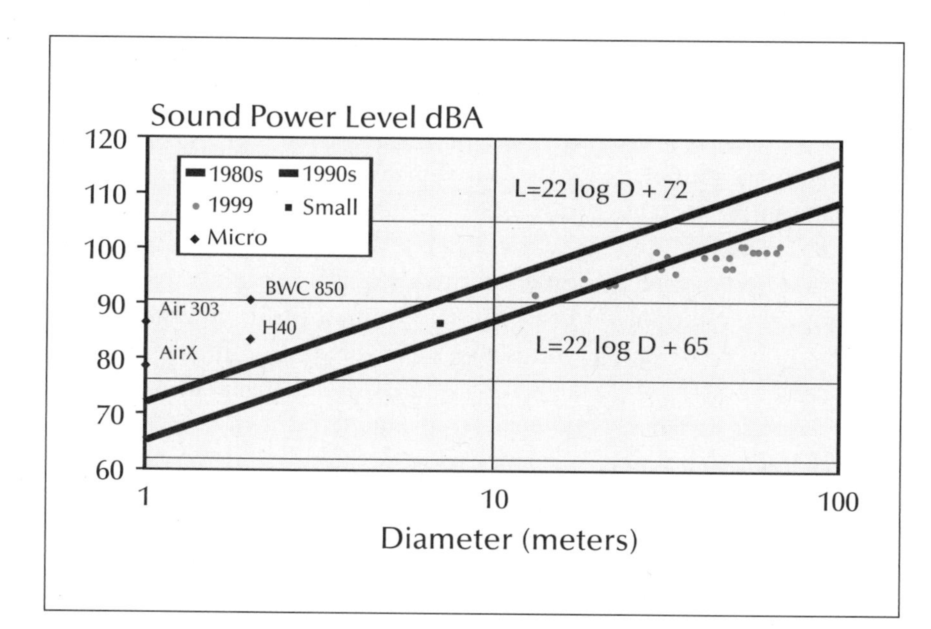

Figure 13-22. Calculated and measured noise emissions. This chart derives from work at ECN by N. C. J. M. van der Borg and W. J. Stam in 1989 on the relationship between source sound power and rotor diameter. Van der Borg and Stam argued that diameter could substitute for tip speed and hence determine sound power. The top line was derived from data on experimental large turbines developed in the late 1970s and early 1980s. The bottom line was derived from data on commercial wind turbines being installed in the late 1980s. Published data for commercial turbines in use from the 1990s through 2000 has been added. Noise emissions from small turbines at the Wulf Test Field and other test sites are also included.

bines are relatively more noisy than bigger turbines. In absolute terms they aren't, but relative to their rotor diameter small wind turbines are noisier. According to van der Borg's model, the Air 403, BWC 850, and Whisper H40 should emit no more than 70 to 80 dBA. Instead the small turbines are 10 to 15 dBA noisier than would be expected for their size. Fortunately manufacturers of small wind turbines have begun to address the issue.

Lowering Wind Turbine Noise

Advances in airfoils and reductions in tip speeds have essentially decoupled noise emissions from rotor diameter for medium-size wind turbines. Building quieter turbines not only makes wind energy a better neighbor, but also makes good business sense. In Europe, where competition is fierce, manufacturers find that quieter turbines give them an edge over their rivals. Manufacturers with quieter turbines can site them in areas where planning officials would prohibit their competition, and quieter turbines ensure that there are fewer headaches after installation, as well as less bad press eroding support for wind energy.

The most direct way to lower noise emissions is to reduce rotor speed. One means of lowering rotor speed on a constant-speed turbine is to operate the turbine at dual speeds. This permits operating the turbine at a lower rotor speed in light winds, when there is less wind noise to mask noise from the turbine. Variable-speed operation is also effective, enabling designers to program operation to lower rotor speeds at night, when noise sensitivity is greatest.

Mechanical noise often has tonal components. The gearbox's high-speed shaft is the most critical element, says Henrik Stiesdal, chief designer for Bonus wind turbines. Mechanical noise can be reduced by redesigning the gearbox and by adding resilient couplings in the drive train to isolate vibrations. Acoustic insulation can also be installed inside the nacelle cover to reduce propagation of mechanical noise.

Stiesdal insists on totally enclosing the drive train and sealing the nacelle canopy. Even ventilation louvers must be carefully designed as sound baffles, he says, or a significant part of the turbine's machinery noise, especially noise at higher frequencies, will escape the nacelle. Stiesdal agrees with NREL acoustician Neil Kelley that noise must be controlled at the source, because "once it gets out, you don't know where it will go."

Consequences

As Graham Chiu found, once noise does get out, the consequences can be costly. One example of the consequences was encountered by Danish manufacturer DanWin. Because of the proximity of one neighbor to a project at Kynby, DanWin took special precautions when building and installing the 21-turbine wind farm. It mounted the 180 kW nacelles on rubber dampers, sharpened the trailing edges of blades on the eight nearest turbines, mounted sand-dampening chambers on four towers, and reduced generator speed to 1,000 rpm from the typical 1,200 rpm.

If there's any doubt as to whether or not your wind turbine might disturb nearby residents, be a good neighbor and contact them in advance.

Despite these precautions, the noise at the nearest residence, a farmhouse 220 meters (720 ft) away, exceeded permissible levels and included a pure tone component from the gearbox. After four years of work, the turbines' noise emissions were reduced from 97 to 102 to 95 dBA, resulting in an acceptable noise level at the dwelling. DanWin's successors achieved this by redesigning the gear teeth and adding further noise treatment. The engineers found they could gain 4 dBA simply by sharpening the trailing edge of each blade, providing one of the most convincing demonstrations that trailing edge thickness is a significant factor in aerodynamic noise. The cost? DKK $4.5 million ($750,000).

Wisconsin Public Service faced a similar dilemma. After installing 14 Vestas V47 turbines in 1999 the utility began receiving noise complaints from neighbors in Lincoln Township, an area experiencing spillover growth from suburban sprawl east of Green Bay. WPS conducted a series of noise studies and eventually offered to buy six homes. Two homeowners accepted, costing the utility about 2 percent of the project's initial investment.

Neighborhood reaction to small turbine noise can also affect how or when owners use their turbine. As in the case of Chiu in New Zealand, public authorities can order the turbine removed. Equally damning could be an order not to operate the turbine at night or in winds above a certain speed. In either case the operation of the turbines would be so marginalized as to dictate its removal.

The Danish windmill owners' association takes a strong stand on wind turbine noise. The association's members not only are the chief advocates of wind energy in Denmark, but also own most of the wind turbines. Many can literally see wind machines outside their windows. They can speak with authority as people who both want wind energy and demand that it be a good neighbor. Their position is clear: Noisy turbines are unacceptable. Noisy machines should either be soundproofed or moved. The goal of the owners' association, one that should be the goal of all wind turbine manufacturers, is to avoid the problem from the start. They have found that once people have been bothered by noise, they remain disturbed, even after the noise has subsequently been abated.

Consideration

Our perception of what constitutes noise is affected by many subjective factors. If your neighbors object to your wind machine because you never invite them to dinner, they're more likely than you are to find the sound produced by it objectionable. On the other hand, if your community has fought rate increases with the local utility, the sound of your wind machine whirring overhead may warm their hearts.

Bergey Windpower suggests that if there's any doubt as to whether or not your wind turbine might disturb nearby residents, be a good neighbor and contact them in advance. Advise them of your plans, and ask for their comments. Answer their questions as forthrightly as you can, and try to incorporate their concerns when designing your installation. Bergey has found that the community's reaction to the noise from a small wind turbine declines after people have had a chance to acclimate to it.

This is equally sound advice for those installing medium-size wind turbines. Be considerate. Newer, quieter wind turbines can be good neighbors—when sited with care.

TV and Radio Interference

Neighbors sometimes worry that a new wind turbine will disrupt their radio and television reception. There have been a few cases in which medium-size turbines have caused ghosting of weak television signals in rural areas. In one case in the early 1980s, Westinghouse's Mod-0A on Rhode Island's Block Island generated complaints as well as electricity. The problem was alleviated by installing cable television on the island.

Interference is a rare phenomenon, and there have been no reported cases due to small wind turbines. Even in the few cases in which interference or ghosting has been documented, the effects have been localized.

There are thousands of wind turbines lining the ridges of the Tehachapi Pass, a major corridor for telephone links between northern and southern California. The turbines surround the microwave repeater stations but are excluded from the microwave path. This provision is sufficient to prevent any interference.

Small wind turbines are used extensively worldwide to power remote telecommunications stations for both commercial and military uses. The turbines would never have been selected if there had been any hint of interference. Unfortunately some wind turbine operators have sought additional revenue by renting space on their towers for telecom dishes and antennas, a practice that detracts from the appearance of the wind turbine (see figure 13-23, Interference, no; ugly, yes).

Figure 13-23. Interference, no; ugly, yes. Vestas V47 with awkward telecom antennas in Germany. While the wind turbine is obviously not interfering with telecommunications here, the antennas and their mounting hardware give this Vestas turbine an undesirable industrial appearance.

Shadow Flicker and the Disco Effect

Shadow flicker occurs when the blades of the rotor cast shadows that move rapidly across the ground and nearby structures. This shadow can disturb some people in certain situations, such as when the shadow falls across the window of an occupied room.

Small turbines are too small and operate too fast to create a significant shadow. Medium-size wind turbines can cause shadow flicker, however, and it can be a nuisance in higher-latitude winters, when the low angle of the sun casts long shadows. It can also be more troublesome in areas with high population densities or

Shadow Flicker

Many North Americans smile when Europeans begin discussing a phenomenon called shadow flicker. Most Americans have never heard of it and can't imagine what the fuss is about. That was my reaction until one fall when I lived near a 75 kW turbine at the Folkecenter for Renewable Energy in Denmark. One morning while working at my desk I felt uneasy. Something was bothering me. I kept looking up from my work, scanning the room for what was wrong. Finally I got up from my desk. Then I noticed it: a shadow repeatedly crossing the room. It was still a few moments before I realized I was a victim of shadow flicker. I flipped on the light, and went back to work.

where neighbors are close enough to be affected by the shadows.

Near Flensburg in Schleswig-Holstein, German researchers examined the effect and found that flicker, under worst-case conditions, would affect neighboring residents a total of 100 minutes per year. Under normal circumstances the turbine in question would produce a flickering shadow only 20 minutes per year.

There are few recorded occurrences of concern about shadow flicker in North America. Ruth Gerath, however, notes that the flickering shadows from the turbines on Cameron Ridge near Tehachapi have startled her horse and those of others in the local equestrian club. Except for the flickering shadows, she says, the turbines seem to have no effect on the horses. The shadows simply cause the horses to stop briefly, until their riders urge them on.

> The charge that wind turbines produce more dead birds than electricity is false.

While few communities have standards regulating shadow flicker, it's wise to be considerate of your neighbors. If there's any question whether nearby residents will be affected, analyze the likely impact before installation.

Professional wind turbine siting software (see the appendixes for details) often includes provisions for calculating shadow flicker. The technique used by these programs is extremely conservative and will project worst-case conditions (bright sun, cloudless sky).

The disco effect is a related phenomenon first noticed in sunny Palm Springs, California. Sunlight glints off the reflective gel coat of the fiberglass blades of the wind turbines in the San Gorgonio Pass. When the blades move, this causes a flash similar to that of a strobe light. As the rotor spins, the flash repeats with a rhythm akin to that of the flashing lights in a discotheque.

To prevent the disco effect from annoying neighbors, Riverside County prohibits reflective blade coatings. The surface finish also dulls after several years in the harsh desert sun, reducing the blades' reflectivity over time.

Birds

Wind energy's chief attribute is its environmental benefits. When sited with care, wind energy is relatively benign. The key is sensitive siting and a frank acknowledgment that wind turbines do have some environmental impact. Though wind turbines have little or no impact on most plants and animals, they can and do kill some birds. Notably, large arrays of medium-size wind turbines have killed birds in the Altamont Pass and near the Straits of Gibraltar. There's no benefit in sugarcoating that fact. Nonetheless, the charge that wind turbines produce more dead birds than electricity is false.

Much has been written about birds and wind turbines. For a more complete account of the problem, see *Wind Energy Comes of Age.* Numerous studies on the topic have been conducted in Europe and North America. Summaries of this research are available from most national wind energy associations.

No single environmental issue pains wind energy advocates more than the effect wind turbines might have on birds. Clearly wind turbines should not kill birds, and we should do everything in our power to ensure that they don't. This is the kind of hot-button issue that elicits strong emotional responses that could, if not addressed honestly, derail the use of wind energy.

That some wind turbines kill birds some of the time should come as no surprise. Most tall structures kill birds to some degree, as do most sources of energy. This should never become an excuse for ignoring the issue, but it does help put it into perspective.

Wind turbines anywhere are capable of killing birds, explains Dick Anderson, a biologist at the California Energy Commission. But nowhere else in the world is the problem as severe as in California's Altamont Pass.

Wind turbines in the Altamont Pass, says Anderson, kill 100 to 300 raptors per year, of which 20 to 50 are golden eagles *(Aquila chrysaetos)*. Golden eagles are a protected species in North America, but are not rare or endangered. While the death of any bird is unfortunate, biologists prefer to place the death in the context of the total population rather than focus on the number of individual deaths, according to Tom Cade, founder of the Peregrine Fund and director of the World Center for Birds of Prey. The deaths, regrettable as they are, "may really have no biological significance," says Cade.

The number of birds killed in the Altamont Pass could be significant for a species, such as the golden eagle, that has suffered population declines throughout its range in California due to urban encroachment. The state's raptors, or birds of prey, are fast losing their habitat to an exploding human population. In the San Joaquin Valley alone, more than 95 percent of wildlife habitat has already been converted to other uses. Consequently wildlife becomes increasingly dependent on the remaining "islands" of undeveloped land. Some of this land remains undeveloped because high winds make it hostile to human habitation. Thus there is the potential for increasing competition between raptors and large-scale wind development for the same resource. The population of golden eagles in the Altamont Pass appears stable, says the CEC's Anderson. The state hasn't yet been able to determine if the number of golden eagles being killed is having a negative effect on the breeding population. Meanwhile, biologists are continuing their fieldwork.

Anderson confirms that wind turbines lining the Tehachapi and San Gorgonio Passes in southern California are also killing raptors, but it's much less of a concern to the state because the numbers killed are significantly lower than those in the Altamont. Fortunately no rare or endangered birds such as bald eagles *(Haliaeetus leucocephalus)*, peregrine falcons *(Falco peregrinus)*, or California condors *(Gymnogyps californianus)* are known to have been killed by wind turbines or their power lines anywhere in California.

Despite the problem among the thousands of medium-size turbines in the Altamont Pass, there's little data on the impact from single medium-size turbines, small clusters of machines, or small wind turbines. It's reasonable to assume that small wind turbines or clusters of larger machines kill birds in proportion to the turbine's size and number. The question of whether small wind turbines also kill birds does arise (see figure 13-24, Birds and small wind turbines).

Consider the case of the Western Pennsylvania Conservancy and Audubon of Western Pennsylvania. They manage a nature center in a Pittsburgh suburb and operate a small wind turbine as part of a display on solar energy. During the mid-1980s they found a dead duck at the base of the tower. Greatly disturbed, they called the dealer. He was speechless. The next day he inspected the wind turbine for any telltale signs. A bird the

Figure 13-24. Birds and small wind turbines. Bird perching on an anemometer boom beneath a Marlec 910F at the Wulf Test Field. As this scene illustrates, small wind turbines are not immune to concerns that they pose a hazard to birds.

size of a duck would have severely damaged the 1 kW Bergey turbine. The dealer found the turbine unscathed.

A few days later a neighbor called the nature center searching for his pet peacock. Meanwhile visitors had begun sighting a fox on the grounds. These reports prompted the center's naturalist to reexamine the dead bird, and the mystery was soon solved. He concluded it was the missing peacock and not a duck, after all. And after finding signs of the fox near the tower, the center concluded that the fox, not the wind turbine, was the culprit.

Still, there are anecdotal reports of collisions between birds and the guy cables of small wind turbines. "Birds collide with just about everything," says NREL's Karin Sinclair. Anytime a structure, whether a house, a skyscraper, or a wind turbine, is raised above ground level, it will pose a hazard to birds; and that includes small wind turbines.

More damaging to wind energy's reputation than the numbers of birds being killed in the Altamont Pass is the manner in which they die. Two-thirds of the golden eagles were killed after colliding with wind turbines or their towers. This lends itself to blaring headlines and self-styled investigative reports revealing the "true story" behind one green technology.

BioSystems, in a report for the California Energy Commission on the problem with wind turbines in the Altamont Pass, tried to put the issue in perspective by noting that 5 to 80 million birds die annually in the United States from collisions with structures ranging from picture windows on homes to cooling towers on power plants.

Birds are killed not only in the production of electricity, but also in its transmission and distribution. Birds die by striking overhead power lines (and telephone lines) or by electrocution. While it's difficult to prevent birds from flying into power lines, most deaths by electrocution are avoidable and can be prevented by modifying transmission line towers.

Ornithologists can only speculate on what happens as birds fly near wind turbines. Flying is hazardous, especially for immature birds. "It's a tricky business to be a fast-flying animal at low altitude," said the University of Pittsburgh's late Melvin Kreithen. "They make mistakes."

The job of the wind industry should be to make flying around wind turbines less hazardous. But there's no panacea or silver bullet for eliminating the problem. Painting splashy stripes on the blades and adding noisemakers have been found wanting. The most effective method is avoiding the problem altogether by siting wind turbines where there are no large concentrations of birds that might collide with the turbines.

Wind companies—large and small—must avoid the fortress mentality evoked by the issue of birds crashing into wind turbines. Some companies respond by trying to control the damage instead of trying to solve the problem. As Exxon found with the *Valdez,* "damage control" may cause as much damage to the company's interests as the disaster itself.

A better approach than damage control is to engage the environmental community before a project is proposed. Environmentalists, including bird lovers, generally support wind energy—when given a chance.

Take Rich Ferguson, for example. Ferguson, energy chair of Sierra Club California, labels the situation in the Altamont Pass "tragic and unacceptable." Nevertheless, he believes the issue is less than black and white and wants to know how many dead birds, specifically golden eagles, are too many? This position doesn't prevent Ferguson from supporting wind energy. When a project was proposed to repower an existing Altamont wind farm with newer turbines, Ferguson urged approval of the project.

One essential step for projects with large numbers of turbines is to study the proposed site beforehand. Ornithologists can determine the level of risk to particular species if the project proceeds. Public authorities and the environmental community must then weigh what risks do exist against the environmental benefits the project provides. Once a large project is in operation, it's also necessary to conduct a postconstruction survey to verify that any impacts are within the range expected.

Though the overall impact on bird populations from wind energy may be slight, the fact that there is an impact at all illustrates, once again, that there are costs to all energy choices. "There's no free lunch," says the CEC's Anderson.

Some birds, including eagles, will fly into wind turbines regardless of mitigation measures. An unpleasant thought, yes. Yet, to some extent, unavoidable. Those who think otherwise are deluding themselves. "Zero kill?" says Tom Cade of the Peregrine Fund. "That's not ever going to happen."

Case Studies

Where it exists, criticism of wind energy results largely from fear of the change this new technology may bring to the community. Just as we grew to accept—and now demand—the utility's intrusion on the landscape, gradually we will grow to accept wind machines, in much the same way and for many of the same reasons.

Though it may fear this technology, the community should not apply more stringent standards to wind machines than it applies to any other similar structure or device now standing. Proponents of wind turbines need not ask for special treatment of wind energy, but they are at least entitled to equal treatment.

Whatever you do, don't bypass the permitting officials. You have a responsibility to comply with the community's wishes, even if you don't agree with them.

In New Cumberland, Pennsylvania, an unthinking homeowner bought a wind machine to install in his backyard. Then, to his chagrin, his application for planning approval was rejected. Not only was wind energy not permitted in his residential neighborhood, but also his lot was physically too small. He hired an attorney and engaged in a lengthy and expensive appeal. His neighbors objected vociferously. Then, amid the glare of television lights and a packed hearing room, his appeal was denied—again. His troubles didn't end there. The dealer then refused to buy back the wind machine and the homeowner had to sell it at a loss. He didn't do his homework, and it cost him dearly.

This unfortunate homeowner can be excused because of his enthusiasm for wind energy and his ignorance of the planning process. The same can't be said for some so-called wind farm developers who have committed similar blunders. The difference is in the sums of money involved: not thousands, as in the homeowner's case, but hundreds of thousands.

One group of self-styled professionals was planning to erect several unreliable wind turbines in a New Jersey residential neighborhood—without planning approval. They were about to begin construction when the local news media broke the story. (There was an exciting mix of New Jersey–style backroom politics involved.) The scheme was quickly killed in a boisterous public hearing.

These cases illustrate how not to install a wind machine. There are literally thousands of examples in which the appropriate approvals have been obtained in an orderly and businesslike manner and the wind turbine successfully installed. Consider the example of an upper-income suburb of Pittsburgh.

Fox Chapel Township has a reputation for strict interpretation of its zoning ordinances. "They'll never let you put one here," some said. Yet the dealer, Bill Hopwood of Springhouse Energy Systems, and the client, the Western Pennsylvania Conservancy, were both respected and thoroughly prepared. (They had to receive approval to erect their anemometer, so they were familiar with the process.) They answered all questions forthrightly, allayed the fears of planning officials, and, to the surprise of cynics, won approval. The wind machine, a Bergey 1000, was installed without incident and has operated successfully for nearly two decades.

14

Installation

He loosened the last bolt. The generator was now ready to swing free.

"All ready?" he yelled.

"Yeah, let 'er rip," replied the ground crew.

"You sure that pulley's secure?" he asked, his voice less certain now.

"Yeah, it's not going anywhere. Let's get this one down and go for a beer."

The old generator rocked on its saddle. Slowly it rolled off toward the gin pole. Suddenly there was a loud *twang* and the squeal of steel cable over pulleys as the 400-pound mass of copper and iron whizzed by . . . to crash through the platform next to him.

He looked about in dazed silence.

"Are you all right?" they asked from below.

He glanced at his feet. Yep, still there. Then to his hands. They were, too, as were all his fingers. *Lucky this time,* he thought.

"I'm okay. What the hell happened anyway?"

"That pulley broke loose from the tower."

This incident actually took place. It happened to an experienced crew working professionally. Though it occurred while removing rather than installing a wind machine, it illustrates what can happen without thorough planning, preparation, and—equally as important in this case—execution.

If you're handy with tools and don't mind hard physical labor, this chapter will offer you guidance on installing a small wind turbine yourself. You gain by replacing the skill, time, and expense of the dealer-installer with your own. You will develop a sense of accomplishment in doing it yourself, and you will learn more about your wind system than in any other way. You will know its strong points and also what can go wrong and where, and how much effort it will take to fix it. The process will also give you an appreciation for the effort and skills required to reliably produce your own electricity.

Wind turbines of less than 3 meters (10 ft) in diameter may be installed by the homeowner or hobbyist with basic construction skills. The introduction of lightweight, tilt-up masts for machines of this size makes such installations easier than ever before. The work can still be hazardous, but no more so than other projects around the home or farm. With proper respect for the hazards involved and close attention to detail, you can safely install a small wind turbine.

As the Alternative Energy Institute's Ken Starcher points out, risk is proportional to size. Larger turbines entail proportionally more risk. The components are heavier

Thoughts on Doing It Yourself

When I first wrote this chapter in 1982, I believed most homeowners with a modicum of tool skills and common sense could safely install a household-size wind turbine themselves. I figured, *Heck, if I can do it, anyone can.* I've since learned that's not true. I can't say whether this conclusion is due to a decline in our collective knowledge about how to use hand tools or work around machinery, or to my becoming more cautious over the decades. I've certainly made my share of mistakes. I was the person on the tower in the anecdote that opens this chapter. I was in charge and I was ultimately responsible. Unfortunately my position in the wind industry does make me aware of the mistakes others—including some professionals—have made, and the injuries that have resulted. As a consequence I now believe that homeowners should only attempt installing wind turbines less than 3 meters (10 ft) in diameter on lightweight tilt-up guyed masts. Products that fit this description are the many micro turbines on the market, as well as Bergey's XL1, Southwest Windpower's H40, and Proven's WT600. Homeowners should avoid installing larger turbines, freestanding truss towers, or heavy-duty guyed towers without hands-on training. Most lack the skills, specialized tools, and safety equipment necessary. The tools can be purchased, and the skills needed can be learned. Workshops, such as those that Mick Sagrillo teaches, or installer training programs offered by manufacturers are worth the money and are the best way to learn how to install wind turbines safely. A book is no substitute for the hands-on learning that's required.

and may require special equipment and techniques unfamiliar to most do-it-yourselfers. Similarly, the installation of medium-size wind turbines requires the skilled use of heavy machinery beyond the ability of even the most resourceful farmer. It's usually best to leave the installation of these machines to professionals.

The following sections provide general information required by any installer of small wind turbines in North America. Although the materials suggested may differ on other continents, the principles and techniques remain the same.

Always consult the manufacturer's installation manual for more detailed descriptions of anchoring and installation techniques. If you plan to install the turbine yourself, the information in this chapter will help you select the tower, anchors, and erection methods that best suit your talents and the conditions at the site. After reading this chapter you may opt to hire a contractor instead of installing it yourself. The information gained, however, will enable you to track the progress and evaluate the performance of the contracted installer.

Whatever route you choose, thorough planning is essential. You must anticipate what will be needed at each step along the way, the problems you may encounter, and how to respond to them. You must coordinate the schedules of your subcontractors, suppliers, and erection crew to keep the project moving smoothly. If you lack any of the required skills, you must find someone who has them.

Pace yourself. Assume it will take twice as long as you expect. A skilled two-person crew can install a micro or mini wind machine in one day. It may take novices much longer.

Prepare for the installation by collecting the parts, fittings, and tools for the job. Learn how components will be assembled and what tools are needed. Make sure you have met all legal requirements and that you're insured for any accidents that may occur. If you're installing the wind system yourself it's a good idea to check whether your insurance will cover hospitalization and liability for friends who lend you a hand.

Without proper execution all your planning and preparation may be for naught. You may know the right way to do a task and have the right tools to do so, but if you don't follow through under the press of time and conditions, the results can be disastrous. Consider the anecdote that opened this chapter. You may be tired, and a trip down the tower to check a pulley may seem unnecessary. It isn't. That's the time to take extra care to do the job right.

One final caveat: Build to local codes even when it's not required. You'll be glad you did. Doing so makes for a sounder, safer, and more serviceable installation.

Parts Control

Installation can be hindered and proper operation of the wind turbine prevented by components damaged in shipment. Before accepting delivery from the freight carrier, examine the invoice or billing form to determine the number of crates shipped. Make sure all are present, and then carefully examine the crates for external damage. If damage is found, open the crates and look at the contents. The crates are designed to take some abuse while still protecting the product inside. Note any damage as precisely as possible and immediately contact the dispatcher at the freight company. (Instant or digital photographs with a date stamp can be helpful in verifying claims.)

Tower sections and sensitive electronic components are the most easily damaged during shipment. Damage to control boxes and inverters is much harder to determine. The best you can do is identify any loose parts.

Catalog the serial numbers on the wind generator, blades, control panel, inverter, and tower. If you have to make a warranty claim, the numbers are much easier to find in your files than at the top of the tower. Serial numbers will also aid in troubleshooting if problems develop.

Make an inventory of all parts received as soon as possible. Many manufacturers provide a parts checklist for this purpose. Use it. The time to realize that an essential bolt is missing is prior to installation.

For those wind systems where the manufacturer doesn't also build the tower, the wind

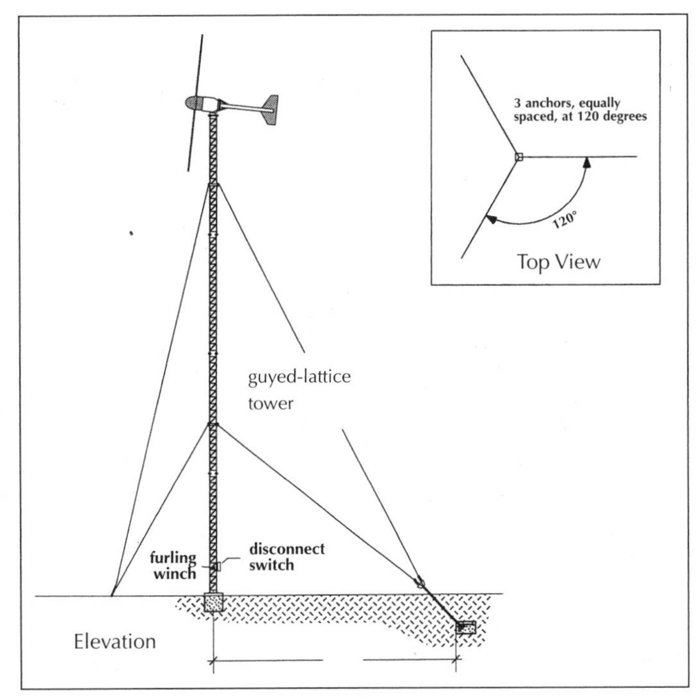

Figure 14-1. Guyed-lattice mast. A typical guyed tower for household-size wind turbine in North America. (Bergey Windpower)

turbine and tower will be shipped separately. Often they will be delivered by separate carriers. Unless you have a special reason for removing the contents from the crates (for an inventory, possibly), leave them as delivered until you're ready for the installation.

Foundations and Anchors

The tower and guy cables (where used) must be kept clear of vines, trees, and shrubs. It may be necessary to clear the site of any plants that could eventually interfere with the tower or guy cables. The site doesn't have to be level, so there's no need to grade the site to bare earth.

Anchors are used to prevent guyed towers from overturning (see figure 14-1, Guyed-lattice mast). Anchors resist uplift. Piers, on the other hand, resist loads in compression. On a guyed tower, for example, the anchors hold the tower upright and resist the forces trying to knock the tower over. The pier beneath the central mast supports the weight of the tower and wind turbine and resists the reactive forces from the guy cables trying to drive the mast into the ground. On freestanding towers, the legs act alternately as piers and as anchors, depending on the direction of the wind.

The type of anchor or pier used is contingent on the tower and the site. If you plan to install a guyed tower, there are several anchoring options to choose from: concrete, screw, expanding, and rock anchors. The best choice for your site is determined by the engineering properties of the soil, the depth to bedrock, and the power equipment available in your area. For a freestanding tower, the choice is limited to concrete.

Anchors

Anchors must withstand the static and dynamic loads acting on the wind system, under all weather conditions, for the life of the system. They must do so without appreciable creep toward the surface or settling. The holding power of anchors depends on the area of the anchor, its depth, the soil in which it is embedded, and the soil's moisture content. Weight is a factor, as well, but it's not as important as you might think.

Soils vary in their ability to resist creep. Resistance to creep is controlled by the soil's shear strength: the resistance of soil particles to sliding over one another. Shear strength is a function of soil type and whether the soil is wet or dry. Shear strength ranges from a maximum in solid rock to a minimum in mucky or swampy soils.

One anchor manufacturer divides the shear strength of soils into two broad categories: cohesive and noncohesive. Cohesive soils, such as those with a high clay content, stick together; the particles cling to each other. These soils have a high shear strength. Noncohesive soils are generally those with a high sand content. In such soils the soil particles slide right by each other. Wind system manufacturers specify that their standard anchor designs are intended only for soils that are cohesive under normal conditions; those with a high clay content.

Anchor holding capacity also decreases as the moisture content increases. Creep can be troublesome in saturated soils because the soil particles become fluid and tend to flow around the anchor. Water also increases the buoyancy of the anchor. The holding capacity of anchors can be reduced 50 percent in wet soils. Wherever possible anchors should be placed below the level of periodic saturation from heavy rains but above the water table.

Frost heave causes similar problems. When soil freezes it expands slightly, just as ice occupies a greater volume than water. If the anchor is not below the frost line, the cycle of freezing and thawing will heave or jack the anchor toward the surface. This is more of a problem for anchors than for piers, because the existing load acts to pull the anchor out of the ground. The forces on piers act counter to frost heave. The frost line varies from year to year and depends on the severity of the winter and the soil cover. Bare soil freezes more quickly and to a greater depth than a soil with a grass cover. (The grass and the organic soil it grows in act as an insulator, slowing the soil's winter heat loss.)

To determine the soil's holding capacity at your site, you can test the soil with a probe or examine nearby road cuts. Better yet, talk to people who work with soil. In the United States, the Soil Conservation Service (SCS), the county extension agent, or the office of the conservation district should be able to help. Explain your plans to them. Describe what it is you want to know and why it's important. (You don't want the anchor pulling out of the ground.) They will be able to tell you not only what kind of soil you will be working with, but also the depth to the water table and the average frost penetration.

Local excavation companies are another good source. They have a feel for subsurface conditions since they work with them daily. They are in business to make money, though,

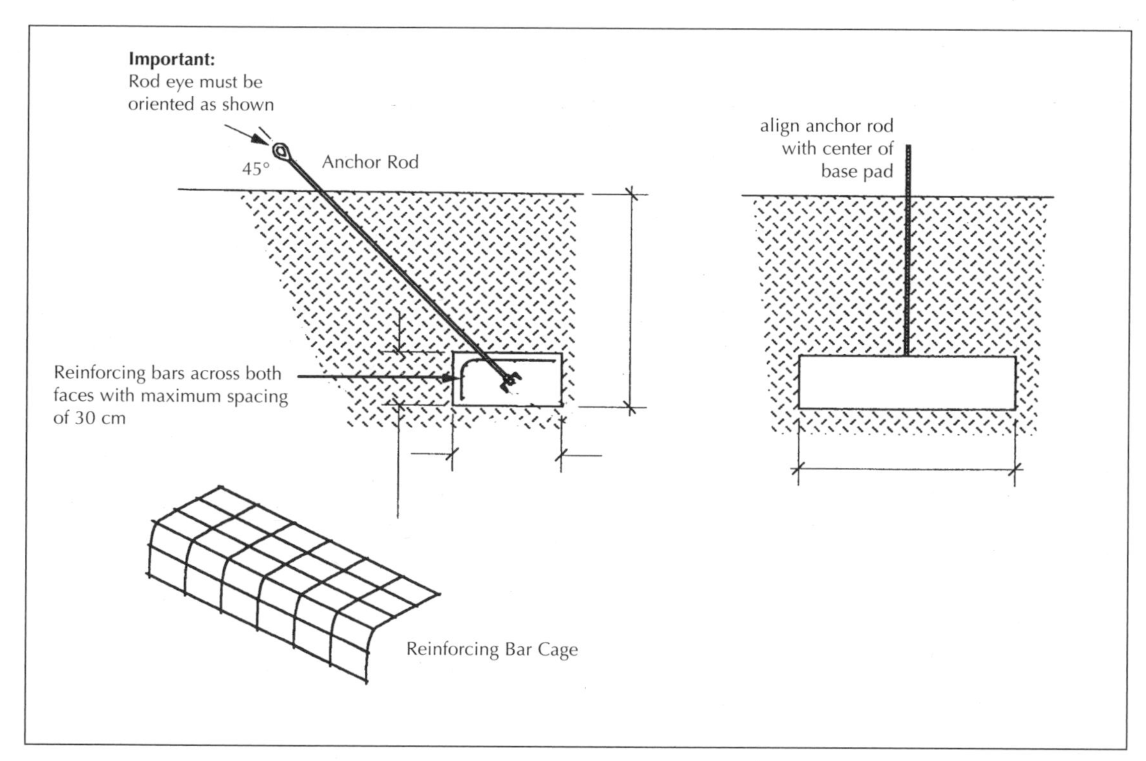

Figure 14-2. Concrete anchor. A detail of a typical concrete anchor for a guyed tower suitable for a small wind turbine. (Bergey Windpower)

not to give out free information. If you want their help, you should hire them.

The requirements for piers are less stringent than those for anchors. Most soils are strong in compression. With an adequate bearing surface, concrete piers of standard dimensions are used throughout North America.

Working with Concrete

The most common method for anchoring a tower or constructing a pier is to excavate a hole and partially—sometimes completely—fill it in with reinforced concrete.

Concrete is literally manufactured rock; conglomerate, to be specific. It's strong in compression, weak in tension. Thus it works well as a pier, or foundation, but poorly as a beam. Tensile strength is improved by reinforcing the concrete with steel rods commonly called rebar (reinforcing bars).

Concrete is rated by its compressive strength. In North America concrete is rated as obtaining its minimum, ultimate strength after curing for 28 days. Strength is a function of the water–cement ratio and the degree to which curing has taken place. The lower the water–cement ratio (the more cement in the mixture), the stronger the concrete. Strength also increases with curing time.

Curing is rapid in the first few days. (Concrete sets or becomes rigid within an hour of adding water.) Hydration doesn't go forward if too much water evaporates in hot weather, or if the concrete becomes too cool in cold weather. Curing should take place for a minimum of seven days before any load is placed on the concrete. The concrete will continue to gain strength if moisture and temperature conditions remain favorable for complete hydration. If you heed the above precautions, concrete can be placed year-round.

Installation drawings invariably show nice neat anchors and piers that look like they were made with a cookie cutter (see figure 14-2, Concrete anchor). Except where an anchor or pier is exposed at the surface, this precision isn't necessary. Where the soils are stiff and will not collapse into the hole, the concrete can be poured in place. For anchor blocks below the surface, this is superior because the concrete acts directly on undisturbed soil. Forms are necessary where the hole is larger than the anchor or pier desired, where the concrete will extend above the surface, or when the anchor is in sandy soil.

The concrete is placed over a grid or cage of rebar, to give the concrete the necessary tensile strength. The rebar is tied together with wire so it won't move when the concrete is placed over it.

All rebar must be covered by at least 3 to 4 inches (about 100 mm) of concrete. When the rebar is closer to the surface of the concrete, acid-laden water can enter the concrete, corroding the steel rebar. As the rebar corrodes, it expands slightly, causing the concrete to spall or chip. For long life, the concrete must seal the rebar from corrosion.

To ensure that the rebar stays where you want it when the concrete is placed, it should be staked down in the excavation or tied to the forms so it won't "swim" around. Pieces of rock or brick can be used to keep the rebar cage off the bottom of the excavation. This keeps the rebar from being too close to the concrete's bottom surface.

Forms can be built from heavy plywood and a wooden frame. The frame, when staked to the side of the excavation, will hold the plywood form in place while the concrete is hardening. Where the soil is stiff and the excavation is no larger than the pier desired, a short form can be made from wooden planks set in the excavation to a depth that gives the finished pier the desired height above the surface. Cylindrical forms can be purchased for placing concrete columns.

Before placing concrete, moisten the forms or the excavation to prevent them from absorbing water from the surface of the concrete mix and reducing its strength. Avoid placing concrete from a height greater than 4 feet (1 m) or the aggregates will begin to separate. Once in the form, work the concrete to eliminate air pockets by poking a board or shovel into the concrete. Work it around the rebar and along the sides of the forms. Don't overwork, or the aggregates begin to settle. Special gasoline-powered vibrators can be rented for working the concrete after it has been placed.

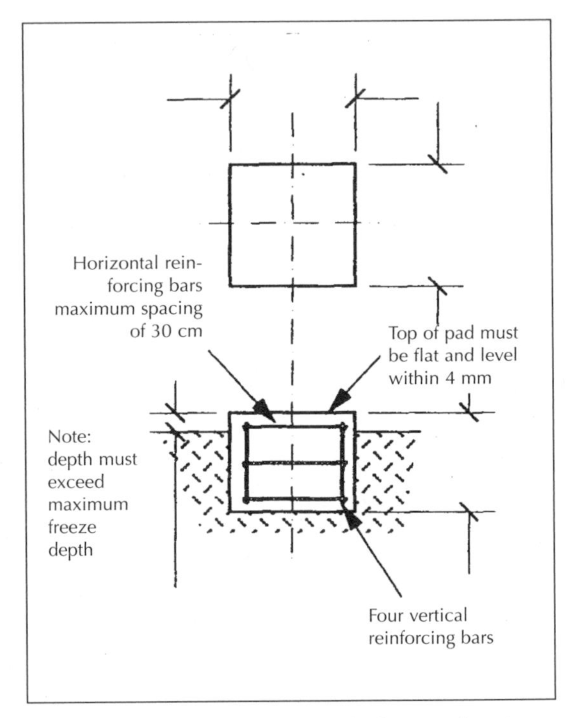

Figure 14-3. Concrete pier. A detail of a typical concrete pier for a guyed tower suitable for a small wind turbine. (Bergey Windpower)

You can simplify the whole process by hiring a contractor to excavate the hole and pour the concrete. Get firm quotes before you do, and make sure they understand what the concrete will be used for. The fact that a lot is riding on their work may discourage them from cutting any corners.

Guyed Towers

Most small wind turbines are installed on guyed towers. And many of those are installed on fixed (non-tilt-up) lattice masts. To support a guyed lattice mast, you'll need a pier for

Table 14-1
Concrete Piers and Anchors for 80-foot (24-m) Guyed Tower for Furling Wind Turbines

	Wind Turbine Rotor Diameter		Pier–Anchor Dimensions		Depth	
	(m)	(ft)	(m)	(ft)	(m)	(ft)
Piers						
	3	10	0.6 x 0.6 x 1.2	2 x 2 x 4	-	-
	5	16	0.8 x 0.8 x 1.2	2.5 x 2.5 x 4	-	-
	7	23	1 x 1 x 1.2	3 x 3 x 4	-	-
Anchors						
	2.5	8	0.8 x 0.8 x 0.3	2.5 x 2.5 x 1	1	3
	3	10	1 x 1 x 0.3	3 x 3 x 1	1.2	4
	5	16	1 x 1 x 0.5	3 x 3 x 1.5	1.2	4
	7	23	1 x 2 x 0.6	3 x 6 x 2	2	6

Note: For normally cohesive soils. For taller towers and less cohesive soils, refer to manufacturer's specifications.

the mast and at least three anchors to keep the mast erect.

The pier is simple to construct (see figure 14-3, Concrete pier) by excavating a hole of sufficient size and depth (see table 14-1, Concrete Piers and Anchors). In weak soils or with an extremely tall tower, a larger pier than specified by the manufacturer may be needed. Similarly, it may also be necessary in some areas to extend the pier deeper than normal to get below the frost line. Strengthen the pier with a rebar cage before placing the concrete.

Guyed towers also need a means to keep the mast in place so it won't scoot off the pier. Before the concrete sets, insert a pin or threaded rod into the center of the pier. Most masts have a base plate that slips over this pin.

Anchors are a little more complex. The type of anchor used depends on several factors: soil strength, the depth to bedrock, and your access to power equipment. In normally cohesive soils you can choose from concrete, expanding, and screw anchors. Concrete anchors are the most popular, followed by screw anchors.

Though screw anchors are widely used by electric utilities, building inspectors and contractors are more familiar with concrete. The equipment needed to excavate the hole for a concrete anchor is also more readily available than the utility line truck often used to drive screw anchors. Concrete anchors can also be adapted to weak or soft soils by simply expanding the anchor's bearing surface—that is, by driving a larger-diameter anchor deeper. If you have any doubts about the holding capacity of screw or expanding anchors, opt for concrete.

Excavators work best at digging trenches and are well suited for making the excavations for concrete anchors. Once the rebar cage has been positioned, the concrete can be placed.

An easier and quicker method than using an excavator is to auger the holes for both pier and anchors. Where soils are not rocky and bedrock is well below the surface, a truck-mounted power auger, like that used by linemen to set wooden utility poles, can drill holes for the pier and anchors in a matter of minutes.

For the pier, auger a hole of the dimensions needed. Set a short form into the top of the hole so that it extends above the surface, place the rebar, and cast the concrete in place.

After the pier is set, the auger can then be used to drill holes for expanding anchors (see figure 14-4, Expanding anchor installation). Expanding anchors work much like toggle bolts and similar fasteners used in plaster walls. Once the arrow-shaped anchor is inserted into the hole and expanded, the barbs resist being pulled back out.

The strength of expanding anchors is controlled by the soil type, the size of the anchor,

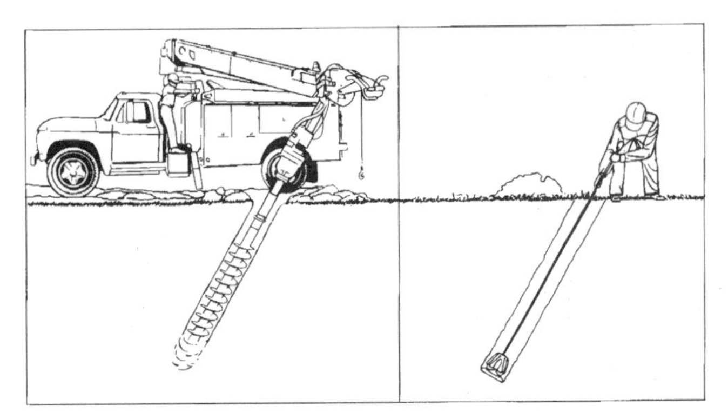

Figure 14-4. Expanding anchor installation. Augering a hole for the anchor, inserting the anchor, expanding the blades of the anchor with a tamping bar, backfilling the hole, and tamping. (A. B. Chance)

Table 14-2
Expanding Anchors for 80-foot (24-m) Guyed Tower for Furling Wind Turbines

Wind Turbine Rotor Diameter		Anchor Diameter		Length	
(m)	(ft)	(mm)	(in)	(m)	(in)
3	10	0.2	8	1.7	66
5	16	0.3	12	1.7	66

Note: For normally cohesive soils. For taller towers and less cohesive soils, refer to manufacturer's specifications.

Figure 14-5. Screw anchor installation. Truck-mounted power augers make quick work of driving screw anchors. (A. B. Chance)

and the firmness of the backfill. Expanding anchors hold better in heavy, stiff soils, less well in sandy or swampy soils. The larger the diameter of the anchor, the greater its holding power. To reach its rated strength, the anchor must be fully expanded into undisturbed soil, and the hole backfilled properly (see table 14-2, Expanding Anchors).

After drilling a hole at the correct angle, set the anchor at the bottom. Strike the anchor with a heavy bar to force the leaves into undisturbed soil. Once the anchor is fully expanded, attach the anchor rod. Gradually backfill the hole, and compact the soil with a tamping bar. That's all there is to it.

The development of screw anchors has further simplified the installation of anchors for guyed towers (see figure 14-5, Screw anchor installation). It's not unusual to install the three anchors needed for a small wind turbine in less than 30 minutes. Many truck-mounted augers have been adapted to drive screw anchors by replacing the auger bit with a special tubular wrench. The hydraulic boom controls both the angle and the rate at which the anchor enters the soil. Unfortunately screw anchors can't be used everywhere. Rocky soils, in particular, can retard the anchor from advancing.

Screw anchors are sized by the diameter of the screw, the number of helixes on the anchor shaft, and the length of the anchor

Table 14-3
Screw Anchors for 80-foot (24-m) Guyed Tower for Furling Wind Turbines

Wind Turbine Rotor Diameter		Anchor Diameter		Length	
(m)	(ft)	(mm)	(in)	(m)	(in)
1	3	152	6	1.7	66
2.5	8	152	6	1.7	66
3	10	203	8	1.7	66

Note: For normally cohesive soils. For taller towers and less cohesive soils, refer to manufacturer's specifications.

rod. Their holding strength is once again based on the cohesiveness of the soil (see table 14-3, Screw Anchors). NRG's Dave Blittersdorf warns against using screw anchors in noncohesive soils such as sand, silt, and peat. Nevertheless, NRG has shipped more than 4,000 of its tilt-up towers over a period of nearly two decades, and nearly all have been installed with screw anchors.

To test the soil yourself, use a technique that soil scientists use. Ball dry soil in your hand. If it doesn't stick together, it isn't cohesive. Pull-out strength is a function of the soil's cohesiveness, or its ability to stick together. Dense, hard clay soils have the highest pull-out strength, and sandy soils the least.

If using the anchor in a cold climate, ensure that the anchor helix or plate is below the frost line, recommends Blittersdorf. In Vermont, where NRG has its tower plant, the frost line is 3 feet (1 m) below the surface. If the anchor helix is within the frost zone, frost heave will gradually lift the anchor toward the surface. Eventually the anchor will pull out, bringing the tower down.

Soil strength greatly affects the holding capacity of any anchor. For a small wind turbine 3 meters (10 ft) in diameter in a normally cohesive soil, the 8-inch (203 mm) screw anchor has a holding capacity more than three times greater than the expected maximum load on the anchor. But in weaker soils, the safety factor drops to less than two. In heavy soils the 8-inch anchors are more than sufficient for this turbine. In weaker soils, or if there's any doubt about the holding capacity of the soil at your site, check with the wind turbine manufacturer.

Where necessary you can double the anchors at each guy. This was the approach taken by Fayette Manufacturing on the 1,500 turbines that once operated in Altamont Pass. Fayette installed its 10-meter (30 ft) turbine on guyed pipe towers using screw anchors. Because of soil conditions in the area, Fayette used two 12-inch (305 mm) screw anchors at each guy point. It was one of the few things Fayette did right.

If you're unfortunate enough to encounter solid rock at or near the surface, none of the preceding anchoring methods can be used. You'll have to drill a hole at the proper angle with an air drill and compressor (see figure 14-6, Rock anchor installation). A rock anchor

Figure 14-6. Rock anchor installation. Drilling the hole, inserting the anchor, and expanding the anchor by torquing the anchor rod. (A. B. Chance)

and rod are then inserted down the hole and wedged in place. These anchors have a high holding capacity, but installation is time consuming and expensive.

For small wind turbines one anchor rod is sufficient for all guy levels in most installations. To minimize bending, the anchor rod must depart the ground at an angle that coincides with the resulting angle of tension in the guy cables. The angle of departure depends on the height of the tower and the guy radius, and it's usually 45 to 70 degrees. In the field it's easiest to use a 45-degree angle of departure by measuring a rise of one over a run of one, but it's always better to follow the manufacturer's recommendations on the appropriate guy angle.

Freestanding Towers

Freestanding towers—whether truss towers or tapered, tubular towers—require an excavation and the placement of concrete. The easiest method, where the depth to bedrock permits, is to use a power auger.

For tubular towers a large-diameter hole is drilled for a pier to support the entire tower. This is the preferred technique for even the largest wind turbines. Massive augers excavate holes for medium-size wind turbines that are large enough to swallow pickup trucks—and have. At one wind farm site near Tehachapi, a pickup truck disappeared and was later found at the bottom of one of the augered excavations. Fortunately no one was injured.

On truss towers holes are augered for a pier at each leg of the tower. In sandy or swampy soils piers are insufficient for truss towers. Instead, like those of the medium-size turbines on the sandy Whitewater Wash near Palm Springs, California, the tower must rest on a massive concrete pad.

Knowing the soil conditions at a site is just as critical when using freestanding towers as when using guyed towers. In 2001 a number of wind turbines larger than 1 MW were installed on tubular towers in Texas using a modified pier commonly used on wind farms in North America. Within a few months of installation, several of the multimillion-dollar wind turbines were lurching off the vertical. The problem? Voids in the limestone bedrock at some turbines didn't provide sufficient overturning resistance. Correcting the massive "leaning towers of Texas" was no simple task.

On household-size wind turbines atop truss towers, the piers for each leg may be up to 3 feet (1 m) in diameter and up to 10 feet (3 m) deep. Often the excavations are left unfinished or in their circular form. The rebar and anchor bolts for anchoring each tower leg are then added, and the concrete is placed.

You must make certain that the base tower sections or anchor bolts used in the foundation don't "swim" around when the concrete is placed. They must also accurately fit the foundation template provided with the tower. Otherwise you could have a rude awakening when you go to set the tower on the base.

Novel Foundations

If power-installed screws work well as anchors, why can't they also be used for the pier supporting the mast of a guyed tower? In theory, at least, they can. The foundations for light standards and transformers at substations have been installed this way for years. No one, however, has adapted this technology to wind systems. When someone does so, an installer will be able to drive the anchors and pier in just minutes instead of waiting days for the concrete in the pier to cure. The whole wind system could then be erected in a day or less.

Using power-installed screw anchors to secure the legs of a freestanding truss tower is another possibility. There may be engineering limitations, particularly in weaker soils, but the advantage of quick and easy installation justifies a look into whether it's possible.

Kits for tilt-up guyed towers from Bergey Windpower and Southwest Windpower for their micro and mini wind turbines use a base plate instead of a pier. The base plate is simply staked to the surface of the ground to prevent it from sliding out from under the tower.

One major innovation in foundation design was developed during the 1990s for tubular towers installed in California. Rather than construct the pier with a solid mass of concrete and rebar, the new design uses a cylinder of reinforced concrete the diameter of the tower that it supports. The hollow core of the cylinder is subsequently filled with the excavated soil. This foundation substantially reduces the amount of concrete needed over that of the traditional method.

Assembly and Erection of Guyed Towers

Two types of guyed towers are used for small wind turbines: fixed, guyed masts; and hinged, tilt-up towers. Fixed, guyed masts use three anchors. Tilt-up towers typically use four anchors.

The guyed lattice masts commonly used for household-size wind machines in North America can be assembled a section at a time with a tower-mounted gin pole, or the entire tower can be assembled on the ground and hoisted into place with a crane. The method you use depends on whether a crane is available and can get to your site. In either case the guy cables must first be cut to length and attached to the guy brackets that fit around the mast.

Guy Cables

Steel cable is shipped in coils or on reels. Avoid damaging kinks in the cable by rolling the coil along the ground to lay out the cable. If the cable is on a spool, use a spool stand to unreel the cable, or carefully walk the spool along the ground to unreel the cable.

If the manufacturer doesn't specify the length of the guy cables in the installation manual, calculate cable length by using the Pythagorean theorem:

$$\text{Guy Length} = (GR^2 + GH^2)^{1/2}$$

where GR is the guy radius and GH is the guy level or height above ground. Give yourself plenty of extra cable to allow for sag and for slight errors in the position of the anchors. You'll need three lengths of cable for each guy level on a fixed tower; four lengths for a hinged, tilt-up tower.

Mark the length of each guy cable on the ground. Unreel the cable and cut it—carefully—to length with bolt cutters. When doing so make sure that both ends of the cable can't whip around. Working with wire rope is like working with an unwound spring.

Guy Cable Attachments

How you attach the cable to the guy brackets on the mast may vary. Tower kits provided by manufacturers use swaged connections. Where swaged connections are not provided, wire rope clips or grips will often be used.

North American utility companies sometimes use strand vises and preformed cable grips for this task on guyed utility poles. In a strand vise, the guy cable is passed through the vise and, when tensioned, wedges the cable in the grip of the vise. These are unsuited for use with wind turbine towers. Because the towers sway with the wind, guy cables are not always in full tension. Consequently strand vises, and other cable attachments like them, can release the guy cable, causing the tower to crash to the ground. Preformed cable grips have been in use by utilities since the 1950s. These attachments use a fine grit embedded in an ahesive to grip the cable. Tension on the grip pulls the strands tighter together. Preformed cable grips allow quick adjustment in the guy cable during installation by being easy to remove and reapply. They were used with guyed towers for Storm Master turbines installed in California on Bergey wind turbines during the 1980s (see figure 14-7, Preformed cable grip). Some wind turbine installers still use them. However, most

Figure 14-7. Preformed cable grip. Legs or strands of the preformed wire grip are wrapped around the cable. Tension in the legs and a fine grit grips the cable. Note the turnbuckle (lower left), and the thimble through the turnbuckle eye.

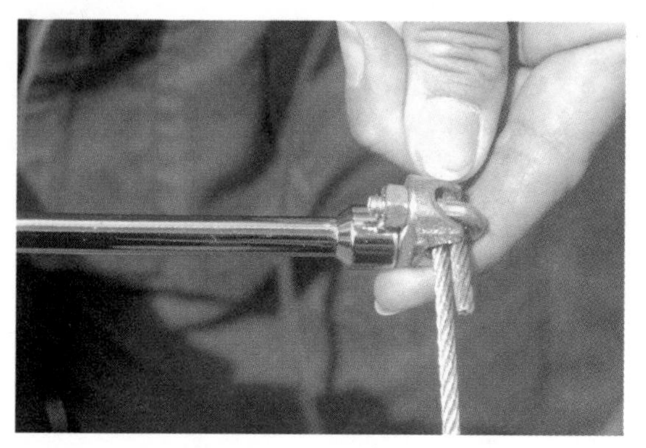

Figure 14-8. Never saddle a dead horse. The forged "saddle" of the wire rope grip must act on a "live" guy cable. The U-bolt acts on the "dead" end of the guy cable.

Figure 14-9. Always use a thimble. Improper attachment of a guy cable at a university test field. The sharp radius of the turnbuckle eye has caused the guy cable to flatten. The weakened cable could fail catastropically. A metal thimble of the proper size should have been used here.

Figure 14-10. Anchor knuckle. The guy cable must not pass over too sharp a bend. The large-diameter anchor eye used on some installations doesn't require a cable thimble.

installations of small wind turbines will use wire rope clips.

Wire Rope Clips and Thimbles

Wire rope clips are simple to use, but they must be used correctly. The clip's forged saddle must bear on the live end of the guy cable—the portion of the cable connecting the tower to the anchor. The U-bolt must bear on the dead end, or the "tail" of the cable after it passes through the anchor eye. As Mick Sagrillo admonishes, "Never saddle a dead horse" (see figure 14-8, Never saddle a dead horse).

Always use a thimble when attaching wire rope. The thimble distributes the load on the cable over a large radius. A sharp bend in the cable will severely weaken it (see figure 14-9, Always use a thimble). Some guy brackets, particularly those used in the utility industry, have built-in thimbles. Similarly, the eye of some anchors has a large knuckle obviating the need for a thimble (see figure 14-10, Anchor knuckle).

Where thimbles are necessary, the guy cable is wrapped around the thimble and passed through two or more clips. The clips prevent the cable from slipping off the thimble (see figure 14-11, Cable thimble). Thimbles can be awkward to fit onto the anchor eye. Hugh Piggott, in his helpful

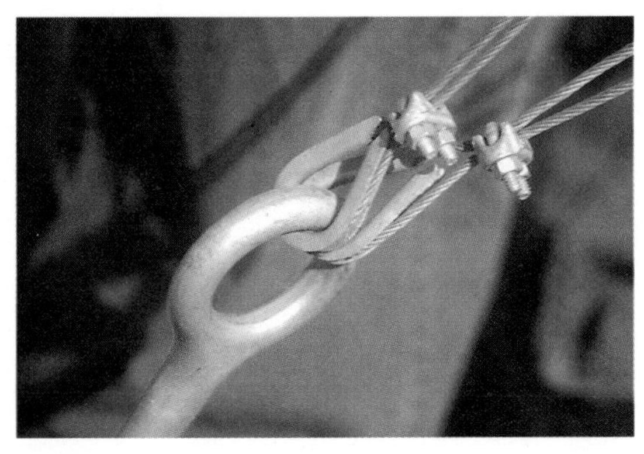

Figure 14-11. Cable thimble. A thimble must be used to distribute the load from the guy cable over a sufficiently large radius to avoid damage to the cable.

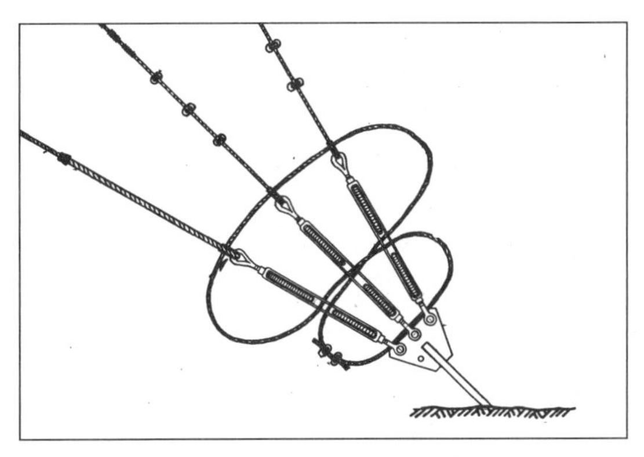

Figure 14-12. Turnbuckle safety cable. Use of a cable through the center of the turnbuckles prevents the turnbuckles from unscrewing and releasing the tower. It's also cheap insurance should a turnbuckle fail. (Bergey Windpower)

book *Windpower Workshop,* describes the simplest way to open the throat of a stamped thimble. Rather than prying the thimble ends apart, twist the ends past each other. After passing the thimble over the anchor eye, simply twist the ends back together. Pliers on either leg of the thimble, says NRG's Dave Blittersdorf, provide a little leverage, making this a snap.

Manufacturers will specify a minimum turnback distance for the tail of the guy cable and the minimum number of clips. Since the wire rope clips are critical to safety, use at last three of them even if the manufacturer says two are sufficient. Apply one clip at the minimum turnback distance. Apply another clip near the thimble sufficient to keep the cable from slipping off the thimble. Place the other clips equidistant between these two. Tighten the nuts on the clips evenly, alternating until the desired torque is reached.

Retighten the clips after the load has been applied. With load, the guy cable will stretch slightly and shrink in diameter, loosening the clip's grip on the cable. Periodically retighten as needed.

On fixed, guyed towers, turnbuckles are used to mechanically tension the cables. It can be just as dangerous to overtension the guy cables, causing the tower to buckle, as to allow too much slack in the cables. There are several methods for measuring the amount of tensioning in a guy cable. It's best to check with the manufacturer for the method it prefers. Where turnbuckles are used, add a safety cable (see figure 14-12, Turnbuckle safety cable).

Tilt-up towers for micro and mini wind turbines seldom use turnbuckles. The cables are tightened by hand. Never, however, release all the cable clips at one time when the cable is under tension, or you may find your turbine and tower slamming into the ground. Instead loosen the topmost grip, pull out the slack, then retighten the grip. Continue until you've worked the slack through all the clips (see figure 14-13, Taking up slack).

Always use cable and anchor fittings designed for the purpose. Anchors, turnbuckles, and wire rope clips are no place to cut corners (see figure 14-14, Substandard turnbuckles). Beware substandard clips, warns Jon Powers, an experienced Tehachapi windsmith. Sometimes there may be telltale excess flashing around the forged saddles. If so, throw them out. They're not worth the risk. Michael Klemen has found such defective clips—after the fact. His Whisper H900 slammed to the ground when he was lowering the tower, and it may have been due to a defective cable clip. It's an expensive lesson, says Klemen.

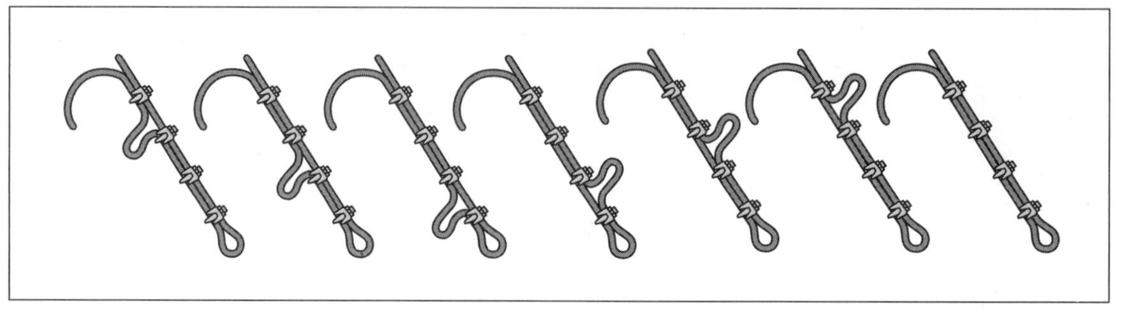

Figure 14-13. Taking up slack. Never release all the wire rope clips when a cable is under tension. Instead loosen the topmost grip, take up the slack, retighten the grip, and proceed until the slack has passed through all the grips. (Vergnet)

Using a Crane

Either the tower and wind turbine can be lifted into place together with a crane or each can be lifted separately. If you're lifting a complete assembly the crane doesn't have to be taller than the tower. For household-size wind turbines, the combined tower and wind turbine can be lifted at some level below the top of the tower. Because the lattice mast is somewhat frail, a nylon sling should be used to spread the lift over the entire cross section of the tower and not act on one leg alone. Lifting a complete assembly also requires that the wind machine must be fully assembled. This may require lifting the tower off the ground slightly to attach the blades and tail vane (where used).

If a crane is to be used bolt all tower sections together, making sure that the section with the guy brackets and cable is in the right position. Attach the wiring conduit and thread the power cable through it as discussed in the subsequent section on wiring. While the crane holds the tower upright, position it over the base plate or pier. Rest the tower on the pier and connect the guy cables to the anchors, pulling them taut. Once this is done—and only then—can the crane be released and final adjustments made to the guy cables.

Figure 14-14. Substandard turnbuckles. Improper use of turnbuckles that were not designed for wind turbine applications at a university test field. The open jaw of the turnbuckle could release the guy cable, causing a catastrophic failure. The installer recognized the problem and welded a gate across the open jaw. Welding the forged turnbuckle could reduce the metal's strength, however, itself contributing to failure of the guy cable. The clevises used here should also include a cotter pin or device to prevent the clevis screw from backing out. All in all, the installer should start over and do it right.

Using a Gin Pole

For reasons of safety, the trend has been away from using gin poles to raise components on fixed, guyed towers. For household-size turbines, installation with a crane is preferred. For micro and mini turbines, manufacturers predominantly use tilt-up guyed towers. Still, the following technique may be used in special circumstances for raising a guyed lattice mast of Rohn 25G or 45G sections. If a gin pole is used, bolt two tower sections together and attach temporary guys. Tip the two sections up onto the pier and tie off the temporary guys to the anchors. They will hold the tower in place until the first guy level is reached. Someone must now climb the tower and bring up the gin pole.

Crane Installation of a Household-Size Turbine

This sequence illustrates the installation of Bob and Ginger Morgan's Bergey Excel on a guyed tower near Tehachapi, California. The tower's location and height were chosen by the Morgans in consultation with the installer, Specialized Turbine Services. The concrete pier for the mast and the three concrete anchors were placed some weeks prior to scheduling the crane. This allowed ample time for the concrete to cure.

Prior to the crane's arrival, the turbine and tower had been assembled and two of the three blades had been bolted to the turbine. STS used barrels and wooden blocks to support the wind turbine and tower during assembly.

The crane arrives.

The crew attaches a nylon sling to the tower. The sling attaches to a crane hook.

Initial lift to clear the ground for attaching the third blade.

Attaching the nose cone or spinner.

The crane raises the turbine and tower as the ground crew guides the mast toward its pier.

The mast is lowered onto the pier. Note the disconnect switch mounted on the tower.

The ground crew attaches guy cables to the turnbuckles. Note the equalizer plate attached to the anchor rod, and the grounding conductor connected to a ground rod.

A windsmith climbs the tower to unhook the crane's boom from the sling, but only after all guy cables are secured to the guy anchors and are tensioned.

The turbine is installed. The crane prepares to leave the site as the windsmith descends the tower.

The gin pole is a boom, or davit, that extends above the top of the tower. It permits tower sections or the wind turbine itself to be lifted up the tower and set in place without the use of a crane. Gin poles need be nothing more than a long section of pipe strong enough to handle the expected loads and having some means of being attached to the tower. Some gin poles are a little more sophisticated and incorporate a horizontal arm that allows the load to be centered on the tower. The gin pole used to lift the modular tower sections of Rohn's lattice mast is simply a 12-foot (4 m) length of aluminum pipe with clamps and a pulley at the top. The clamps are designed for attaching the gin pole to one leg of the lattice mast. This same gin pole has been used numerous times to hoist small wind turbines, though its strength limits it to machines no larger than 3 meters (10 ft) in diameter.

Pulleys are used to direct the hoisting rope over the gin pole to the load. Never use a gin pole without first routing the hoisting rope through a pulley at the base of the tower. This pulley permits the hoisting crew to stand clear of the tower and be well away from any falling objects. It also prevents any unnecessary bending of the gin pole. With a base pulley in place, the hoisting tension acts directly on the gin pole from below. This minimizes the bending forces on the gin pole.

An improperly attached base or down-tower pulley caused the accident described earlier. The lack of such a pulley at the base of the tower killed Robert Skarski in front of his wife in 1993 when the tower he was on buckled due to the lateral loads on an improperly rigged gin pole.

Similar mishaps have occurred when the attachment of the gin pole to the tower has failed. It's paramount not only that the gin pole be firmly attached to the tower, but also that it not move laterally when the load is applied. The manufacturer's recommendations for the materials used in the gin pole, and for its attachment to the tower, should be followed scrupulously. Erecting a wind turbine can be dangerous. Always use extreme caution. When in doubt, consult the manufacturer.

With the gin pole now in place, the next tower section is brought up. The third section will usually have the lower guy bracket attached. Once the section has been bolted down the guy cables can be strung to the anchors. These cables are tensioned by hand until the three assembled tower sections are vertical and the tower is straight.

The gin pole is then released and moved to the top of the topmost section. A new section is hoisted up and bolted into place . . . and so on, until the tower is completed. After each set of guy cables is attached, the tower should be checked for plumb and twist. If the lower sections have been aligned properly it is possible to simply sight along the tower to check the alignment of the upper levels. A transit can be used to be certain. The tower must be vertical for proper yawing of the wind turbine.

Erecting a wind turbine can be dangerous. Always use extreme caution. When in doubt, consult the manufacturer.

Normally the guys on the Rohn towers can be tensioned by simply turning the turnbuckles. For household-size wind turbines it may be necessary to mechanically tension the guys with a coffing hoist or come-along attached to the guy cable with cable grips. The tension is then measured by a dynamometer in line with the hoist. Follow the manufacturer's directions for attaching the hoist and for the amount of tension required.

Always attach the guy cable directly to the anchor or to a turnbuckle that is attached to the anchor. The coffing hoist or come-along acts in parallel with the guy cable and is used only for tensioning the cable, not for connecting the cable to its anchor. Tensioning hoists and the grips that they use can, and have, failed—disastrously. Steen Aagaard was seriously injured when such a tensioning system inadvertently released and the tower he was on crashed to the ground.

After the guys are in place, the tower is aligned, and the guys are properly tensioned, you're ready to raise the wind turbine. Do so only on a calm day. Any wind at all will make it more difficult to position the turbine once it's atop the tower.

For small wind turbines it's best to do as much of the assembly as possible on the ground, since it's awkward to work at the top of the tower. Household-size turbines may require the use of a machine stand or cribbing. This permits attachment of the blades and other components.

Attach the hoisting cable and one or two tag lines. The wind turbine may have an eye bolt used for lifting, or it may require a special lifting jig. Whatever's used, it's important that the hoisting line lift at the wind turbine's center of gravity. If not, the wind turbine will be a lot more difficult to handle on its way up the tower, and once you're on top you'll have a heck of a time mounting it to the tower. The turbine may be easy to move around on the ground, at the end of a long lift line, but it's a lot more difficult at the top of the tower. Everything you do on the tower is more difficult, and more dangerous, than when you do it on the ground.

Use the tag lines to keep the machine from banging into the tower and tangling with the guy cables. The tag lines must be longer than the tower is high. Keep the pull on the tag lines moderate (just enough to prevent the machine from hitting the tower), particularly as the turbine nears the top of the tower. It's easy to buckle a gin pole if too much force is used on the tag line.

Once the turbine is mounted atop the tower, string the conduit and fish the power cable to the generator. Torque all tower fasteners to the specified value and apply locking nuts where required.

Freestanding Towers—Assembly and Erection

Technically truss towers can also be erected with a tower-mounted gin pole. But no one does it in practice. Only in rare cases such as remote sites in Alaska or other inaccessible areas are gin poles used to assemble truss towers. The sections are much heavier and more awkward to work with than those on a guyed tower. Most installers simply call in a crane.

Each individual member on a truss tower is so heavy that by the time the first section is bolted together you're not going to move it anywhere without heavy equipment. Ideally you'd like the crane to simply drive up to the tower, raise it, in one lift set the tower on its foundation, and leave. You don't want the crane to move sections of the tower around the site because you didn't thoroughly plan the assembly. This is particularly important if the wind turbine has been mounted on the tower. The more moving that's required, the greater the likelihood of damaging the turbine.

The tower can be raised first and the wind turbine mounted on top in a second lift. When using this method the boom of the crane must extend above the top of the tower to allow for the crane hook and the lifting jig to clear the top of the tower.

A crane with a shorter boom can do the same job if it raises the wind turbine and tower at one time (see figure 14-15, Freestanding tower installation). When lifting the turbine and tower together, the lift should be made some distance below the top of the tower yet well above the tower's center of mass. As the tower is raised, its weight keeps the bottom sections on the ground while the upper sections move toward the vertical. If the tower has been positioned correctly during assembly, the bottom will slide across the ground toward the foundation. Because the lift is being made below the top of the tower, the rotor blades are able to clear the boom as the tower nears the vertical.

With the skillful use of hand trucks, dollies, and come-alongs, even the heaviest towers can be fully assembled without power equipment. But it's wise to assemble the tower where it will make the crane operator's job as simple as possible. Begin by bolting the lowest (the heaviest) section together. Its placement will determine the location of the remaining tower sections, and it won't be moved again until the crane arrives.

Truss towers are like giant erector (meccano) sets, and will be puzzling until a few of the cross-girts are bolted into place. Lightly tighten the bolts on the cross-girts, but don't over-tighten. As you move along the tower section, you'll find that some of the pieces won't fit easily. There are always a few pieces that need some "convincing" before they fall into place.

Erection wrenches (also called spud wrenches) and drift pins are helpful in these situations (see figure 14-16, Tower assembly tools). An erection wrench has a long tapered shaft that's used by ironworkers to solve stubborn alignment problems between two pieces of metal. The shaft is inserted into the holes, and it's used to lever the pieces into position with the help of a little muscle. A bull pin is a similar device. It, too, has a long tapered shaft, but instead of a wrench on one end it has a striking face. The bull pin is dropped into the holes needing a little nudge, and then driven with a hammer until the holes are aligned. Both of these tools are well suited for aligning holes on the flange plates between each tower section.

Bolt all section together until the tower is fully assembled. The wind machine is then mounted on the tower and the conduit for the power cable installed. You'll find that strapping the conduit to the tower while it's on the ground and fishing through the conductors is much easier than trying to do it after the tower is erected. Once you have erected the tower, tighten the bolts to the desired torque.

An old trick from the aerospace industry is to install bolts on truss tower legs upside down. If the nut loosens, the bolt falls free,

Figure 14-15. Freestanding tower installation. Raising a Proven wind turbine at a workshop sponsored by the Midwest Renewable Energy Association. Note that the Proven wind turbine was not designed for this type of tower, which complicated the installation.

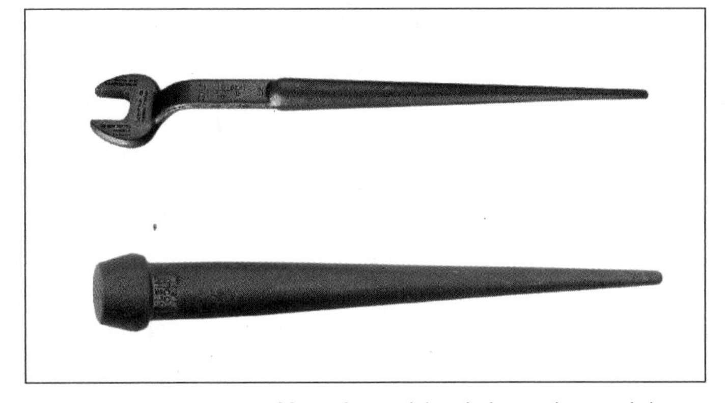

Figure 14-16. Tower assembly tools. Useful tools for working with lattice towers. Erection wrench (top) and bull pin (bottom). (Klein Tools)

Figure 14-17. Tapered tubular tower. This Bergey Excel was installed on a tapered tubular tower at a mobile telephone station at Pont-au-Père, Quebec. Such towers make a clean installation but require a crane for servicing the turbine.

notifying the user that the bolt needs to be replaced. While the towers may look massive, constant vibration from the turbine can loosen fasteners. Because of these vibrations, the nuts on all bolts must be prevented from loosening. Use self-locking nuts or add special locking nuts after the nut and bolt have been tightened. These are easiest to install while the tower is still on the ground. Don't overlook this step. Locking nuts are specified for a reason: Towers have failed without them.

The tower is now ready for the lift. Attach the lifting sling so that the stress is distributed onto tower members strong enough to take the load. Don't, for example, wrap the sling around a tower cross-girt. Instead use a tower leg; better yet, use two legs. Note that the sling must not slide along the leg as the tower is being lifted.

The crane will slowly set the tower down on the flanges or threaded bolts in the foundation. On foundations with flanges, align the holes with the drift pin and judicial use of a pry bar. Drop in the bolts once the holes are aligned. Before removing the sling, level the tower.

There are two ways to level the tower. On lighter towers, shims are forced in between the section flanges. The heavier towers use threaded bolts with adjusting nuts between the foundation and the lower tower section. The nuts are used to level the tower. When the tower is level, tighten the mounting bolts and install the locking nuts. The sling can now be safely removed and the crane sent on its way.

Tubular Towers

During the 1980s some small wind turbines were installed on freestanding tubular towers. These towers relied on a slip fit between sections. For these towers the first section is placed on the foundation and leveled with adjusting nuts. The next section is slipped onto the first, and so on. Gravity does the rest. The wiring run is then strung inside the tower and the wind machine installed. There are still a few small wind turbines installed on tapered tubular towers (see figure 14-17, Tapered tubular tower).

In contrast, nearly all medium-size wind turbines use large-diameter tubular towers. While a few medium-size turbines have been installed on lattice towers, some up to 130 meters (426 ft) tall, tubular towers are preferred for their clean lines, low maintenance, and ability to shelter service personnel during inclement weather.

For these towers the first section is raised upright, often with two cranes. One crane lifts the top of the section toward vertical; the second lifts the bottom of the section so that it just clears the ground. This prevents the bottom of the first section from scooting across the ground, damaging the flange upon which it will rest. Once the section is vertical, the crane positions it over the bolts protruding from the foundation. Additional sections are added in a similar manner, and the sections are bolted together.

When raised, the interior components of the tubular sections are nearly complete. The ladders, cable raceways, lighting, and in some systems even the fall protection rails are in place.

The nacelle, sans rotor, is then lifted onto the top of the tower, followed by the rotor. Typically the rotor is fully assembled on the ground and lifted as one piece, again using two cranes in the same manner as for raising the tower sections (see figure 14-18, Raising the rotor). The enormous blades on some multimegawatt turbines preclude raising the rotor fully assembled, and the blades are individually hoisted into position and bolted to the hub in a delicate aerial ballet.

Figure 14-18. Raising the rotor. Two cranes are used to lift the rotor and hub on this 2 MW turbine. The small crane keeps the lowest of the three blades off the ground. For larger turbines, the nacelle may be mounted on top of the tower, the drive train components and hub added, then each blade raised separately. (Bonus Energy)

Tilt-up Towers—Assembly and Erection

All towers for small wind turbines, whether guyed or freestanding, can be hinged and tipped upright into the vertical position, eliminating the need for a crane. Hinged towers simplify assembly because all tower and wind turbine components can be added while they are on the ground. They simplify service and repair for the same reason. Rather than climbing the tower and manhandling awkward components in the air, you can lower the tower and do the job on the ground. Avoiding the need for a crane also reduces the cost of installation and service. Additionally, tilt-up towers allow turbines to be lowered in cyclone-prone regions, such as the Caribbean, for protection during hurricanes. With a gin pole these towers can be raised by a heavy-duty industrial winch, tow vehicle, griphoist, or small crane (see figure 14-19, Tilt-up tower).

The hinges do add to the cost and complexity of the installation, and they also introduce a potentially weak link in the tower structure. Worse, many a hinged tower has been dropped by an installation or service crew unfamiliar with the loads involved. Even experienced crews have dropped turbines.

Installation of a Medium-Size Wind Turbine

These photos illustrate installation of a 750 kW turbine on a wind farm in the San Gorgonio Pass near Palm Springs, California. The turbine was one of many installed to replace hundreds of smaller machines that previously had been removed.

Foundation for a tubular tower to support a medium-size wind turbine. Note the positioning frame over the rebar cage to ensure alignment of the tower flange.

A detail of the threaded rebar used for tensioning of the rebar cage in this low-cost foundation design. Unlike traditional systems, the interior of the excavation is only partially filled with concrete.

Placing the first tower section. (Neal Emmerton)

Placing the top two tower sections in one lift. (Neal Emmerton)

Lifting the nacelle on this NEG-Micon turbine. (Neal Emmerton)

Assembling the rotor. Note the slotted holes on the fixed-pitch hub and the blade bolts protruding from the blade root.

Raising the rotor. A small crane prevents the lower blade from striking the ground. (Neal Emmerton)

Almost there. Windsmiths, not visible inside the nacelle, bolt the rotor to the mainshaft. (Neal Emmerton)

Ready for operation. This is one of nearly 70 similar turbines installed on this wind farm in 1998. (Neal Emmerton)

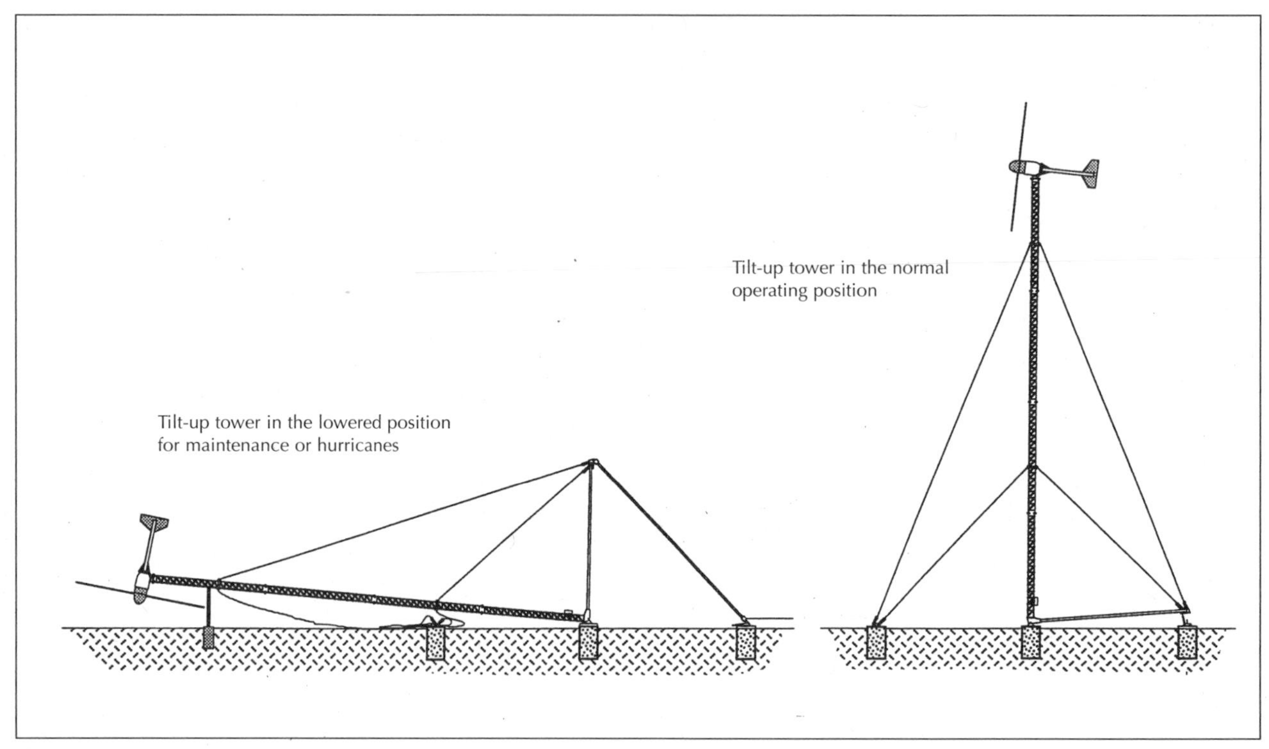

Figure 14-19. Tilt-up tower. Erecting a hinged tower with a gin pole. (Bergey Windpower)

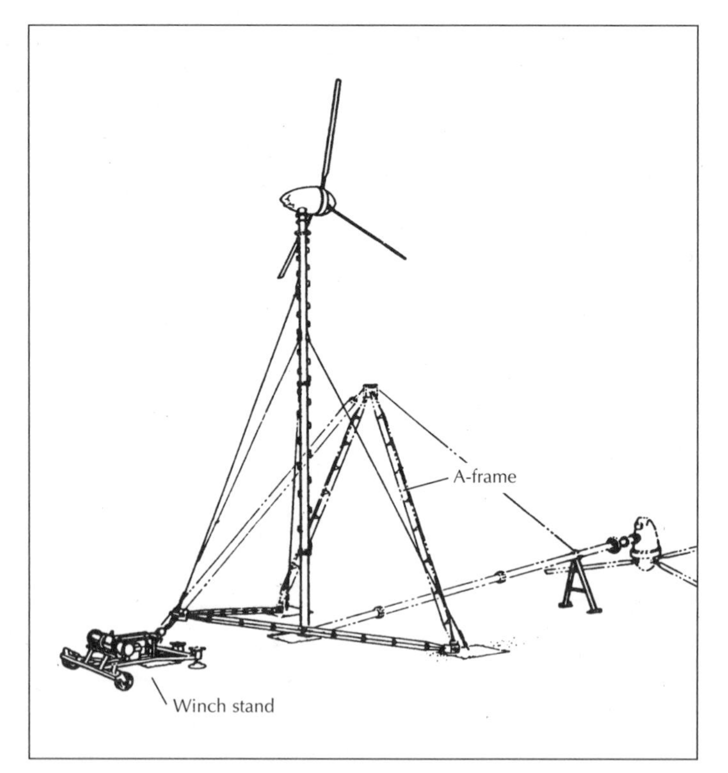

Figure 14-20. A-frame gin pole. An A-frame gin pole is used to stabilize a hinged tower that is guyed at only three anchors. Only a few of these Windworker 10 turbines were installed before the manufacturer ceased production.

Some have watched helplessly while their tow vehicle was dragged across the ground and a tower with an expensive wind turbine atop it headed inexorably to earth. Despite these limitations, hinged towers make sense when designed and operated properly.

The advantages of hinged towers dictate that they'll continue to be used for small wind turbines and even some larger machines. French manufacturer Vergnet uses guyed tilt-up towers for its entire product line. It offers one model, a 220 kW turbine with a rotor 26 meters (85 ft) in diameter, on a tilt-up mast 50 meters (164 ft) tall.

Vergnet's towers feature a guyed, tubular mast with accompanying gin pole. The tower is prevented from moving laterally during erection by two guys at right angles to the lift. The free guy cable is routed to the top of the gin pole. The tower is raised by drawing a block and tackle together between the gin pole and the free anchor or, in Vergnet's case, with a powered griphoist.

Towers with three guy cables can also be

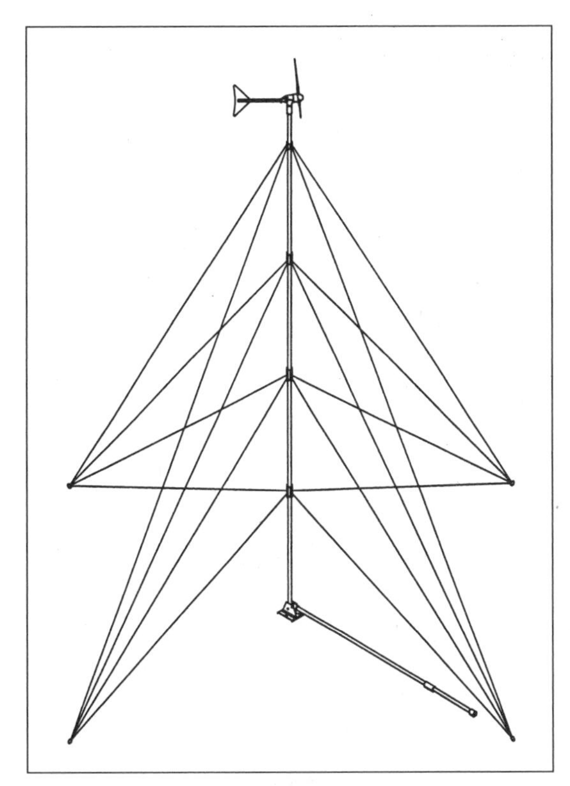

Figure 14-21. Mini wind turbine on a tilt-up tower. A Whisper H40 on a guyed tilt-up tower from Southwest Windpower's kit. The gin pole for raising the tower rests at a right angle to the tower. (Southwest Windpower)

Sample Tool List for a Micro Turbine

When studying the installation manual for a wind turbine, highlight the tool list. If there isn't one, ask for it. Knowing what you need in order to do the job right can save a lot of time—and aggravation—in the field. This is especially true at remote sites. You can lug around a complete toolbox, of course, but this isn't always practical. Southwest Windpower's installation manual for its AirX, for example, includes a nearly complete tool list for mounting the turbine on the tower, even though its turbine is shipped with many of the tools needed.

- 5/16-inch hex key (included)
- 3/16-inch hex key (included)
- 5/32-inch hex key (included)
- Torque wrench (spanner)
- Electrical tape or heat-shrink tubing
- Wire strippers
- Wire crimpers or electrician's tool

Substitute an electrician's (narrow blade) screwdriver for the crimping tool and use block connectors instead of the crimped connections suggested. Block connectors (see the sidebar Up-Tower Block Connectors for Micro Turbines) simplify removal of the turbine for repairs. SWP surprisingly also suggests solder, and a soldering iron or propane torch. Soldered connections are impractical and dangerous in the field, especially in fire-prone arid regions, though they do reduce voltage losses across the junction.

used, but there's no lateral restraint on the mast during the lift. This lateral motion can be prevented by using a two-piece gin pole in the shape of an A-frame. The A-frame can be built inexpensively from four sections of lattice mast, but the bases must be hinged to allow the gin pole to move with the tower (see figure 14-20, A-frame gin pole).

Truss towers have been raised in the same way. One manufacturer uses the A-frame gin pole; others use a single pole. Either the gin pole is attached to the tower's foundation with a hinge or it stands separately. When standing apart from the tower the gin pole must be guyed to prevent it from moving laterally out from under the load.

Freestanding tubular towers have also been erected with a gin pole mounted at the base of the hinged tower. In the case of both the truss tower and the tubular tower, the hoisting cable is passed over the gin pole to the top of the tower. Once upright the raising cable can either be removed or left in place for service calls.

In Europe several small wind turbines have been installed atop hinged, tubular towers with the use of a powerful hydraulic cylinder (see figure 7-6 Hinged, freestanding tubular tower).

Tilt-Up Guyed Towers

For micro and mini wind turbines, a tilt-up, guyed tower is both cheaper and easier to install than any other tower choice (see figure 14-21, Mini wind turbine on a tilt-up tower). While installing such a tower isn't a snap, it's

Using a Griphoist to Erect a Guyed Tilt-up Tower

The following sequence illustrates installation of a Southwest Windpower Air-model micro turbine on a 45-foot (13.7-meter) NRG mast at the Wulf Test Field in the Tehachapi Pass. Using Tractel's Super Pull-All model griphoist, only two people are necessary to safely raise and lower the turbine. Three elements make the NRG tower system easy to use at remote sites such as the Wulf Test Field: screw anchors that can be driven by hand, lightweight mast sections, and precut lengths and swaged fittings on the lifting cables.

Taking delivery of an Air-brand micro turbine and NRG tower kit.

Taking inventory. Make sure all the components needed are on site.

Assembling the mast's base plate. The base plate on the NRG guyed mast substitutes for a pier in a fixed-mast, guyed tower.

Carefully align anchor positions relative to the base plate. Precision at this step allows raising and lowering the tower with minimal adjustments.

Screw anchor. The holding power of screw anchors is a function of soil conditions, the diameter of the helix, and the depth that the helix is driven into the ground. The NRG tower system relies on screw anchors.

Driving the screw anchors by hand. This can be good exercise, especially in dense soil.

Assembling the NRG mast. Sections slip together.

Unreeling guy cable. The NRG tower and similar tower kits from other suppliers include all the guy cables and attachments that are needed.

Gin pole and rigging. Yellow ropes stay the gin pole while raising the tower. They are later removed. Lifting cables (with red tags) are clipped into the snap hook shown.

Raising the tower without the turbine. This is a necessary step to ensure that the guy cables are properly adjusted before raising the wind turbine and tower. Always attach the rear guy cable before raising the tower!

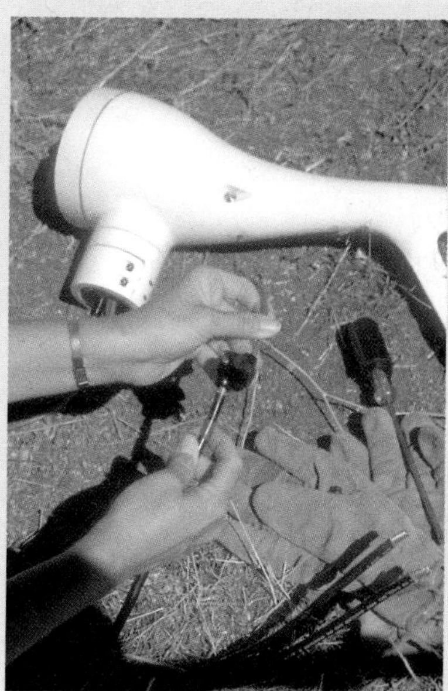

Making connections.

Strain relief. A wire net suspends the tower conductors from an eye bolt in the tower wall. Cable connections (terminations) are not rated for supporting the tower conductors.

Mounting the wind turbine.

Using a griphoist to raise the turbine and tower.

Note the tag line on the rear guy cable. The second person stands at a right angle to the tower's fall line, and well outside the fall zone.

The turbine is installed.

much simpler than erecting a freestanding lattice tower with a crane, for example. There's also the advantage of being able to raise and lower the tower whenever you want to service the turbine. And some turbines have to be serviced far more frequently than others.

Raising a guyed tilt-up tower for a micro or mini wind turbine can be safely accomplished with as few as two people, but is not risk-free. Consider the admonition never to walk under the tower when it's being raised. Wind experimenter Michael Klemen, who dropped one of his towers, explains why this precaution is important. Klemen calculates that an 80-foot (24 m) tower would hit the ground within two seconds after the raising cable was released. By the time you realize there's a problem, Klemen says, there's no time to get out of the way. For his part, Mick Sagrillo likens a falling guyed tower to a "Wisconsin cheese slicer."

Tilt-up tower kits, like those of NRG Systems, Southwest Windpower, or European suppliers, use a gin pole to raise the mast. The gin pole is effectively a second, shorter mast at a right angle to the tower. The gin pole acts as a lever to reduce the lifting loads.

NRG's towers consist of thin-walled tubing and are widely used in the wind industry for meteorological masts. They have packaged tower kits for Bergey's 850 and its replacement, the XL1. For the Bergey XL1, NRG specifies its 3.5-inch (89 mm) mast system for towers up to 42 feet (13 m) in height. These are shipped in 7-foot (2.1 m) lengths, allowing transport by parcel delivery services worldwide. Bundles of three sections in this size are easy to work with. Taller towers for the XL1 require NRG's 4.5-inch (114 mm) tube, sections of which are shipped in 10-foot (3 m) lengths in bundles of three pieces each. These require transport by motor freight, and the cartons are hefty. Though the cartons require some effort to move around, the individual sections are easily manageable.

In the NRG tower system, five anchors are needed: one each for the four guy cables, and the fifth for raising the tower. Because NRG's tower sections slip together and the gin pole comprises several sections, the fifth anchor must be directly below the gin pole when the tower is fully upright.

Griphoists

Everyone who has chosen a tilt-up tower to support a small wind turbine has had to face the difficult question of how to raise it. The most common technique among do-it-yourselfers in the United States is to raise tilt-up towers with a truck or tractor. Unless you're skilled with hand signals and with working around heavy equipment, however, raising a tower with a truck or farm tractor can be a ticklish and dangerous operation. Robert Skarski was killed when the truck driven by his brother pulled the tower over before anyone had time to react.

Properly using a vehicle for raising a tilt-up tower demands a large crew. Wisconsin's Mick Sagrillo recommends two on the truck (one to drive and one to watch for hand signals) and one for each anchor, or a minimum of six people. Gathering together such a large group always presents a challenge. When the inevitable glitches arise it puts you in the awkward position of either asking everyone to come back another day or forging ahead and taking chances you shouldn't. You can quickly wear out your credit with friends and family if the tower raising doesn't go as planned. You don't want a bunch of your friends standing around asking, "Hey, are we going to install this windmill or not?"

Communal tower raising can be a rewarding experience, like Amish barn raising, bringing people together for a common purpose. But barns last indefinitely. You put it up and it stays up. Not so with a wind turbine. Whether we like it or not, small wind turbines do need repairs and we have to bring them down before we can haul them off to the local repair shop. Some are up and

Figure 14-22. Griphoist. A Super Pull-All model griphoist with wire rope, snap hook and keeper, and lever operating handle is used to raise mini wind turbines on NRG tilt-up towers at the Wulf Test Field. The griphoist is also a handy tool around the farm or ranch.

down a lot. Gathering six people together every time you want to raise or lower your turbine quickly becomes tiresome.

Fortunately griphoists are an alternative common where cranes are either too expensive or too difficult to use. Hand-operated griphoists are used throughout Europe and Canada for a variety of applications that include raising wind turbines and meteorological masts (see figure 14-22, Griphoist).

Scoraig Wind Electric's Hugh Piggott uses a griphoist to install wind turbines in Scotland. A griphoist is "hard to beat for erecting tilt-up towers," says Piggott, "because it's slow and fail-safe." Unlike the driver of a truck or other vehicle being used to raise a tower, the operator of the winch has full control of the operation, and there's no dependence on hand signals or risk of missed cues.

To Piggott, this tool is a *tirfor*. Tractel, the world's largest manufacturer of griphoists, officially calls it a griphoist-*tirfor-greifzug* product. In English, the word *griphoist* says it all. But the tool was originally sold as a *tirfor*, which in French says much the same. *Greifzug* is the German equivalent.

The griphoist pulls a few inches of cable through the tool's body on each stroke of the operating lever, both on the back stroke and on the forward stroke. Because it's a simple mechanical device, you can actually feel the tension in the cable. This gives the operator a tactile sense of the load. The heavier the load, the more difficult it is to move the lever. The loads in tower raising are greatest when the tower is on the ground and least as the tower nears the vertical. Operating the griphoist takes the most effort when the tower first begins to leave the ground. For loads nearing the limit of a particular size, operating the griphoist will take some effort (see table 14-4, Griphoists).

Table 14-4
Griphoists (*Tirfor* or *Greifzug*) and Electric Hoists Suitable for Micro and Mini Wind Turbines

Type	Model	Maximum Capacity		Wire Rope Diameter		Weight	
		(kg)	(lbs)	(mm)	(inches)	(kg)	(lbs)
Griphoist	Pul-All	300*	700	4.72	3/16	1.8	4
Griphoist	Super Pull-All	500*	1,500	6.5	1/4	3.8	8
Griphoist	T-508	800*	2,000	8.3	5/16	6.6	17
Griphoist	T-516	1,600*	4,000	11.5	7/16	13.5	35
Griphoist	T-532	3,200*	8,000	16.3	5/8	24	58
NRG winch	X1**	907	2,000	3.18	1/8	12.3	27
NRG winch	S6**	2,722	6,000	6.35	1/4	58	128

Notes: *Derated for European market.
**Does not include cost or weight of battery, but does include pulley blocks and battery cables.

Figure 14-23. Raising a tilt-up tower with an electric winch. *Left:* The crew from NRG Systems raised a 4.5-inch (114 mm) mast with an electric winch. *Right:* The tower was then lowered, a Bergey 1.5 kW turbine was mounted on top, and the tower was raised again.

Another hoist option, the one used by some American meteorologists to install anemometer masts, is an electric winch (see figure 14-23, Raising a tilt-up tower with an electric winch). They typically power the winch from the battery of a truck or haul in special-purpose batteries. But Zephyr North's Jim Salmon prefers a griphoist to raise NRG towers in Canada. "[Griphoists] are easier to control" than either winches or vehicles, he says, "and in some cases much safer than [electric] winches."

Unlike electric winches, griphoists are readily portable. You can lug a griphoist into areas where you would never consider hauling an electric winch and battery, or even driving a truck, for that matter. Endless Energy's Harley Lee swears by the heavy-duty griphoist he used to raise a 130-foot (40 m) anemometer mast on a rugged Maine mountaintop. Lee says the griphoist and winch cable, though heavy, were easier to carry up the mountain than the batteries, electric winch, and backup generator that would have otherwise been necessary.

Griphoists are not come-alongs; the latter is a lightweight tool found in North American hardware stores that uses a spool for coiling a short length of wire rope. Ranchers, for example, use come-alongs to tighten fencing. And for that purpose they don't need a long cable.

It's the spool or drum that sets come-alongs, as well as winches in general, apart from griphoists. Technically griphoists are not winches but hoists. Winches use a drum to spool the hoisting cable, like the large drum on a crane. Griphoists, in contrast, pull the hoisting cable directly through the body of the hoist. Tractel likens the locking cams inside the griphoist to the way we take in a rope, "hand over hand." To use a griphoist

Up-Tower Block Connectors for Micro Turbines

One challenge facing installers of mini and micro turbines is making good terminations or connections between the conductors on the wind turbine and the conductors running down the tower. Often the space available is cramped, making good connections problematic. Some use wire nuts. These are less than ideal because they can work loose from the wind turbine's vibration. "We never use 'wire nuts' over here," says Hugh Piggott in Scotland.

Some manufacturers suggest split-bolts. These are cumbersome to use, though, inside the 1.5-inch pipe for the AirX. Bergey Windpower recommends NSI-brand insulated connectors for up-tower terminations. While costly, these connectors are rated for both copper and aluminum and include an antioxidant compound and a plug for sealing the set screw, which clamps the conductor in place. Best of all, these connectors are easy to use in the field, requiring only a screwdriver. A less expensive option are "blocs" used for electrical connections in France. (These are known in Anglophone countries as blocks.) Like the NSI-brand connectors, blocs are available in a range of sizes, are insulated, and use two set screws to secure the conductors. Blocs come in strips; you simply break off the number of connectors you need. Blocs are not certified for use in the United States or Canada, but should be suitable for the low-voltage wiring used in micro and mini wind turbines, when used within the limitations of the connector: The set screws must be tightened securely, checked periodically, and protected from the elements. None of these connectors is designed for supporting the weight of the conductors dangling down the tower. All must be used in conjuction with strain relief.

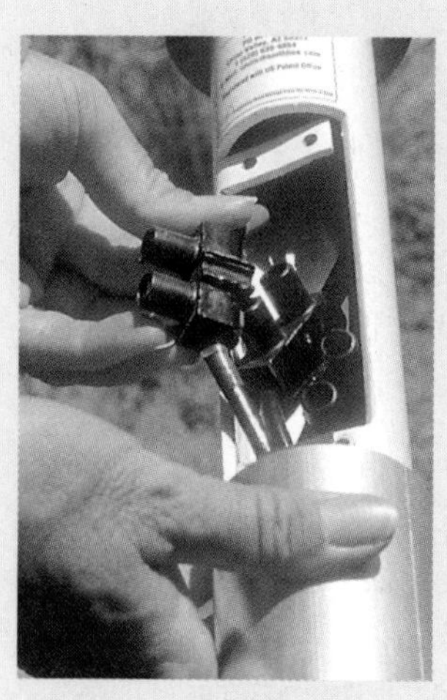

Up-tower connectors. The block connectors used with a Creative Designs up-tower junction box simplify installation of this micro turbine.

you move a lever forward and back. This lever pulls the cable through the tool.

The hoisting cable for a griphoist can be any length, since there is no need to spool the cable on a drum. (Capstan winches can also use cables of any length, but they pass the cable over a drum.) When specifying the size of griphoist needed, don't overlook the amount of cable required. The standard lengths shipped with most griphoists are insufficient for raising tilt-up towers. Simply order the correct length required from the griphoist manufacturer. Don't try to skimp by using guy cable in the griphoist.

There are several brands of griphoists on the market. They come in a range of capacities suitable for most small wind turbine applications. The manufacturer of either the wind turbine or the tower will specify the capacity (in pounds or kilograms) needed to raise a tilt-up tower of a given height, and the amount of "run" or cable necessary.

As with any winch, the griphoist and the hoisting cable should have safety keepers or gates on all lifting hooks. Gates ensure that the hooks stay coupled to the load when there's unintended slack in the cable. They're absolutely essential.

Since little has been written about griphoists, it's surprising the number of people who have used or are now using them. Bergey Windpower, for example, has been using griphoists for remote installations since 1993. It recommends griphoists to its overseas clients, says Pieter Huebner, Bergey's field technician. Vergnet even offers griphoists as an accessory to its tower kits. If you have to buy any one tool for your off-the-grid wind system, says Hugh Piggott, it should be a griphoist.

Wiring

The bible on wiring in the United States is the National Electrical Code (NEC). It's the rulebook for what can and can't be done, but the final say in electrical matters is in the hands of your local electrical inspector, fire underwriter, or code enforcement officer.

Licensed electricians in your area will be familiar with the local application of the codes. They won't be familiar, however, with wind generators and their specific requirements, especially as they relate to interconnection with the utility. This section will be helpful to your electrician and essential if you plan to wire the wind system yourself. In either case, when unsure of your next step, check with your local electrical inspector before you begin work.

Why is the approval of the electrical inspector or fire underwriter so important? Because your fire insurance may be void without it. And in some communities you can't sell your home without electrical wiring that's "up to code."

The electric utility is responsible for all wiring from the nearest transformer to the service drop (where the conductors are secured to your house above the kilowatt-hour meter). Your responsibility begins at this point.

The conductors from a wind system are much like the power lines from a small generating station. The lines can be strung aboveground or buried. They enter your home or business in a manner similar to those of the utility. Let's look first at wiring on the tower; then we'll consider how we're going to deliver the power to your farm, home, or business.

In all cases, the leads from the generator must be connected (terminated or spliced, in the jargon of electricians) to the wires (conductors) running down the tower. These connections must be permanent and weatherproof. How this is accomplished depends on the kind of wind turbine and on the manufacturer.

When the generator is stationary within the tower, as in a vertical-axis wind turbine or in some horizontal-axis machines with right-angle drives, the generator leads can be directly connected with the conductors on the tower. On more conventional wind turbines, though, there must be a mechanism for transferring power from the moving platform of the wind machine to the stationary tower. Slip rings and brushes usually perform this task on small wind turbines.

On larger household-size wind turbines and medium-size commercial wind turbines, however, the power cable from the generator hangs freely through the center of the tower. As the machine yaws, or turns to face the wind, the pendant conductors twist. The cable is permitted to twist several revolutions before it must be unwound. Contemporary medium-size and larger turbines monitor how much the nacelle yaws in any direction, and when a set number of twists in the pendant cable are counted, the rotor is braked to a halt. The nacelle then yaws to unwind the cable. Once completed the rotor is released and the turbine returns to service.

Good terminations between conductors are those that are mechanically tight, electrically insulated, and corrosion resistant. Terminations may be made in a number of ways: with split-bolts, wire nuts, crimp (compression) connectors, or insulated termination blocks. Each has its merits.

Split-bolts are easy to use and come in sizes suitable for even the heaviest power cable. The stripped ends of the conductors are inserted into the jaws of the bolt and the faces tightened with wrenches (spanners). These connections need an insulated boot and may loosen under a wind turbine's constant vibration.

Wire nuts are popular connectors for home wiring because of their ready availability and ease of use. Nearly every home handyman is familiar with them. They are not well suited to wind turbine applications. They, too, may loosen under vibration and need to be secured with electrical tape. Moreover, they're most practical on solid, rather than stranded, wire and only on the smaller wire sizes. Wire nuts for heavy-gauge conductors are hard to find and are more difficult to use.

Crimp or compression connectors are the

best all-around termination. They are mechanically sound and will not loosen as a result of vibration. Some include an insulated covering, but all can be insulated with a boot of heat-shrink tubing slipped over the connector. Their chief disadvantage is the need for a special crimping tool; also, in the larger wire sizes this tool is both expensive and awkward to use.

The connection between the leads from the slip rings and tower conductors should be made inside a junction or J-box. The conductors are then run down the tower inside metal or plastic conduit.

In the past, some installers have taken unapproved shortcuts, using twist-lock plugs at the top of the tower and then lacing the power cable down a tower leg. They would attach the cable to the tower with electrical tape or with nylon cable ties. This technique doesn't meet building or electrical codes in many communities, and presents a potential electrical hazard, to say nothing of creating future maintenance problems.

The turbine's power conductors should be protected from physical damage. Conductors on the exterior of the tower should be protected within conduit. Either electrical metallic tubing (EMT) or plastic tubing can be used. Metal conduit is preferable because it's strong and provides some shielding against voltage transients and lightning. Each 10-foot (3 m) section of EMT is joined by weatherproof compression couplers and mounted on the tower with conduit hangers. The hangers should be spaced two per section, with one at each end near the coupler. If you're using a PVC (polyvinyl chloride) conduit, make sure it's rated for electrical use. Because PVC flexes, use three or four conduit hangers per section. Feeding the conductors down the inside of tubular towers and guyed tubular masts provides the necessary protection.

Aboveground and Buried Cable

Almost no one runs cable aboveground today. In Denmark the electric utilities bury all distribution lines. Only high-voltage lines remain aboveground. It's just simpler to excavate a trench and lay direct-burial cable, and that's how it's done on wind farms around the world. For small wind turbines, though, it is better to lay PVC conduit in the trench and pull or "fish" the conductors through the conduit. This allows replacement of the conductors, should that ever be necessary. There's also less likelihood that the conductors will be damaged as the trench is backfilled.

In many cases you'll want to install a disconnect switch in the line between the wind turbine and the wind system's control panel. The disconnect switch is necessary to permit isolation of the wind system from the load center or control panel during emergencies or when maintenance of the wind turbine is required.

While a disconnect switch may seem like a redundant safety device on interconnected wind systems (because the wind turbine automatically disconnects from the utility line during power outages), it's not. The switch gives linemen or windsmiths a positive mechanism for protecting themselves. When you throw the switch off and insert a "lockout" (a metal tag warning that someone is working on the line) through the switch handle, service personnel know with certainty at which point the circuit is broken.

From the utility's viewpoint, a lockable disconnect switch for a wind turbine should be located near its service entrance or kilowatt-hour meter. This is where fire crews will want to see the disconnect switch, as well. For practical purposes, a second disconnect switch should be located at the base of the wind turbine tower. This is useful for isolating the wind turbine when servicing the machine.

The wind generator's output is then wired to a control panel or inverter, depending on the type of wind system used. In no case should DC output ever be connected directly with the AC service panel. Generators producing DC must incorporate a synchronous inverter before they can be interconnected with utility-supplied AC. Those wind systems producing "wild" AC—that is, AC of varying voltage and varying frequency—must also use an inverter.

For household-size wind turbines intended for interconnection with the utility, conductors from the control panel or inverter are wired to a dedicated circuit in the building's existing service panel. Output from the wind system must not be plugged into a wall outlet or wired to a circuit that's already supplying a load in the building, such as a refrigerator or a series of receptacles. Instead a circuit breaker must be "dedicated" to the wind system.

Strain Relief of Tower Conductors

Always support the power cables on the tower with some form of strain relief to carry the weight of the cables. Disregard the suggestion by some manufacturers that the leads from their micro turbines are strong enough to resist being yanked out of the generator. It's always good practice, and often required by electrical codes, to use strain relief on the tower conductors. Otherwise the leads will eventually pull out of the generator, causing big headaches later.

On guyed tubular towers, thread the power cables down the inside of the mast. Support the weight of the conductors with a strain relief wire net. The net works like a Chinese finger puzzle to grip the cable bundle. Hang the strain relief from an attachment point at the top of the tower. The strain relief can make or break a wind turbine installation.

For the connections inside the tower you can use compression connectors for making vibrationproof splices between the wind turbine leads and the cables carrying power down the tower. These connectors are often used for connecting the power supply to submersible well pumps, and can be found in farm supply and plumbing stores. Pros such as Mick Sagrillo and Hugh Piggott warn against using split-bolts, a common alternative, as they can work loose from vibrations in the tower. Compression terminals can obviate the need for heavy-duty and expensive crimping tools when you're working with heavy-gauge cable. Barrel connectors are another handy choice for terminating heavy, stiff conductors to terminal blocks or circuit breakers.

When you buy the power cable for inside the hinged tubular towers of Westwind Turbines, the Australian company includes end terminations and the suspension strap necessary to support the power cable. That would make even a hard-nosed buyer like Mick Sagrillo happy. Sagrillo argues that manufacturers should ship termination kits and strain reliefs with their turbines to ensure that the job gets done right the first time. France's Vergnet, for example, offers components with its turbines and towers that most other manufacturers leave out. Vergnet includes 35 meters (115 ft) of power cable with all its turbines. It also includes junction boxes, grounding, and lightning protection with its tower kits.

Conductors and Conductor Sizing

The electricity produced by a wind generator is seldom of benefit at the top of the tower where the turbine is located. The electricity must instead be transmitted to the point of use before it provides any benefit. Thus the cables connecting the wind turbine to the load are as integral to a wind power system as the tower supporting the turbine.

No practical conductor transmits all the electricity that is passed through it. Some is lost due to resistance. The length, diameter, and material of the cables connecting the wind turbine to the load—whether a service panel in your cellar or batteries in the barn—determine the amount of power and energy transmitted and, conversely, the amount lost.

These losses are proportional to the type of material (copper has less resistance than aluminum), diameter (thick cable has less resistance than thin cable), and distance to the batteries (short cables have less resistance than long ones). These resistive losses are reflected

American Wire Gauge to Metric Conversion

In the United States manufacturers of small wind turbines specify conductor size in American Wire Gauge (AWG). The rest of the world uses cross-sectional area in square millimeters (mm^2). Converting from AWG to mm^2 is, unfortunately, not straightforward. U.S. manufacturers sometimes offer the equivalent metric diameter. This is not cross-sectional area. They leave it up to the user to make the conversion. If in doubt use a metric size that's larger than the AWG equivalent and check whether the rated ampacity for the cable in the application you intend meets local codes. Note that the resistance (ohms per measure of distance) is rated at greatly different temperatures. The rated resistance for AWG sizes is from the National Electrical Code in the United States. Increasing the temperature increases resistance.

American Wire Gauge to Metric Cable Conversion and DC Resistance in Ohms

AWG	Strands	Area (cmils)	(in^2)	Resistance Copper @ 75° C (Ohms/1000 ft)	Area (mm^2)	Nearest Metric Size* (mm^2)	Resistance Copper @ 20° C (Ohms/km)
10	7	10,380	0.0082	1.2900	5.3	6	3.080
8	7	16,510	0.0130	0.8090	8.4	10	1.830
6	7	26,240	0.0206	0.5100	13.3	16	1.150
4	7	41,740	0.0328	0.3210	21.1	25	0.727
3	7	52,620	0.0413	0.2540	26.7	35	0.524
2	7	66,360	0.0521	0.2010	33.6	35	0.524
1	19	83,690	0.0657	0.1600	42.4	50	0.387
0	19	105,600	0.0829	0.1270	53.5	70	0.268
2/0	19	133,100	0.1045	0.1010	67.4	70	0.268
3/0	19	167,800	0.1318	0.0797	85.0	95	0.193
4/0	19	211,600	0.1662	0.0626	107.2	120	0.153
250	19	250,000	0.1963	0.0535	126.7	150	0.124

Note: 1 inch = 25.4 mm; direction of conversion is from AWG to metric sizes; reversing the direction could result in undersize conductors.
Source: AWG, NEC 1999 Edition; metric, Folkecenter for Renewable Energy, Denmark

in the voltage drop between the wind turbine and the load.

Wind turbine manufacturers specify the cable size and material for a range of wind-turbine-to-load distances that will allow their products to perform as designed. For example, a BWC 850 was installed 150 feet (45 m) from its load at the Wulf Test Field in the Tehachapi Pass. Bergey Windpower recommended using #8 AWG copper wire for the 24 V battery-charging system. These specifications assume that a portion of the electricity produced by the wind turbine will be lost in the conductors.

Operators of commercial wind power plants fully understand this phenomenon and account for it. Most consumers do not, and are unpleasantly surprised when their shiny new wind turbine doesn't produce quite as much as advertised in the glossy brochures. While there are many reasons why small wind turbines sometimes underperform, one is undersizing the conductors from the generator to the load (see figure 14-24, Power lost in conductors).

Resistance in the conductors is seen as a voltage drop between the wind turbine on the tower and the load, for example, at the batteries. The voltage drop not only affects the amount of power transmitted, but, if it's

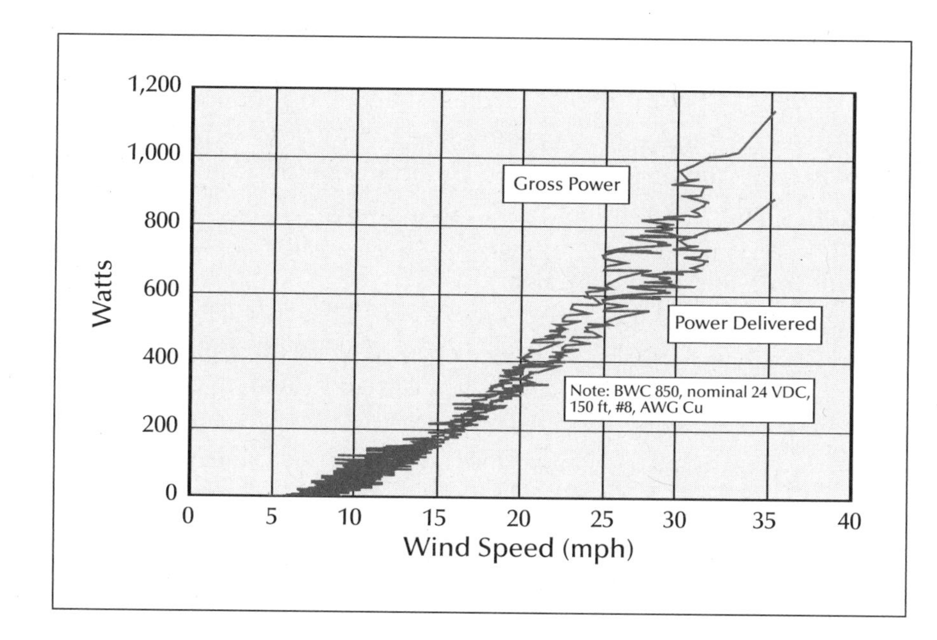

Figure 14-24. Power lost in conductors. Measured power delivered to the batteries and an estimate of the gross power generated by a Bergey 850 at the Wulf Test Field as a function of wind speed. The difference is the amount of power lost in the copper conductors between the wind turbine and the batteries. At the rated wind speed and above, 15 to 20 percent of the power produced by the wind turbine is lost in the conductors specified by the manufacturer.

severe enough, can also affect the operation of electrical equipment.

On interconnected wind systems, the voltage drop between the wind turbine and the load is critical to proper operation. Take the case of a vocational-technical school that installed a 6-meter (20 ft) diameter wind turbine driving an induction generator. The school was responsible for wiring the system into its service panel. Students and their instructor mapped out the conduit run and laid the conductors. When all was finished the installer flipped the switch—and nothing happened. He checked the wind turbine. Everything seemed fine. He then checked the students' wiring; nothing wrong there, either. As they sat and scratched their heads, some wise guy suggested they measure the voltage.

The problem? Low voltage. The wind turbine's control system sensed a voltage below its disconnect value and wouldn't energize the generator. (Manufacturers incorporate this feature to detect a power outage so a wind generator cannot energize a downed line and potentially kill a lineman.)

The dealer and the students then went back to the books. The wiring was sized according to the installation manual, or so it seemed. Their mistake was failing to take into account that the panel where the wind system was interconnected with the school's service was a long way from the utility's entrance to the school. The wind turbine's conductors were sized properly for the run from the service panel in the outlying classroom. But the long distance from the utility's service entrance to the wind turbine's control panel caused an excessive voltage drop. They remedied the situation by installing heavier-gauge conductors.

Acceptable voltage drops range from 1 percent for interconnected wind systems using power electronics to 3 percent for those with induction generators.

Using Ohm's law we can calculate the voltage drop in a circuit and, subsequently, estimate the amount of power lost due to resistance:

$$I = V/R$$

where I is current in amps (for *intensité,* as Ampère referred to it), V is voltage in volts, and R is resistance in ohms (Ω). Solving for volts:

$$V = RI$$

The voltage drop increases in direct proportion to resistance and current. An increase in current increases the voltage drop through the conductor. If we wanted to use a 1.2-kilowatt wind generator in a 24-volt battery-charging system, we'd need to pump 50 amps through the conductors. If we sized the conductors so they would handle 5 amps at 24 volts with only a 1 percent voltage drop (0.24 volt lost), we'd encounter a 10 percent voltage drop (2.4 volts) at 50 amps.

If we increase the voltage of the 1.2-kilowatt system, say from 24 to 48 volts, the conductors will carry only half the current at rated power as before: 25 amps. Consequently the voltage drop at rated power would be only 5 percent.

Now consider the amount of power that would be lost in both systems. Where P is power:

$$P = VI$$

Substituting the value for volts from Ohm's law:

$$P = (RI)I$$
$$P = RI^2$$

While doubling system voltage cuts the current in half, it reduces the power lost in the conductors by a factor of four.

For this reason system voltage increases with the size of the wind turbine. Micro turbines operate at 12 to 24 volts, mini wind turbines at 24 to 48 volts, household-size turbines at 48 to 240 volts, medium-size wind turbines at 400 to 600 volts, and some megawatt-scale turbines in the thousands of volts.

The total resistance, R_T, in a circuit is:

$$R_T = (\Omega/\text{unit of length}) \times 2L$$

Where L is the wire run or distance from the wind turbine to the load, and 2 is the number of conductors for a DC circuit. For a three-phase circuit:

$$R_T = (\Omega/\text{unit of length}) \times 3L$$

Many micro and mini wind turbines rectify AC at the alternator and transmit DC to the load; thus calculating the voltage drop is straightforward. Other small wind turbines, such as the BWC 850 mentioned previously, transmit three-phase AC to a rectifier near the batteries. Unfortunately calculating resistance or power loss due to current flow in three-phase AC-rectified circuits is somewhat less straightforward.

Scoraig Wind Electric's Hugh Piggott explains that in rectified three-phase circuits, the AC phase current is about 0.82 of the current on the DC side of the rectifier for low-impedance alternators. For high-impedance alternators, the AC phase current is nearer 0.74 of the DC current.

Measurements on the BWC 850 at the Wulf Test Field confirmed this. At high wind speeds and high current, the AC phase current was 0.72 times the DC current. At low wind speeds and low current flows, AC phase current was more than 0.8 DC current.

For the three-phase BWC 850 at 300 feet in the example, the voltage at the load at rated power is:

$$V = [(P/V) \times (0.72/\text{phase})] \times R_T$$
$$V = [(850\text{ W}/24\text{V})\ (0.72/\text{phase})]$$
$$\times [(3\text{ phases}) \times (0.0008090\ \Omega/\text{ft}) \times (150\text{ ft})]$$
$$V = 9$$

Or a drop of more than 20 percent, for the nominal 24-volt system.

The power lost at rated power is:

$$P = I^2R\text{, and}$$
$$P = [(850\text{ W}/24\text{V}) \times (0.72/\text{phase})]^2$$
$$\times [(3\text{ phases}) \times (0.0008090\ \Omega/\text{ft}) \times (150\text{ ft})]$$
$$P \sim 240\text{ W}$$

If the turbine is designed to deliver 850 W to the batteries at rated power then it must generate:

$$P_T = 240\text{ W} + 850\text{ W}$$
$$\sim 1{,}090\text{ W}$$

Consequently at rated power this BWC 850 would lose about 22 percent of the power produced (240 W/1,090 W).

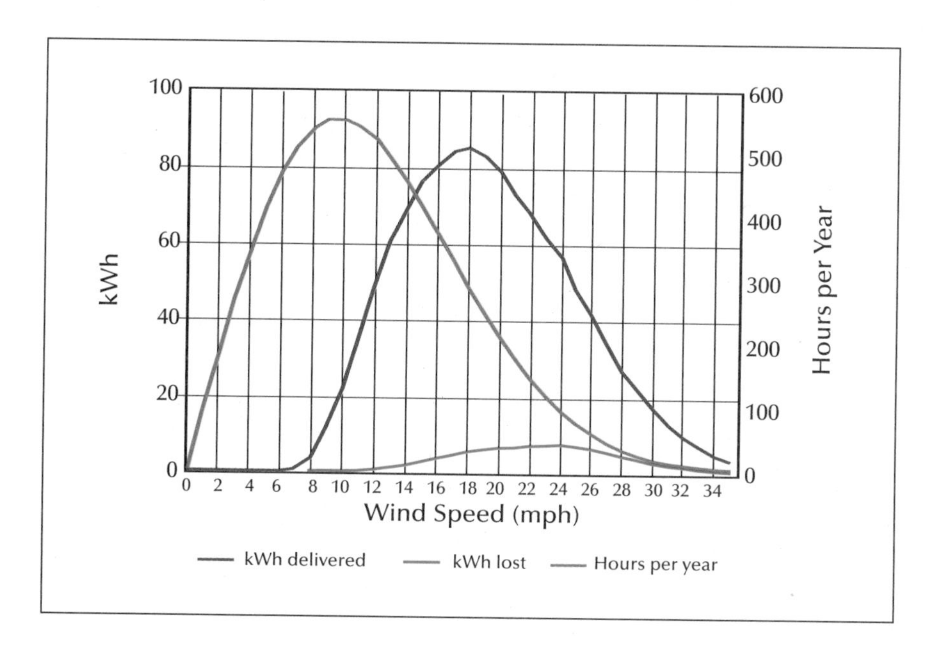

Figure 14-25. Energy lost in conductors. Calculated energy delivered to the batteries and the energy lost in the conductors for a 12 mph (5.4 m/s) annual average wind speed with a Rayleigh distribution. The calculation is based on actual power measurements of a Bergey 850 at the Wulf Test Field. Though as much as 20 percent of the power is lost at higher winds, only 10 percent of the annual energy generation is lost to resistance because the wind occurs at these higher speeds only a few hours per year.

In the real world, the conductors don't operate at 75°C. Typically they are buried in the ground, and their starting temperature would be much lower, closer to 20°C. Temperature plays an important role in resistance losses in the conductors. Resistance can be adjusted for temperature of copper conductors where R_1 is the initial resistance and R_2 is the resistance at the new temperature:

$R_1 = 0.8090\ \Omega/1{,}000$ ft

T_2 = Ground temperature, 20°C

$R_2 = R_1[1 + \alpha(T_2 - 75)]$,

where α for copper = 0.00323

$R_2 = 0.8090 \times [1 + 0.00323\ (20 - 75)]$

$R_2 = 0.8090 \times 0.822$

$R_2 = 0.665\ \Omega/1{,}000$ ft

Substituting the resistance at ground temperature into the earlier equation for power loss,

$P \sim 200$ W

Total power at the turbine then is:

$P_T = 850$ W (delivered) + 200 W (lost)

$\sim 1{,}050$ W

Power lost in the conductors at 20°C is:

$P_{Lost} = 200\ \text{W}/1{,}050\ \text{W}$

$\sim 19\%$

Consequently possibly as much as 20 percent of the total power produced by the wind turbine could be lost in the conductors at the rated wind speed.

Fortunately this isn't the whole story. At many sites wind turbines operate very little of the time near their rated power (see figure 14-25, Energy lost in conductors). This diminishes the effect of the power lost in the conductors on the total energy delivered by the wind turbine to the load. On any typical wind site, says Gordon Proven of Proven Engineering in Scotland, most of the energy is generated in winds that produce about half of the wind turbine's rated power. Consequently the resistive losses are less significant than at higher wind speeds. For the BWC 850 in this example at a 12 mph (5.5 m/s) site, less than 10 percent of the energy produced is lost in the conductors.

While a 10 percent loss in a battery-charging system, such as the BWC 850, isn't critical, losses of this magnitude can make or break commercial wind power plants. For this reason conductor sizing is an important element of project design.

Conduit Fill

The size of the wire you use will determine the size of conduit needed for protected cable runs, such as on the outside of a tower or in a trench. For practical reasons there's a limit to

Junction or J-Boxes

One challenge facing do-it-yourselfers wiring a small wind turbine is how to work with the heavy-gauge conductors needed on long runs from the turbine to the load. Disconnect switches, for example, are rated by their current-carrying capacity and not by the size of conductor they can accept. If you use a switch of the correct current rating, it will be too small for heavy-gauge conductors. On the other hand, if you buy a switch large enough it will be much more costly than necessary. You can solve the problem by using junction boxes at each end of a long cable run. The heavy-gauge wire is joined with a wire sized appropriately for the control panel or disconnect switch. It's important to make good splices when using junction boxes, using insulated split-bolts or appropriately rated block connectors. This prevents shorts or grounds and keeps the overall resistance of the wiring run to a minimum.

Junction box. The junction box (bottom) allows connection of the heavy-gauge conductors used for long wire runs to the terminals in the disconnect switch (top).

the amount of cable that can be stuffed into conduit. Electrical codes also limit the size and number of conductors used inside conduit of a given size.

Electricians generally prefer a conduit larger than the size necessary for meeting code requirements. The larger diameter eases the task of pulling the conductors through the conduit. Bergey Windpower, for example, recommends conduit one to three sizes larger than that required by electrical codes to simplify long wire runs. When you're pulling heavy, stiff conductors a long distance through conduit, you'll quickly appreciate why. It may cost a few cents more but it makes your life a lot easier.

For short runs you can simply push the cables through the conduit. On longer runs, such as through a buried trench from the tower to your service panel, you can push a metal fish tape through the conduit first. The fish tape has a woven basket or cable grip to which the conductors are attached. The tape is then used to pull the wires through the conduit. Another technique is to feed a nylon rope through each section or "stick" of conduit as you assemble the sections. You then pull the conductors through the assembled conduit with the rope.

Surge Protection

Wind turbines, like most electrical appliances, are susceptible to damaging voltage spikes. Electrical systems are grounded to limit voltage surges from nearby lightning strikes. Grounding also ensures the prompt operation of fuses, circuit breakers, and protective devices from other electrical faults.

Lightning is only the most obvious of several sources of voltage spikes; passing clouds and the rapid opening and closing of switches on the utility's distribution system are others. Many wind turbine owners have found that voltage surges caused by the utility's lines are more frequent than those caused by lightning.

Lightning can short-circuit a wind generator in less than a second. Though it occurs almost instantaneously, lightning can be of sufficient voltage to break down thick insulation and arc over insulators. We try to minimize the effects of lightning by using lightning arresters and ground rods.

Lightning or surge arresters furnish a path to the ground when a greater-than-normal voltage exists in a conductor. This drains off the excess voltage. After the voltage has returned to normal, the flow to the ground ceases. Though there's no foolproof protection against lightning, arresters do provide some degree of protection to power lines and other electrical components.

Equipment and buildings can also be protected by raising the effective ground level with a static line or lightning rod. The static line on the utility's distribution system and the lightning rods on farm buildings are attempts to do just that. Lightning rods, such as those at utility substations, offer a 45-degree cone of protection beneath them. When lightning strikes a lightning rod, the lightning passes directly to the ground without first going through the object shielded below. For this reason some medium-size and larger wind turbines sport lightning rods above the nacelle.

Medium-size turbines incorporate lightning protection in the blades as well as the nacelle. This may not prevent all damage during a direct strike, but it does minimize damage. Manufacturers estimate that megawatt-scale turbines in Europe will be struck at least once per year.

Most systems rely on thorough grounding to drain off any static charge. Reducing the buildup of static electricity during a storm minimizes the possibility of a direct strike. Lightning doesn't always strike the tallest object. There are many documented cases in which telecommunications towers, because they are thoroughly grounded, have been spared a direct hit, while nearby trees have been incinerated. Proper grounding lessens the possibility of a direct strike and minimizes the damage if one does occur.

To ground the tower drive several copper-clad ground rods deep into the soil. When bedrock is near the surface the ground rods can either be driven in at an angle or buried in a trench.

On freestanding towers use two or more ground rods. On guyed towers drive a ground rod near the mast and near each concrete anchor. Connect the ground rods to the tower or guy cables with heavy-gauge copper wire with brass or bronze clamps.

In areas of high lightning incidence or where the soil is dry and sandy, Jim Sencenbaugh recommends installing a ground net. On guyed towers the mast and each anchor are wired to their own ground rod. Then all the ground rods are tied together with a buried ground wire. On freestanding towers there should be a ground rod for each leg of the tower. Sencenbaugh also electrically bonded together each tower section with a jumper wire on his installations. This provided a continuous path down the tower to ground.

Ground Nets

The passage on the grounding technique used by Jim Sencenbaugh was originally written in 1982. Arrogance—and ignorance—knows no limits. In the mid-1990s, Zond Systems, the sole U.S. manufacturer of medium-size wind turbines, installed a small wind farm in a lightning-prone area of Texas. The project, partially subsidized by the U.S. Department of Energy, was plagued by lightning-induced outages. After extensive studies on the cause of the problem, the company installed a ground net much like the one Jim Sencenbaugh had used on his wind turbines in the late 1970s.

Additional Notes on Wiring

Electrical codes in the United States require that all metal electrical enclosures, disconnect switches, control panels, and inverters be grounded so that any fault (short circuit) between a live or hot conductor and the enclosure will be conducted safely to ground. Grounding causes the circuit's protective devices to function—fuses to blow, circuit breakers to trip. This prevents the metal enclosure from becoming energized and presenting a shock hazard.

Control boxes and switches must be properly mounted, and all holes or cut-outs not in use must be sealed. There must also be sufficient clearance in front of any panel or junction box for safely servicing the circuit. Control panels and inverters with ventilation louvers must be located so as to allow for free air circulation.

All terminals, connectors, and conductors must be compatible. Poor connections and the use of dissimilar materials such as copper and aluminum are a major cause of electrical fires. Aluminum is particularly troublesome. It oxidizes when exposed to the atmosphere and forms a highly resistant crust. This increased resistance causes the connection to heat up under heavy loads and is believed to be responsible for numerous fires. Whenever aluminum is used, terminal blocks, split-bolts, and other connectors must be rated for aluminum (Al) or for copper and aluminum (CO/AL, or Cu-Al). Aluminum connections should also be coated liberally with an antioxidizing compound. (Don't overdo it, though. These compounds are conductive and could cause a short circuit if you get sloppy.)

Once the installation is complete you're ready for the final check of your wind system before you begin operation, the subject of the next chapter.

15

Operation, Performance, and Maintenance

"Wind turbines are not toasters. You can't just plug them in and walk away," says Wisconsin's Mick Sagrillo. And therein lies the problem. Even seemingly simple devices, such as micro wind turbines, require periodic attention—some would say care and nurturing—in order to perform reliably over long periods of time.

Once in service wind turbines operate automatically, whether charging batteries or delivering megawatts of electricity to the utility network. Small wind turbine rotors typically freewheel and, when there's sufficient wind, spin up to speed and begin generating electricity or pumping water with seeming effortlessness. In high winds the rotor furls or regulates without any user intervention.

Similarly, the much more complex medium-size wind turbines think for themselves. Each turbine monitors environmental conditions—for example, wind speed and direction—and internal parameters—such as the presence of utility power, voltage, frequency, generator temperature, and oil pressure. The controller—the turbine's brain—compares its measurements with its programming and, when conditions are acceptable, signals the turbine to begin or continue operation. When any condition exceeds the manufacturer's programmed limits, the controller orders the turbine to begin the sequence to take the turbine out of service: to pitch the blade tips and apply the parking brake on stall-regulated turbines, or to feather the blades and apply the brake on pitch-regulated machines. If this is an "emergency stop" the controller dials the owner or responsible windsmith and transmits a message of what went wrong. The operator can tell the controller to restart the turbine, or the operator can send someone to investigate.

This ability to operate for days, weeks, and months on end without human intervention can give a false impression: first, that the turbines are indestructible—they are not—and second, that they need little or no maintenance. Small wind turbines are particularly misleading in this regard.

For micro and mini wind turbines, there are very few "owner serviceable" components. Nevertheless, careful periodic inspection can catch problems before they become serious. Household-size turbines may require more actual maintenance; less for designs using direct drive, and more for those using transmissions, blade pitching mechanisms, or blade tip brakes.

Medium-size and larger turbines require the same degree of maintenance as any piece of heavy machinery, such as a farm tractor or large diesel engine. Again careful inspection pays dividends in detecting problems before they become costly.

The ruggedness of modern wind turbines also suggests that initial start-up, or "first rotation," is less critical than it really it is. Most defects, in both large and small wind turbines, occur during the first few months of operation. This is a critical period in the life of any rotating machine. The start-up procedure for a large wind turbine resembles the preflight checklist for an aircraft. A lot's riding on getting it right the first time. Fortunately the steps are well defined, and the professionals who install these turbines have the experience to do it right.

Small Wind Turbine First Rotation

Installation of small wind turbines is sometimes more haphazard, and the installers less experienced. For this reason try not to let the thrill of starting your new wind turbine get the better of you. As Jack Park advised more than two decades ago, avoid "fire-'em-up-itis," a disease sometimes fatal to wind turbines. True, small wind turbines are ruggedly built to withstand the elements. They are, however, still electrical machines. As such they're sensitive to proper installation. As with most electrical components, you don't get a second chance. An improperly wired inverter or control panel can lead to costly repairs before the wind system has generated its first kilowatt-hour.

Take your time and look over the entire installation. Go over the manufacturer's checklist: all bolts correctly torqued, cables secured, wires connected to the proper terminals, and so on. If everything meets your satisfaction, go through the suggested start-up procedure. Avoid starting the machine for the first time in a high wind. If a problem develops, light wind minimizes the potential for damage and makes it easier to bring the turbine under control. It's also a good idea to keep a watchful eye on the entire wind system for the first few days, to make certain that it operates as expected.

Interconnected Wind Systems

You should notify the utility—in writing—several months before installation that you plan to interconnect a wind turbine with its lines. This gives your letter ample time to move through the utility's bureaucracy and get the proper clearances. If the utility requires any special switches or metering, it's better to find out as early as possible in order to minimize costly modifications once the turbine is in place. Expedite the process by calling first and finding the person responsible for small power producers or independent generators. Include in your notification the brand name and model number of the wind turbine, the maximum power output in kilowatts, the operating voltage, the number of phases, and a line drawing of the proposed installation, including any disconnect switches.

The line drawing should show the location of all disconnect switches and protective relays (contactors) relative to the service panel, the kilowatt-hour meter, and the point where the utility's service enters the building or property. Describe how the wind system functions under both normal and emergency conditions (such as a power outage on the utility's lines). The utility wants to know how the design of the wind turbine guards against energizing a downed line and endangering its personnel.

Your letter should address the utility's concerns. It should explain how the wind turbine's controls will automatically isolate the wind turbine from the utility's lines whenever a fault on the wind turbine side of the interconnection is detected, a fault on the utility's line is detected, or abnormal operating voltage, frequency, or phase relationship is detected.

Additional information on the wind system's power factor or VAR (volt-ampere-reactance), current harmonics, and maximum in-rush currents may be helpful to the utility, but is unnecessary to guarantee a safe interconnection. The turbine's manufacturer will provide this information to you if the utility insists on it. Most utilities in North America and Europe should have enough experience with wind turbines—of all sizes—so that these questions have been answered many times over.

The utility is responsible for determining whether the interconnection poses a safety hazard to its personnel or to you, and whether it will interfere with its service to other nearby customers. To determine this, it will want to inspect the installation before the wind system begins operation. Don't panic; this is a reasonable action. In some cases the utility will accept the inspection report of the local fire underwriter or code enforcement officer, and issue you an approval to begin operation until its own personnel can get out to your site themselves. Sometimes the utility engineer can offer valuable advice on how the wind turbine can best meet both your needs and those of other utility customers.

Battery-Charging Wind Systems

Stand-alone power systems are much more complex than interconnected wind machines. There are more components (batteries, inverters, and backup generators), more cables, and more connections. There's a lot more room for error than in interconnected systems. Each component demands special attention and should be carefully checked before initial start-up. Consult the manufacturers' service or operation manuals for a start-up checklist on each component. Again follow the checklists carefully.

Monitoring Performance

One key to overseeing any wind turbine is to regularly monitor its performance. In short, does it do what it's supposed to do? Is it producing the amount of electricity, or lifting the amount of water expected? Has its performance degraded over time? Operators of commercial wind turbines in wind power plants use automatic data-acquisition systems for this purpose. They pore over their digital records looking for any sign of abnormality, like a doctor examining X-rays. And as with a deadly disease, early detection and treatment can prevent more serious consequences later.

For interconnected wind turbines, both large and small, monitoring performance can be as simple as reading the kilowatt-hour meter on a regular basis. Records of monthly and annual generation are the bare minimum needed to determine trends. When you're examining the production of a specific wind turbine, it's useful to compare it to like machines in similar wind regimes. While there's little published data on the performance of small wind turbines, there are reams of data on medium-size and larger turbines. Private wind turbine owners in Denmark regularly record monthly production from their wind turbines and mail the results to a central clearinghouse. Using this and other sources, the quarterly trade magazine *WindStats* reports on the production from wind turbines around the world. The most reliable data is reported by owners of wind turbines in Denmark, Germany, and the Netherlands. The Bundesverband Windenergie's annual *Windenergie Marktübersicht (Wind Energy Market Survey)* is another good source for turbines operating in Germany. The California Energy Commission periodically also summarizes production data from the state's giant wind farms in its Wind Project Performance Reporting System.

Data from these sources can't be used to say with certainty how well a specific wind turbine will perform at a specific site during a specific period; however, such historical information remains invaluable. Analysts use this data to estimate a range of production that a given model might reasonably be expected to generate. Such comparisons are particularly helpful in cutting through the marketing hype surrounding some wind turbines. At the least, this data can be helpful in determining the upper limits of what can be expected.

For example, such data can illustrate that a manufacturer's claims about a new wind turbine are outlandish in comparison to all the wind turbines that have gone before it. The data doesn't prove that the new wind turbine can't deliver as much electricity as promised, but suggests that this is unlikely. Similarly, if the manufacturer asserts that the new wind turbine will produce orders of magnitude more electricity than other wind turbines of comparable swept area, the data suggests that the claims are wildly exaggerated.

Small Wind Turbines

The simplicity of small wind turbines has one drawback. When they're running properly—with no unusual sounds or vibrations—it's difficult to know how well they're performing. Unlike their larger brethren, small wind turbines have little or no metering to indicate their performance. You can make up for this deficiency by installing your own monitoring system and periodically checking the wind system's performance. At a minimum, every interconnected wind system should include a dedicated kilowatt-hour meter. (Surprisingly, some don't.)

You will also want to install an anemometer on the tower near the rotor (but not so close that the rotor interferes with the anemometer). And you'll also need some means to "log" the data from the anemometer. The expense of household-size and larger wind turbines justifies the purchase of commercial data loggers, such as Second Wind's Nomad used at the Wulf Test Field. Resourceful experimenters such as Michael Klemen in North Dakota and Hugh Piggott in Scotland adapt secondhand computers for the task. Either approach works. At a minimum what you want is the ability to collect wind speed data and average it over a period of time. For more sophisticated analysis you will want to install a watt transducer to measure instantaneous power.

One simple test is to observe the wind speed when the turbine first starts generating. You can observe when it reaches peak power, and subsequently when it begins to furl or otherwise curtail generation. You're seeking only a gross approximation. First, the wind measured by the anemometer will never be the same as that striking the rotor. Second, wind speeds are always fluctuating. There's a lag between the time when a gust hits the rotor and when the wind turbine responds. Likewise, there's a lag when the wind ebbs and the rotor begins to slow down. Don't be alarmed if your initial observations of wind speed and power don't match the manufacturer's specifications. You need to analyze a great number of observations to make any sense of what's happening.

The data logger stores, processes, and summarizes these electronic observations into something meaningful. What you want the data logger to do is tally numerous measurements of the average of both power and wind speed over either 1-minute or 10-minute periods. The logger then sorts these average measurements by wind speed, which can then be compared to the manufacturer's advertised power curve.

On interconnected wind systems, you can improvise by using a kilowatt-hour meter to sum the total generation for a given period. You can then combine this with the average wind speed for the same period and compare it to the manufacturer's projection of annual or monthly generation at that average wind speed. While less sophisticated than producing a power curve with a watt transducer and data logger, this method remains a useful tool.

Maintenance

"Take good care of your tools, and they will take care of you," Mike Barker's grandfather once told him. Barker, a budding wind turbine installer, believes it's a good practice to live by.

Reading a Kilowatt-Hour Meter

Kilowatt-hour meters measure the amount of electrical energy that passes through them. Knowing how to read these meters is useful for monitoring either your electricity consumption or the production from a wind turbine. If you're cutting your consumption with energy-saving features in your home or business, you can read the meter daily to gauge your progress. Because the meters are inexpensive and widely available, every wind turbine producing line-quality power should have one installed. With the meter in place, you can track the wind turbine's performance: whether or not it's working and, if so, how well. By measuring the rate at which energy is produced, you can calculate average power as well.

Kilowatt-hour meter. This meter registers 6,918 kWh.

There are several versions in use. Meters with digital displays, while easier to read, are uncommon in North America. Those with a row of clock faces are more prevalent. Reading the meter's registers is little different from telling time with a traditional clock. But be aware that some of the registers rotate clockwise, others counterclockwise. When the pointer is between two numbers, the lower number is the number that has been passed (see sibebar photo of a Kilowatt-hour meter).

You can also use the kilowatt-hour meter to measure average power. Within the kWh meter is a metal disk that spins in response to the flow of electricity. On the meter's face is a number labeled kh. This factor is in watt-hours of energy that pass through the meter per disk revolution. If kh = 1, the disk must revolve 1,000 times before 1 kWh registers on the meter. To measure average power, watch the meter for one minute and count the number of revolutions (there's a black mark on the disk for this purpose), then multiply by kh and 60 minutes per hour:

$$P_{avg} = \text{rev/min} \times \text{kh (watt-hr/rev)} \times 60 \text{ min/hr}$$

You now have the one-minute average power passing through the meter. For example, assume the timing disk makes 10 revolutions in one minute and kh = 3.6. The rate at which electricity passes through the meter is:

$$P_{avg} = 10 \text{ rev/min} \times 3.6 \text{ (watt-hrs/rev)} \times 60 \text{ min/hr}$$

$$P_{avg} = 2{,}160 \text{ W} = 2.16 \text{ kW}$$

The adage can certainly be applied to maintaining your wind turbine. As Mick Sagrillo tells his workshop students, "Life expectancy [of a wind turbine] is directly related to your involvement with the machine."

Small Turbines

There's no concise answer to the question of how much maintenance is required on small wind turbines. Some wind machines are marvels of simplicity and appear nearly maintenance-free. Others are more complex, and the level of service required is more obvious. There are some small wind turbines that never seem to work right, or for very long, without a repair. Others operate day in, day out with no problem. The amount of maintenance required depends on the type of wind machine, its size, and the approach of its designers.

Here are a few guidelines. Rotors with fixed-pitch blades require less maintenance than those using variable-pitch governors. Machines using direct drive require less maintenance than those using transmissions. Freewheeling drive trains require less maintenance than those where the rotor must be motored up to speed. And those turbines using passive yaw to orient the rotor require less maintenance than those using active yaw drives. If minimizing maintenance is one of the designer's top priorities, it's reflected in the final product. Wind machines destined for remote, battery-charging applications, where maintenance is not only infrequent but also costly, are designed to be as maintenance-free as possible. On the whole it can be said that small, integrated, direct-drive wind turbines require far less maintenance than medium-size wind machines. However, the cost of maintenance on a small wind turbine can represent a much greater portion of the small turbine's revenue than that for medium-size wind turbines because small turbines produce so many fewer kilowatt-hours.

Many small wind turbines are nearing a state of hands-off operation. But we're not there yet. In an age where we're accustomed to automatic chokes and "idiot lights" on our cars, we should hardly be expected to run up and down a tower in foul weather carrying a grease gun. Yet after developing autos for 100 years, we still service them regularly. We shouldn't expect a machine operating in an environment as punishing as the wind to be as maintenance-free as a refrigerator in your kitchen.

Mick Sagrillo warns his students against being lulled into the false notion that small wind turbines are trouble-free. "Sure, you can go out only once a year and look at your turbine," he says, tongue in cheek. "If it's not on the top of the tower, it's on the ground, and you definitely have a problem." Danish wind turbine designer Claus Nybroe agrees. "Of course it has to be maintained, and even repaired a bit from time to time. Even the best [wind]mill will experience some trouble."

You should inspect the turbine and tower at least twice each year, says Sagrillo: once in spring, after the turbine has withstood winter storms, and once in fall, in preparation for winter. This gives you the opportunity to detect any problems that need to be corrected. "This is an essential consideration when it comes to small turbines," says Nybroe, since owners often have to do everything themselves. For example, Robert Gutowski has operated his Bergey 1000 nearly 20 years at his home in Northampton, Massachusetts. Gutowski installed the turbine and tower himself, and his experience has come in handy. He's had to remove the turbine from his 125-foot (38 m) tower seven times during the past two decades to replace rotor bearings and re-epoxy the turbine's magnets.

If you like, you can do this inspection in a cursory manner from the ground, on a calm day, using binoculars. Check the rotor for symmetry. See if all blades look alike. If they don't, you obviously have a serious problem. Watch how the turbine changes direction as the wind shifts. Note if the turbine yaws smoothly or abruptly. Erratic yawing can be due to turbulence, in which case the only solution is to install a taller tower. Erratic yawing can also occur if the tower is not vertical. (You can check the plumb of the tower with a level, plumb bob, or transit.) While on terra firma, check that the tower is still properly grounded and that the guy cables (where used) are tensioned correctly. If the turbine was installed on a hinged tower, lower the tower to ground level for a more detailed inspection. Look for cracks, worn fittings, and excessive grease or oil.

Check that all bolts on the turbine and tower are snug. If any bolts or nuts are missing, replace them immediately. (Usually if any are missing, you will already know about it.) Check the yaw assembly and whether the turbine can yaw freely or binds in one position. If it does bind up you may need to grease or replace the yaw bearings.

Maintenance may include little more than tightening loose bolts. Occasionally it may also entail cleaning slip rings, where used, or

Proven on the Falkland Islands

There are few places where off-the-grid wind power systems make more sense than on the Falkland Islands. The high cost of flying in diesel fuel to remote telecom and military sites translates into $21,000 per hour of diesel-generator operation. At such sites, says Tim Cotter of the Islands' Development Corporation, small wind turbines can pay back their installation costs in a few hours.

There are also few places harder on small wind turbines than the Falkland Islands. The average wind speed of more than 8 m/s (18 mph) beats lightweight turbines into scrap metal in a matter of months. "What's important for us is to have a wind turbine that is tough and reliable. The Proven has worked very well in this respect," says Cotter.

Proven on the Falklands. Tough places require tough turbines. Regular replacement of springs and blade hinges keeps these Provens in service on the Falkland Islands. (Tim Cotter)

The Islands' Development Corporation's goal is to power three-fourths of the islands' farms with wind. It installed its first Proven turbine in late 1995 and now operates more than four dozen of the machines. "The relatively low rpm [of the rotor] means no turbine has suffered blade erosion, whereas a GRP [glass-reinforced polyester] blade on an 800-watt high-speed turbine lasted only 12 months. We have suffered some minor failures, but with farmers claiming up to 80 percent reduction in [backup] fuel consumption, this has been an acceptable expense. It compares with tire wear on a four by four. The fold-down tower makes it easier for the farmer to lower it safely to change blades and replace springs," explains Cotter. The Proven "is a very tough piece of kit."

replacing worn components—the springs on Proven's flapping rotor, for example. If the wind machine has grease fittings, they will have to be greased semiannually or quarterly. Micro and mini wind turbines, however, use sealed bearings and bushings that are designed to last the life of the machine.

If the turbine uses a gearbox, as many household-size and larger turbines do, the oil will have to be regularly changed. This can be messy, but it needn't be. If it's required make sure it gets done. Commercial wind farm operators have developed techniques that enable you to change transmission oil in a safe and environmentally sound manner. Manufacturers can specify which system works best with their particular product.

Visually check for corrosion and secure connections at all wiring terminations on the turbine, in junction boxes, at disconnect switches, and in the control panel. Use extreme caution anytime you open the control panel, synchronous inverter, or disconnect switch. If the connections look good, leave well enough alone. When closing the door on any electrical enclosure make sure you don't pinch any conductors between the door and the box.

Balance of Remote Systems

In stand-alone systems batteries are the single most maintenance-intensive component, generally followed by the backup generator. The batteries' state of charge should be routinely monitored. If the state of charge indicates that the batteries are low, remember that lead-acid

Figure 15-1. Torquing tower bolts. Windsmiths tighten the bolts on a truss tower supporting a 750 kW wind turbine near Palm Springs, California. Though this turbine is only a few months old, there is grease on the blades from leaky pitch bearings. If the problem worsens the blades will need to be cleaned.

batteries should never be discharged to more than 20 percent of their capacity; when they are, permanent damage can result.

Make sure that the backup generator operates properly in order to recharge the batteries when needed. For this reason it may be necessary to start the backup generator occasionally to keep it well lubricated and to ensure that it will work when you need it.

Periodically it will be necessary to produce an equalizing charge on lead-acid batteries, to balance out any voltage discrepancies between cells. Most battery chargers are capable of providing an equalizing charge. Occasionally top off battery electrolyte fluid with distilled water. Protect battery terminals from corrosion, and keep battery tops clean. Also keep the battery storage area tidy, and don't store anything above the batteries. This will prevent metal objects from falling across battery terminals.

For more information on proper handling of batteries and their maintenance in a remote power system, read *The New Solar Home Book* by Joel Davidson. *Home Power* magazine also carries articles with helpful tips on how to get the most use out of batteries and inverters (see the appendixes for details).

Medium-Size Wind Turbines

Medium-size wind turbines require more, and more highly skilled, maintenance than small wind turbines, but less today than in the 1980s. Mike Kelly, operations manager for Zilkha Renewable Energy, explains that early commercial turbines required quarterly maintenance while turbines now need only biannual inspections and service. Kelly, who has managed wind turbines in California, Italy, and Iowa, says that scheduled maintenance ranges from simple visual inspection and general housekeeping to more sophisticated measures, such as analysis of the vibration of critical bearings and examination of pitting on the teeth of transmission gears. Kelly adds that good housekeeping, while occasionally overlooked, is critical. "If there's oil everywhere, you can't find the leaks" that need to be addressed. "On a million-dollar machine, housekeeping more than pays for itself."

With the advent of sophisticated monitoring equipment, notes Kelly, managers are now able to predict the life remaining on critical components. These components can then be replaced before they fail, on a schedule that minimizes lost production. Such advanced techniques transform costly, unscheduled maintenance into routine maintenance.

Unscheduled maintenance results when a fault takes the turbine out of service, or when the operator detects a problem with the potential to seriously damage the turbine. During a period of strong winds, lost production from unscheduled maintenance costs the operator lost revenue. Wind plant managers, such as Kelly, schedule maintenance to minimize lost production, ideally when the turbines are idle in light winds.

Though manufacturers provide a maintenance schedule, says Kelly, each wind farm operator modifies it based on actual experience with the wind turbines. He pays special attention to high-wear components in the braking and yaw systems and to visual inspection of the blades both from the ground and from the nacelle. More rigorous annual blade inspections—for example, of the pivoting blade tips on some machines—require specialized equipment, cranes, lifts, or rigging that allow both examination and minor repairs with the blade on the rotor.

Kelly says his crews test the torque on 10 percent of all bolts in a tower annually. They test the flange bolts between sections of tubular towers, for example, as well as the foundation bolts. If all meet specifications he takes no further action. He argues that truss towers, properly designed and assembled, should not require any more frequent bolt tightening than tubular towers (see figure 15-1, Torquing tower bolts). With more than two decades working in the wind industry, Kelly is the first to admit that many truss towers have not met this standard.

Figure 15-2. Tvindkraft. Tvind turbine near Ulfborg, Denmark, circa 1980. For scale, note the door at the base of the tower. An elevator takes service personnel up to the nacelle inside the hollow, reinforced concrete tower.

Tvindkraft

There are now many wind turbines that have been operated and maintained for more than 20 years. Yet few have quite the amazing background of the turbine at the Tvind school in the northwest corner of Denmark's Jutland Peninsula. From 1975 to 1978, Tvind built one of the world's largest wind turbines (see figure 15-2, Tvindkraft). There is little about the school, its students, and the construction of its wind turbine that is traditional.

At the time, most wind energy development was directed toward multimillion-dollar behemoths constructed by the likes of Boeing, General Electric, and Messerschmit-Bölkow-Blohm (MBB). Though well funded, these massive turbines seldom performed as expected or operated very long before being scrapped, their builders moving on to other, more lucrative ventures.

Tvind students, faculty, and a dedicated group of volunteers built their turbine for less than $1 million—a fraction of the cost of the large wind turbines of the day—by doing much of the work themselves and using secondhand components. They salvaged the

Bushland Enertech

Even some of the old turbines from the 1970s and early 1980s—when well maintained—will continue to spin out the kilowatt-hours. Like its counterparts at Canada's Atlantic Wind Test Site, the USDA's experiment station near Bushland, Texas, operates an early version of the Enertech E44 (see figure 12-12, Long-lived Enertech). The USDA's E44—a moniker designating the rotor diameter in feet—has been operating since 1982, logging more than 100,000 hours of operation.

During its more than two decades surviving the winds of the Llano Estacado (staked plains) of the Texas Panhandle, the turbine has operated in three different configurations: first 20 kW, then 40 kW, and finally 60 kW. Every five to seven years, says former USDA staffer Eric Eggleston, they would take down the turbine and replace the seals on the main shaft, just like changing the seals on an automotive crankshaft. They've also had to replace a worn yaw assembly. The household-size turbine is typically available for operation 97 percent of the time. It's out of service about 1 percent of the time (about 90 hours per year) for weather-related events, 1 percent for scheduled maintenance, and another 1 percent for unscheduled maintenance.

gearbox from an ore-crushing plant in Sweden, the generator from a Swedish paper mill, and the main shaft from a supertanker. Tvind's "mill group" built the mammoth blades on the 54-meter (180 ft) diameter rotor by hand, and photographs of young Danes carrying the blades from shop to construction site appeared in newspapers around the world. They did so as a political statement that wind energy need not be the sole domain of the aerospace giants.

They succeeded. For more than two decades the 1-megawatt turbine has generated electricity for the school and the surrounding area. While not without its share of problems, the turbine has operated more than 82,000 hours and generated nearly 14 million kWh through 2001.

To create a "modern" turbine the designers opted to depart from Danish practice and place the rotor downwind of its concrete tower. But the students and faculty were not the Luddites that some pictured them to be. The school received valuable technical assistance from engineers, some of whom would eventually lead Risø National Laboratory's test station for small wind turbines. This was beneficial to both: Tvind was able to deal with some thorny technical problems, and Tvind's advisers gained experience and hands-on knowledge of large wind turbines at the school's expense.

According to Peter Karnøe at the Copenhagen Business School, it was experience gained on the Tvind project that eventually led to the development of the Danish wind turbine blade industry. Subsequently boat manufacturer LM Glasfiber adopted the technology and began producing fiberglass blades for the ever-increasing numbers of commercial Danish wind turbines. In 1993 one of Tvind's blades failed. Subsequently all the blades were replaced and the turbine was again returned to regular operation. Allan Jensen, who has maintained the Tvind turbine since 1982, is probably the longest-serving windsmith on a single turbine in the world. Along with his coworker Britta Jensen (no relation), Allan performs daily, weekly, quarterly, biannual, and annual inspections. Though they have detected some pitting on the gear teeth, the gearbox has never needed repair or replacement, nor has the generator. This is striking. Many much smaller wind turbines in commercial service in California as well as in Denmark, and other countries have needed their drive train components repaired or replaced more than once.

Each day one of them dons bright blue Danish coveralls, rides the elevator to the nacelle, and makes a simple visual inspection to ensure that all's well. In their weekly inspections, they check that all lights and meters are functioning. They also inspect the gearbox filter, and from inside the hub they look for cracks in the blades.

Every quarter they test all safety systems

and perform an emergency stop. They also grease some fittings. Biannually they check bolt torque as part of a scheduled 10-year rotation. That is, every bolt is inspected and torqued once in ten years.

Each year Allan and Britta inspect all critical bolts for corrosion and fatigue. They also inspect and clean each blade and paint metal components as needed. Every three years they inspect hidden areas for corrosion and check the lightning rods on the nacelle. "Everything has been very reliable," says Britta.

Tvind has taken care of its history-making turbine and the turbine has taken care of them. In more than two decades, the school has only had to replace the blades and blade pitch bearings.

Blade and Tower Cleaning

One surprising lesson learned in California during the mid-1980s was the necessity of periodically giving the rotor a good bath. Many early wind turbines installed in California performed poorly, even when operating as the manufacturer expected. The search for lost production reads like a good detective novel. Eventually engineers identified the culprit: dirty blades.

Nearly all wind turbines had been designed in rainy climates: Denmark, Germany, or the northeastern United States. But it seldom rains in sunny southern California. In the explosion of new life that takes place every spring, millions of insects hatch and fly off to find a mate. Some find the leading edge of a wind turbine blade instead. And like the windshield of your car as you barrel down the highway at high speed, splattered insects begin to accumulate, leaving a sticky goo. On your windshield this goo obscures your vision. On a wind turbine blade's leading edge it disturbs the airflow. A good hard rain washes the accumulated dust and dead bugs off the blade as the rotor turns.

In arid environments, such as southern California, there's not enough rain to clean the blades. Once this was realized regular maintenance began to include washing the blades. On some wind turbines dirty blades cut production nearly in half. Thus washing the blades became a necessity.

Some companies developed spray bars that were permanently attached to the tower. Water trucks would then be used to pump large volumes of water through the spray bars while the turbine was in operation, simulating rain. Some firms specialized in washing wind turbine blades. They adapted the high-pressure nozzles used in the utility industry to clean insulators on electric lines, so that they sprayed a powerful stream of water at the blade's leading edge. Other firms specialize in rappelling down the blade from the hub with a high-pressure wand, cleaning the blade as they descend (clearly not a job for the faint of heart). Today's airfoils are less sensitive to soiling than they were 20 years ago. Nevertheless, blades need periodic cleaning, especially if hydraulic fluid or grease leaks onto them.

Towers may need cleaning, as well. Some wind turbines' designs—there's no delicate way to put this—are incontinent. Mitsubishi's 250 kW model (of which there are more than 600 near Mojave, California) is notorious for its leaky drive train and the oil-stained towers that result. Leaking fluids pose a potential environmental hazard if they soak into the soil. Leaks also pose a risk to technicians if they make access to the turbine hazardous. While fluid leaks are a design defect in this Mitsubishi model, other turbines occasionally suffer from a spill and need cleaning as well (see figure 15-3, Regular bathing). It's always cheaper to prevent leaks in the first place, or provide drip pans (for collecting unavoidable leaks), than it is to clean a wind turbine tower.

Painting

At coastal sites where corrosive salt-laden air envelopes the turbine and tower every day, metal surfaces may require periodic painting. Many tubular towers for commercial wind

Figure 15-3. Regular bathing. Cleaning a tower on a large wind turbine. Note use of carabiners with locking gate, gloves, helmet, and eye protection. (Rope Partner)

turbines are shipped with special coatings to ward off corrosion. Some older towers, and all freestanding lattice towers, use galvanizing to protect the underlying steel. Unfortunately the coatings on all towers are not of equal quality, and the effectiveness of the best coatings are compromised when nicked or scraped to bare metal. You need only see the lattice and tubular towers in India's Gujarat state on the Gulf of Kachchh to see how quickly metal coatings deteriorate in corrosive environments. Even in California's relatively benign Altamont Pass, the marine air has disfigured the cheaply painted towers of some turbines.

Under normal conditions galvanized metal should rarely need treatment (see figure 15-4, Painting). Painted or coated surfaces may periodically need touching up, especially if they have been washed repeatedly to remove grease or oil. At some sites, or with lower-quality coatings, towers may need to be repainted during their lifetime.

Cost of Maintenance

Always budget for maintenance. Commercial wind developers understand this. It's the job of managers like Mike Kelly to devise complex budgets to keep their turbines in service at minimum expense. Yet small wind turbines and even medium-size turbines in distributed applications have repeatedly fallen prey to the misperception that the wind is free. Indeed, the wind is free, but the turbine and the cost of maintaining it are not.

Many a wind turbine has been installed under a foreign-aid or publicly supported demonstration program, only to operate intermittently. Grants pay for the turbine and its installation but seldom for its maintenance. Often it isn't clear who bears responsibility for the turbine and who will operate it and pay for its upkeep. Consider a cautionary tale from New England.

The high school in Hull, Massachusetts, won a state government grant for an Enertech E44 to help with the school's electricity bill. The school installed the household-size turbine at the appropriately named Windmill Point near the entrance to Boston Harbor in early 1985. The Enertech operated sporadically until mid-1996, when it was removed. During this period, the turbine only twice produced as much electricity as expected. The turbine's poor performance was attributed in part to the harsh marine environment and in part to poor maintenance by the school. Maintenance and repair of the E44 cost Hull High School $17,000 during its decade of fitful operation. Prorated, this is equivalent to about 2 percent of the turbine's installed cost, annually, but was insufficient to maintain the turbine properly.

Fortunately Hull's experience with the E44 didn't sour the community on wind energy. On December 27, 2001, John MacLeod, the manager of the Hull Light Plant, and a group of community activists dedicated a new Vestas V47 at the same site. They intended for the 660 kW turbine to provide municipal lighting for the town of 10,000. The dedication took place during the darkest week of the year, symbolically saying to the community at large that the lights of Hull would shine brightly without the burning of oil, like the lamps in the temple at Jerusalem celebrated during Hanukkah, the festival of lights.

Hull's V47, one of the few urban wind turbines in North America, has become the pride of the community. Hull Light's MacLeod reports that tourists to Windmill Point have increased threefold since the commercial wind turbine was put into service. The Vestas turbine not only powers all Hull's streetlights, traffic lights, and other municipal lighting needs—more than 400,000 kWh annually—but also generates more than sufficient revenue, says MacLeod, to pay for its maintenance.

Figure 15-4. Painting. Treating galvanized surfaces on a freestanding lattice tower near Montefalcone, Italy. Note the use of a boom or aerial lift. Contrast this with figure 15-1, where the windsmiths are suspended from their harnesses inside the lattice tower.

Small Wind Turbines

As Claus Nybroe explained, most owners repair their own micro and mini wind turbines. Consumers internalize the cost in the form of their time and effort. Hiring a professional for simple tasks can cost a significant proportion of the benefits these wind turbines generate.

On the other hand, the greater generation from household-size turbines, as well as their greater complexity, often justifies professional help. In a 1988 report Wisconsin Power & Light summarized the results from monitoring four household-size turbines in the mid-1980s. Most of the turbines had been operating for three years; one for as long as six. Though not a statistically valid sample, the data does indicate a range of maintenance costs that can be expected. The utility found that during the three-year period, repairs cost as little as $100 for one turbine and as much as $6,000 for another.

The cost of operation and maintenance is often given in units of currency relative to kilowatt-hours produced. Some costs are fixed: They remain constant, regardless of how hard the wind turbine works or how much electricity it generates. Other costs vary with how much time the wind turbine is in operation and how hard it works when it is. Costs per kilowatt-hour are often higher in areas of moderate winds and lower in areas with more wind (and thus more total generated kilowatt-hours). In the modest wind regime of Wisconsin, the cost of repairs to the turbines ranged from a low of $0.003 per kWh to as much as $0.09 per kWh.

Medium-Size Wind Turbines

During the mid-1980s the cost for operating and maintaining well-designed wind turbines in California averaged $0.01 per kWh. Modern technology has cut this cost nearly in half during the first five years of operation. Danish authority BTM Consult estimates that operation and maintenance costs for wind turbines in the 55 kW class—the first commercial medium-size wind turbine—have averaged about 0.03 euro per kWh ($0.03/kWh) during their two decades of operation. BTM says turbines in the 200 to 500 kW class cost about 0.01 to 0.015 euro per kWh ($0.01–0.015/kWh) in the decade they have been in service.

Danish statistical clearinghouse Energi og Miljødata (EMD), in a study for Risø, expressed the O&M costs in a slightly different manner (see table 15-1, Operation and Maintenance Costs in Denmark). It calculated that annual operation and maintenance costs averaged from 1 to 7 percent total installed cost. EMD's estimates include a reinvestment in the turbine after year 10 of 20 percent of the total installed cost prorated over the second decade of the turbine's life to account for replacement of major components.

Table 15-1
Operation and Maintenance Costs in Denmark (as % of Total Investment)

	Years 1–10	Years 10–20*
150–300 kW	1–3%	4–7%
500–600 kW	1–2%	4–5%

Note: *Includes prorated 20% reinvestment after year 10.

Source: IEA, Energi og Miljødata.

Life Expectancy

Some small wind turbines, notably lightweight machines, may survive for as few as five years before requiring replacement. Others, with regular maintenance and the occasional repair, have been operating for two decades.

For commercial wind turbines, the rule of thumb has been a 20-year life span, and many machines have demonstrated that this is a realistic expectation. BTM Consult reports that many early turbines in Denmark are being replaced before they reach the end of their useful lives. As the technology rapidly advances, explains BTM Consult, a turbine's economic life becomes shorter than its technical life. Many used turbines from California and Denmark have found new homes, demonstrating that these early machines have many more years remaining to spin wind into electricity.

16

Safety

We obey the law to stay in business,
but we obey the laws of physics to stay alive.
—Anonymous windsmith

The capture and concentration of energy—in any form—is inherently dangerous. Wind energy exposes those who work with it to hazards similar to those in other industries. Of course there are the hazards that, taken together, are unique to wind energy: high winds, heights, rotating machinery, and the large spinning mass of the wind turbine rotor. Wind energy's hazards, like its appearance on the landscape, are readily apparent. Wind energy hides no latent killers; no black lung, for example. When wind kills it does so directly, and with gruesome effect.

In this chapter we'll first examine the record and glean what we can from fatal accidents with wind energy. Then we'll turn to the tools and practices necessary for working safely with the technology. Unpleasant as the accounts described may be, they emphasize the need to work safely—because your life quite literally depends on it.

Fatal Accidents

Death in the maw of a wind machine is nothing new. H. C. Harrison recounts in *The Story of Sprowston Mill* how his great-grandfather, Robert Robertson, was killed in 1842 after becoming entangled in the sack hoist on his English windmill. There are historical accounts of similar deadly accidents in France, and no doubt like tales can be found in other countries where wind energy has been used.

Since its rebirth in the 1970s wind energy has directly or indirectly killed 20 people worldwide. The first was Tim McCartney, who fell to his death near Conrad, Montana, in the mid-1970s while trying to salvage a 1930s-era windcharger. There are few details on McCartney's death other than that his broken body was found near the tower. News reports said simply that he fell during high winds. McCartney was followed a few years later by Terry Mehrkam, a pioneering Pennsylvania designer and manufacturer of wind turbines. Mehrkam was killed in late 1981 near Boulevard, California. Unfortunately Mehrkam was not the last.

A short while later there was a spate of electrocutions. Pat Acker was killed while constructing the foundation for a wind turbine near Bushland, Texas, and Jens Erik Madsen was electrocuted while servicing a wind turbine in Denmark.

In 1983 Canadian Eric Wright rode an experimental Darrieus wind turbine to his death when it fell over during installation near Palm Springs, California.

Terry Mehrkam

BOULEVARD, CA (UPI)—Terrence Mehrkam, 34, owner of a Hamburg, Penn., windmill manufacturing company, was struck and killed by the blade of one of his own windmills at a "wind farm" in this San Diego County community.

The coroner's office said Mehrkam was struck by one of the blades after falling from a platform . . .

Sadly, Terry Mehrkam's death in 1981 wasn't the first; nor was he the last person killed working with wind energy. His accident and those of others should serve as constant reminders of the danger inherent in working on a power plant high above the ground. California's Department of Occupational Safety and Health (CalOSHA) concluded that Mehrkam climbed to the top of the tower without using any form of fall protection and either fell or was thrown off the tower to his death.

Terry Mehrkam. Terry Mehrkam stands atop the nacelle on one of his 40-foot (12.2 m) diameter wind turbines near Hamburg, Pennsylvania. He was thrown from a similar wind turbine and killed in 1981. At the time, Mehrkam was not wearing a work belt, climbed to the nacelle, and tried to manually stop a "runaway" rotor. Here, Mehrkam has a rope tied around his waist.

The year 1984 was deadly. Ugene Stallhut was driving a tractor as a tow vehicle when it flipped over and crushed him on a farm in Iowa. Then Art Gomez was killed while servicing a crane in California's San Gorgonio Pass. The same year J. A. Doucette was crushed to death while unloading a container of tubular towers in the Altamont Pass.

The simple medical description of John Donnelly's death found in the files of his company's insurer fails to describe the horror of his fate. Death by "multiple amputations" sanitizes a truly grisly accident in 1989, a nightmare witnessed by his coworker, who watched helplessly as Donnelly was drawn inexorably into the nacelle's slowly spinning machinery. What made Donnelly's accident even more terrifying for windsmiths everywhere was its cause: Donnelly's lanyard, a device designed to prevent falls, became entangled on the revolving main shaft and dragged him to his death.

Not long after Donnelly's accident near Palm Springs, Dutch homeowner Dirk Hozeman was killed in a like manner. Against professional advice he climbed to the nacelle of his Polenko turbine in a vain attempt to stop the runaway rotor from destroying itself in a violent winter storm. Tragically the turbine had been inoperative for two years and had just recently been returned to service. After squeezing into the small nacelle, Hozeman, like Donnelly, became snagged on a turning shaft. Rescue crews retrieved his body the next day, after the wind subsided.

Also in 1989 two men were killed in a single accident on the Danish island of Lolland. Three men were suspended from a crane in a basket when the rotor they were servicing unexpectedly began to move. Two, Leif Thomsen and Kai Vadstrup, were thrown to the ground. The third man dangled from his lanyard 30 meters (100 ft) aboveground until he could be rescued.

Then in 1991 Thomas Swan, a crane operator, was electrocuted near Tehachapi when the boom on his crane snared a 66,000-volt power line.

Richard Zawlocki fell to his death in 1992 while descending a tower near Palm Springs.

Robert Skarski died in 1993 while installing a small wind turbine at his Illinois home. He was killed when the tower he was on buckled and fell to the ground.

Mark (Eddie) Ketterling was nearly cut in half by a chunk of ice knocked off the interior of a tubular tower in Minnesota in 1994.

The series of deadly accidents continued in 1997. Crane operator Randy Crumrine was crushed when his crane collapsed in Sibley, Iowa. Ivan Sørensen fell to his death near Lemvig on Denmark's Jutland Peninsula while removing a wind turbine from its tower. And Bernhard Saxen was crushed inside the nacelle of a wind turbine when it flew off the top of its tower at the Kaiser-Wilhelm-Koog test center in Germany.

In a bizarre year 2000 accident, a young parachutist crashed into a wind turbine on the German island of Fehrmarn.

Hazards

Falling from the tower is the single most apparent occupational hazard of working with wind energy. Industry practice and what some would argue to be common sense suggest that McCartney, Mehrkam, and Zawlocki all made the same fatal mistake: They did not use any form of fall protection.

Falls

While fall protection terminology can be arcane, the principles are not. Fall protection comprises both tools to prevent falls and devices to arrest falls that do occur. Where falls can occur, for example, in the absence of a work platform and railing, a fall protection system must include three elements: body support, lanyard, and anchorage. The lanyard—short sections of rope or webbing—connects the body support to a sturdy attachment on the nacelle or tower. Body or work belts and lanyards in conjunctin with a suitable anchorage are tools used to restrain windsmiths from falling while allowing them to position themselves to perform a given task. If a windsmith does fall, a fall-arresting system is designed to prevent serious injury. Like fall restraint, a fall-arresting system includes a lanyard and a suitable anchorage. Unlike fall restraint, a fall-arresting system requires a full-body harness.

Little is known about how McCartney died. He obviously was not using a fall-arresting system. But Mehrkam's death was investigated by members of California's Department of Occupational Safety and Health. They concluded that Mehrkam climbed to the top of the tower without any form of fall protection and either fell or was thrown off the tower to his death. Unlike Mehrkam, Zawlocki was wearing his work belt when he fell while descending the tower. He was not, however, using his positioning lanyard as protection against a fall from the tubular tower's interior ladder. His lanyard was later found atop the tower—holding the nacelle cover open.

Zawlocki's death is more troubling for what it says about the human factor in all accidents. It is evident that Zawlocki was aware of the risk of working on the tower because he was wearing his work belt. But he failed to use his positioning lanyard. We will never know why. We do know that when safety equipment is inconvenient or uncomfortable, there is a tendency to avoid using it.

All medium-size and larger wind turbines include a fall-arresting system designed specifically to prevent the kind of accident that killed Zawlocki. This system uses a metal sleeve that slides along a steel cable or rail that runs the length of the tower. When ascending or descending the tower, windsmiths attach their body harness to this sleeve. Should they slip, the sleeve grips the cable, arresting their fall.

The tower Zawlocki was descending had recently been installed in a repowering project for which turbines had been moved from northern California to the San Gorgonio Pass in southern California. The tower lacked a fall-arresting cable or "ladder-safety device" in the jargon of the trade. As an alternative technicians were instructed to attach their lanyards to the ladder when climbing the tower. This entails climbing a few rungs, removing, and then reattaching their lanyards. Anyone who has used this system knows that it is awkward and time consuming or, in other words, inconvenient. As a consequence this technique is more honored in the breech than observed in practice, which led to Zawlocki's death.

Spinning Rotors

The second lesson to be learned from these accidents is that no one should ever work atop the tower when the wind turbine rotor is turning, especially during high winds. This simple proscription would have prevented the deaths of Terry Mehrkam in California, Dirk Hozeman in the Netherlands, and Bernhard Saxen in Germany.

Mehrkam died when he tried to stop the rotor on one of his wind turbines. The brakes had failed in high winds. Subsequently the rotor "ran away" and went into uncontrolled overspeed. This is the worst scenario imaginable for going anywhere near the wind turbine. Under these conditions the rotor on Mehrkam's machine would have been a blur. As insane as it seems now, it was Mehrkam's practice with runaways such as this to climb the tower and manually brake the rotor to a halt by wedging a pry bar into the brake calipers.

Terry Mehrkam made two mistakes: He climbed the tower while the rotor was spinning uncontrollably, and he did so without any fall protection, not even a work belt and lanyard. He must have been frantic to save his machine, because he had used a rope restraint in the past. The rope restraint, crude even by the lax standards of the day, could have prevented his fall and saved his life. It was utter foolhardiness to mount the turbine and straddle the nacelle like Slim Pickins riding a nuclear bomb to its target in the movie *Dr. Strangelove.* He should have simply walked away, cleared the site, and waited for the wind to subside.

Similarly, Dirk Hozeman should have walked away from his wind turbine and waited for the storm to pass. There is little that you can do when a wind turbine rotor becomes unloaded in high winds, as when the grid goes down and the overspeed control devices fail.

In the industry's "wild west" days there are accounts of windsmiths shooting ropes through the rotors on runaway wind turbines in the San Gorgonio Pass. Such attempts were often unsuccessful, and wind companies eventually abandoned efforts to rescue turbines in overspeed as being too risky. As longtime Tehachapi windsmith Jon Powers explains, "Wind turbines are replaceable; people are not."

Another lesson from these accidents is the danger inherent in working around poorly designed wind turbines. The Mehrkam turbine and its clones were notorious for self-destructing in California's windy passes. The Polenko was only slightly better.

On occasion early Danish turbines also failed catastrophically. Yet Danish turbines failed far less often than the flimsy U.S.–designed machines of the day. The pitchable blade tips on Danish turbines could typically be depended on to protect the rotor from destroying itself. "We rely on the centrifugal tips on our stall-regulated turbines to account for brake, generator, and drive train failures," says Powers. When the wind subsides and the turbine is safe to approach, the nacelle is yawed out of the wind and the rotor eventually stops.

In contrast to Danish designs, Mehrkam's turbines had no aerodynamic means of overspeed control. If the drive train or brakes failed and the winds were high, his machines often self-destructed. Though the turbine that killed Bernhard Saxen during a storm at the German test center was built by a reputable manufacturer and was a far cry from the crude contraptions built by Mehrkam, it remained experimental. By its nature an experimental turbine has not proven that it is safe to operate under all conditions and as such poses a potential hazard.

On medium-size and larger turbines that have a proven record of reliability, there may be occasions when it is acceptable for a technician to remain inside the nacelle while the turbine is operating, say site managers. The caveats are that no one enter or exit the nacelle while the turbine is operating, that there is sufficient space inside the nacelle to work safely, that all rotating shafts are covered, and that this never be attempted during high winds or storms when the control systems are under stress. Saxen should not have been in the nacelle of an experimental turbine during a storm. Whatever his reason for doing so, he took an unnecessary risk.

All contemporary commercial wind turbines have shaft guards. Modern megawatt-class turbines and many—but not all—medium-size turbines have sufficient space inside the nacelle for someone to safely stand and observe the turbine's operation. Regardless of the turbine's size, site managers warn that no one should enter or leave the nacelle unless the turbine has been brought to a full stop and the nacelle is prevented from yawing. Access to even the megawatt-scale nacelles is tight and often requires some scrambling.

Some newer turbines have automatic emergency stop switches on the upper access hatch leading into the nacelle. Should anyone reach the upper platform and raise the hatch while the turbine is operating, the switch activates the control system and brings the turbine to a halt. Such switches are intended to prevent potentially hazardous entry into the nacelle while the wind turbine is operating.

When anyone is working on the rotor or drive train the rotor must be locked in place. How this is done varies from turbine to turbine, but typically involves placing a pin, suitably rated for the task, through a rotating component of the drive train. Such a locking pin would have prevented the accident in

Dynamic Braking of Small Turbines

One fundamental rule of working with wind turbines is to never go near a spinning rotor. Period.

Unfortunately most small wind turbines, even some household-size turbines, lack a mechanical brake that can stop and hold the rotor. If you have to work near a small wind turbine that doesn't have a mechanical brake, furl the rotor or apply a dynamic brake, and then only approach the rotor in calm weather. Neither furling nor dynamic braking will entirely stop the rotor from spinning, but these measures can bring the rotor under a semblance of control.

Manufacturers of mini and micro turbines seldom provide the ability to manually furl the rotor. Dynamic braking, effected by shorting the phases in the stator of permanent-magnet alternators, is the only means available for controlling the rotor on many small turbines. However, dynamic braking can bring the rotor to a near halt only in light winds.

Southwest Windpower includes such a "brake" switch with the control panels on its Whisper line. Bergey provides a similar switch on the control panels of its XL1 model. You can order a suitably rated switch for Southwest Windpower's Air series.

But as Mick Sagrillo warns, dynamic braking may not work when you need it most: during a storm's high winds. Strong wind may overpower the generator, says Sagrillo, spinning the rotor and causing potentially damaging current in the generator's windings.

Alternators large enough to stop the rotor with dynamic braking under any wind conditions, says Bergey Windpower's Mike Bergey, would have to be several times larger than otherwise needed. Like Sagrillo, Bergey warns that dynamic brakes or "stop switches" must be used with extreme care. If the rotor doesn't reduce speed quickly the switch must be reopened immediately to avoid damage to the alternator. Scoraig Wind Electric's Hugh Piggott suggests putting an ample dump or resistive load directly on the turbine output instead of shorting the phases together. Piggott says this will provide more braking torque than a dead short. Once the rotor begins to slow down the brake switch can then be applied to short the windings.

Denmark, where the brake was inadvertently released, allowing the rotor to begin turning and catch the basket with the three windsmiths. A locking pin would have also prevented John Donnelly's death near Palm Springs

During a seemingly calm day, Donnelly climbed the turbine to repair a damaged brake. The turbine was off line, but without the brake the rotor was able to freewheel. A slight breeze started the rotor turning, catching Donnelly off guard. By the time he realized what was happening his lanyard was snagged and it was too late to react.

Locking pins themselves are not foolproof. There are limits to their effectiveness. Manufacturers may specify wind speeds above which such devices should not be used. In one case two windsmiths were servicing the hub on a large wind turbine. They placed the locking pin in position as instructed, but they disregarded the onset of increasing wind speeds. While they were inside the hub the locking pin failed, and the rotor began to spin, trapping them inside the hub. With their tools flying about their heads, they must have thought they were on a deadly Mr. Toad's Wild Ride. Eventually their coworkers saw that something was amiss and braked the rotor to a halt. If the blades had not been pitched to feather, and the rotor only capable of spinning slowly, this humorous tale would have taken a deadly turn.

Electrical

Next to falls and spinning shafts, the most serious hazard with wind energy is working around electricity. The most common form of serious, nonfatal accidents in California wind plants is injuries from electrical burns. In one case a Tehachapi woman was maimed when she touched energized equipment inside a transformer cabinet. Again the hazards are similar to those in the electric utility industry, and the precautions developed during the past century for safely generating and transmitting electricity are applicable. If they had been followed they would have prevented the Tehachapi accident and others like it.

Construction

Thirteen of those killed died in construction-related accidents. McCartney was dismantling a turbine, and Mehrkam was trying to rescue a turbine he had just recently installed. Swan was driving his crane to move a turbine to a better site, and Zawlocki was completing the reinstallation of a turbine that had been moved from a site in the Altamont Pass to Palm Springs. Similarly, Sørensen was removing a turbine so it could be replaced with a more modern machine. Skarski was installing a small turbine. Ketterling and Crumrine were on crews installing medium-size turbines when they were killed; Acker was positioning a rebar cage for a wind turbine's foundation when the forklift holding the rebar contacted overhead power lines. Construction is not an ongoing activity, and the risks associated with it normally occur only twice in the life of the wind turbine: during installation and during removal.

Analysis

The deaths in the wind energy industry, while alarming, may not accurately reflect what can be expected from a mature technology. Many of the hazards encountered in building and operating a wind turbine are not unique to wind energy.

Most of the deaths have occurred during construction or construction-related accidents. Five have died during operation or maintenance of the turbines.

One-half the deaths have occurred on or around turbines of the size installed during the great California wind rush of the mid-1980s. Seven men have been killed working with larger turbines, and at least three have been killed working with small turbines.

The preponderance of those killed were Americans: 12 U.S. citizens and one

Canadian. The mortality rate in the United States and the Netherlands is twice the international average of those working with wind energy. The majority of those killed in the United States were killed during construction: installing, moving, or removing wind turbines. The high number of deaths in the United States may be connected to the typically frantic, year-end, tax-subsidy-driven installation booms that have come to characterize commercial U.S. wind development. Germany, in contrast, has one of the lowest mortality rates in the wind industry despite the phenomenal growth of wind energy in the country since 1990.

Yet no passerby has ever been injured or killed by a wind turbine. The German parachutist, though a member of the public, was not a passerby: someone who walks or drives by a wind turbine and is inadvertently injured. The wind turbine that killed the parachutist could easily have been a building, tree, or cell phone tower.

There remains some, albeit minor, risk for neighbors and passersby. For example, some wind turbines have thrown their blades. Few have, it is true, but as turbines have grown larger, the consequences of any accident have grown as well.

Despite their hazards, wind machines are no more dangerous than many other aspects of modern life. We have all grown to accept the hazards of the electricity and natural gas that flow through our homes. Yet accidents with these common energy sources, though not frequent, are certainly not rare. Common do-it-yourself projects, such as painting the eaves or repairing the family car, are just as dangerous as working on a wind machine.

Treat wind systems with the same respect you would give any machine, and work as though your life were at stake—because it is. Safety equipment must be used, and used properly, before it's of any value. A safety harness, lanyard, or hard hat is no good when left on the ground.

Tower Safety and Climbing Gear

Work on the tower poses the most risk to those who want to install or service their own wind turbine: The possibility of accidents is greatest, and the severity of possible injuries is highest. Anyone who has hung from the top of a slender guyed tower in a strong wind readily appreciates this. So let's turn to work on the tower.

The most reliable way to avoid accidents is to avoid the hazard. Manufacturers of small wind turbines should strive to eliminate working on the tower altogether. Wherever possible, for example, assembly should be completed on the ground and the wind turbine and tower erected as an entire unit. Similarly, performing maintenance on small wind turbines while they are atop the tower should be avoided as much as possible by either using a hinged tower that can be lowered to the ground or by using a "boom lift."

Raising and lowering hinged towers create their own not-insignificant hazards. But tilt-up towers and integrated micro and mini wind turbines with few moving parts are beginning to make the goal of minimal maintenance on top of a tower a reality. Still, there are some small wind turbines installed on fixed guyed towers or on freestanding lattice towers, requiring inspection—if not maintenance—at the top of the tower.

Positioning Belts and Full-Body Harnesses

The term *safety belt* is often used incorrectly to describe any belt or harness used where a fall hazard exists. The so-called safety belts once used on construction sites and lineman's belts are more correctly labeled positioning or work belts (see figure 16-1, Working on the tower). These belts free the hands and, when used with care, can restrain falls.

Work belts were once made from wide leather straps; today they use synthetic webbing buckled around the waist. On each side of the belt are large metal D-rings. One or

Figure 16-1. Working on the tower. The Alternative Energy Institute's Ken Starcher (left) teaches the author (right) how to replace the brushes on a rebuilt Jacobs in 1979. The positioning belts used here were intended to restrain the wearer from falling. These belts were not designed to safely arrest a fall, if one occurred, but rather to aid working at heights. Today full-body harnesses and energy-absorbing lanyards would be used in addition to the positioning belts shown here. A work platform would have made this job much easier—and safer.

two lanyards link the wearer of the belt to the tower via attachments at the D-rings and to anchorages on the tower. When you're using a positioning belt your legs carry most of your weight. These belts were never designed to safely stop or arrest a fall, however. They are simply a tool to help perform the task at hand. Should a fall occur you'll need more protection than a work belt can provide.

Two decades ago many of us didn't know the difference between a belt and a harness. We thought a work belt was sufficient for working safely work at heights. Since then we've learned that during a fall a tremendous shock is transmitted though the lanyard or lifeline to the work belt and, hence, to the pelvis. An unconscious wearer may flip upside down after a fall and slip out of the belt completely.

European windsmiths have been using full-body harnesses for many years. Today anyone who works around wind turbines professionally should use a full-body harness (see figure 16-2, Full-body harness). Harnesses have been required as part of fall-arresting systems in the United States since 1995.

Full-body harnesses are designed to safely arrest a free fall, using leg and chest straps to distribute the shock of the fall over the entire torso instead of only the pelvis. These harnesses incorporate D-rings on the back and on the chest. In combination with a lanyard and appropriate anchorage, the D-ring on the back is designed to keep the wearer upright after a fall. The chest D-ring is used when climbing a ladder.

Combination harnesses include both fall-arresting and positioning components. Work on the towers of small wind turbines often requires standing for uncomfortably long periods on a narrow ladder rung or cross-girt. Some harnesses include a seat sling or strap across the buttocks that allows hanging par-

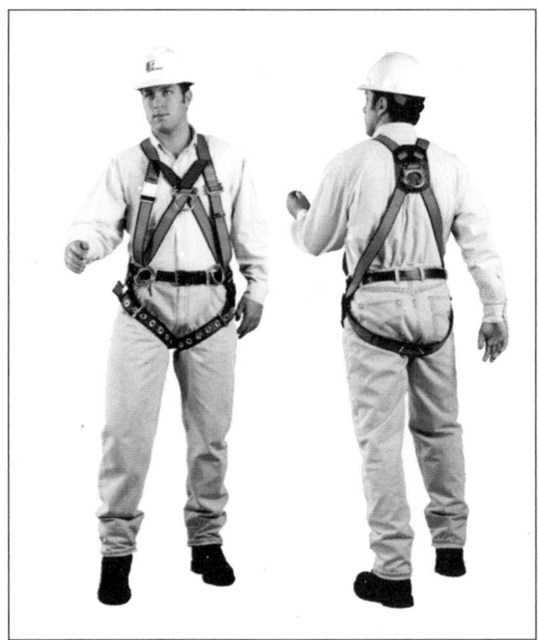

Figure 16-2. Full-body harness. This is a combination full-body harness for fall protection (back and chest D-ring) and positioning belt (side D-rings). The chest D-ring is used when ascending and descending a ladder. (MSA)

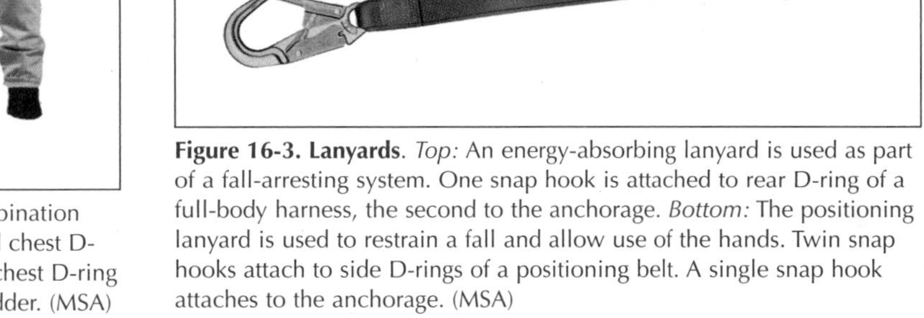

Figure 16-3. Lanyards. *Top:* An energy-absorbing lanyard is used as part of a fall-arresting system. One snap hook is attached to rear D-ring of a full-body harness, the second to the anchorage. *Bottom:* The positioning lanyard is used to restrain a fall and allow use of the hands. Twin snap hooks attach to side D-rings of a positioning belt. A single snap hook attaches to the anchorage. (MSA)

tially suspended from the positioning lanyards. These positioning belts take some of the load off the legs and make working at height more tolerable.

In the United States, the Occupational Safety and Health Administration (OSHA) sets standards for work belts and personal fall-arresting systems. The University of South Carolina explains that OSHA compliance requires a fall-arresting system consisting of a full-body harness, lanyard, connectors (snap hooks), and lifeline or anchorage capable of supporting a 5,000-pound (2,500 kg) load. Sit harnesses used for sport climbing have been used by homeowners and do-it-yourselfers while servicing their wind turbines. Sit harnesses for sport climbing are rated at 3,000 to 4,000 pounds (1,400 to 1,800 kg) breaking strength, less than that required by OSHA. Professionals should use only OSHA-rated harnesses.

Lanyards, Lifelines, and Anchorages

For a full-body harness to safely arrest a fall or for work belts to free the hands, they must of course be connected to a suitable anchorage on the wind turbine or tower. To do so often requires a lanyard (see figure 16-3, Lanyards). To comply with OSHA in the United States, a fall-arresting system must be capable of limiting the forces on the body to less than 1,800 pounds (820 kg) when stopping a free fall. Such a system must be used when the hazard of falling can't be controlled by railings or floors. To meet this standard lanyards should use an energy-absorbing element.

Unlike lanyards, which can be used either for positioning or to arrest a fall, lifelines are used only for arresting a fall. On a wind turbine a lifeline runs the length of the tower and is anchored either at the top or at both the top and the bottom of the tower.

For maximum fall protection make sure that lanyards and lifelines are always at their peak strength. Rope and cable serve many functions during the installation and service of a wind system. For this reason note that lifelines and lanyards used for fall protection

Figure 16-4. Work platform and lanyard anchorage. Niels Ansø servicing the Folkecenter for Renewable Energy's 7 kW turbine, used for domestic heating. Ansø has parked and tied off the rotor so it won't rotate. He has attached his lanyard to the anchorage ring at the top of the tower, just below the nacelle. Though there is no railing, this turbine includes a sturdy work platform, fall-arresting system, and clearly identifiable lanyard attachments.

should be dedicated to that application and not used for any other purpose.

Repeated use, dirt, chemicals, sunlight, oil, and grease all weaken rope and synthetic webbing. Keep lanyards and harnesses clean and replace them regularly.

Lanyards or lifelines must be attached to a fixed structural element rated for fall protection. On modern medium-size and larger wind turbines, anchorages for fall-protection lanyards are easily identifiable and appropriately labeled (see figure 16-4, Work platform and lanyard anchorage).

Note that on some megawatt-scale wind turbines, a low rail surrounds the top of the nacelle; similar rails can be seen on the hubs of some machines, as well. The rails on some turbines, while possibly useful for positioning, are not anchorages rated for fall protection. Never assume that a rail is rated for fall protection unless the manufacturer has assured that it will arrest a fall. Even then inspect the anchorage first before using it. German technicians have found severely corroded bolts in some anchorages on wind turbines at coastal locations. The German windsmiths concluded that some of these bolts would not arrest a free fall.

Wind plant managers instruct their windsmiths that they must be connected to an anchorage whenever there's a fall hazard. On top of the nacelle of a modern megawatt-scale turbine, for example, the windsmith must use two lanyards, says Zilkha Renewable Energy's Mike Kelly. When moving from one position to the next on top of the nacelle, for example, one lanyard always remains attached to an anchorage. An aide-mémoire, says Mick Sagrillo, is "One [lanyard] for me, one for the wife and kids."

Small wind turbine towers, except for those built in Denmark and Germany, seldom offer ready lanyard anchorages. Ian Woofenden, a senior editor at *Home Power* magazine, argues that small wind turbines in North America, when not installed on tilt-up towers, often use guyed lattice masts or freestanding lattice towers, which offer ample anchorages. "The whole top of the tower is an anchorage," says Woofenden. Possibly, but give careful thought to how you use those anchorages. Woofenden is a skilled arborist who has learned how to work safely at heights.

Avoid the temptation to simply wrap your lanyard around the top of the tower like a lineman on a utility pole. Should a fall occur, you'll slide down the mast until you reach the first set of guys. This is a common hazard to linemen on wooden utility poles in North

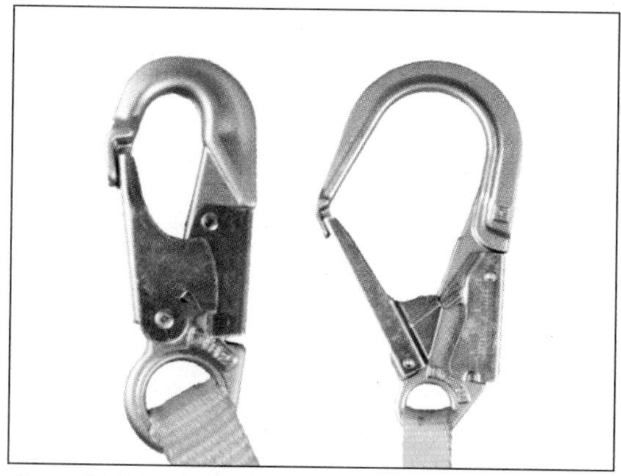

Figure 16-5. Locking snap hooks. Snap links (carabiners) and snap hooks with locking gates can be applied and released with one hand. The lock prevents the gate from unexpectedly opening and releasing. *Left:* A double-acting snap hook for attaching to an anchorage ring. *Right:* A rebar snap hook. (MSA)

America when their climbing gaffs "cut out" and they drop straight to the ground, with wooden splinters flying. Make sure you pass your lanyard through the tower—not around it.

Snap Hooks, Carabiners, and Slings

As the name implies, snap hooks are shaped like hooks and can be snapped onto an anchorage, a tower leg, the cross-girt of a lattice tower, or the D-ring of a body harness. Riggers, windsmiths, and others in the construction industry use snap hooks for positioning and fall protection. Carabiners or snap links may also be used if they're rated for fall protection. Carabiners are useful for clipping gear onto a work belt. Anytime a rope must be fastened and unfastened a number of times, a carabiner can make the task easier.

All snap hooks and carabiners used for positioning and fall protection should have both a spring-loaded gate (keeper) and a latch or fail-safe lock, says CalWind's Jon Powers (see figure 16-5, Locking snap hooks). The fail-safe or "double-acting" lock prevents the gate from opening inadvertently and releasing from the anchorage during a fall.

Sewn loops of synthetic webbing or "slings" were developed after climbers learned that a great deal of valuable rope was being lost when constructing rope harnesses. Slings can be used to carry equipment up the tower, tie down the rotor on small wind turbines, and perform any number of other tasks. Wider and stronger polyester slings are used extensively in rigging. Where erection jigs or fixtures are not handy, slings (appropriately rated) are a good choice for lifting loads such as small wind turbines or a completely assembled turbine and tower.

Fall-Arresting Systems

The most dangerous activities involved in working on a wind system are ascending and descending the tower. Amateurs often compound the risk by scaling the tower, then securing their lanyards after they have reached their work station. Such a practice is even more dangerous when descending. After several hours on the tower you're tired and your timing can be off, particularly in winter when biting winds quickly sap your strength.

Several manufacturers offer devices designed to mitigate this climbing hazard. These fall-arresting systems employ a sleeve that slides along a taut steel cable or synthetic rope that spans the height of the tower (see figure 16-6, Fall-arresting cable and sleeve). You attach your harness directly to the sleeve with a snap hook or carabiner. The sleeve rides up the cable as you climb the tower. Should you slip it locks onto the cable, arresting your fall.

Fall-arresting systems on wind turbine towers typically use steel cable. Because the tower sections on medium-size and megawatt-scale turbines are so large, the towers are not fully assembled on the ground. Each section is lifted independently into position. Consequently the fall-arresting cable hasn't been terminated at the base of the tower.

When Vision Quest Windelectric erected its Vestas turbines near Pincher Creek in the Canadian province of Alberta, Mike Bourns's first task was to ascend the towers and complete installation of the fall-arresting system. The fall-arresting cable had previously been attached to the top tower section before the tower section

Figure 16-6. Fall-arresting cable and sleeve. Windsmiths attach the chest D-ring on their safety harness to the carabiner before climbing the ladder inside this Vestas V44 tower in Traverse City, Michigan. When windsmiths ascend, the sleeve is below their harness (the lever points up) and slides freely up the cable. Should a windsmith fall the sleeve locks onto the cable (the lever points down, as shown here), arresting the fall. Many fall-arresting cables and rails are positioned in the center of the ladder, and the chest D-ring is attached directly to the sleeve with the carabiner. All towers that must be climbed to service the wind turbine should have a fall-arresting system. Note that the carabiner uses a locking gate.

Figure 16-7. Climbing a tower with a fall-arresting system. Niels Ansø is descending the tower of a household-size turbine at the Folkecenter for Renewable Energy, with two lanyards. He has clipped the large snap hook of one lanyard to his positioning belt. He used this lanyard while he was on the work platform at the top of the tower. He has attached the chest D-ring on his full-body harness to the fall-arresting sleeve that slides along a steel cable (not visible).

had been hoisted into position. The remainder of the cable lay coiled on the platform of the top tower section. Bourns scaled the tower using two lanyards in what's known in the trade as the 100 percent tie-off technique.

As Bourns climbed the ladder, he made sure that one lanyard was always clipped to a ladder rung. When he reached the platform of the top tower section, he lowered the fall-arresting cable to the ground, where it was terminated at the base of the ladder. He and others could now safely use the ladder by linking the chest D-ring on their body harnesses to a carabiner attached to a sleeve sliding along the cable (see figure 16-7, Climbing tower with a fall-arresting system).

Fall-arresting systems may also use synthetic rope. Some use a steel rail. The latter has the advantage that it can be installed on tower sections along with the access ladder before the tower sections are raised into place. As the tower sections are bolted together, the rails are also joined, providing continuous protection for those installing the tower and turbine. A fall-arresting rail permanently a part of tubular tower sections could have prevented Richard Zawlocki's death near Palm Springs. No fall-arresting system was in place when he descended the tower, and he failed to use his lanyard to protect himself.

In practice these fall-arresting systems protect windsmiths as they climb the ladder to the

Figure 16-8. Servicing a medium-size turbine. A windsmith services a Vestas V27 on Alta Mesa near Palm Springs, California. The nacelle of this 225 kW turbine is large enough that all work on the drive train is performed inside, though the low head height requires opening the rooftop doors.

enclosed work platform found on most medium-size wind turbines and the enclosed nacelle on larger machines (see figure 16-8, Servicing a medium-size turbine). Household-size turbines used in commercial applications also use cable and sleeve devices, but these are rarely found on most small wind turbines in North America.

As Bourns demonstrated, the use of two lanyards can be an alternative to fall-arresting cable systems. When ascending, one lanyard is attached above as far as possible. When it is reached, the second lanyard is attached above and then the first removed, and so on. This ensures that you are always tied to the ladder rung or tower cross-girt—even when reattaching a lanyard. Still, a fall-arresting cable and sleeve system is always preferable because it is so much easier to use—and therefore more likely to be used.

The two-lanyard technique illustrates a good overall safety practice—always keep one lanyard attached. "You must never rely on your hands alone," says Zilkha's Mike Kelly, who has climbed his share of towers. Like other fall-arresting lanyards today, these lanyards must be capable of absorbing the energy of a fall.

When servicing most small wind turbines, because they often lack fall-arresting systems you will need to use the 100 percent tie-off technique.

Fall-arresting systems are used to prevent serious injury should you fall. They should not be used as a work tool for positioning, say fall safety experts. Where a lanyard is used for positioning, then a separate fall-arresting system should be employed. Work tools and fall-arresting systems should be independent. F. Nigel Ellis explains in his book *Introduction to Fall Protection* (see the bibliography for further details) that "if you put your weight on it, it's for positioning."

To recap, for tower work, positioning lanyards and a work belt allow you tie off, freeing

Tower Work and Do-It-Yourselfers

Any wind turbine and tower that cannot be safely lowered to the ground for servicing should have a fall-arresting system for ascending, descending, and working atop the tower, a sturdy work platform, and safe, clearly identifiable anchorage points for attaching your lanyard. Homeowners and do-it-yourselfers should stay off towers of any type unless they've received training in tower safety.

your hands. You'll still need a full-body harness and fall-arresting system, such as a self-retracting lifeline anchored above you, to work safely. While these principles are acknowledged among professional windsmiths, homeowners, farmers, and even some small wind turbine manufacturers often remain unaware of their relevance to working safely with wind energy.

Figure 16-9. Wooden work platform. Even the most rugged wind turbines in the most out-of-the-way places need occasional service, and a work platform makes the job easier. Here an old Aerowatt stands on a makeshift tower at a ranger station guarding rare Magellanic penguins in Chile's Patagonia region. (Nancy Nies)

Work Platforms

Most older medium-size wind turbines incorporate a work platform at the top of the tower to aid in servicing the machine (see figure 8-2, Home built, but well built). Contemporary medium-size and larger turbines are large enough that most work is performed from safely inside the nacelle. On these turbines the body harness and energy-absorbing lanyard are used principally to protect against a fall while ascending and descending the tower. They are also used when you're working on the top of the nacelle or wherever there's the risk of a fall.

Small wind turbines, especially those manufactured in North America, seldom provide a work platform at the top of the tower. As a result, working on the tower of a small wind turbine demands a combination harness that can safely arrest a fall, while also freeing the hands for assorted tasks. Working at the top of the tower without a platform also demands stamina, as anyone who has done it can attest.

Ty McNeal knows. He's installed a host of wind turbines in Iowa, from household-size machines like the Bergey Excel and Jacobs 20 kW to medium-size Danish wind turbines. But McNeal has tired of installing and servicing the smaller turbines, in part because they lack a work platform. "I just don't want to hang on to a toothpick anymore," he says. "At least with the 65 kW [Danish] turbines there's a solid platform to stand on."

Household-size turbines manufactured in Europe often do include work platforms, modest as they may be (see figure 7-8, Guyed pipe tower). When they don't, enterprising owners make their own (see figure 16-9, Wooden work platform).

In the absence of a work platform, a boom or aerial lift provides a comfortable and secure working environment (see figure 16-10, Boom or aerial lift). Lift platforms include a guardrail and lanyard anchorages. These lifts have become increasingly common for servicing wind turbines of all types and sizes. They also eliminate hazards associated with climbing the towers of small wind turbines.

Figure 16-10. Boom or aerial lift. A technician services an Atlantic Orient AOC 15/50 at the National Renewable Energy Laboratory near Golden, Colorado. The tower for the AOC 15/50 includes a fall-arresting cable, but no work platform. The absence of a work platform necessitates using a boom lift in order to comfortably service the drive train. As here, the platform on the boom lift includes a railing and lanyard anchorages.

Ladders

As with work platforms and fall-arresting systems, small wind turbines installed in North America on guyed lattice masts lack a ladder. You're expected to use the mast's cross-girts to scale the tower. Freestanding lattice towers, on the other hand, provide climbing pegs.

Tubular towers on medium-size and larger wind turbines include an interior ladder. Vindmølleindustrien spokesman Søren Krohn notes that some Danish manufacturers place the access ladder at a certain distance from the tower wall. Windsmiths climb the ladder with their back to the wall. This enables them to rest by leaning back and placing their shoulders against the inside wall of the tower.

Some manufacturers of commercial turbines also provide rest platforms at regular intervals. This detail is much appreciated when you're climbing a 50-meter (160 ft) tower. Some, such as Tvind's tower in Denmark, include both a ladder and an elevator.

For more on fall protection, lanyards, snap hooks, work platforms, and ladders, see the Rigging section of the bibliography.

More Tower Tips

Besides combination harnesses and lanyards, there are other pieces of gear that can make tower work both safer and more comfortable. Boots are one; gloves, another. Always wear boots with firm, nonslip soles. Your feet tire less and are less likely to slip from a girt or ladder rung than with street shoes. Gloves do more than protect the hands—they help you get a better grip, and a good grip is paramount. Leather is best. The galvanizing used on towers forms droplets on the steel before it cools. These droplets can be sharp as a knife, and cut through cloth gloves with ease.

Hard hats or helmets are also essential attire. Admittedly they are uncomfortable—particularly in winter—and they're difficult to wear in a high wind unless fitted tightly or used with a chin strap. Their value becomes apparent, however, when you're working around small wind turbines that lack parking brakes, because the blades may hit you.

Most micro and mini wind machines do not have parking brakes. Even when furled or dynamically braked, the rotor may still spin slowly. Those blades may not look like much, but they can easily knock you off the tower. The rotor drive train contains a lot of inertia when it's turning, and this inertia can drive a lightweight blade with damaging force.

A similar problem is the unexpected yawing of the turbine in gusty winds. Just when you think you're clear of all that machinery, the wind will change direction and bring everything swinging your way. It's then that a hard hat and a fall protection system are truly important. Larger, commercial turbines feature both a parking brake for the rotor and a brake to prevent the nacelle from yawing unexpectedly.

Never work alone. Always have someone nearby who can go for help if you need it. Because even a slight breeze makes it hard to hear commands from the top of a tower, handheld radios are a useful tool for talking with your ground crew. Better yet are voice-activated or "hands-free" radios.

If the tower doesn't have a cable and sleeve device, climb exterior ladders on the windward side whenever possible. The wind will force you into the tower, not off it. Never climb the tower in high winds. At California wind plants, for example, no work on exposed lattice towers is performed in winds above 25 to 30 mph (12 m/s).

Keep the base of the tower clear, in case a tool or some lost parts come hurtling to earth. No one should work at the base of the tower while someone is working above. One California windsmith—who wasn't wearing a hard hat—received a serious head injury when his colleague dropped a bolt from the top of the tower!

As for tools, always carry them in a tool belt, keep them on a tool loop, or keep them in a bucket. All other items should be hoisted up with a rope once you are safely in place. Take a handline up with you on your first trip. Then use a nylon or canvas bucket to ferry small parts up and down the tower or to hold parts while you're working.

When around rotating machinery, whether it's a wind machine at the top of a tower or a bicycle in the garage, don't wear rings, watches, necklaces, loose clothing, or long hair (tuck it under your hard hat if need be). In *Windpower Workshop* Hugh Piggott recounts a "hair-raising" tale told by Mick Sagrillo of the encounter between Mick's former ponytail and the slowly turning shaft of a "Jake" in an Alaskan shop.

Never climb a guyed tower that is not properly secured to its anchors. This may seem patently obvious, but Steen Aagaard's accident shows that it's not. Stay clear of the tower during ice storms or freezing rain. If operating, the rotor will shed ice by throwing it to the ground. Ice projectiles typically strike directly below the wind turbine, and can be a hazard in the immediate vicinity of the tower. Never climb a wet or ice-covered ladder or tower. Remember, the wind turbine's expendable; you're not.

NRG Systems' Dave Blittersdorf warns that raising and lowering hinged towers demands your full attention. His rules are no bystanders, no news media, no distractions. With tilt-up towers, such as those made by NRG Systems, the lifting loads are greatest when the tower is near the ground, explains Blittersdorf. This is when any component that can fail, will. Though this is advantageous when you're raising the tower, as it provides the opportunity to test all components under full load, it can be disastrous when you're lowering the tower.

Members of an experienced field crew at the Alternative Energy Institute at West Texas A&M were lowering a 25 kW Carter turbine at their test field outside Canyon, Texas. They had done so many times before. They were good at it, but this time was different. There was a miscommunication. Worker 1 thought the turbine was ready to be lowered after checking that the

Steen Aagaard's Crippling Fall

"Two Hurt in Turbine Accident" blared the headline in the *Bakersfield Californian.* "The tower toppled while it was being erected, horrifying onlookers," continued the article by Jill Hoffman, a correspondent at the scene. The article was accompanied by photos of the crumpled tower, a crushed pickup truck, and a close-up of Steen Aagaard, 38, clinging to the slender tower as it fell over.

The details of what happened remain sketchy, but the general outline of events is clear from eyewitness accounts and from an accident report prepared by CalOSHA.

Aagaard was installing a Bergey Excel on an 80-foot (24 m) guyed lattice tower. The Excel uses a 7-meter (23 ft) rotor and weighs 1,000 pounds (460 kg). Two sets of three guy cables and anchors are designed to hold the lattice mast of solid steel rod upright.

The turbine, tail vane, and rotor had been mounted while the tower was on the ground. Aagaard, an experienced Tehachapi windsmith, was directing a rented crane that was raising the assembled turbine and tower when the accident occurred.

Aagaard made sure the mast had been placed on its pier and the guy cables attached, as he had done in the past. He then climbed the tower to release the lifting sling from the crane boom. Prior to climbing the tower, he personally inspected the guy cables, Aagaard told CalOSHA. He climbed the tower wearing a full-body harness and clipped his positioning lanyard to the tower. Aagaard then released the crane without incident. While remaining on the tower, he directed his ground crew in tensioning the guy cables.

At this point one of the cables "came loose from the come-along device" being used, says CalOSHA. The remaining guy cables "caught and held the tower precariously for a moment," says the newspaper account, before "the tower pivoted through the air and crashed to the ground."

Erik Slocum, a member of Aagaard's ground crew, was tensioning one of the guy cables with a "come-along tool" when it released, says CalOHSA's report. Slocum instinctively grabbed the guy cable and was pitched 15 feet (5 m) into the air. He suffered minor injuries and was taken to Kern Medical Center, where he was treated and released.

The guy cables were never directly attached to the guy anchors or the guy anchor turnbuckles. Instead the guy cables were attached to a cable grip. This grip was then attached to a tensioning device—what CalOSHA calls a "come-along device." This tensioning tool was then attached to the guy anchor turnbuckle.

This arrangement is common in rigging to allow quick and frequent take-up of slack in a cable or wire rope. However, it is not intended for use where someone would be at risk should the grip unexpectedly release. In other words, it is not rated for human loads.

The CalOSHA report notes only that the guy cable "came loose" from the tensioning tool. It's not clear whether the cable slipped through the "come-along device" or the cable grip used to hold the cable. In either case the guy cable was never securely attached to the anchor's turnbuckle before Aagaard climbed the tower. The turnbuckles are normally used to tension the guy cables after the tower has been set on its pier, but before anyone ascends the tower.

Bergey Windpower's installation manual explains the sequence to be used. Item 16 states, "Attach each of the guy wires to its turnbuckle . . ." Subsequently item 17 advises using "the turnbuckles to move the tower towards vertical and set tension in the guy wires." Finally item 18 says, "After the guy wires are secure and adjusted, the crane rigging can be released."

Other factors may have played a part. Aagaard was experienced in servicing commercial wind turbines. Installing household-size turbines was a sideline. He and his crew may have been overconfident; the Bergey turbine was a fraction of the size of the turbines they normally serviced. The news media was present, and there was a host of onlookers. Even the most experienced crew can be distracted by curious passersby. And when the media is present, it takes willpower not to "perform."

The onlookers were also too close to the tower and the installation crew. "Many of those present had to rush clear of the tower as it fell," the *Californian* reported. No one except the installation crew should ever be within the tower's fall zone. Item 1 under tower safety in Bergey's installation manual advises that "Persons not involved in the installation should stay clear of the work area."

Aagaard survived his fall, but suffered disabling injuries. The fall broke his back in two places, paralyzing him below the waist. Eight months after the accident, CalOSHA fined Aagaard's company $450 for violations of its regulations. One of these was not following the manufacturer's installation instructions.

hoisting cable was secured to the gin pole and the tow vehicle, and proceeded to release the turnbuckle connecting the gin pole to the anchor. Meanwhile worker 2 decided to replace the pin connecting the hoisting cable to the tow vehicle without telling worker 1. As a consequence, when the turnbuckle was released the guy cable was not attached to its anchor, and the hoisting cable was not attached to the tow vehicle. And to make matters worse, there was a photographer in the path of the tower. The tower whizzed by inches from the photographer's head and crashed to the ground. No one was hurt, but there were lots of deep breaths and red faces. The photographer quickly left Texas and never returned.

Small Turbine Electrical Safety

When servicing the control panel, disconnect the power supply from the turbine. Install a fused disconnect switch for this purpose (see figure 16-11, Disconnect switch). It will come in handy. In a battery-charging system, disconnect the batteries as well. If interconnected with the utility, also disconnect utility power.

Permanent-magnet alternators produce a voltage whenever the rotor turns—even when disconnected from the load or control panel! Open-circuit voltage can be up to five times nominal voltage; the hazard to safety and the potential damage to sensitive electronic equipment are thus both considerably greater.

Before you disconnect the turbine from the control panel or the load, check with the turbine manufacturer. Unlike Bergey Windpower's designs, few small turbines were ever intended to operate unloaded. Scoraig Wind Electric's Hugh Piggott warns that waiting for a calm day may not alone be sufficient. Restrain the rotor physically or by shorting the alternator's phases together to prevent the rotor from generating damaging voltages.

Off-the-grid power systems can also experience high current draws and high charging rates. Both conditions require that for safe operation all cabling be amply sized and the connections terminated correctly.

To protect against overcurrent, fuse all power sources in any installation, off-the-grid or interconnected. In a hybrid wind and solar system, fuse both sources and fuse AC and DC loads as well as connections to the batteries. Several manufacturers offer pre-engineered, preassembled power panels that include all necessary fusing or circuit breakers for both the DC and AC side of off-the-grid power systems. These panels are part of an encouraging trend toward more standardized and professional DC to AC power systems. Use them.

If you have any doubts about how to properly fuse a part of your power system or how to make sound terminations, consult the manufacturer or supplier of the component.

Wind generators produce high voltages. Use extreme caution anytime you open the control box or the nacelle cowling, or work around the slip rings. Always turn off the power from all sources before working around electrical components. In any wind system, there is power both from the wind turbine side and from either the utility side or the batteries.

Use insulated tools whenever possible. Remember, electricity can kill. But if you're working on the tower, the shock itself may not be the greatest danger. Electric shock can cause you to lose your grip. Even if you fall only a short distance, you could be seriously injured.

Before poking your insulated screwdriver into a control panel, check the circuits with a multimeter. You could have thrown the wrong switch by mistake, or you could wrongly assume that someone else has de-energized the circuit. This is a particular hazard during the installation of multiple turbines, when there is a lot of activity on the construction site and it may not be clear which turbines have been energized. This is also a problem in wind–PV hybrids with multiple power sources. Always test first.

Take your time, and think about what you're doing. Never wear metal jewelry when working around electricity. Don't wear rings—even inside gloves—or watches or necklaces.

Avoid constructing a tower near utility lines. If you have any doubts about clearance between the tower or the boom of a crane and a power line, call the utility company before you start to erect the tower. This precaution could have prevented Pat Acker's death in Texas.

Stay clear of the tower if a storm, especially an electrical storm, is threatening. A lightning

Figure 16-11. Disconnect switch. Throwing the switch to "off" and putting a lock through the handle ensures that no current is flowing from the switch to the load. This disconnect switch (left) was installed by Dave Blittersdorf on the guyed tilt-up tower supporting the Bergey Excel at his home (right) near Burlington, Vermont. Because the Bergey Excel uses a permanent-magnet alternator, there is a potential electrical hazard whenever the rotor is turning. Note the junction box beneath the switch box, flexible conduit from the tower to the disconnect switch, bare copper grounding wire connecting the tower to a buried ground rod, and "hitch pins" on the tower hinge. Instead of the "hitch pins" shown here, use a through-bolt with a locknut.

Figure 16-12. Insulated tools. Mick Sagrillo demonstrates how he insulates his metal tools when working around batteries.

strike anywhere near the tower will energize all metal components. Static buildup before a storm can produce a similar effect.

Batteries

Always use extreme caution when working around batteries. Use substantially more precautions than you would use when working near an automotive battery, advises Mick Sagrillo. Because batteries store energy in a concentrated form, be extremely careful when working around or servicing batteries in a stand-alone power system. An accidental spark can ignite the explosive hydrogen gas given off when batteries are fully charged. Rapid discharge by a short circuit—for example, from a tool falling across the terminals—can cause batteries to explode.

Ventilate the battery storage area to avoid the accumulation of hydrogen gas, and thoroughly fuse battery cables to prevent unintended discharge. Wear goggles whenever working near batteries to protect your eyesight from battery acid spray.

Beware of dropping metal tools onto exposed battery terminals. Some pros recommend insulating metal tools for working around batteries (see figure 16-12, Insulated tools). Cap battery terminals with plastic covers to reduce the hazard.

Vent the batteries adequately to the outdoors to prevent concentrations of explosive hydrogen gas. Avoid any source of sparks or open flame around the batteries.

Rope and Cable

Rope is one of the most essential tools for installing and servicing small wind turbines. Because of its importance and its many uses, rope has reached a state of complexity that's almost baffling. To pick the right rope or cable for the job, an understanding of a rope's inner workings is helpful.

Rope is made from fibers. These can be natural or synthetic. Hemp, manila, cotton, and sisal are examples of natural rope. Manila is the most common. Synthetic rope is made of nylon, polyethylene, or polyester. Fibers are twisted to form yarns, and strands are wound from the yarn. Several strands are then combined to form rope. The role of polyester rope is increasing over that of nylon, says DBI Sala's Scott Paul, because of its superior properties.

Strands of steel cable, as opposed to those of rope, are wound from wires instead of yarn. The wire is made from various grades of steel, depending on their intended use. For example, stainless steel may be used when corrosion protection is desired.

Europeans have developed a jacketed rope that has become a climber's standard accessory. In kernmantle construction a core (kern) of braided or twisted strands is covered by a protective, braided sheath (mantle). No strands are exposed at the surface. Consequently there's less opportunity for debris to work between the strands and damage the rope. The sheath can take a lot of abuse while still preserving the integrity of the core. Ropes of this type can be found in wind turbine applications.

Manufacturers indicate the breaking strength of rope or cable on the spool. You should look for this when you purchase rope and cable because it will partially determine the rope you choose for a particular job. Remember that breaking strength applies only to new rope; it decreases with age and use. Breaking strengths vary between North American and European manufacturers. American manufacturers list the average breaking strength; European manufacturers list the minimum breaking strength, which can be from 10 to 15 percent less than the average breaking strength.

The force acting on a rope can greatly exceed the simple weight of a load when the lift is uneven or the load drops suddenly. Knots can reduce the breaking strength of rope by as much 45 to 75 percent. As with the webbing of harnesses and lanyards, soiling, grit, sunlight, and use can reduce a rope's strength even further.

When choosing the type and size of rope for a specific job, buy it according to its working strength, not its breaking strength. Working strength allows a reserve or margin of safety. This will account for any reduction in strength due to knots or sudden loads on the rope. Dave Blittersdorf, for example, uses a safety factor of 5:1 in the cables he specifies for raising NRG Systems' tilt-up towers.

In general, nylon has twice the strength of manila rope. Nylon stretches; manila does not. Polyethylene also stretches but is made from a coarser fiber than nylon and is more resistant to abrasion. Polyester is more resistant to sunlight than the other synthetics, but like manila it doesn't stretch. These features determine which rope is better suited for a particular job. Nylon is preferred over manila or polyester for lanyards, for example, because it stretches.

Like rope, not all steel cable is created equal. Guy strand is stiff and heavy. Winch cable, in contrast, is flexible so it can be wound around the spool of a winch or threaded over a pulley. Guy strand comes in several grades, from common or utility grade to extra-high strength. Guy strand is also galvanized. Other cable may be of stainless steel or simply of untreated steel.

The construction of steel cable is given by the number of strands and the wires per strand. Almost all wire rope has at least six to eight strands. For example, a "6 x 7" cable has six strands and seven wires per strand. What you use depends on what you want to do with the cable. For a hoisting line, winch cable is superior. Tractel's Super Pull-All model griphoist (North American version) is shipped with 0.25-inch (6.5 mm) diameter, 7 x 19 wire rope.

Inspect rope regularly for signs of abrasion. Natural rope usually shows clear signs of fatigue and wear. It becomes limp, surface fibers become soft and fuzzy, and the inner fibers rot. Synthetic rope becomes limp and soft with frayed surface fibers. No rope should be used for more than five years, even when properly treated.

When using steel cable for a hoisting line, check for nicks or kinks that may weaken the cable. When cutting wire rope, keep a grip on both ends so it doesn't recoil and whip around. The cable should not be pulled over a sharp radius or tied in a knot like a fiber rope. Cable should be looped over a thimble and secured with wire rope clips as explained in chapter 14.

Pulleys

Pulleys serve several functions: to change the direction of the force applied to a hoisting line, to provide a mechanical advantage, and to reduce the rate at which a load is moved. They make the job of hoisting wind turbine components easier and safer. Pulleys allow you to lift a load from the security of ground level, rather than putting yourself in the absurd and dangerous position of carrying up tower and wind turbine components by hand.

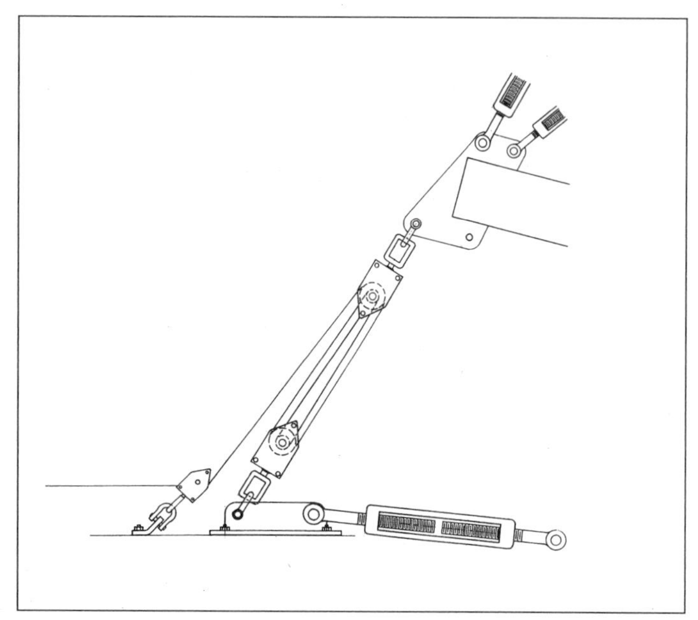

Figure 16-13. Pulleys. Pulley blocks are used to raise a Bergey Excel tower with a ginpole. The winch line is on the left; the turnbuckle for gin pole on the right. The pulleys shown here provide a 5:1 mechanical advantage. (Bergey Windpower)

Carrying the generator for a small wind turbine up a tower may seem ludicrous, but it's been done. By fixing a pulley on a gin pole at the top of a tower and a snatch block (a pulley with a locking snap hook) at the base, you can hoist components while standing a safe distance from the tower.

Fixed pulleys alone provide no mechanical advantage. To gain a mechanical advantage, there must be a pulley that moves with the load. The number of lines between the movable pulley, or block, and the fixed pulley determines how much mechanical advantage you gain. If there are two lines between the movable block and the fixed block, you need only one-half the force to lift the load. To gain this advantage, though, you must use twice as much rope. From high school physics, you may remember that work equals force times distance. To perform the same amount of work as before when we reduce the force by half, we must double the distance through which it acts (see figure 16-13, Pulleys).

Winches and Hoists for Small Wind Turbines

Heavy loads, such as complete tower assemblies and small wind machines, can be hoisted up a tower using a tow vehicle or a winch. Similarly, a tow vehicle or winch can be used to raise tilt-up towers with a gin pole.

Using a tow vehicle, either a truck or tractor, to raise a wind turbine is a ticklish operation. Whenever a vehicle is used you lose a degree of control over the lift, no matter how good the driver. Ugene Stallhut was killed when the tractor he was using as a tow vehicle flipped over and crushed him. Robert Skarski was killed when his brother, driving the tow vehicle, literally pulled the tower over with Robert on it.

Winches are preferred over vehicles for raising wind turbines and heavy tower components. Most vehicle-mounted winches have insufficient spool capacity for raising towers. Electric winches selected for the purpose, such as those used by NRG Systems to raise their tilt-up towers, have the required spool capacity.

Winches should have a brake that locks the spool in either direction. The brake should automatically engage if power is lost. The winch should be mounted on its own foundation or anchorage. More than once installers have been shocked to see their "tow" vehicle being pulled across the ground as a hinged tower headed toward its natural state of repose—on the ground.

The griphoist, as explained in chapter 14, is an extremely valuable tool for raising small wind turbines. The griphoist does not spool cable on a reel; instead it pulls or releases the cable through the tool body. Thus the cable can be of any length. Griphoists offer superior control, and hence safety, over tow vehicles and electric winches for raising and lowering small wind turbines.

Loss Prevention

If there is any possibility that the tower will be scaled by thrill seekers, children, or vandals, install anticlimb guards. These can be purchased from the tower supplier, or you can improvise. On freestanding truss towers, you can remove the lower rungs of the ladder or climbing pegs. On guyed lattice masts, you can wap the lower section with sheet metal. This is nearly always sufficient. There's no need to fence wind turbine towers. Period.

Protect manual controls at the base of the tower by removing winch handles or chaining them down. The massive doors on the tubular towers of medium-size wind turbines include locks for preventing unauthorized entry. Some doors, especially in vandal-prone California, incorporate a blind metal cover protecting the lock, discouraging even the most ardent troublemaker.

Because the attachments of the guy cables on guyed towers are so tempting to vandals, treat the threads of bolts to prevent the nuts from being removed. (One way to do this is to peen the exposed threads.) Also install a safety cable through the turnbuckles. This prevents both vandals and normal vibrations from loosening the turnbuckles and releasing the cable.

Similarly, ensure that the hinges on tilt-up towers can't work free. Use hinge shafts or bolts with a bored hole in both ends. Retain the hinge shaft with washers and a through-bolt with locknut. Avoid using the retaining pins found on trailer hitches. They were designed to be quickly and easily removed, which is the last thing you want for a pin holding up your valuable wind turbine.

Avoid placing guy anchors in pathways, but where you must do so, consider planting low shrubs or bushes around the anchors. People tend to detour around hedgerows rather than charging through them. Shrubs can also soften the line between the tower and the anchor. Slip fluorescent plastic guards over the guy cables to make them more visible.

When tubular towers will be installed during winter, they should be shipped with covers to prevent ice and snow from accumulating inside. This simple precaution would have prevented Eddie Ketterling's death on an icy winter day in southwestern Minnesota.

Manufacturers or operators of large wind turbines must also ensure that all large-diameter openings in the nacelle or hub, such as access to the interior of a blade root, are covered. The blade roots on megawatt-scale turbines are so cavernous that there is a very real hazard that someone could tumble into the blade while servicing, say, the pitch bearings. If a door covers a hatch in the floor of the nacelle, it must be capable of supporting the weight of an adult without failure. There are harrowing tales of floor hatches—used to raise small components to the nacelle—giving way.

All fall-arresting components must be inspected prior to use by the end user and should be periodically inspected by somone trained to detect defects. Harnesses and lanyards should always be replaced after they have arrested a fall. Harnesses and lanyards soiled with grease or oil should also be replaced. And periodically inspect bolts supporting fall-arresting anchorages for corrosion or missing fasteners.

This chapter provides general guidelines for working safely around wind turbines, large and small. This material is by no means exhaustive—entire books have been written on fall protection alone. If you're prudent and cautious you should be able to install, operate, and maintain a wind turbine in relative safety. If you have any doubts about your ability to perform tasks safely, don't hesitate to seek professional help or attend a workshop where safe working practices are part of the class syllabus.

17

Looking to the Future

He that will use all winds, must shift his sail.
—John Fletcher, "Faithful Shepherdess"

In the preceding chapters we've learned that wind energy works. It may not be as simple or as straightforward as we would like, but wind energy does work, and it works reliably. Three decades after the modern wind revival began, we can say with certainty that wind energy has come of age as a commercial generating technology, and that it's here to stay.

No miracles or media-grabbing breakthroughs are necessary before we put wind energy to productive use pumping water, powering remote homesteads, or generating electricity in parallel with that from the utility, whether we do it with one wind turbine or thousands. As Green Energy Ohio's Bill Spratley argues, we don't need more studies, we need to seize the day or, as he puts it, "Carpe Ventum"—seize the wind—and put more of this technology to use.

As wind energy continues to grow, new applications are appearing that at one time seemed far-fetched, the province of dreamers like naval architect Bill Heronemus, who envisioned wind turbines riding offshore platforms, or others who saw a future with wind-powered electric cars, or communities powered by their own wind turbines.

Still, there remain many challenges facing wind energy, some due to its remarkable success. And there remains plenty of room for improvements in the technology, especially among small wind turbines.

Expanded Applications

As we've seen, small wind turbines have been traditionally called upon to provide valuable services beyond the reach of utility lines. Such battery-charging applications, while important, have been overshadowed by the development of large central stations—wind power plants—connected to the grid. These massed arrays can contain hundreds and sometimes, particularly in North America, thousands of wind turbines. While enormously successful, both economically and technically, wind farms are only one way that wind energy can be used interconnected with the utility network. Another potentially important role, one suitable to large and small turbines alike, is the use of wind turbines interconnected with the grid, but at the end of the utility's distribution lines (see figure 17-1, End of the line).

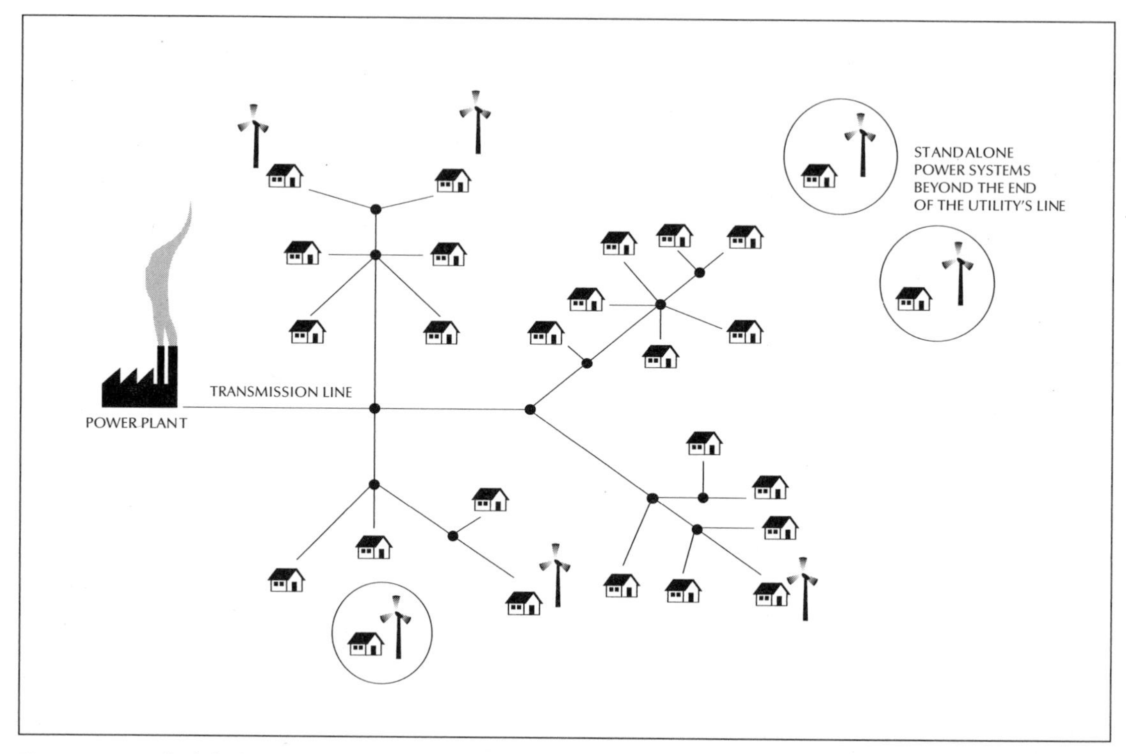

Figure 17-1. End of the line. Small battery-charging wind turbines traditionally have been used beyond the reach of utility lines. They and their medium-size brethren, however, can also be useful in distributed applications providing services at the end of a utility's distribution system.

Distributed Wind

Unlike conventional power plants, wind energy is modular; each turbine is a power plant unto itself. One of wind's attributes is that it can be dispersed across the landscape and employed where the energy is needed: in a backyard, for example, or near a factory, as at Schafer Systems in Iowa. In this way less energy is lost in transmission. Further, distributed sources of generation make the network more resilient.

Carl Weinberg, the former chief of Pacific Gas & Electric's research division, points out that large wind plants connected to the high-voltage network are no different from other centralized power plants. That's part of wind's success story: Wind power plants now appear much like conventional sources to utility managers. Nonetheless, such concentrations of wind turbines sacrifice the benefits of decentralization that individual wind turbines can provide.

Utilities, such as Pacific Gas & Electric, have found that by reducing the loads at the end of a heavily used line, they can avoid construction of expensive new transmission capacity. Loads can be reduced through conservation, or by installing modular sources of generation such as wind turbines. Dispersed sources of generation help the utility meet growing demand at lower cost than by traditional methods of expanding the distribution system. Further, says Canadian Doug Woodward, geographic dispersal of wind turbines—whether small, medium, or large machines—evens the ebb and flow of wind-generated supply across a distribution network as weather systems move through a region.

French manufacturer Vergnet has adopted the strategy for marketing its small and medium-size turbines "in proximity" to the load, or end user. The wind turbines are sited locally for local benefit. The grid serves, in

Figure 17-2. Distributed wind. One example of the many distributed wind turbines on farms across Denmark.

Vergnet's case, to distribute the energy from the wind turbine to the consumer. As in distributed wind, several turbines can be connected at distribution voltages, or one turbine connected at low voltage on the consumer's side of the distribution transformer. Vergnet hopes that farmers and others will want to develop their own wind resources with clusters of medium-size turbines, instead of leasing their land to commercial wind farm developers.

Distributing wind turbines throughout the utility network in this way takes full advantage of wind's modularity. The utility and its ratepayers benefit by beefing up the distribution system and serving new customers at lower cost. If rural residents are compensated sufficiently, wind energy can become a new source of farm revenue, helping preserve rural communities (see figure 17-2, Distributed wind). Society gains a nonpolluting, renewable resource. Everyone wins.

Community Wind

Wind can be—and is today—a producer of commodity or bulk electricity, but wind energy can be so much more. Wind energy can become a tool for rural economic development. Equally as important, wind energy can offer ownership, and with it a sense of control, to farmers and villagers buffeted by globalization and the industrialization of agriculture. Wind could offer a ray of hope to dying rural communities.

Community wind, as Mike Mangin calls it, is wind development on a human scale. Mangin, a thought-provoking Wisconsin maverick, promotes small clusters of medium-size turbines, like those found Denmark and Germany. To Mangin, community wind signifies local turbines, locally owned. They may be owned individually, cooperatively, or collectively through municipal governments. The key, says Mangin, is for the community to identify the turbines as its own. Doing so

MinWind—Farmer Owned

When giant wind farms began rising in their midst, a group of innovative Minnesota farmers asked whether there was another way. Could they do it themselves, or would they, too, have to lease their land on the state's famed Buffalo Ridge to out-of-state wind developers? Was there a way to develop wind energy and yet keep as many of the benefits as possible flowing through the local economy?

After two years of struggle a group of farmers in the southwest corner of the state proved that community wind development could indeed be done. Like farmers elsewhere in North America, they found that the biggest challenge was not the technology, not the wind resource, and not raising the money. It was finding a utility that would buy the electricity at a fair price.

The stubborn farmers wouldn't surrender their dream, says Mark Willers. They eventually persevered, erecting four 950 kW NEG-Micon turbines on the land of one of the participating farmers.

Willers, who raises corn (maize) and soybeans near Beaver Creek, Minnesota, was one of the farmers instrumental in bring the project to fruition. He and his neighbors could see that farming in the Midwest had become, in business school jargon, a mature industry. They were competing with each other to produce a commodity product at ever-lower cost. The farmers needed another source of revenue if they were to survive. And if they couldn't survive, the communities of which they were a part would also die.

From their experience with the area's cooperatively owned ethanol plants, they knew the energy business and could see that wind energy looked attractive. Wind energy was increasingly competitive with other sources, and, because of its modularity, offered the possibility of local ownership.

According to Willers, "There's a difference between local ownership," and what it can offer the community, and absentee ownership in which the landowner—the farmer—receives only a royalty. Willers and the group of 66 local investors made a point of using local contractors and consultants to maximize local benefits.

The farmers created two partnerships with two turbines each, dubbed MinWind I and II, and structured them to be open to members of the community who are not farmers themselves. Owners include two grocers and a newspaper editor, as well as farmers in the area.

The partnerships are managed essentially as a cooperative. Willers and his colleagues created them to take advantage of both the federal tax credit and Minnesota's incentive payment of $0.015 per kilowatt-hour for wind projects up to 2 MW, a program specifically designed to help the state's farmers.

Farmers have both the financial resources and the entrepreneurial spirit needed to make projects such as MinWind happen, says Willers proudly. The farmers shocked observers when they raised the equity portion of the $3.5 million project in only 12 days.

Willers and the civic-minded group of farmers who made MinWind a reality are now working with other groups to replicate elsewhere their model of local ownership.

---

may avoid the all-too-common conflicts encountered when developers, viewed as outsiders, propose projects that primarily benefit absentee owners.

Mangin cites controversial proposals for wind farms in Wisconsin as an example of what can go wrong. He characterizes the bumbling of some Wisconsin utilities as being "like sumo wrestlers teaching each other ballet." Long accustomed to pushing projects through over local objections, the companies made one blunder after another, he says, damaging wind energy's credibility in the state.

Community wind development can never guarantee total community acceptance, but it does offer one more way in which wind energy can be used to best advantage.

Offshore Wind

In the early 1970s an obscure professor at the University of Massachusetts proposed an outlandish scheme, or so it was thought at the

Figure 17-3. Offshore. Bonus 2 MW turbines on the Middelgrunden sand bank outside Copenhagen. This is one of several offshore projects operating off the coast of northern Europe. (Bonus Energy)

time. Bill Heronemus, who as a young engineer helped construct nuclear submarines, broke ranks with his former colleagues, concluding that plans to lace Boston Harbor with submerged nuclear power plants were not only too costly but also too dangerous. Heronemus instead proposed fleets of offshore wind turbines.

At a time when nuclear power was seen as the wave of the future, Heronemus was decried as an unrealistic, wild-eyed dreamer. Then nuclear faltered, and all the king's men couldn't put it back together again. Meanwhile wind energy proved itself, and Heronemus's vision has taken form—not in Boston Harbor as yet but in the shallow waters off Denmark. By 2040 the Scandinavian country plans to provide 50 percent of its electricity from renewable sources of energy—in large part from offshore wind turbines (see figure 17-3, Offshore). Unlike those envisioned by Heronemus, these offshore wind plants were not built on floating platforms but were firmly rooted to the ocean floor—and to reality.

While the offshore market has been slow to develop, small near-shore projects were completed in waters off Denmark and Sweden in the late 1990s. Because offshore turbines require more expensive infrastructure, such as waveproof towers and foundations, projects have longer lead times. They must also be larger than those on land to justify the expense.

One large project of twenty 2 MW turbines was built in 2000 on the Middelgrunden banks in the straits between Copenhagen and Sweden. The turbines, half of which are owned by a cooperative of the capital city's residents, are visible from the Christiansborg Palace, the seat of the Danish parliament. Middelgrunden is the first of many to come. In 2002 Denmark began installing more than 300 MW offshore under a government-sponsored program: the 160 MW Horns Rev project in the North Sea west of Esbjerg, and the 158 MW Rødsand project in the Baltic Sea.

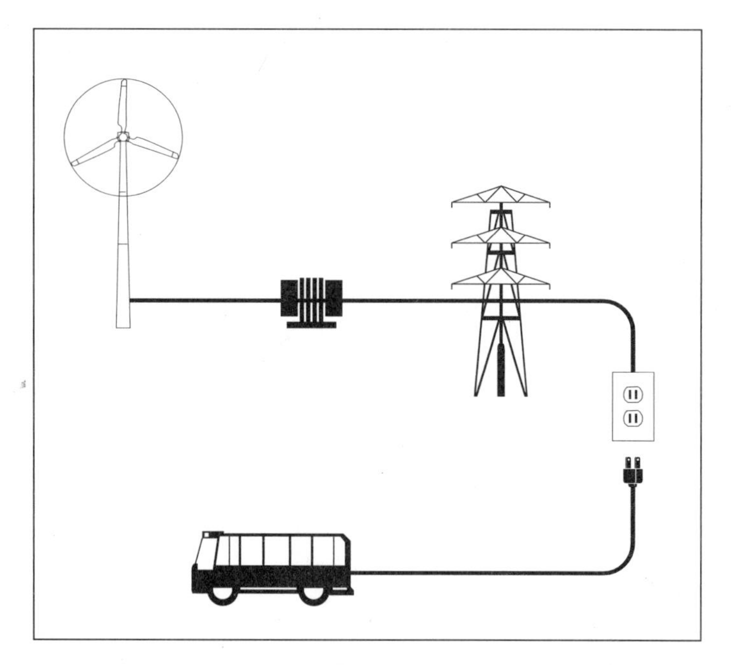

Figure 17-4. Electric vehicle charging. While large-scale storage of bulk wind-generated electricity in batteries remains impractical, storing wind in electric vehicle batteries is conceptually attractive. Such a system would enable wind turbines to provide fuel for transportation.

Offshore projects have been proposed for off the coast of the Netherlands, Germany, and France, as well as for the Irish Sea between Ireland and Great Britain. Some projects have also been proposed for along the eastern seaboard of North America. Offshore development has the potential to produce large quantities of bulk electricity near urban markets.

Electric Vehicle Charging

Using wind and solar energy to somehow reduce our seemingly insatiable appetite for liquid fuels in transportation has long been a dream of renewable energy advocates. Just as offshore wind plants were once considered a pipe dream, refueling electric vehicles with wind energy may still seem fanciful (see figure 17-4, Electric vehicle charging).

Electric vehicles produce fewer nitrogen oxides, carbon monoxides, and reactive organic gases than gasoline-powered vehicles, even when taking into account emissions from the conventional power plants that would normally generate the electricity to charge them. Of course, they would be even cleaner if the electricity came from the wind.

Transportation opens up a potentially huge new market for wind generation. Electric cars could absorb large amounts of generation that utilities might otherwise have difficulty using. When wind energy is abundant and demand weak, excess generation could be directed into electric vehicle batteries: a kind of mobile storage system.

Current electric vehicles consume about 0.5 kilowatt-hour per mile (1.6 km) traveled. If the typical North American vehicle travels 15,000 miles (25,000 km) per year, it would need some 7,500 kWh. One medium-size wind turbine of 700 kW capacity at a site with a class 4 wind resource of 7 m/s (16 mph)—as is found in many parts of the world—could alone power 300 vehicles. Communities in Denmark, Germany, and California are now doing just that, albeit on a modest scale (see figure 17-5, Tehachapi rides on the wind).

Other forms of transportation could also become a potential new market for clean sources of electricity. Canada's Vision Quest Windelectric is supplying wind-generated electricity for Calgary's light rail system. Calgary's "ride the wind" program uses 26 million kWh of wind-generated electricity annually, the production from 12 of Vision Quest's turbines. Similar projects could be used to offset the consumption of subway systems in major cities, or of high-speed electrified rail systems such as those in France and Germany, or the magnetic levitation rail line in Shanghai, China.

Better Technology

Observations made at the Wulf Test Field, and by users of small wind turbines worldwide, reveal that the technology of commercial or wind farm turbines has progressed much more rapidly than that of small turbines. Commercial medium-size and megawatt-class wind turbines

Figure 17-5. Tehachapi rides on the wind. Like their counterparts in Denmark and Germany, Tehachapi civic leaders decided to tap the area's abundant winds—and its abundance of used wind turbines—to power a budding fleet of municipal vehicles. Here an early-1980s vintage Nordtank turbine spins out wind electricity for the city of Tehachapi.

are technological marvels. They set a standard of reliability and performance that the small wind turbine industry should strive to reach.

Small Wind Turbines

To achieve this objective first requires a frank appraisal of the technology. Trying to market small wind turbines as though they were as simple and easy to use as solar panels, says Home Power's senior editor Ian Woofenden, is a mistake. It only leads to dissatisfied consumers charging that small wind turbines don't work. Rather, Woofenden says, we must build robust equipment that will stand up to the harsh conditions it's likely to encounter.

Others suggest that small wind turbine manufacturers must also recognize that there are places where small wind turbines make sense and places where they don't. Emphasis on sales of turbines alone leads to inappropriate applications and, again, dissatisfied customers. "What we [the small wind turbine industry] should be about is producing renewable electricity—and not just selling equipment," urges Wisconsin's Mick Sagrillo.

Yet Sagrillo also admonishes consumers to acknowledge that reliability, low maintenance, and longevity all come at a price. These values are front-loaded in a wind turbine. Consumers must "be wary of Internet marvels" that promise far more than they can deliver, Sagrillo says.

With rotor efficiencies now nearing 40 percent (about 70 percent of the theoretical limit), no one expects any startling technological leaps that will revolutionize small wind turbines. In keeping with the steady progress of the past decade, analysts expect continuing incremental improvements. Advances in aerodynamics have already resulted in quieter, more efficient airfoils.

Better instrumentation, now seen in electronic inverters, could be a boon to owners of small turbines, said Christopher Meier in an Internet discussion group. Instrumentation and diagnostic software could help in troubleshooting problems and simplifying after-sales service calls. Imagine calling the manufacturer's support line and being able to read off a list of parameters that the turbine's electronics have measured, or e-mailing a service file downloaded from a port on the inverter.

Nearly all commercial-size turbines today have their own phone lines and internal diagnostics. When something goes wrong, or even when something may go wrong, the turbine dials the service desk at the manufacturer, the owner, or both, and lists its ailments. Turbine owners and operators regularly poll their turbines to view "real time" performance, sensor readings, and operational characteristics such as blade pitch, contactor closures, and so on.

Windsmiths who service dozens of turbines every day often begin by scrolling through several screens of operating parameters on the turbine's controller before taking any action. Such electronic diagnostics on small wind turbines could greatly help users and manufacturers alike.

Rare-earth magnets, such as neodymium-iron-boron, are finding increasing application in small wind turbines. By incorporating these magnets with their high flux density in new generators, wind turbine designers believe they will squeeze more power out of low wind speeds.

Power electronics are also advancing rapidly. New electronic switching devices are driving a revival of static and synchronous inverter technology. The effects have already been seen in cheaper, more reliable inverters for the stand-alone power systems introduced during the late 1980s. Advances in power electronics for medium-size turbines are filtering down to the small turbine industry. Improvements are just ahead in interconnected wind systems, as well.

Another development that could broaden the appeal of small wind turbines is the further integration of the components (turbine, tower, and inverter) with installation. According to Sagrillo, "The market is moving toward turnkey systems"—in which wind turbines would be sold and installed like any other major household appliance. Mike Bergey agrees. He sees continued refinement of installation with screw anchors, and simple, easily erected guyed towers. Collectively these refinements will permit erection of small turbines in a matter of hours, not days.

Altogether these enhancements will make small wind turbines more accessible—and more cost-effective—to a greater number of users. Mike Bergey estimates that manufacturers could potentially reduce the overall cost of electricity produced by small wind machines by one-third if there was sufficient demand.

Mick Sagrillo, whose dream of "a wind generator in everybody's backyard" was dashed during the 1980s, remains hopeful. "It's no longer just tinkerers," he says, who are interested in wind energy. Sagrillo and others are convinced that environmental concern will repower the small wind turbine market in North America, as it has for larger turbines in Europe. He believes small wind turbines will eventually find their place in the sun.

Medium-Size and Larger Wind Turbines

The cost of electricity from commercial wind turbines continues to decline through economies-of-scale as the turbines grow ever larger. Savings also accrue in large projects of multiple turbines by spreading infrastructure cost, such as for roads and cabling, over the greater number of kilowatt-hours that big projects produce. Countering this pressure is the limited number of windy sites near high-voltage transmission lines where massive projects can be built.

Turbines continue to increase in size, especially for offshore where bigger turbines justify the greater infrastructure cost. Surprisingly, the push for bigger turbines on land has come not from utilities—with the exception of some prestige projects of the *my-windmill's-bigger-than-yours* variety—but from German farmers. Where planning restrictions have limited farmers to a single turbine, they've pushed for the biggest turbine possible, to maximize their revenues.

Despite the march toward multimegawatt turbines, some manufacturers are reexamining the medium-size class for applications where the megawatt-scale turbines are simply too large. Enercon, always an innovator, introduced its 300 kW E30 not long after the market-dominating success of its 500 kW E40. When the company introduced a 4.5 MW direct-drive turbine for offshore, it was pushing forward with its 100 kW E20 model, a size that harks back to the mid-1980s.

Enercon and other German manufacturers have also spearheaded the increasing use of power electronics in variable-speed turbines. For the megawatt-class machines, Danish and German manufacturers have all moved to full-span pitch control. While nearly all German manufacturers have opted for variable-speed operation, some Danish firms have continued using fixed-speed operation, a technology that has served them well.

Direct drive remains a select fraternity with Enercon in Germany, Jeumont in France, and Lagerwey in the Netherlands. The remaining manufacturers have stayed with what they know best: gearboxes.

As the turbines grow larger, so do the costs and risks of turbine development. Consequently fewer and fewer new manufacturers are able to break into the exclusive club. General Electric opted to buy its way in with the purchase of Enron's German and U.S. operations, and no doubt other multinational conglomerates will follow this example.

Challenges Facing Wind

Wind energy's success, its steady technological progress, and its rapidly growing deployment from India in the South to Canada in the North, and from Japan in the East to Europe in the West, can lure the industry into a false sense of security. There are challenges facing wind energy—public acceptance, for example—that could derail wind's march to the future. Most dangerous of all is the lure of panaceas.

Bigger Is Better

Where wind energy must compete with commodity prices, or where land-use plans limit owners to one turbine, there is an overpowering push for larger and larger machines. Don Smith, a veteran of both large and small turbines, is quick to point out that for every organism as well as machine, there is an optimal size. What that size is for wind turbines will be in part determined by their application, whether on land or offshore, urban or rural. All admit there is a limit. That limit has yet to be determined, but clearly larger and larger wind turbines are no panacea for wind energy's competitiveness with fossil fuels.

Green Power

During the difficult period of utility deregulation in North America, wind energy's proponents in the environmental community were forced to innovate. With no state or federal programs to support renewable energy, with no electricity feed laws, the only choice was to market so-called green power directly to consumers in those states and provinces where it was permitted. This allowed conscientious consumers in Michigan, for example, or in Colorado to pay a slight premium for their electricity so the utility would install a few wind turbines. In some regions, such as Michigan, it was the only way forward.

Unfortunately green power makes ethical consumers pay for benefits that everyone enjoys: clean air. At the same time, the existing system makes everyone, consumers and nonconsumers alike, pay for air pollution from fossil-fired power plants. Thus green power is inherently unfair. Worse, some of the firms involved have questionable track records that have given green power a bad name. Overall, little wind capacity has been constructed under green power programs worldwide in comparison to other development strategies. Even for urban dwellers, for whom green power seems most suited, installing a wind turbine as a member of a cooperative, such as those in Toronto or in Copenhagen, offers more rewards than paying for overpriced electricity from a marketing company. Whether or not green power is a good buy, it's clearly not a panacea for the absence of programs that allow wind energy to flourish.

Offshore

Planners have long looked offshore as a means of reducing siting conflicts between wind turbines and their neighbors in densely populated Europe. In the midst of sometimes protracted planning battles, wind companies and developers eye offshore wistfully as an answer to all their ills. *Out of sight, out of mind,* they muse. Not so. No place is out of sight or out of mind in today's interconnected world. If wind energy doesn't solve problems with public acceptance on land, the problems will only be transferred to another venue. We need only examine the vicious fight over Cape Wind's project in Massachusetts's Nantucket Sound to know that offshore is no cure-all for the problems wind energy faces on land.

Environmental

Wind's explosive expansion has assured us that the technology is here to stay, but this very success has also shone a light on one of wind energy's most striking features: the rapidity with which wind turbines can be added to the landscape. One moment there are none, then within days there can be tens of machines, and within weeks, hundreds of them. Wind energy can change the landscape literally overnight.

Regardless of our zeal to see wind energy succeed, we should never overlook wind's warts, such as the rapaciousness of an irresponsible developer or the fraudulent hype of an Internet swindle. Honest advocacy will move us farther in the direction we want to go than public relations puffery.

Wind energy is no environmental panacea. It's just one of many technologies we must use to build a sustainable future. Wind energy is a relatively benign technology, when used with care and developed with sensitivity to the community of which it will be a part. Wind energy, more than other technologies, depends on public acceptance. We ignore this at our peril.

Too Cheap to Meter

One unfortunate aspect of the on-again, off-again U.S. love affair with renewable energy has been the desperate attempt by the Department of Energy to curry favor with Congress when renewables are out of the limelight. To justify budgets for the development of new wind energy technology—some of which is already widely used in Europe—DoE has resorted to projecting costs for wind energy far into the future. Using DoE-developed technology, wind will be the lowest-cost energy option for the nation, or so the department's studies say. This is a dangerous game. Congress typically responds by saying, in effect, *Well, if wind energy will be that cheap in the future, why do it now? We'll develop wind energy later, when it's more cost-effective.* Worse, it forces DoE to project increasingly lower costs each time it holds its hat out for funding.

To those with a historical perspective,

DoE's wind projections sound eerily like those of nuclear's proponents in the 1950s. When the chairman of the U.S. Atomic Energy Commission, Lewis Strauss, promised that nuclear power would become "too cheap to meter," he created an expectation that could never be met. The failure to reach that unrealistic target called into question the basic tenets of the nuclear program. For wind to avoid this trap, proponents must never oversell the technology. Wind will never be "too cheap to meter," in part because it produces a higher-value product than that of fossil and nuclear fuels.

What's Needed

To go from where we are today to where we want to be in the years to come, there are several steps we can and must take. The most immediate is to pay a fair price for the premium product that wind-generated electricity represents. As a means to greater acceptance, we also need to encourage greater participation in the use and development of wind energy through community ownership. And we need better-informed and more aggressive advocacy in order for wind energy to reach its full potential.

Wind Turbine Owners' Associations

To start the debate in countries without a feed law, advocates of wind energy need to align themselves with wind turbine owners or potential wind turbine owners. Advocates must realize that trade associations typically represent manufacturers and wind developers. They understandably cater only to their members' needs, and not necessarily to those of the greater wind community.

Associations of wind turbine owners have a greater stake in feed laws and will put them at the top of the agenda, like their counterparts at the Bundesverband Windenergie in Germany. And like the Danish owners' association, Danmarks Vindmølleforening, they can demand that manufacturers guarantee that their turbines will perform as advertised.

In Ontario, Doug Woodard sees a need for a international consumers' union among wind turbine owners, to facilitate the sharing of ideas and experiences across political and cultural frontiers. This kind of sharing is like that now provided internally among members of the owners' associations in Germany and Denmark. Perceptively Woodard suggests an international association modeled after the Danish and German organizations, not only to foster development of wind technology with a human face, but also to foster the political developments that can make it happen.

Making Wind Go Farther

To Irish wind developer Michael Layden, "Each wind farm, each turbine is a psychic victory, giving us a warm feeling that we're making progress." But the world's ballooning population and increasing per-capita energy consumption erode the gains made. Layden echoes landscape architect Christoph Schwans's view from Germany, where wind turbines have become a feature of the landscape, that "we'll only make real progress if we also cut consumption."

One objective of using renewables is to improve the overall environmental efficiency of energy supply. Minimizing losses in transmission should be a priority—as should minimizing overall consumption of electricity through continued improvements in end-use efficiency and the more responsible use of energy. North Americans, in particular, are profligate. If society chooses to make greater use of wind energy a top priority, we should be, as Buckminster Fuller would say, doing more with less. In doing so we get the most out of our wind investment, both in economic and in environmental terms. For example, wind turbines provided 1 percent of California's electricity in 2002. If Californians cut their consumption in half, which they could easily do, wind's contribution would double to 2 percent.

Restructuring the Grid

The utility network was built to distribute supply from a few central sources. The challenge is to rebuild the grid to collect and balance contributions from many different, dispersed sources. This requires a fundamental conceptual shift in how we view the electric utility network, as well as a willingness to invest in the infrastructure to make it happen. The grid then becomes a means of collecting generation as much as a means of distributing it.

Likewise, the train wreck that is utility deregulation has eliminated nearly all oversight of this essential public service—and with it planning for future transmission capacity—across North America. Already transmission constraints are strangling growth of the technology in windy states of the upper Midwest. Activists are toying with once unthinkable proposals for breaking the logjam. Schemes such as large-scale hydrogen generation, with pipelines carrying the wind-produced hydrogen from windy regions like Buffalo Ridge to urban areas like Minneapolis–St. Paul, may no longer be the stuff of science fiction.

Fair Price for a High-Value Product

The problem with using household-size wind turbines—or even medium-size and larger wind turbines—at distributed homes, farms, and businesses lies not with the technology but with politics, says wind researcher Eric Eggleston. It always has been so.

Doug Woodard, from his perspective in Canada, observes that much of the interest in and advocacy of small wind turbines in North America grows out of political and cultural conditions. American advocates of wind energy have had little choice. They either support a large wind farm or they install a small wind turbine in their backyard. There has been no "third way" where they, too, can use commercial-size wind turbines.

In Europe, where wind energy is valued much more highly than in North America, the situation is quite different and small wind turbines are often relegated to their traditional role of powering remote sites. Farmers, businesses, and communities in Europe are freer to choose the technology that serves their interests best, regardless of size, than their counterparts across the Atlantic. They can do so because wind pays.

The greatest roadblock to expanded use of wind energy remains payment for its true value. To Hugh Piggott of Scoraig Wind Electric, "Wind energy is [more] expensive because it is environmentally superior to cheaper, polluting, nonsustainable sources of energy."

It makes sense to pay more for wind-generated electricity than for electricity from coal, oil, or even natural gas. It's an old idea, but one only now being implemented, mostly in Europe. Because wind turbines produce electricity cleanly without air or water pollution, use a renewable resource, provide generating diversity, and can be dispersed throughout a utility's distribution system, the electricity they produce is worth more than that produced by conventional power plants, which have none of these attributes.

If an "environmental premium" were added to the current commodity value of wind-generated electricity, the use of wind energy, whether through small or medium-sized wind turbines, would expand as rapidly as it has in Germany and Spain. After numerous legal challenges European courts have ruled that such above-market payments from electricity feed laws are, as in Spain's case, justified compensation for wind's environmental benefits.

The Bundesverband Windenergie's Joachim Twele observes that the world market for wind energy can be divided into a commodity-driven segment and an environmentally driven segment. European markets are environmentally driven, and wind energy is growing apace because Europeans are willing to pay for wind's benefits. During the 1990s, says Twele, wind energy moved from niche player to competitive industry not because of government-sponsored research programs, but because of the dynamic market created by premium payments.

Commodity-driven markets force manufacturers and developers to compete on price, and price alone. "Attempts by the wind industry to make wind cheaper usually result in poor-quality wind turbines," says Piggott. "The result is noisy, unreliable machines which give the impression that wind energy does not work." While Piggott is referring specifically to small wind turbines, his observation is equally applicable to larger machines. Environmentalists add that efforts to cheapen wind also leads wind companies to take shortcuts, and when they do the environment sometimes suffers.

What's holding back greater adoption of feed laws is the defeatist attitude found among renewable energy advocates in some countries. They can be heard saying, "Feed laws, great idea, but it can't be done here." To say *it can't be done here* is a prescription for inaction. The Danes did it, the Germans are doing it, and the Spanish and French are doing so, as well. It's helpful to remember that at one time feed laws, and the development of wind energy in general, were considered impossible in the French context, given the French state's longstanding support of nuclear power.

We must "live in hope," says writer Mark Herstgaard. "Vaclav Havel, Nelson Mandela, and Mikhail Gorbachev did the impossible because they lived in hope." They didn't shrink from what seemed an overwhelming task.

German parliamentarian Hermann Scheer didn't flinch from the challenge, either. "We didn't wait for a consensus," in the Bundestag, says Scheer about the struggle to protect Germany's feed law from an assault by some of the most entrenched and powerful electric utilities in the world. "We pushed."

They also organized. Scheer and the German wind energy association, with help from trade unionists and conservative farmers, organized a massive—and ultimately successful—demonstration in Berlin to demand legislative support for the feed law. Scheer, addressing the crowd, cited a Chinese proverb: "If the winds of change blow, some build walls, others build windmills."

Energy for Life

The 19th-century Danish theologian N. F. S. Grundtvig expounded a philosophy that has had a profound and lasting effect on Danes and the cooperative movement in Denmark. More than a century ago Grundtvig said that all public policy, in fact all human endeavors, should be life affirming. Denmark's development of wind energy is built upon this foundation. It's not just a question of abstract economics. Danes believe using wind energy is the right thing to do.

We need to envision an energy system that is sustainable, a system that meets the needs of the people of today as well as those of tomorrow, and a system that is built upon sufficiency for all, equitably distributed (see figure 17-6, Energy for life). We must envision a system that enhances the quality of life for all people, rich and poor alike. Such a system is built upon services rendered and needs met, not upon a constant and neverending growth of supply. We must envision an energy system for people, or, as Grundtvig would say, an energy system for life. Wind energy can help us reach that dream.

Figure 17-6. Energy for life. Wind energy can and should be a celebration of life, for the people of today and for those of tomorrow.

Appendix

Constants

ICAO Standard Atmosphere Air pressure at sea level (p)	=	29.92 in Hg
	=	760 mm Hg
	=	1013.25 mb
	=	1.01325 x 105 N/m^2 or Pa
Temperature (T)	=	15°C
	=	288.15 K
	=	59°F
Acceleration due to gravity (g)	=	9.807 m/s^2
Gas constant for dry air (R)	=	287.04 J/kgK
Absolute temperature	=	273.15 K
Normal atmospheric lapse rate	=	6.5°C/1000 m
Density of dry air, standard atmoshpere (D)	=	1.225 kg/m^3

Conversions

Speed

1 mph = 0.447 m/s
1 mph = 0.870 knots
1 mph = 1.61 km/h
1 knot = 1.15 mph
1 knot = 0.514 m/s
1 knot = 1.85 km/h
1 m/s = 2.24 mph
1 m/s = 1.94 knots
1 m/s = 3.60 km/h
1 km/h = 0.621 mph
1 km/h = 0.54 knots
1 km/h = 0.278 m/s

Length

1 inch = 2.54 centimeters
1 centimeter = 0.394 inches
1 meter = 3.28 feet
1 foot = 0.305 meters
1 kilometer = 0.620 miles
1 mile = 1.61 kilometers

Area

1 square kilometer = 0.386 square miles
1 square kilometer = 100 hectares
1 square mile = 2.59 square kilometers
1 hectare = 10,000 square meters
1 hectare = 2.47 acres
1 acre = 0.405 hectares
1 acre = 4,047 square meters

Volume

1 cubic meter = 35.3 cubic feet
1 cubic meter = 264 U.S. gallons
1 cubic foot = 0.028 cubic meters
1 liter = 0.264 U.S. gallons
1 U.S. gallon = 3.78 liters
1 U.S. gallon = 0.004 cubic meters

Flow Rate

1 liter/second = 0.264 U.S. gallons/second
1 cubic meter/minute = 264 U.S. gallons/minute
1 gallon/minute = 3.79 liters/minute
1 gallon/minute = 0.063 liters/second
1 gallon/day = 0.158 liters/hour

Weight

1 metric ton = 1.10 short tons
1 kilogram = 2.21 pounds
1 pound = 0.454 kilograms

Power

1 kilowatt = 1.34 horsepower
1 horsepower = 0.746 kilowatts

Torque

1 Newton-meter = 0.738 foot-pounds
1 foot-pound = 1.356 Newton-meters

Temperature

degrees Fahrenheit = 9/5°C + 32
degrees Celsius = 5/9 (°F - 32)
degrees Kelvin = 273.15 + °C

Energy Equivalence of Common Fuels

1 kWh = 3413 Btu
= 3.41 cu ft of natural gas
= 0.034 gallon of oil
= 0.00017 cord of wood
= 860 kilocalories
1 Therm = 100,000 Btu
= 100 cu ft of natural gas
= 1 gallon of oil
= 29.3 kWh
= 0.005 cord of wood
= 25,194 kilocalories
1 gallon of oil = 100,000 Btu
1 cord of wood = 20,000,000 Btu
1,000 cu ft of nat. gas = 1,000,000 Btu

Approximate Primary Energy Offset by Direct Generation of Electricity

1 kWh = 10,000 Btu
1 kWh = 2,520 kilocalories
1 kWh = 0.1 therm
600 kWh = 1 barrel of oil equivalent

Scale of Energy Equivalents

Approximate size wind turbine to provide the same annual energy output by Rotor Diameter*

Energy (kWh)	Rotor Diameter (m)	(ft)	
1			auto battery, 100 watt light for 10 hours
10			electric space heater for 10 hours
50	0.25	1	average per capita consumption in India
1,000	1.5	5	
3,500	2.5	8	average residential consumption in Northern Europe
4,000	3	10	
6,500	4	13	average residential consumption in California
12,000	5	16	average residential consumption in Texas
15,000	6	20	
20,000	7	23	typical electric home consumption in Oklahoma
75,000	10	33	
250,000	18	60	
500,000	25	80	
1,000,000	35	110	
2,000,000	45	150	
3,000,000	55	180	
4,000,000	65	210	
6,000,000	80	260	
8,000,000	90	300	
10,000,000	100	330	

Note: * Class 4 wind resource, 50 meters (164 feet) hub height, 7.5 m/s (16.8 mph).

Scale of Equivalent Power

Rotor Diameter (m)	(ft)	Typical Wind Turbine Peak Power Rating (kW)	Equivalent
1.5	5	0.25	⅓ horsepower electric motor, electric drill
2	7	0.50	⅔ horsepower electric motor
3	10	1	hair dryer, electric space heater
7	23	10	garden tractor
10	33	25	
18	59	100	passenger car engine
25	82	250	
40	131	500	heavy truck engine
50	164	1,000	race car engine, small diesel locomotive
100	328	3,000	diesel locomotive
		500,000	coal-fired generator, small nuclear reactor
		1,000,000	large nuclear reactor

Air Density Correction Spreadsheets

Spreadsheet Notation

Note that the following tables are presented in the form of spreadsheets (worksheets). The formulas for calculating the values in critical cells are given at the bottom of the spreadsheet. By keying these formulas into your spreadsheet software, you can produce these tables. The formulas use conventions found in Microsoft Excel. For use with Lotus 123 or Corel Quattropro substitute a "+" for the initial "=" (equals) sign, and insert "@" (at) before function commands such as SUM, EXP, or WEIBULL. For example, Microsoft's "=SUM(J28:J47)" is equivalent to "@SUM(J28..J47)". For other differences, see your software's help menu.

Air Density Correction Spreadsheet for Temperature

	A	B	C	D	E	F
1	Change in Air Density with Temperature at Sea-Level Pressure					
2						
3			Air	Change		
4	Temperature		Density	Relative		
5	°C	°F	kg/m³	to 15°C		
6	-20	-4	1.394	1.14		
7	-15	5	1.367	1.12		
8	-10	14	1.341	1.10		
9	-5	23	1.316	1.07		
10	0	32	1.292	1.05		
11	5	41	1.269	1.04		
12	10	50	1.247	1.02		
13	15	59	1.225	1.00		
14	20	68	1.204	0.98		
15	25	77	1.184	0.97		
16	30	86	1.164	0.95		
17	35	95	1.146	0.94		
18	40	104	1.127	0.92		
19	45	113	1.110	0.91		
20	50	122	1.092	0.89		
21	Cell C6: =101325/(287.04*(273.15+A6))					

Air Density Correction Spreadsheet for Altitude

	A	B	C	D	E	F
1	Change in Air Pressure and Density with Elevation above Sea-Level					
2	at Constant Temperature (15°C)					
3						
4	Elevation			Air	Relative	
5	above		Air	Density	Change	
6	Sea Level		Pressure	to 15°C	in Density	
7	m	ft	mb	kg/m³		
8	0	0	1013.25	1.225	1.00	
9	500	1,640	954.93	1.155	0.94	
10	1,000	3,280	899.96	1.088	0.89	
11	1,500	4,920	848.15	1.025	0.84	
12	2,000	6,560	799.33	0.966	0.79	
13	2,500	8,200	753.32	0.911	0.74	
14	3,000	9,840	709.96	0.858	0.70	
15	3,500	11,480	669.09	0.809	0.66	
16	4,000	13,120	630.58	0.762	0.62	
17	Cell C8: =C$7*(EXP((-A8*9.807)/(287.04*288.15)))					
18	Cell D8: =(C8*100)/(287.04*288.15)					

Air Density Correction Spreadsheet for Normal Lapse Rate (Standard Atmosphere)

	A	B	C	D	E	F
1	Change in Air Pressure and Density with Elevation for Normal or					
2	Environmental Lapse Rate (6.5°C/1000 m)					
3					Air	Change
4	Elevation		Temperature	Pressure	Density	Relative
5	m	ft	K	mb	kg/m^3	to SL 15°C
6	0	0	288.15	1013.25	1.225	1.00
7	500	1,640	284.90	954.60	1.167	0.95
8	1,000	3,280	281.65	898.74	1.112	0.91
9	1,500	4,920	278.40	845.55	1.058	0.86
10	2,000	6,560	275.15	794.94	1.007	0.82
11	2,500	8,200	271.90	746.81	0.957	0.78
12	3,000	9,840	268.65	701.06	0.909	0.74
13	3,500	11,480	265.40	657.62	0.863	0.70
14	4,000	13,120	262.15	616.38	0.819	0.67
15	Cell C7: =288.15-(6.5/1000)*A7					
16	Cell D7: =1013.25*((288.15-(6.5/1000)*A7)/288.15)^(9.807/(287.04*(6.5/1000)))					
17	Cell E7: =D7/(287.04*C7)*100					
18						

Air Density Correction Spreadsheet for a Given Temperature and Elevation

	A	B	C
1	Air Pressure and Density for a Given Temperature and Elevation		
2			
3			Example
4		Sea Level	from
5		at 15°C	Text
6	Temperature		
7	°F	59	90
8	°C	15	32.22
9	Elevation		
10	ft	0	4,920
11	m	0	1,500
12	Pressure (mb)	1,013.25	857
13	Density (kg/m^3)	1.225	0.977
14	Change relative to SL @ 15°C	1.00	0.80
15	Cell B8: =(+B7-32)*5/9		
16	Cell B11: =B10/3.28		
17	Cell B12: =1013.25*(EXP((-B11*9.807)/(287.04*(273.15+B8))))		
18	Cell B13: =B12/(287.04*(273.15+B8))*100		

Wind Speed Distributions

You need only the average wind speed to determine the Rayleigh speed distribution. Use the following table for average wind speed in both mph and m/s. Average speed is in the top row. The distribution is the column below the average speed. The wind speed bin corresponds to the horizontal axis on graphs of wind turbine power curves. The Probability of Occurrence is the percentage of time the wind occurs within that wind speed bin.

Rayleigh Wind Speed Distribution

	A	B	C	D	E	F	G	H	I	J	K
1	Rayleigh Wind Speed Distribution For Average Annual Wind Speed										
2											
3	Wind	Probability of Occurrence at Annual Average Wind Speeds									
4	Speed										
5	Bin										
6		3	4	5	6	7	8	9	10	11	12
7	0	0.0000	0.0000	0.0000	0.0000	0.0000	0.0000	0.0000	0.0000	0.0000	0.0000
8	1	0.1599	0.0935	0.0609	0.0427	0.0315	0.0242	0.0192	0.0156	0.0129	0.0108
9	2	0.2462	0.1613	0.1108	0.0800	0.0601	0.0467	0.0373	0.0304	0.0253	0.0213
10	3	0.2387	0.1893	0.1421	0.1076	0.0833	0.0659	0.0533	0.0439	0.0367	0.0312
11	4	0.1728	0.1790	0.1520	0.1231	0.0992	0.0807	0.0664	0.0554	0.0468	0.0400
12	5	0.0985	0.1439	0.1432	0.1264	0.1074	0.0903	0.0761	0.0645	0.0552	0.0476
13	6	0.0453	0.1006	0.1217	0.1194	0.1080	0.0947	0.0821	0.0710	0.0617	0.0538
14	7	0.0170	0.0620	0.0943	0.1049	0.1023	0.0942	0.0844	0.0748	0.0661	0.0585
15	8	0.0052	0.0339	0.0673	0.0864	0.0919	0.0895	0.0834	0.0760	0.0686	0.0616
16	9	0.0013	0.0166	0.0444	0.0671	0.0788	0.0817	0.0796	0.0748	0.0691	0.0631
17	10	0.0003	0.0072	0.0272	0.0492	0.0645	0.0719	0.0735	0.0716	0.0678	0.0632
18	11	0.0000	0.0028	0.0154	0.0343	0.0507	0.0612	0.0660	0.0668	0.0651	0.0620
19	12	0.0000	0.0010	0.0082	0.0226	0.0383	0.0503	0.0576	0.0608	0.0612	0.0597
20	13	0.0000	0.0003	0.0040	0.0142	0.0278	0.0401	0.0490	0.0542	0.0563	0.0564
21	14	0.0000	0.0001	0.0019	0.0085	0.0194	0.0310	0.0406	0.0472	0.0509	0.0524
22	15	0.0000	0.0000	0.0008	0.0048	0.0131	0.0233	0.0328	0.0402	0.0452	0.0480
23	16	0.0000	0.0000	0.0003	0.0026	0.0085	0.0170	0.0259	0.0337	0.0394	0.0432
24	17	0.0000	0.0000	0.0001	0.0014	0.0053	0.0120	0.0200	0.0276	0.0338	0.0383
25	18	0.0000	0.0000	0.0000	0.0007	0.0032	0.0083	0.0151	0.0222	0.0285	0.0335
26	19	0.0000	0.0000	0.0000	0.0003	0.0019	0.0056	0.0111	0.0175	0.0237	0.0289
27	20	0.0000	0.0000	0.0000	0.0001	0.0011	0.0036	0.0080	0.0136	0.0194	0.0246
28	21	0.0000	0.0000	0.0000	0.0001	0.0006	0.0023	0.0057	0.0103	0.0156	0.0207
29	22	0.0000	0.0000	0.0000	0.0000	0.0003	0.0014	0.0039	0.0077	0.0123	0.0171
30	23	0.0000	0.0000	0.0000	0.0000	0.0002	0.0009	0.0026	0.0057	0.0096	0.0140
31	24	0.0000	0.0000	0.0000	0.0000	0.0001	0.0005	0.0017	0.0041	0.0074	0.0113
32	25	0.0000	0.0000	0.0000	0.0000	0.0000	0.0003	0.0011	0.0029	0.0056	0.0090
33	26	0.0000	0.0000	0.0000	0.0000	0.0000	0.0002	0.0007	0.0020	0.0042	0.0071
34	27	0.0000	0.0000	0.0000	0.0000	0.0000	0.0001	0.0004	0.0014	0.0031	0.0055
35	28	0.0000	0.0000	0.0000	0.0000	0.0000	0.0000	0.0003	0.0009	0.0022	0.0042
36	29	0.0000	0.0000	0.0000	0.0000	0.0000	0.0000	0.0002	0.0006	0.0016	0.0032
37	30	0.0000	0.0000	0.0000	0.0000	0.0000	0.0000	0.0001	0.0004	0.0011	0.0024
38	31	0.0000	0.0000	0.0000	0.0000	0.0000	0.0000	0.0001	0.0003	0.0008	0.0018
39	32	0.0000	0.0000	0.0000	0.0000	0.0000	0.0000	0.0000	0.0002	0.0005	0.0013
40	33	0.0000	0.0000	0.0000	0.0000	0.0000	0.0000	0.0000	0.0001	0.0004	0.0009
41	34	0.0000	0.0000	0.0000	0.0000	0.0000	0.0000	0.0000	0.0001	0.0002	0.0007
42	35	0.0000	0.0000	0.0000	0.0000	0.0000	0.0000	0.0000	0.0000	0.0002	0.0005
43	36	0.0000	0.0000	0.0000	0.0000	0.0000	0.0000	0.0000	0.0000	0.0001	0.0003
44	37	0.0000	0.0000	0.0000	0.0000	0.0000	0.0000	0.0000	0.0000	0.0001	0.0002
45	38	0.0000	0.0000	0.0000	0.0000	0.0000	0.0000	0.0000	0.0000	0.0000	0.0002
46	39	0.0000	0.0000	0.0000	0.0000	0.0000	0.0000	0.0000	0.0000	0.0000	0.0001
47	40	0.0000	0.0000	0.0000	0.0000	0.0000	0.0000	0.0000	0.0000	0.0000	0.0001
48	41	0.0000	0.0000	0.0000	0.0000	0.0000	0.0000	0.0000	0.0000	0.0000	0.0000
49	42	0.0000	0.0000	0.0000	0.0000	0.0000	0.0000	0.0000	0.0000	0.0000	0.0000
50	43	0.0000	0.0000	0.0000	0.0000	0.0000	0.0000	0.0000	0.0000	0.0000	0.0000
51	44	0.0000	0.0000	0.0000	0.0000	0.0000	0.0000	0.0000	0.0000	0.0000	0.0000
52	45	0.0000	0.0000	0.0000	0.0000	0.0000	0.0000	0.0000	0.0000	0.0000	0.0000
53	46	0.0000	0.0000	0.0000	0.0000	0.0000	0.0000	0.0000	0.0000	0.0000	0.0000
54	47	0.0000	0.0000	0.0000	0.0000	0.0000	0.0000	0.0000	0.0000	0.0000	0.0000
55	48	0.0000	0.0000	0.0000	0.0000	0.0000	0.0000	0.0000	0.0000	0.0000	0.0000
56	49	0.0000	0.0000	0.0000	0.0000	0.0000	0.0000	0.0000	0.0000	0.0000	0.0000
57	50	0.0000	0.0000	0.0000	0.0000	0.0000	0.0000	0.0000	0.0000	0.0000	0.0000
58	Cell B7: =1*((PI()/2)*($A7/(B$6^2))*EXP(-(PI()/4)*($A7/B$6)^2)), where the width of the wind speed bins is 1;										
59	0 to 0.49 bin not included.										

Weibull Wind Speed Frequency Distribution

	A	B	C	D	E	F
1	Weibull Wind Speed Frequency Distribution					
2	Probability of Occurrence with k and C For Tera Kora, Curaçao,					
3	and Helgoland, Germany					
4						
5			Tera Kora	Helgoland		
6	Avg. Speed (V_{avg})		7.2	7.2	m/s	
7	Shape Factor (k)		4.5	2.09		
8	Scale Factor (C)		8	8	m/s	
9						
10	Wind Speed					
11	Bin					
12	m/s	mph				
13	0	0.0	0	0		
14	1	2.2	0.000388	0.026734		
15	2	4.5	0.004386	0.054557		
16	3	6.7	0.017946	0.078857		
17	4	9.0	0.047569	0.097035		
18	5	11.2	0.096231	0.107633		
19	6	13.4	0.156255	0.110364		
20	7	15.7	0.203712	0.106001		
21	8	17.9	0.206932	0.096109		
22	9	20.2	0.155347	0.082661		
23	10	22.4	0.080143	0.067661		
24	11	24.6	0.025934	0.052826		
25	12	26.9	0.004717	0.039402		
26	13	29.1	0.000424	0.028110		
27	14	31.4	0.000016	0.019197		
28	15	33.6	0.000000	0.012559		
29	16	35.8	0.000000	0.007874		
30	17	38.1	0.000000	0.004733		
31	18	40.3	0.000000	0.002728		
32	19	42.6	0.000000	0.001508		
33	20	44.8	0.000000	0.000800		
34	21	47.0	0.000000	0.000407		
35	22	49.3	0.000000	0.000199		
36	23	51.5	0.000000	0.000093		
37	24	53.8	0.000000	0.000042		
38	25	56.0	0.000000	0.000018		
39	26	58.2	0.000000	0.000007		
40	27	60.5	0.000000	0.000003		
41	28	62.7	0.000000	0.000001		
42	Cell C15: =(C$7/C$8)*(($A15/C$8)^(C$7-1))*EXP(-($A15/C$8)^C$7), or					
43	Cell C15: =WEIBULL(A15,C7,C8,0)					

Winds on Helgoland approximate those of a Rayleigh distribution.

Map of Wind Resources in the United States

The accompanying map, U.S. Annual Average Wind Power, is taken from *Wind Energy Resource Atlas* of the United States. The map was published in 1987 by Battelle's Pacific Northwest Laboratory for the U.S. Department of Energy. This map and its companion state maps incorporate the most up-to-date information on the wind resource in the United States available to the general public. Unlike the previous survey, published in the early 1980s, this update includes data from both private and public sources never before assimilated into one comprehensive assessment of wind energy in the United States.

The Battelle wind atlas is the first place to turn for an overview of wind energy's potential anywhere in the United States. The 1987 atlas contains valuable background information on the power in the wind, terrain features, and meteorological conditions that affect the wind resources within each region of the country. The complete atlas and new high-resolution digital maps from the National Renewable Energy Laboratory can be found on the Web at www.nrel.gov/wind/wind_map.html.

Battelle's maps present a range of possible values of wind power density as wind power classes as shown in the following table. Commercial wind power plants have been developed in class 4 resources, though the most productive are sited in areas with class 6 or better.

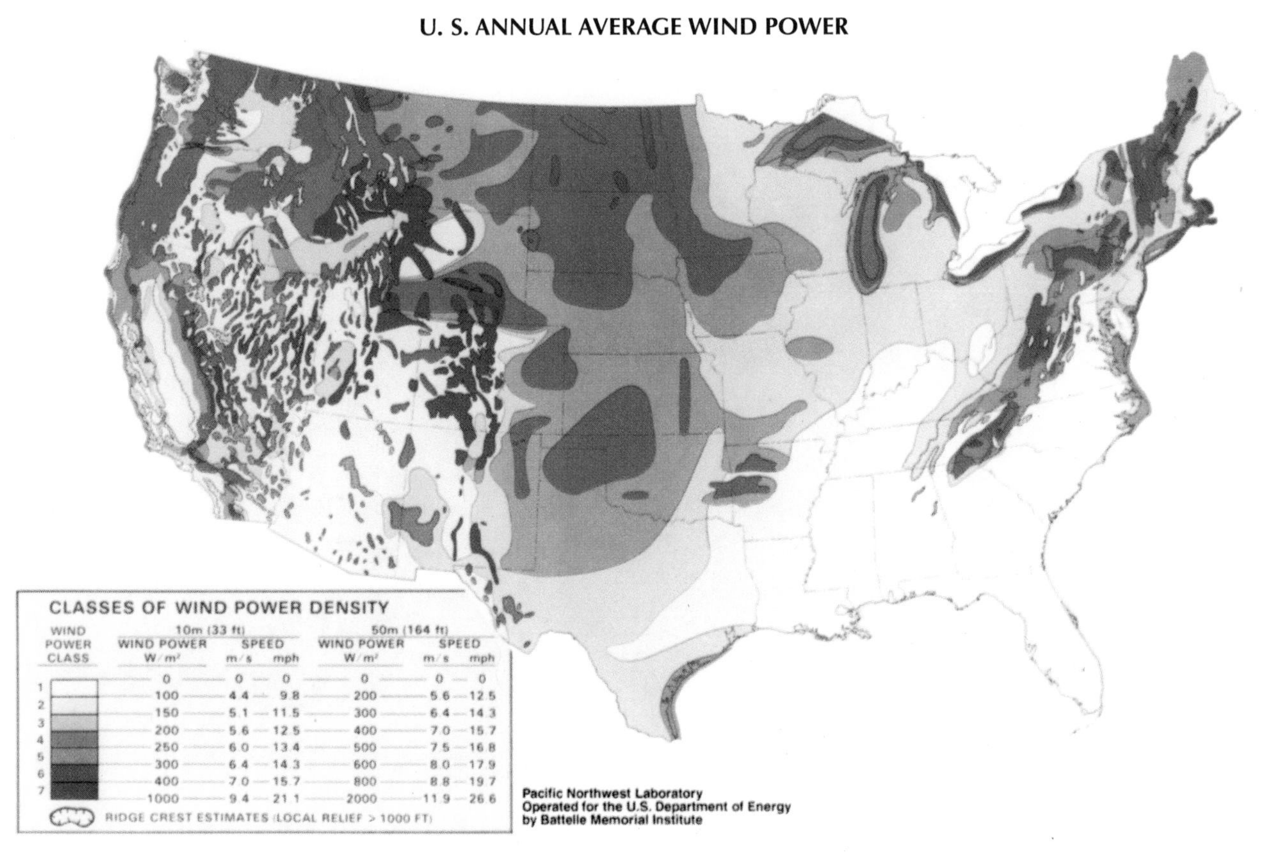

WIND POWER CLASS	10m (33 ft) WIND POWER W/m²	10m (33 ft) SPEED m/s	10m (33 ft) SPEED mph	50m (164 ft) WIND POWER W/m²	50m (164 ft) SPEED m/s	50m (164 ft) SPEED mph
	0	0	0	0	0	0
1						
	100	4.4	9.8	200	5.6	12.5
2						
	150	5.1	11.5	300	6.4	14.3
3						
	200	5.6	12.5	400	7.0	15.7
4						
	250	6.0	13.4	500	7.5	16.8
5						
	300	6.4	14.3	600	8.0	17.9
6						
	400	7.0	15.7	800	8.8	19.7
7						
	1000	9.4	21.1	2000	11.9	26.6

Source: *Wind Energy Resource Atlas of the United States*

Battelle Classes of Wind Energy Flux

	Wind Speed and Power at 10 m (33 ft)			Wind Speed and Power at 30 m (100 ft)			Wind Speed and Power at 50 m (164 ft)		
	Power Density (W/m²)	Speed (m/s)	(mph)	Power Density (W/m²)	Speed (m/s)	(mph)	Power Density (W/m²)	Speed (m/s)	(mph)
	50	3.5	7.8	80	4.1	9.2	100	4.4	9.9
1									
	100	4.4	9.9	160	5.1	11.4	200	5.5	12.3
2									
	150	5.0	11.5	240	5.9	13.2	300	6.3	14.1
3									
	200	5.5	12.3	320	6.5	14.6	400	7.0	15.7
4									
	250	6.0	13.4	400	7.0	15.7	500	7.5	16.8
5									
	300	6.3	14.1	480	7.4	16.6	600	8.0	17.9
6									
	400	7.0	15.7	640	8.2	18.4	800	8.8	19.7
7									
	800	8.8	19.7	1,280	10.3	23.1	1,600	11.1	24.9
8									
	1,200	10.1	22.6	1,920	11.8	26.4	2,400	12.7	28.4
9									
	1,600	11.1	24.9	2,560	12.9	28.9	3,200	13.9	31.1

Note: Increase in speed and power with height assumes 1/7 power law.

Maps of World Wind Resources

The following maps depict wind resources worldwide. The maps were prepared by Battelle Pacific Northwest Laboratories in the early 1980s. Since these maps were prepared, many countries have performed their own national surveys.

AUSTRALASIA—WIND ENERGY RESOURCE

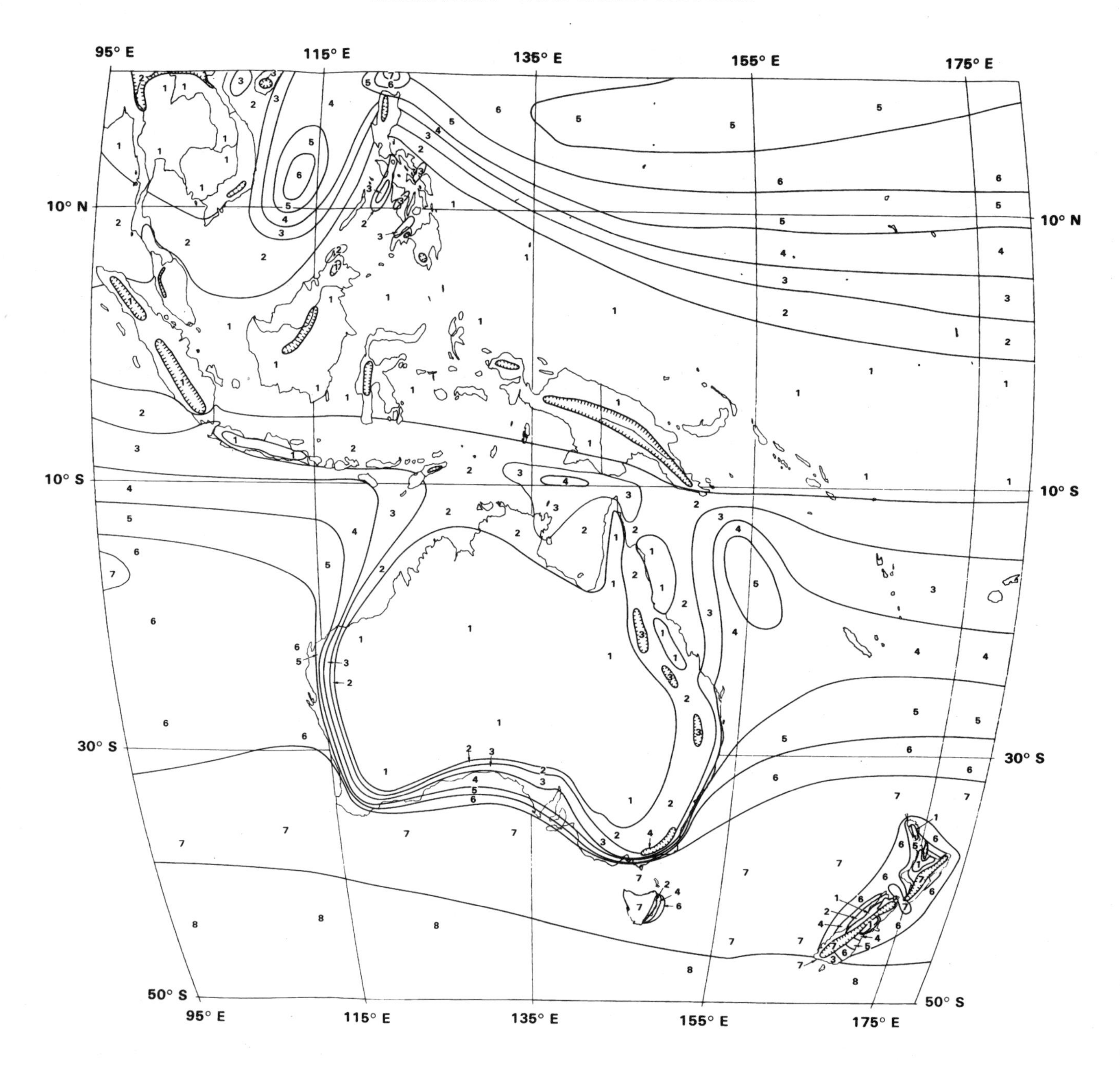

AFRICA—WIND ENERGY RESOURCE

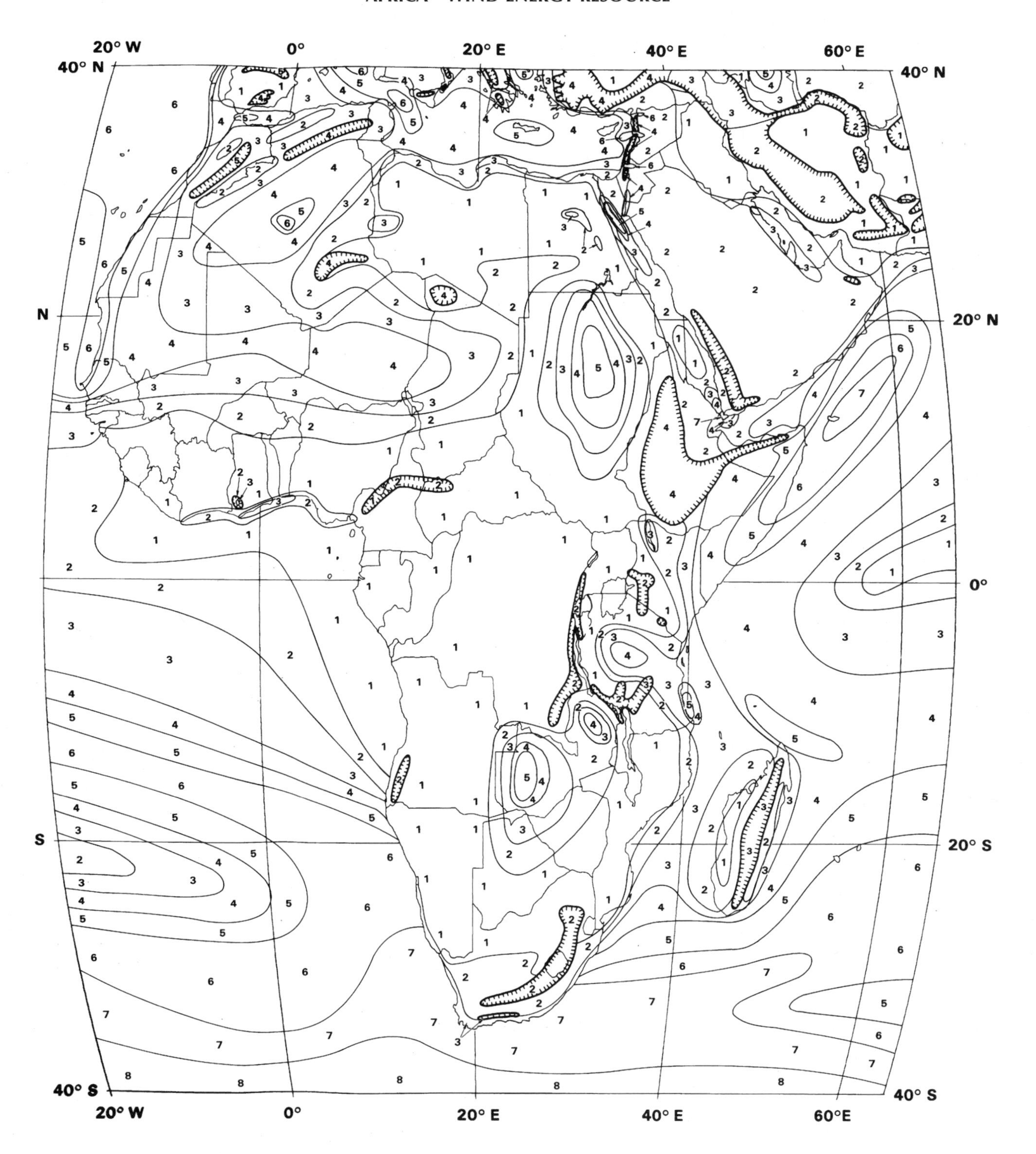

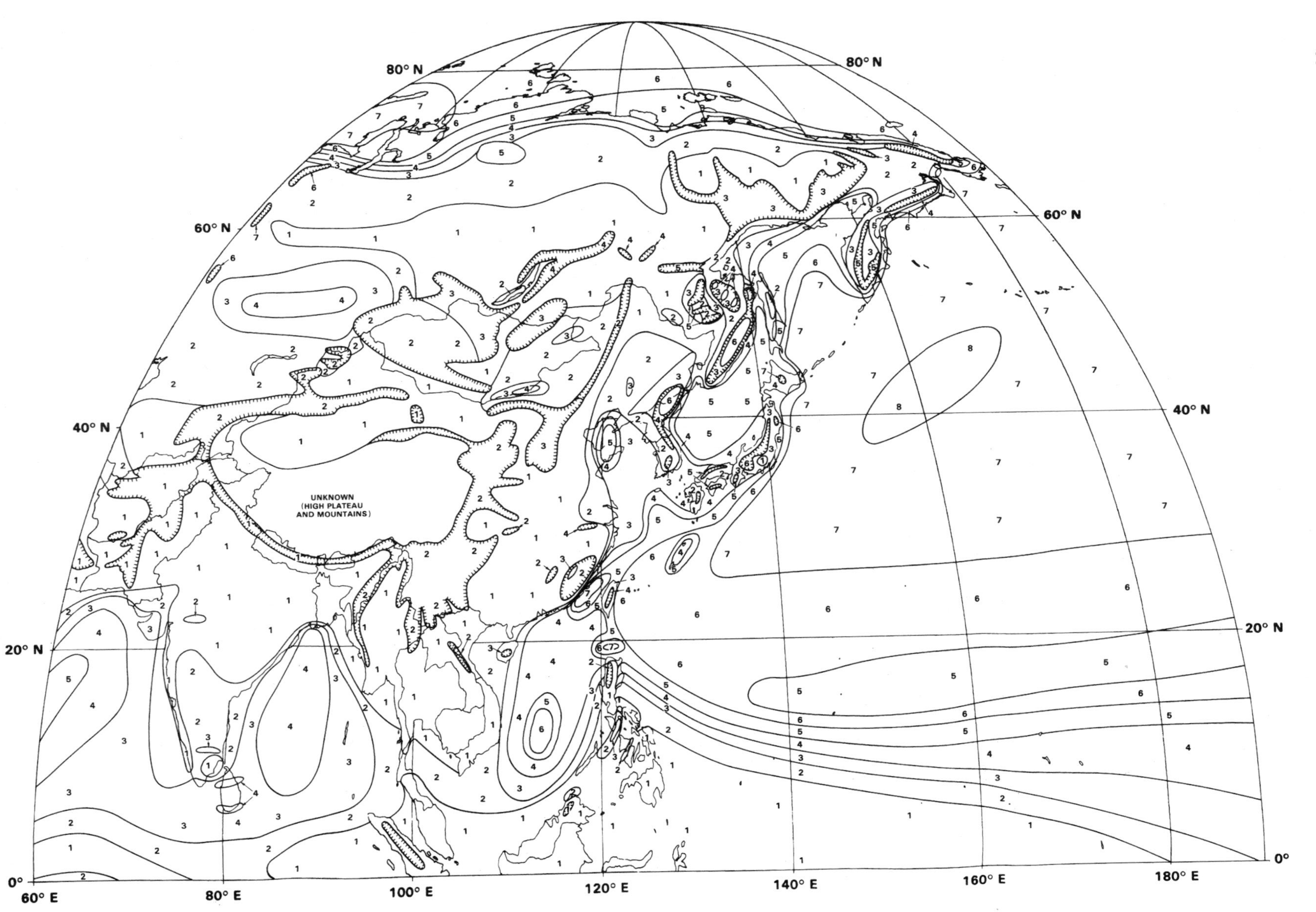
EASTERN ASIA—WIND ENERGY RESOURCE
UNKNOWN
(HIGH PLATEAU
AND MOUNTAINS)
80° N
60° N
40° N
20° N
0°
60° E
80° E
100° E
120° E
140° E
160° E
180° E

EUROPE AND WESTERN ASIA—WIND ENERGY RESOURCE

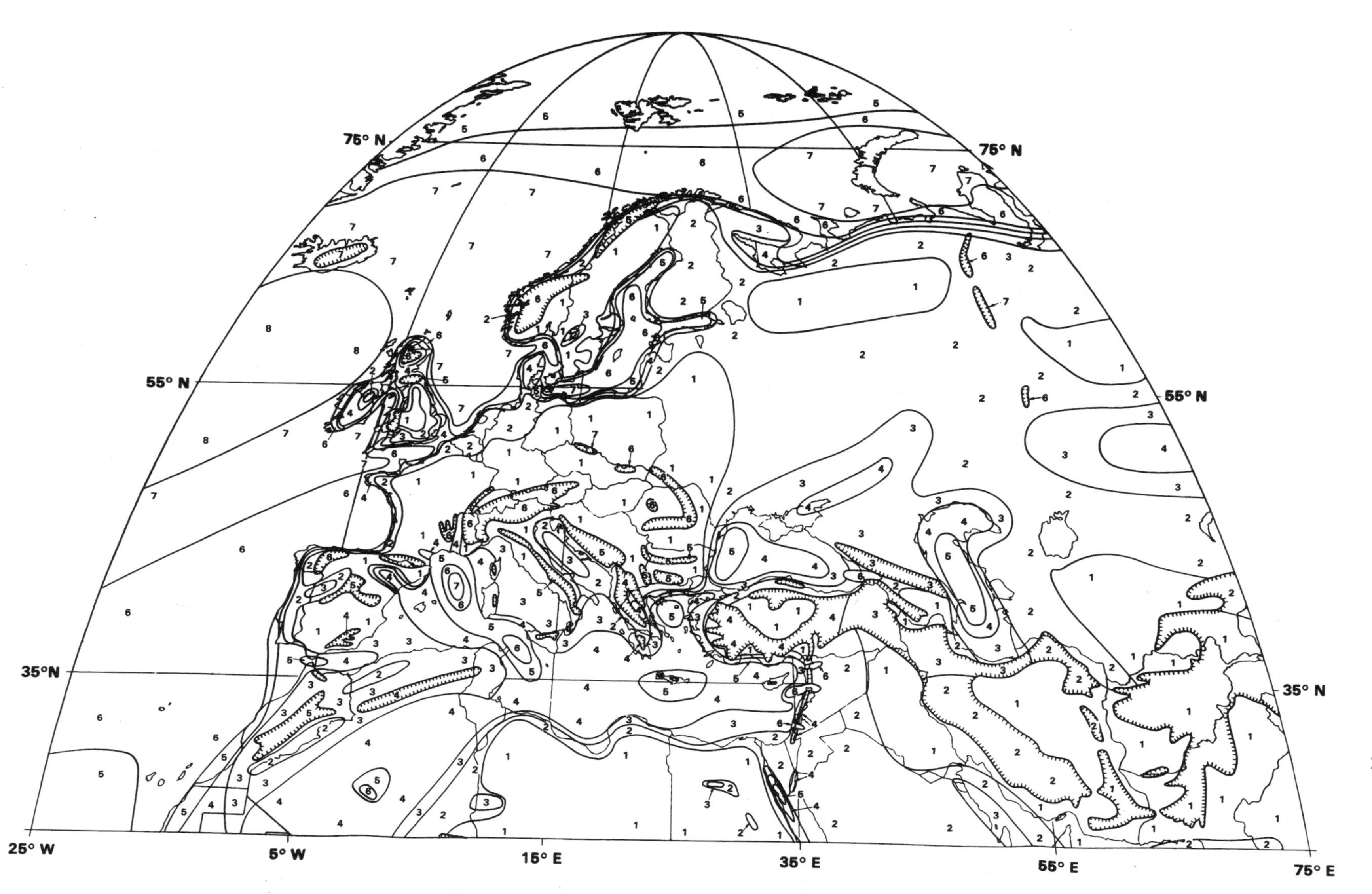

SOUTH AMERICA—WIND ENERGY RESOURCE

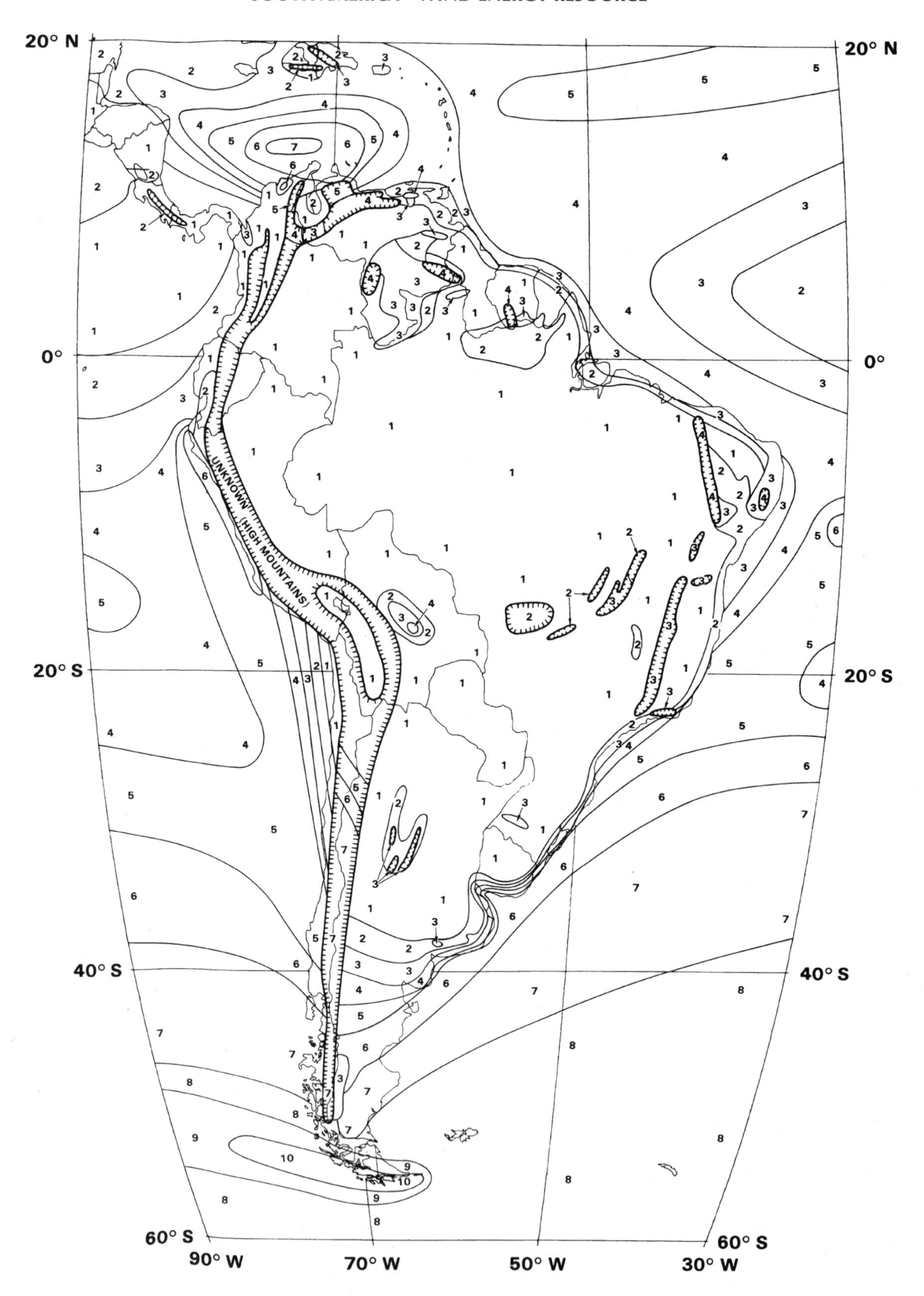

Estimates of Annual Energy Output

The following tables estimate the amount of energy a wind turbine of a given size will produce in a given wind regime using the swept area method. The table assumes a Rayleigh distribution of wind speeds. The overall conversion efficiency assumed is given in the column labeled Total Effic. The assumed efficiencies have been derived from a survey of product literature of manufacturers worldwide. The results are approximations and won't necessarily correspond to estimates by the manufacturer for the same conditions using a power curve and a specific wind speed distribution.

These estimates represent gross energy generation. They do not account for energy lost in the cables between the wind turbine and the load. Nor do these estimates account for energy that may be spilled or dumped in battery-charging systems when the batteries are fully charged.

There is one table for small wind turbines and another for medium-size and large wind turbines. The first table uses typical conversion efficiencies found in small wind turbines. These turbines are designed for different applications than medium-size turbines, and their performance reflects this. The second table incorporates conversion efficiencies typically found on the medium-size and larger wind turbines used in commercial applications such as wind power plants.

Estimated Annual Energy Output at Hub Height in Thousands of kWh/yr for Battelle Power Classes, Small Wind Turbines

	A	B	C	D	E	F	G	H	I	J	K	L	M	N	O	P	Q	R	S
1	Estimated Annual Energy Output for Battelle Wind Power Classes at 30 m (100 ft) Hub Height in Thousand kWh/yr																		
2	Small Wind Turbines																		
3	Battelle Power				Wind Speed and Power					Rotor Diameter									
4	Class at 10 m (33 ft)				at 30 m (100 ft) Hub Height														
5		Power			Power														
6		Density	Speed		Density	Speed		Total	m	1	2	3	4	5	6	7	8	9	10
7		W/m^2	m/s	mph	W/m^2	m/s	mph	Effic.	ft	3.3	6.6	9.8	13.1	16.4	19.7	23.0	26.2	29.5	32.8
8		50	3.5	7.8	80	4.1	9.2	0.265		0.1	0.6	1.3	2.3	3.6	5.3	7.2	9.3	11.8	14.6
9	1																		
10		100	4.4	9.9	160	5.1	11.4	0.236		0.3	1.0	2.3	4.2	6.5	9.4	13	17	21	26
11	2																		
12		150	5.0	11.5	240	5.9	11.5	0.207		0.3	1.4	3.1	5.5	8.6	12	17	22	28	34
13	3																		
14		200	5.5	12.3	320	6.5	14.6	0.1925		0.4	1.7	3.8	6.8	11	15	21	27	34	42
15	4																		
16		250	6.0	13.4	400	7.0	15.7	0.178		0.5	2.0	4.4	7.8	12	18	24	31	40	49
17	5																		
18		300	6.3	14.1	480	7.4	16.6	0.1635		0.5	2.2	4.9	8.6	14	19	26	35	44	54
19	6																		
20		400	7.0	15.7	640	8.2	18.4	0.149		0.7	2.6	5.9	10.5	16	24	32	42	53	66
21	7																		
22		1,000	9.5	21.3	1,600	11.1	24.9	0.12		1	5	12	21	33	48	65	85	107	132
23	8																		
24		1,200	10.1	22.6	1,920	11.8	26.4	0.12		2	6	14	25	40	57	78	101	128	159
25	9																		
26		1,600	11.1	24.9	2,560	12.9	28.9	0.12		2	8	19	34	53	76	104	135	171	211
27	Note: Assumed efficiency based on published data. Actual performance will vary. Increase in speed and power with height assumes one-seventh power law.																		
28	Cell C8: =ROUND((B8/(0.6125*1.93))^(1/3),1)																		
29	Cell E8: =(B8/(1/3)^(3*1/7))																		
30	Cell F8: =ROUND((E8/(0.6125*1.93))^(1/3),1)																		
31	Cell J8: =(((J$6/2)^2*PI())*$E8*8760*$H8)/1000/1000																		

Estimated Annual Energy Output for Battelle Wind Power Classes at 50 m (164 ft) Hub Height in Thousand kWh/yr

Medium-Size Wind Turbines

	Battelle Power Class at 10 m (33 ft)			Wind Speed and Power at 30 m (100 ft) Hub Height				Rotor Diameter					
	Power Density (W/m²)	Speed (m/s)	(mph)	Power Density (W/m²)	Speed (m/s)	(mph)	Total Effic.	(m) (ft)	20 66	25 82	30 98	35 115	40 131
	50	3.5	7.8	100	4.4	9.9	0.25		70	110	150	210	280
1													
	100	4.4	9.9	200	5.5	12.3	0.3		170	260	370	510	660
2													
	150	5.0	11.5	300	6.3	14.1	0.3		250	390	560	760	990
3													
	200	5.5	12.3	400	7.0	15.7	0.3		330	520	740	1,010	1,320
4													
	250	6.0	13.4	500	7.5	16.8	0.28		390	600	870	1,180	1,540
5													
	300	6.3	14.1	600	8.0	17.9	0.28		460	720	1,040	1,420	1,850
6													
	400	7.0	15.7	800	8.8	19.7	0.25		550	860	1,240	1,690	2,200
7													
	1,000	9.5	21.3	2,000	11.9	26.7	0.22		1,210	1,890	2,720	3,710	4,840
8													
	1,200	10.1	22.6	2,400	12.7	28.4	0.22		1,450	2,270	3,270	4,450	5,810
9													
	1,600	11.1	24.9	3,200	13.9	31.1	0.22		1,940	3,030	4,360	5,930	7,750

Note: Assumed efficiency based on published data. Actual performance will vary. Increase in speed and power with height

45	50	55	60	65	70	75	80	90	100
148	164	180	197	213	230	246	262	295	328
350	430	520	620	730	840	970	1,100	1,390	1,720
840	1,030	1,250	1,490	1,740	2,020	2,320	2,640	3,340	4,130
1,250	1,550	1,870	2,230	2,620	3,030	3,480	3,960	5,020	6,190
1,670	2,060	2,500	2,970	3,490	4,050	4,640	5,280	6,690	8,260
1,950	2,410	2,910	3,470	4,070	4,720	5,420	6,160	7,800	9,630
2,340	2,890	3,500	4,160	4,880	5,660	6,500	7,400	9,360	11,560
2,790	3,440	4,160	4,950	5,810	6,740	7,740	8,810	11,150	13,760
6,130	7,570	9,160	10,900	12,790	14,830	17,030	19,370	24,520	30,270
7,360	9,080	10,990	13,080	15,350	17,800	20,430	23,250	29,420	36,330
9,810	12,110	14,650	17,440	20,460	23,730	27,250	31,000	39,230	48,440

Estimating Annual Energy Output from Manufacturer's Power Curve and Weibull Wind Speed Frequency Distribution Relative to Standard Conditions

	A	B	C	D	E	F	G	H	I	J	K
1	**Estimating Annual Energy Output from Manufacturer's Power Curve and**										
2	**Weibull Wind Speed Frequency Distribution Relative to Standard Conditions**										
3											
4	Adapted from a spreadsheet created by Mike Bergey, Bergey Windpower Co.										
5	**WARNING: Actual performance may vary. See notes below.**										
6											
7	Wind Turbine:		Bergey XL1								
8											
9	**Inputs**					**Output**					
10	Rated Power (kW)			1		Hub Height Avg. Wind Speed (m/s)				5.67	
11	Rotor Diameter (m)			2.5		Altitude Correction				0.85	
12	Swept Area (m^2)			4.9		Temperature Correction				0.98	
13	Avg. Wind Speed (m/s)			5		Annual Energy Output (kWh)				2,137	
14	Weibull K factor			2.0		Avg. Daily Energy (kWh)				6	
15	Altitude (m)			1500		Avg. Monthly Energy (kWh)				178	
16	Avg. Temperature (C)			20		Avg. Power (kW)				0.24	
17	Anemometer Height (m)			10		Avg. Conversion Efficiency				0.24	
18	Tower Height (m)			24		Capacity Factor				0.24	
19	Wind Shear Exponent			0.14		Annual Specific Yield (kWh/m^2/yr)				435	
20	Availability			100%							
21	Turbulence Intensity			0%							
22	Safety Margin			0%							
23											
24	Wind	Gross	Turbine		Turbine		Turbine		Gross	Turbine	
25	Speed	Wind	Power		Corrected	Weibull	Avg.		Wind	Avg.	
26	Bin	Power	Curve	Effic.	Power	Freq.	Power		Energy	Energy	
27	(m/s)	(kW)	(kW)	(%)	(kW)	Dist.	(kW)	(h/yr)	(kWh/yr)	(kWh/yr)	
28	1	0	0.000	0.00	0.0	0.0481	0.00	422	1	0	
29	2	0	0.002	0.08	0.0	0.0894	0.00	783	19	1	
30	3	0	0.022	0.27	0.0	0.1186	0.00	1,039	84	19	
31	4	0	0.060	0.31	0.1	0.1330	0.01	1,165	224	59	
32	5	0	0.125	0.33	0.1	0.1331	0.01	1,166	438	122	
33	6	1	0.230	0.35	0.2	0.1218	0.02	1,067	693	206	
34	7	1	0.375	0.36	0.3	0.1031	0.03	903	932	284	
35	8	2	0.530	0.34	0.4	0.0814	0.04	713	1098	317	
36	9	2	0.700	0.32	0.6	0.0602	0.04	527	1156	310	
37	10	3	0.880	0.29	0.7	0.0419	0.03	367	1103	271	
38	11	4	1.070	0.27	0.9	0.0274	0.02	240	961	216	
39	12	5	1.200	0.23	1.0	0.0170	0.02	149	772	150	
40	13	7	1.230	0.19	1.0	0.0099	0.01	87	574	90	
41	14	8	1.200	0.15	1.0	0.0055	0.01	48	397	48	
42	15	10	1.150	0.11	1.0	0.0029	0.00	25	255	24	
43	16	12	1.095	0.09	0.9	0.0014	0.00	12	154	11	
44	17	15	1.040	0.07	0.9	0.0007	0.00	6	87	5	
45	18	18	0.990	0.06	0.8	0.0003	0.00	3	46	2	
46	19	21	0.940	0.05	0.8	0.0001	0.00	1	23	1	
47	20	24	0.890	0.04	0.7	0.0001	0.00	0	11	0	
48						Total	0.24			2,137	
49	Cell D12: =PI()*(D11/2)^2										
50	Cell J10: +D13*((D18/D17)^D19); Cell J11: (1.2252-(0.0001194)*D15)/1.2252; Cell J12: =(1013250/(2870000*(273.15+D16)))*1000/1.2252; Cell J13: =J48*(D20)*(1-D21)*(1-D22); Cell J14: =J13/365; Cell J15: =J13/12; Cell J16: =G48; Cell J17: =J48/(SUM(I28:I47)): Cell J18: =J13/(8760*D10): Cell J19: =J13/D12										
51	Cell B28: =(A28^3*(1.2252/2)*D$12)/1000										
52	Cell D28: =C28/B28										
53	Cell E28: =C28*$J$11*$J$12										
54	Cell F28: =($D$14/($J$10/0.89))*((A28/($J$10/0.89))^($D$14-1))*(EXP(-((A28/($J$10/0.89))^$D$14)))										
55	Cell G28: =F28*E28										
56	Cell H28: =F28*8760										
57	Cell I28: =B28*H28										
58	Cell J28: =E28*H28; Cell J48: =SUM(J28:J47)										
59											
60	Notes										
61	1. This spreadsheet accounts for changes in air density that may or may not be applicable to wind turbines using pitch regulation.										
62	2. The accuracy of the technique used in this spreadsheet is limited by the validity of the power curve and the wind speed data. The measurement and prediction of power curves, especially for small battery-charging wind turbines, is an arcane art frought with uncertainty. For best results use only power curves verified by an independent testing laboratory.										
63	3. The technique used in this spreadsheet is best suited for estimating annual energy output and is not suited for calculating performance over shorter periods.										
64	4. Turbulence intensity is a measure for incorporating the power-robbing effects of turbulence.										
65	5. Safety margin is a means to account for loss of power due to a host of factors. Bergey Windpower suggests using 5% for hybrid power systems in a residential application, 15% to 25% in a telecommunication application, and 20% to 40% for high priority loads.										
66	6. This spreadsheet uses a Weibull wind speed frequency distribution to calculate annual energy output. The Rayleigh distribution is a special case where Weibull K=2.										
67											
68	How To Use the Spreadsheet										
69	1. Enter the data for the turbine you are planning to use and for the site where it will be used in the column labeled Inputs. (For conventional horizontal axis wind turbines, swept area is calculated using the area of the circle swept by the rotor.) You will need to enter at least the average annual wind speed, the Weibull K factor, the elevation where the turbine will be installed, the height of the tower, and the rotor diameter. Enter a wind shear exponent for the conditions that best represent your site.										
70	2. Enter data for the power curve of the wind turbine you plan to use in the third column by wind speed bin in meters/second.										
71	3. Enter any adjustments for Availability, Turbulence Intensity, or Safety Margin in the Inputs column that are appropriate for your situation.										

Estimates of Water Pumping Capacity of Farm Windmills

The following table estimates the amount of water a traditional American windmill of a given diameter will pump daily from a given depth within different wind regimes. The tables assume that the overall efficiency of the windmill is 5 percent in a Rayleigh wind speed distribution. Actual performance may vary depending on the windmill and whether it's properly matched to the well pump.

To use the table, first find the total dynamic head in the leftmost column. Then find the average annual wind speed at hub height. For example, if the total pumping head is about 100 feet (30 m) at a site with an 11 mph (5 m/s) average wind speed, a farm windmill with 10-foot windwheel will pump about 4,100 gallons (16 m3) per day.

Approximate Daily Pumping Volume of American Farm Windmill in m3/day and Gallons/day

	A	B	C	D	E	F	G	H	I	J	K	L
1	Approximate Daily Pumping Volume of American Farm Windmill in m^3/day and gallons/day											
2												
3	Rotor Diameter		3.05	m	10	ft						
4			Average Annual Wind Speed									
5		m/s	3		4		5		6		7	
6		mph	6.7		9.0		11.2		13.4		15.7	
7												
8	Pumping Head											
9	m	ft	m^3/dy	gals/dy	m^3/dy	gals/dy	m^3/dy	gals/dy	m^3/dy	gals/dy	m^3/dy	gals/dy
10	10	30	10	2,700	24	6,300	47	12,300	80	21,200	128	33,700
11	20	70	5	1,300	12	3,100	23	6,100	40	10,600	64	16,800
12	30	100	3	900	8	2,100	16	4,100	27	7,100	43	11,200
13	40	130	3	700	6	1,600	12	3,100	20	5,300	32	8,400
14	Cell B10: =ROUND(A10*3.28,-1)											
15	Cell C10: =(0.32*5/4)*($A:$C$3)^2*(C$6^3)/$A10											
16	Cell D10: =ROUND(C10*264,-2)											
17												
18	Source: Center for International Development, Research Triangle Institute											

Cash Flow Models

The following tables include the formulas within critical cells for calculating cash flow from the two examples used in the text.

Cash Flow for Single Household-Size Wind Turbine in U.S. Residential Use

	A	B	C	D	E	F	G	H	I	J	K	L
1	Cash Flow for Single Household-Size Wind Turbine in U.S. Residential Use											
2												
3	Assumptions:											
4	Rotor diameter		7	m	Retail rate		0.10	$/kWh	Utility rate escalation		5%	
5	Avg. wind speed		6	m/s	Resale rate		0.02	$/kWh	Inflation rate		3%	
6	Yield		365	kWh/m²/yr	% at retail rate		100%		Down payment		20%	
7	Installed cost		$35,000		Tax credit rate		0.0165	$/kWh	Loan term		10	yrs
8	O&M		0.01	$/kWh	% Tax credit used		100%		Loan interest		8%	
9	Insurance		1%		Tax bracket		35%		Discount rate		6%	
10												
11	Results:											
12	Swept area		38	m²								
13	Annual energy output		14,000	kWh/yr								
14												
15		Gross			Loan	Loan	Tax Value	Revenue	Cumulative			
16	Year	Revenue	O&M	Insurance	Interest	Principal	(Cost)	(Loss)	Revenue			
17	0	-7,000	0	0	0	0	0	-7,000	($7,000)			
18	1	1,400	-140	-350	-2,240	-1,933	$0	-4,663	($11,663)			
19	2	1,470	-144	-361	-2,085	-2,087	$2,991	-1,686	($13,349)			
20	3	1,544	-149	-371	-1,918	-2,254	$3,046	-1,647	($14,996)			
21	4	1,621	-153	-382	-1,738	-2,435	$3,102	-1,607	($16,602)			
22	5	1,702	-158	-394	-1,543	-2,630	$3,158	-1,566	($18,168)			
23	6	1,787	-162	-406	-1,333	-2,840	$3,215	-1,525	($19,694)			
24	7	1,876	-167	-418	-1,106	-3,067	$3,273	-1,485	($21,178)			
25	8	1,970	-172	-430	-860	-3,313	$3,332	-1,444	($22,622)			
26	9	2,068	-177	-443	-595	-3,578	$3,391	-1,403	($24,025)			
27	10	2,172	-183	-457	-309	-3,864	$3,450	-1,363	($25,388)			
28	11	2,280	-188	-470	0	0	$3,508	2,850	($22,538)			
29	12	2,394	-194	-484	0	0	$3,684	3,006	($19,532)			
30	13	2,514	-200	-499	0	0	$3,868	3,169	($16,363)			
31	14	2,640	-206	-514	0	0	$4,061	3,342	($13,021)			
32	15	2,772	-212	-529	0	0	$4,264	3,523	($9,498)			
33	16	2,910	-218	-545	0	0	$4,478	3,714	($5,783)			
34	17	3,056	-225	-562	0	0	$4,702	3,915	($1,868)			
35	18	3,209	-231	-578	0	0	$4,937	4,127	$2,259			
36	19	3,369	-238	-596	0	0	$5,183	4,349	$6,608			
37	20	3,538	-245	-614	0	0	$5,443	4,583	$11,191			
38	Total 20-year revenue							$11,191				
39	Net present value using a discount rate derived from a 20-year mortgage.							($6,000)				
40	Generation used on site offsets consumption at retail rate. Tax credit applies only to sale of electricity.											
41												
42	Cell B17: =-(K6*C7); Cell B18: =(C$13*G$4*G$6)+(C$13*G$5*(1-G$6)); Cell B19: =B18+($K$4*B18)											
43	Cell C12: =($C$4/2)^2*PI(); Cell C13: =ROUND(C6*C12,-2); Cell C18: =-($C$13*$C$8); Cell C19: =C18+(C18*K$5)											
44	Cell D18: =-C$9*C$7; Cell D19: =C18+(C18*K$5)											
45	Cell E18: =IF(A18<=K$7,(IPMT(K$8,A18,K$7,(C$7-(C$7*K$6)))),0)											
46	Cell F18: =IF(A18<=K$7,(PPMT(K$8,A18,K$7,(C$7-(C$7*K$6)))),0)											
47	Cell G18: =(B18*(1/(1-G$9)))+(E18*-G$9)											
48	Cell H18: =C18+D18+E18+F18+G18; Cell H38: =SUM(H17:H37); Cell H39: =ROUND(NPV(K9,H17:H37),-3)											
49	Cell I18: =H18+I17											
50												
51												

Cash Flow for Single Medium-Size Wind Turbine (225–300 kW) in U.S. Business or Farm Application

	A	B	C	D	E	F	G	H	I	J	K	L
1	Cash Flow for Single Medium-Size Wind Turbine (225-300 kW) in U.S. Business or Farm Application											
2												
3	Assumptions:											
4	Rotor diameter		29	m	Retail rate		0.10	$/kWh	Utility rate escalation		5%	
5	Avg. wind speed		6	m/s	Resale rate		0.02	$/kWh	Inflation rate		3%	
6	Yield		720	kWh/m^2/yr	% at retail rate		90%		Down payment		20%	
7	Installed cost		$300,000		Tax credit rate		0.0165	$/kWh	Loan term		10	yrs
8	O&M		0.01	$/kWh	% Tax credit used		100%		Loan interest		8%	
9	Insurance		1%		Tax bracket		35%		Discount rate		6%	
10												
11	Results:											
12	Swept area		661	m^2								
13	Annual Energy Output		480,000	kWh/yr								
14												
15		Gross			Loan	Loan	Depreciation	Income	Tax	Revenue	Cumulative	
16	Year	Revenue	O&M	Insurance	Interest	Principal	Deduction	Tax	Credit	(Loss)	Revenue	
17	0	-60,000	0	0	0	0	0	0	0	-60,000	($60,000)	
18	1	44,160	-4,800	-3,000	-19,200	-16,567	-60,000	14,994	792	16,379	($43,621)	
19	2	46,368	-4,944	-3,090	-17,875	-17,892	-96,000	26,439	816	29,822	($13,799)	
20	3	48,686	-5,092	-3,183	-16,443	-19,324	-57,600	11,771	840	17,256	$3,457	
21	4	51,121	-5,245	-3,278	-14,897	-20,870	-34,560	2,401	865	10,097	$13,553	
22	5	53,677	-5,402	-3,377	-13,228	-22,539	-34,560	1,011	891	11,034	$24,587	
23	6	56,361	-5,565	-3,478	-11,425	-24,342	-17,280	-6,515	918	5,955	$30,541	
24	7	59,179	-5,731	-3,582	-9,477	-26,290		-14,136	946	908	$31,449	
25	8	62,138	-5,903	-3,690	-7,374	-28,393		-15,810	974	1,942	$33,391	
26	9	65,244	-6,080	-3,800	-5,103	-30,665		-17,591	1,003	3,008	$36,400	
27	10	68,507	-6,263	-3,914	-2,649	-33,118		-19,488	1,033	4,108	$40,507	
28	11	71,932	-6,451	-4,032	0	0		-21,507		39,942	$80,449	
29	12	75,529	-6,644	-4,153	0	0		-22,656		42,076	$122,525	
30	13	79,305	-6,844	-4,277	0	0		-23,864		44,320	$166,845	
31	14	83,270	-7,049	-4,406	0	0		-25,135		46,680	$213,525	
32	15	87,434	-7,260	-4,538	0	0		-26,472		49,163	$262,688	
33	16	91,805	-7,478	-4,674	0	0		-27,879		51,775	$314,463	
34	17	96,396	-7,703	-4,814	0	0		-29,358		54,521	$368,984	
35	18	101,216	-7,934	-4,959	0	0		-30,913		57,410	$426,394	
36	19	106,276	-8,172	-5,107	0	0		-32,549		60,448	$486,842	
37	20	111,590	-8,417	-5,261	0	0		-34,269		63,643	$550,486	
38	Total 20-year revenue									$550,486		
39	Net present value using a discount rate derived from a 20-year mortgage.									$214,000		
40	Generation used on site offsets consumption at retail rate. Tax credit applies only to sale of electricity.											
41												
42	Cell B17: =-(K6*C7); Cell B18: =(C$13*G$4*G$6)+(C$13*G$5*(1-G$6)); Cell B19: =B18+($K$4*B18)											
43	Cell C12: =($C$4/2)^2*PI(); Cell C13: =ROUND(C6*C12,-4); Cell C18: =-($C$13*$C$8); Cell C19: =C18+(C18*K$5)											
44	Cell D18: =-C$9*C$7; Cell D19: =D18+(D18*K$5)											
45	Cell E18: =IF(A18<=K$7,(IPMT(K$8,A18,K$7,(C$7-(C$7*K$6)))),0)											
46	Cell F18: =IF(A18<=K$7,(PPMT(K$8,A18,K$7,(C$7-(C$7*K$6)))),0)											
47	Cell G18: =-0.2*C$7; Cell G19: =-0.32*C$7; Cell G20: =-0.192*C$7; Cell G21: =-0.1152*C$7; Cell G22: =-0.1152*C$7; Cell G23: =-0.0576*C$7											
48	Cell H18: =-(+B18+C18+D18+E18+G18)*G$9											
49	Cell I18: =(1-G$6)*G$7*G$8*C$13; Cell I19: =I18+($K$5*I18)											
50	Cell J17: =B17+C17+D17+E17+F17+H17+I17; Cell J38: =SUM(J17:J37); Cell J39: =ROUND(NPV(K9,J17:J37),-3)											
51	Cell K17: =J17+K15											

Nongovernmental Organizations

The following organizations can be contacted for further information. The Alternative Energy Institute conducts field trials of small wind turbines at its Wind Turbine Test Center and is a good source of practical information on small machines. Denmark's Folkecenter for Renewable Energy also tests small wind turbines. Most of the national wind energy associations hold annual conferences on the latest developments in wind turbine technology within their respective countries. These conferences deal mostly with the commercial use of medium-size wind turbines, though most have sessions on small wind turbines, as well.

Outside Anglophone countries there's usually a distinction between trade associations that promote the interests of businesses working with wind energy, such as wind turbine manufacturers, and groups that represent the interests of consumers or the environment. In Germany, Bundesverband Windenergie represents wind turbine owners and Fördergesellschaft Windenergie represents the interests of manufacturers. This distinction is important to know when searching for information. Trade associations exist to promote the interests of their members and not necessarily those of the general public. Don't expect the American Wind Energy Association or the Danish Wind Turbine Manufacturers Association to help you locate information on German wind turbines.

Alternative Energy Institute
West Texas A&M University
Box 60248
Canyon, TX 79016
phone: 806 651 2295
fax: 806 651 2733
aeimail@mailwtamu.edu
www.windenergy.org

American Wind Energy Association
122 C Street NW, #380
Washington, DC 20001
phone: 202 383 2500,
800 634 4299
(fax on demand)
fax: 202 383 2505
windmail@awea.org
www.awea.org

Australian Wind Energy Association
GPO Box 4499
Melbourne, Victoria 3000
Australia
phone: 61 3 9663 2323
fax: 61 3 9663 2040
info@auswea.com.au
www.auswea.com.au

Austrian Wind Energy Association
(Interessengemeinschaft Winkdraft Österreich)
Mariahilfer Strasse 89/22
A-1060 Wien
Austria
phone: 43 581 7060
fax: 43 581 7061
igw@atmedia.net
www.atmedia.net/IGW

British Wind Energy Association
26 Spring Street
London W2 1JA
United Kingdom
phone: 44 171 402 7102
fax: 44 171 402 7107
info@bwea.com
www.bwea.com

Bundesverband Windenergie (German Wind Turbine Owners' Association)
Herrenteichsstrasse 1
D49074 Osnabrück
Germany
phone: 49 541 35 0600
fax: 49 541 35 06030
bwe_os@t-online.de
www.wind-energie.de

Canadian Wind Energy Association
100, 3553 31st NW
Calgary, Alberta T2L 2K7
Canada
phone: 800 9 CanWEA,
403 289 7713
fax: 403 282 1238
canwea@canwea.ca
www.canwea.ca

Centre for Alternative Technology
Llwyngwern Quarry
Machynlleth, Powys Wales
United Kingdom
phone: 44 1654 702 400
fax: 44 1654 702 782
info@cat.org.uk
www.cat.org.uk

Centro Brasiliero de Energia Eolica
Centrol de Tecnologia-UFPE
Cidade Universitaria
Recife, Pernambuco CEP 50740-530
Brazil
phone: 55 81 4532975
fax: 55 81 4534662
www.eolica.com.br

Danmarks Vindmølleforening (Danish Wind Turbine Owners' Association)
Egensevej 24
DK-4840 Nr. Alslev
Denmark
phone: 45 54 43 13 22
fax: 45 54 43 12 02
info@dkvind.dk
www.dkvind.dk

European Wind Energy Association
Rue du Trône 26
B-1040 Brussels
Belgium
phone: 32 3 546 1940
fax: 32 3 546 1944
ewea@ewea.org
www.ewea.org

Fördergesellschaft Windenergie (German Wind Turbine Manufacturers' Association)
Flotowstrasse 41-43
D-22083 Hamburg
Germany
phone: 49 40 27 80 91 82
fax: 49 40 27 80 91 76
FGW-HH@t-online.de
www.wind-fgw.de

France Énergie Éolienne (French Wind Energy Association)
330, Rue de Mourelet, ZI Courtine
F-84000 Avignon
France
phone: 33 4 32 76 03 10
fax: 33 4 32 76 03 28
association-free@wanadoo.fr
www.fee.asso.fr

Irish Wind Energy Association
Arigna
Carrick on Shannon, County Roscommon
Ireland
phone: 353 78 46072
office@iwea.com
www.iwea.com

New Zealand Wind Energy Association
P.O. Box 553
Wellington
New Zealand
phone: 64 4 586 2003
fax: 64 4 586 2004
nzwea@windenergy.org.nz
www.windenergy.org.nz

Nordvestjysk Folkecenter for Vedvarende Energi (Folkecenter for Renewable Energy)
P.O. Box 208
DK-7760 Hurup Thy
Denmark
phone: 45 97 95 65 55
fax: 45 97 95 65 65
energy@folkecenter.dk
www.folkecenter.dk

Renewable Energy Research Laboratory
Engineering Laboratory
University of Massachusetts
Amherst, MA 01003
phone: 413 545 4359
fax: 413 545 0724
rerl@snail.ecs.umass.edu
www.ecs.umass.edu/mie/labs/rerl

South African Wind Energy Association
P.O. Box 22
St. James 7946
South Africa
phone: 27 21 788 9758
fax: 27 21 788 9758
sawea@icon.co.za
www.icon.co.za/~sawea

Suisse Eole
Crêt 108a
CH-2314 La Sagne
Switzerland
phone: 41 32 932 40 23
fax: 41 32 931 18 68
contact@suisse-eole.ch
www.suisse-eole.ch

Vindkfraftförenigen r.f. (Finnish Wind Energy Association)
P.O. Box 124
FIN-65101 Vasa
Finland
phone: 358 500 862 886
fax: 358 6 312 8882
www.vindkraftforeningen.fi

Vindmølleindustrien (Danish Wind Turbine Manufacturers' Association)
Vester Voldgade 106
DK-1552 Copenhagen
Denmark
phone: 45 33 73 03 30
fax: 45 33 73 03 33
danish@windpower.dk
www.windpower.dk

World Wind Energy Association
Eduard-Pflueger-Str. 43
D-53113 Bonn
Germany
phone: 49 228 369 90 83
fax: 49 228 369 90 84
wwindea@wwindea.org
www.wwindea.org

Government-Sponsored Laboratories

These research centers concentrate on highly technical aspects of medium-size and large wind turbine technology. Several have tested small wind turbines, however, including the National Engineering Laboratory (NEL) in Scotland, the National Renewable Energy Laboratory in Colorado, Risø National Laboratory in Denmark, and the U.S. Department of Agriculture in Texas.

Atlantic Wind Test Site
RR 4
Tignish, PEI COB 2BO
Canada
phone: 902 882 2746
fax: 902 882 3823
info@awts.pe.ca
www.awts.pe.ca

CIEMAT, Departameno de Energias Renovables (DER)
Avenida Complutense 22
ES-28040 Madrid
Spain
phone: 34 1 346 6254
fax: 34 1 346 6037
www.ciemat.es

Deutsches Windenergie Institut (DEWI)
Eberstrasse 96
D-26382 Wilhelmshaven
Germany
phone: 49 4421 48080
fax: 49 4421 480843
dewi@dewi.de
www.dewi.de

ECN (Energieonderzoek Centrum Nederlands)
Postbus 1
NL-1755 ZG Petten
Netherlands
phone: 31 224 56 49 49
fax: 31 22 46 32 14
wind@ecn.nl
www.ecn.nl/wind/

Institut für Solare Energieversorgungstechnik (ISET)
Königstor 59
D-34119 Kassel
Germany
phone: 49 561 72 940
fax: 49 561 72 94 100
mbox@iset.uni-kassel.de
www.iset.uni-kassel.de

National Engineering Laboratory (NEL)
East Kilbride
Glasgow, Scotland G75 OQU
United Kingdom
phone: 44 3552 20222
fax: 44 3552 72333
info@nel.uk
www.nel.uk

National Renewable Energy Laboratory
National Wind Technology Center
1617 Cole Boulevard
Golden, CO 80401-3393
phone: 303 384 6900
fax: 303 384 6999
www.nrel.gov/wind

Risø National Laboratory
The Test Station for Wind Turbines
P.O. Box 49
DK-4000 Roskilde
Denmark
phone: 45 4677 5000
fax: 45 4677 5970
vea@risoe.dk
www.risoe.dk/amv

Sandia National Laboratory
P.O. Box 5800
MS 0708
Albuquerque, NM 87185
phone: 505 844 0183
fax: 505 845 9500
www.sandia.gov/wind

U.S. Department of Agriculture in Texas
Agricultural Research Service
P.O. Box 10
Bushland, TX 79012
phone: 806 356 5734
fax: 806 356 5750
www.cprl.ars.usda.gov/wesm.htm

Windtest Kaiser-Wilhelm-Koog
Sommerdeich 14b
D-25709 Kaiser-Wilhelm-Koog
Germany
phone: 49 48 56 9010
fax: 49 48 56 90149
info@windtest.de
www.windtest.de

Internet Sites

There are literally hundreds of World Wide Web sites on the topic of wind energy. For tips on the design of small wind turbines, visit the sites of Scoraig Wind Electric's Hugh Piggott (www.scoraigwind.co.uk) and Windmission's Claus Nybroe (www.windmission.dk).

Most national wind energy associations sponsor Web sites with information on programs in their countries. Many of the European Web sites are offered in several languages, including English.

The most comprehensive site on medium-size wind turbines is that sponsored by the Danish Wind Turbine Manufacturers' Association (Vindmølleindustrien) at www.windpower.dk or www.windpower.org. Information is available in English, German, Spanish, and Danish. The fastest way to browse this site is with the association's companion CD-ROM.

Remember that trade associations are in business to promote their members' products. You won't find information about the German manufacturer Enercon on the Web sites of the American Wind Energy Association or the Danish Wind Turbine Manufacturers' Association.

The best site in French is that sponsored by the University of Quebec at Rimouski, www.eole.org.

There are a number of excellent Web sites in German. Visit the wind energy sites at the University of Münster (www.uni-muenster.de/Energie), or at the Technical University of Berlin (www.ilr.tu-berlin.de/WKA/engwindkraft.html)—one of the earliest Web sites with information on wind energy.

As with any source of information, be wary. Occassionally there is bogus and misleading material at sites advertising mail-order wind turbines. Some of these sites are "here today, gone tomorrow."

Electronic Networks

There are two electronic networks or "news groups" administered by the American Wind Energy Association that deal with wind energy: awea-wind-home and awea-windnet. These electronic conferences are accessible by subscribers worldwide.

Participants in awea-wind-home discuss small wind turbines and issues associated with the use of wind energy at farms, homes, and businesses. Subscribers to awea-windnet are typically professionals who work with wind energy.

To participate in these news groups, you must subscribe by sending a blank message to awea-wind-home-subscribe@yahoogroups.com or by registering on the yahoogroups.com Web site.

Before posting any questions or comments, it's always good "netiquette" to monitor the lists for some time to become familiar with the topics and the participants. It's also wise to read the FAQ, or frequently asked question, files first. Participants on these lists have answered many of the most common questions, so many times they may bristle at seeing them again. Be forewarned that discussions on these lists can be quite lively. AWEA monitors or censors content on awea-windnet; awea-wind-home is more freewheeling.

Workshops

Both Mick Sagrillo and Hugh Piggott offer workshops on small wind turbines. Sagrillo targets sizing, proper siting, and installation. Piggott teaches how to build a sturdy, functioning small turbine from scrap parts.

Mick Sagrillo
Sagrillo Light & Power
E 3971 Bluebird Road
Forestville, WI 54213
phone: 920 837 7523
fax: 920 837 7523
msagrillo@itol.com

Hugh Piggott
Scoraig Wind Electric
Dundonnell, Ross Shire
Scotland IV23 2RE
United Kingdom
phone: 44 18 54 633 286
fax: 44 18 54 633 233
hugh@scoraigwind.co.uk
www.scoraigwind.co.uk

Several organizations offer short courses on wind energy for professionals, including ECN in the Netherlands and DEWI in Germany. AEI also offers a distant learning class via the Internet or by CD-ROM.

Alternative Energy Institute
West Texas A&M University
Box 60248
Canyon, TX 79016
phone: 806 651 2295
fax: 806 651 2733
aeimail@mail.wtamu.edu
www.windenergy.org

Deutsches Windenergie Institut (DEWI)
Eberstrasse 96
D-26382 Wilhelmshaven
Germany
phone: 49 4421 48080
fax: 49 4421 480843
dewi@dewi.de
www.dewi.de

ECN (Energie Centrum Nederlands)
Postbus 1
NL-1755 ZG Petten
Netherlands
phone: 31 224 564705
fax: 31 224 568214
wind@ecn.nl
www.ecn.nl/wind/products/training/

International Wind Energy Consultants

There are dozens of firms worldwide providing professional services to the wind energy industry. These three are the best known. There also a number of planning bureaus in Germany that specialize in siting medium-size wind turbines throughout Europe.

AeroVironment Inc.
222 E. Huntington Drive
Monrovia, CA 91016
phone: 626 357 9983
fax: 626 359 9628
windpower@aerovironment.com
www.aerovironment.com

BTM Consult
I. C. Christensens Alle 1
DK-6950 Ringkøbing
Denmark
phone: 45 97 32 52 99
fax: 45 97 32 55 93
btm@btm.dk
www.btm.dk

Garrad Hassan
St. Vincent's Works
Silverthorne Lane
Bristol BS2 0QD
United Kingdom
phone: 44 117 972 9900
fax: 44 117 972 9901
info@garradhassan.com
www.garradhassan.com

Characteristics of Selected Small Wind Turbines

Small wind turbines is a broad and arbitrary category. This class includes micro wind turbines—the smallest of small turbines—mini wind turbines, and household-size wind turbines. Micro turbines as those from 0.5 to 1.25 meters (about 2 to 4 ft) in diameter. These machines include the Rutland 500 as well as the Air series. Mini wind turbines are slightly larger and span the range between the micro turbines and the bigger household-size machines. They vary in diameter from 1.25 to 2.75 meters (4 to 9 ft) and include such models as the Whisper H40 and the Rutland 1800. Household-size wind turbines (a translation of the Danish term *hustandmølle*) are the largest of the small wind turbine family. As you would expect from the broad range of home sizes available, wind turbines in this class span a wide spectrum. They include turbines as small as the Whisper H80 with a rotor 3 meters (10 ft) in diameter, along with Lagerwey's 18-meter (60 ft) machine, capable of generating 80 kW.

Most small wind turbines are suitable only for battery-charging and similar applications. The larger turbines can be used to produce utility-compatible electricity that can be safely interconnected with the local utility.

The turbines in the table are sorted by swept area, and not by rated power.

Std. Power Rating: This is an arbitrary rating designed for those who insist on comparing wind turbines by power and not swept area. It assumes that each wind turbine is capable of generating 200 watts per square meter of swept area. This will overstate the capability of some turbines and understate the capability of others, especially those with permanent-magnet alternators using neodymium magnets. For example, it will understate the performance of the Air series, but will overstate that of the Rutland 913.

Mfg. Rated Power and Rated Wind Speed: This information is taken from manufacturer product literature; most has not been verified by independent third-party tests.

Mfg. Perf. at Rated Power: The overall conversion efficiency of the wind turbine at the rated wind speed, derived from the rated output claimed by the manufacturer. These claims are also unverified. This is but one of several useful tools in evaluating performance claims. For example, if the manufacturers of most small wind turbines expect their turbines to convert 20 to 30 percent of the energy in the wind to electricity at rated power, then a manufacturer that claims a 40 percent conversion efficiency may either have made a mistake or be inflating its turbine's performance.

Spec. Rotor Loading at Rated Power: The power that a manufacturer expects its wind turbine to produce at its rated wind speed relative to the area swept by the rotor is an indication of how hard the blades have to work. It's also one indicator of potential noise. This measure is another tool when comparing wind turbines. For example, if many small wind turbines produce 100 to 150 watts per square meter of rotor area at rated power, then the rotors on those turbines that produce 200 to 300 W/m2 have to work much harder than competing products.

Means of Control: All small wind turbines designed for use on land should include some aerodynamic means for controlling rotor overspeed. This is a time-tested axiom. Some of the micro turbines listed—those with no aerodynamic overspeed control (nc)—are suited only for use on boats in protected harbors or on land in low-wind regimes.

Characteristics of Selected Small Wind Turbines

Manufacturer	Model	(Axis)	Rotor Height (m)	(ft)	Rotor Dia. (m)	(ft)	Area (m^2)	(ft^2)
Ampair	Dolphin	v	0.43	1.4	0.26	0.9	0.11	1.2
Windside Oy	0.15	v	0.50	1.6	0.30	1.0	0.15	1.6
Windside Oy	0.30C	v	1.00	3.3	0.30	1.0	0.30	3.2
Windside Oy	4	v	4.00	13.1	1.00	3.3	4.00	43.0
Marlec	Rutland 500	h			0.51	1.7	0.20	2.2
LVM	Aero2gen	h			0.58	1.9	0.26	2.8
J. Bornay	G60	h			0.75	2.5	0.44	4.8
LVM	Aero4gen F	h			0.86	2.8	0.58	6.2
LVM	Aero4gen	h			0.87	2.9	0.59	6.4
Marlec	Rutland 910-3F	h			0.91	3.0	0.65	7.0
Marlec	Rutland 913	h			0.91	3.0	0.65	7.0
Ampair	100	h			0.92	3.0	0.66	7.1
MG Plast	Aerplast	h			1.0	3.3	0.79	8.4
Everfair Enterprises	Fourwinds III	h			1.0	3.3	0.81	8.7
Southwest Windpower	Air X	h			1.2	3.8	1.07	11.5
Aerocraft	AC 120	h			1.2	3.9	1.13	12.2
LVM	Aero6gen	h			1.2	4.0	1.17	12.6
LVM	Aero6gen F	h			1.2	4.0	1.17	12.6
Wind Generator Prod.	Windbugger	h			1.3	4.2	1.27	13.6
MG Plast	Aerplast	h			1.4	4.4	1.43	15.4
J. Bornay	Inclin 250	h			1.4	4.4	1.43	15.4
Guangdong Yuehua	F50/6	h			1.5	4.9	1.77	19.0
Atlantis	WB 15H	h			1.5	4.9	1.77	19.0
Southwest Windpower	Windseeker 503	h			1.5	5.0	1.81	19.5
Electrovent	30 amp	h			1.5	5.0	1.83	19.6
Electrovent	15 amp	h			1.5	5.0	1.83	19.6
Hamilton Ferris	WP200	h			1.5	5.0	1.83	19.6
Everfair Enterprises	Fourwinds II	h			1.5	5.0	1.83	19.6
LVM	Aero8gen F	h			1.6	5.1	1.89	20.3
Aerocraft	AC 240	h			1.7	5.4	2.14	23.0
MG Plast	Aerplast	h			1.7	5.4	2.14	23.0
LMW	250	h			1.7	5.6	2.27	24.4
Cataventos	Gerador Kenya	h			1.8	5.9	2.54	27.4
Marlec	Rutland 1803	h			1.8	6.0	2.62	28.2
Guangdong Yuehua	F200/7	h			1.9	6.2	2.84	30.5
A. Harbarth	D 303	h			1.9	6.2	2.84	30.5
Atlantis	WB 20H	h			2.0	6.6	3.14	33.8
Sunrise Solar	Soma 400	h			2.0	6.6	3.14	33.8
Soluciones Energet.	Velter II	h			2.0	6.6	3.14	33.8
J. Bornay	Inclin 600	h			2.0	6.6	3.14	33.8
Guangdong Yuehua	F300/7	h			2.1	6.9	3.46	37.3
Southwest Windpower	Whisper H40	h			2.1	7.0	3.58	38.5
Fortis	Espada	h			2.2	7.2	3.80	40.9
Guangdong Yuehua	F500/8	h			2.2	7.2	3.80	40.9
CSRI Elektropribor	UVE500	h			2.2	7.2	3.80	40.9
LMW	600	h			2.2	7.2	3.80	40.9
E. Schoder		h			2.2	7.3	3.94	42.4
SVIAB	VK240	h			2.4	7.9	4.52	48.7
Qingdao Windmill Fac.	FD2.4-200	h			2.4	7.9	4.52	48.7

*Std. Power Rating= swept area x 200 W/m^2

Key: nc=no control, h=horizontal furling, v=vertical furling, ab=air brake, tb=tip brake, p=pitch to feather, ps=pitch to stall,
GFRP=glass-fiber-reinforced polyester or fiberglass, wood lam.=wood laminate
Poly=glass-fiber-reinforced polypropylene PGFR=polycarbonate glass-fiber-reinforced

Std. Power Rating* (kW)	Mfg. Rated Power (kW)	Rated Wind Speed (m/s)	(mph)	Mfg. Perf. at Rated Power (%)	Spec. Rotor Loading at Rated Power (W/m²)	No. of of Blds.	Blade Material	Means of Control
0.02	0.02	10.0	22	0.32	193	3	GFRP	nc
0.03	0.04	12.0	27	0.22	233	2	GFRP	nc
0.06	0.07	12.0	27	0.22	233	2	GFRP	nc
0.80	0.70	12.0	27	0.17	175	2	GFRP	nc
0.04	0.03	10.0	22	0.20	122	6	nylon	nc
0.05	0.01	9.8	22	0.07	38	5	nylon	nc
0.09	0.06	10.0	22	0.22	136	5	GFRP	nc
0.12	0.07	10.3	23	0.18	121	6	nylon	h
0.12	0.07	10.3	23	0.18	118	6	nylon	nc
0.13	0.09	10.0	22	0.23	138	6	nylon	h
0.13	0.09	10.0	22	0.23	138	6	nylon	nc
0.13	0.05	10.0	22	0.12	73	6	GFRPoly	nc
0.16	0.06	10.0	22	0.12	76	3	GFRP	nc
0.16	0.21	12.8	29	0.20	259	4	GFRP	nc
0.21	0.40	12.5	28	0.31	373	3	CFRP	nc
0.23	0.12	9.0	20	0.24	106	5	GFRP	v
0.23	0.12	10.3	23	0.15	103	6	nylon	nc
0.23	0.12	10.3	23	0.15	103	6	nylon	h
0.25	0.18	28.8	64	0.01	142	3	aluminum	nc
0.29	0.15	9.0	20	0.23	105	3	GFRP	v
0.29	0.25	11.0	25	0.21	175	2	nylon	v
0.35	0.05	6.0	13	0.21	28	3	Poly	v
0.35	0.30	10.0	22	0.28	170	3	CFRE	v
0.36	0.50	12.5	28	0.23	276	3	wood	v
0.37	0.36	15.6	35	0.08	197	2	wood	ab
0.37	0.18	12.5	28	0.08	99	2	wood	ab
0.37	0.24	10.3	23	0.20	132	2	wood	ab
0.37	0.24	12.8	29	0.10	132	2	GFRP	ab
0.38	0.24	12.8	29	0.10	127	3	wood lam.	h
0.43	0.24	9.0	20	0.25	112	5	GFRP	v
0.43	0.30	9.5	21	0.27	140	3	GFRP	v
0.45	0.25	10.0	22	0.18	110	3	GFRP	h
0.51	0.20	10.0	22	0.13	79	2	wood	ab
0.52	0.34	10.0	22	0.21	130	2	GFRP	h
0.57	0.20	7.0	16	0.34	71	3	Poly	v
0.57	0.30	12.0	27	0.10	106	3	nylon	v
0.63	0.60	10.0	22	0.31	191	3	CFRE	v
0.63	0.40	10.0	22	0.21	127	2	GFRP	v
0.63	1.00	13.0	29	0.24	318	3	CFRE	h
0.63	0.60	11.0	25	0.23	191	2	CFRE	v
0.69	0.30	7.4	17	0.35	87	3	Poly	v
0.72	0.46	10.0	22	0.21	129	3	PGFR	h
0.76	0.60	12.0	27	0.15	158	2	GFRP	h
0.76	0.50	8.5	19	0.35	132	3	Poly	v
0.76	0.50	10.0	22	0.21	132	3	GFRP	h
0.76	0.60	12.0	27	0.15	158	2	GFRP	h
0.79	0.50	12.0	27	0.12	127	3	GFRP	v
0.90						3	nylon	h
0.90	0.30	8.0	18	0.21	66	2		p

pt=pitchable tip, mb=mechanical brake

CFRP=carbon-fiber-reinforced polyester

CFRE=carbon-fiber-reinforced epoxy

GFRPoly=glass-fiber-reinforced polypropylene

Characteristics of Selected Small Wind Turbines

Manufacturer	Model	(Axis)	Rotor Height (m)	Rotor Height (ft)	Rotor Dia. (m)	Rotor Dia. (ft)	Area (m²)	Area (ft²)
Baltaruta	BRC mini	h			2.4	7.9	4.52	48.7
Aerocraft	AC 500	h			2.4	7.9	4.52	48.7
Aerocraft	AC 750	h			2.4	7.9	4.52	48.7
Aeromax	OB1	h			2.4	8.0	4.67	50.3
Qingdao Windmill Fac.	FD2.5-500	h			2.5	8.2	4.91	52.8
Qingdao Windmill Fac.	FD2.5-700	h			2.5	8.2	4.91	52.8
Bergey Windpower	XL.1	h			2.5	8.2	4.91	52.8
Proven Wind Turbines	WT600	h			2.6	8.4	5.11	54.9
E. Schoder		h			2.7	8.8	5.64	60.7
A. Harbarth	E 600	h			2.7	8.9	5.73	61.6
Synergy Power Corp.	S-3000	h			2.7	8.9	5.73	61.6
Giacobone	Eolux	h			2.7	8.9	5.73	61.6
Sunrise Solar	Soma 1000	h			2.7	8.9	5.73	61.6
Guangdong Yuehua	F1000/9	h			2.7	8.9	5.73	61.6
Guangdong Yuehua	F1500/10	h			2.7	8.9	5.73	61.6
Solartechnik Geiger	SG 270	h			2.8	9.0	5.94	63.9
J. Bornay	Inclin 1500 neo	h			2.9	9.4	6.42	69.1
W+W Windtechnik	W+W 1200	h			3.0	9.8	7.07	76.0
Southwest Windpower	Whisper H80	h			3.0	10.0	7.30	78.5
Bergey Windpower	1500-PD	h			3.1	10.0	7.31	78.6
Bergey Windpower	1500-24	h			3.1	10.0	7.31	78.6
LMW	LMW 1003	h			3.1	10.2	7.65	82.3
LMW	LMW 1500	h			3.1	10.2	7.65	82.3
Fortis	Passaat	h			3.1	10.2	7.65	82.3
Synergy Power Corp.	S-5000	h			3.5	11.5	9.62	104
Westwind	Light Wind	h			3.5	11.5	9.62	104
Westwind	Standard	h			3.5	11.5	9.62	104
Proven Wind Turbines	WT2500	h			3.5	11.5	9.62	104
African Wind Power	AWP36	h			3.6	11.8	10.2	110
Solartechnik Geiger	SG 400	h			4.0	13.1	12.6	135
Qingdao Windmill Fac.	FD4-1500	h			4.0	13.1	12.6	135
J. Bornay	Inclin 3000 neo	h			4.0	13.1	12.6	135
Vergnet	GEV 4/2	h			4.0	13.1	12.6	135
Agroluz		h			4.1	13.4	13.2	142
Abundant Renewable Energy	Jacobs short case	h			4.3	14.0	14.3	154
Abundant Renewable Energy	Jacobs long case	h			4.3	14.0	14.3	154
Southwest Windpower	Whisper 175	h			4.6	15.0	16.4	177
Guangdong Yuehua	F5000/11	h			4.6	15.1	16.6	179
Solartechnik Geiger	SG 490	h			4.9	16.1	18.9	203
LMW	2500	h			5.0	16.4	19.6	211
Northern Power Sys.	HR3	h			5.0	16.4	19.6	211
LMW	3600	h			5.0	16.4	19.6	211
Fortis	Montana	h			5.0	16.4	19.6	211
Vergnet	GEV 5/5	h			5.0	16.4	19.6	211
Guangdong Yuehua	F7500/12	h			5.0	16.4	19.6	211
Calorius		h			5.0	16.4	19.6	211
Westwind	Light Wind	h			5.1	16.7	20.4	220
Westwind	Standard	h			5.1	16.7	20.4	220
W+W Windtechnik	W+W 3000	h			5.1	16.8	20.6	222

*Std. Power Rating= swept area x 200 W/m²

Key: nc=no control, h=horizontal furling, v=vertical furling, ab=air brake, tb=tip brake, p=pitch to feather, ps=pitch to stall,
GFRP=glass-fiber-reinforced polyester or fiberglass, wood lam.=wood laminate
Poly=glass-fiber-reinforced polypropylene PGFR=polycarbonate glass-fiber-reinforced

Std. Power Rating* (kW)	Mfg. Rated Power (kW)	Rated Wind Speed (m/s)	(mph)	Mfg. Perf. at Rated Power (%)	Spec. Rotor Loading at Rated Power (W/m²)	No. of of Blds.	Blade Material	Means of Control
0.90	0.50	10.0	22	0.18	111	2		h
0.90	0.50	8.5	19	0.29	111	3	GFRP	p
0.90	0.75	9.5	21	0.32	166	3	GFRP	p
0.93	1.00					3	CFRE	nc
0.98	0.50	9.0	20	0.23	102	3		p
0.98	0.70	10.0	22	0.23	143	3		p
0.98	1.00	11.0	25	0.25	204	3	GFRP	h
1.0	0.60	10.0	22	0.19	117	3	Poly	ps
1.13	1.00	12.5	28	0.15	177	3	GFRP	v
1.1	0.50	12.0	27	0.08	87	3	nylon	v
1.1	0.50	9.0	20	0.20	87	3	GFRP	v
1.1	0.60	12.0	27	0.10	105	3	GFRP	v
1.1	1.00	10.0	22	0.29	175	2	GFRP	v
1.15	1.00	9.5	21	0.33	175	3	Poly	v
1.15	1.50	10.5	24	0.37	262	3	Poly	v
1.2	1.00	14.0	31	0.10	168	4	GFRP	h
1.3	1.50	12.0	27	0.22	233	2	CFRE	v
1.4	1.20	12.0	27	0.16	170	3	GFRP	h
1.5	0.86	10.5	24	0.17	118	3	PGFR	h
1.5	1.50	12.5	28	0.17	205	3	GFRP	h
1.5	1.50	12.5	28	0.17	205	3	GFRP	h
1.53	0.75	10.5	24	0.14	98	3	CFRE	h
1.5	1.00	10.5	24	0.18	131	3	CFRE	h
1.5	1.40	16.0	36	0.07	183	3	GFRP	h
1.9	0.85	9.0	20	0.20	88	3	GFRP	v
1.9	1.80	12.5	28	0.16	187	3	GFRP	h
1.9	2.50	14.0	31	0.15	260	3	GFRP	h
1.9	2.20	12.0	27	0.22	229	3	Poly	ps
2.0	0.90	12.0	27	0.08	88	3	GFRP	h
2.5	2.0	14.0	31	0.09	159	4	GFRP	h
2.5	1.0	10.0	22	0.13	80	3	GFRP	p
2.5	3.0	12.5	28	0.20	239	2	CFRE	v
2.5	2.0	12.5	28	0.13	159	2	wood lam.	ps
2.6	2.0	10.0	22	0.25	151	3	GFRP	p
2.9	1.80	12.5	28	0.11	126	3	wood lam.	h
2.9	1.80	12.5	28	0.11	126	3	wood lam.	h
3.3	2.4	10.0	22	0.24	146	2	CFRE	h
3.32	5.00	11.0	25	0.37	301	3	Poly	v
3.8	4.0	14.0	31	0.13	212	3	GFRP	h
3.9	2.0	10.0	22	0.17	102	3	GFRP	h
3.9	3.0	12.5	28	0.13	153	3	wood lam.	v
3.9	3.0	12.0	27	0.14	153	3	GFRP	h
3.9	4.2	16.0	36	0.09	214	3	GFRP	h
3.9	5.0	15.0	34	0.12	255	2	wood lam.	ps
3.93	7.50	12.0	27	0.36	382	3	Poly	v
3.9			0		0	3	GFRP	?
4.1	4.0	12.5	28	0.16	196	3	GFRP	h
4.1	5.0	14.5	32	0.13	245	3	GFRP	h
4.1	3.0	12.0	27	0.14	146	3	GFRP	h

pt=pitchable tip, mb=mechanical brake

CFRP=carbon-fiber-reinforced polyester CFRE=carbon-fiber-reinforced epoxy

GFRPoly=glass-fiber-reinforced polypropylene

Characteristics of Selected Small Wind Turbines

Manufacturer	Model	(Axis)	Rotor Height (m)	(ft)	Rotor Dia. (m)	(ft)	Area (m²)	(ft²)
Aerocraft	AC 3000	h			5.3	17.2	21.6	233
Taawind Australia	Taawin 3500	h			5.3	17.2	21.6	233
Aerocraft	AC 5000	h			5.3	17.2	21.6	233
Taawind Australia	Taawin 5000	h			5.3	17.2	21.6	233
African Wind Power	AWP54	h			5.4	17.7	22.9	246
Proven Wind Turbines	WT6000	h			5.5	18.0	23.8	256
Synergy Power Corp.	S-20000	h			5.8	19.0	26.4	284
J. Bornay	Bk 12	h			5.8	19.0	26.4	284
Vergnet	GEV 6/5	h			6.0	19.7	28.3	304
ATEV		h			6.0	19.7	28.3	304
Atlantic Orient	Windlite	h			6.7	22.0	35.3	379
Gaia Wind	5.5 kW	h			7.0	23.0	38.5	414
Bergey Windpower	Excel R/48	h			7.0	23.0	38.5	414
Westwind	Light Wind	h			7.0	23.0	38.5	414
Eurowind		h			7.0	23.0	38.5	414
Bergey Windpower	Excel PD	h			7.0	23.0	38.5	414
Vergnet	GEV 7/10	h			7.0	23.0	38.5	414
Bergey Windpower	Excel-S	h			7.0	23.0	38.5	414
Fortis	Alize	h			7.0	23.0	38.5	414
Eurowind		h			7.0	23.0	38.5	414
Westwind	Standard	h			7.0	23.0	38.5	414
Wind Turbine Ind.	23-10	h			7.0	23.0	38.6	415
Wind Turbine Ind.	23-12.5	h			7.0	23.0	38.6	415
Wind Turbine Ind.	26-15	h			7.9	26.0	49.4	531
Wind Turbine Ind.	26-17.5	h			7.9	26.0	49.4	531
ENEL	MiniWind E15	h			8.0	26.2	50.3	541
Taawind Australia	Taawin 15 kW	h			8.0	26.2	50.3	541
Solartechnik Geiger	SG 800	h			8.0	26.2	50.3	541
Arsenal	WPP 2	h			8.0	26.2	50.3	541
Wind Turbine Ind.	29-20	h			8.8	29.0	61.4	661
Wind Turbine Ind.	31-20	h			9.5	31.0	70.2	755
Vergnet	GEV 10/15	h			10.0	32.8	78.5	845
Kramer Winturbinen	K15	h			10.0	32.8	78.5	845
Vergnet	GEV 10/25	h			10.0	32.8	78.5	845
Cellpart	JBA-15	h			11.0	36.1	95.0	1,022
Arsenal	WPP 15	h			12.0	39.4	113	1,217
Enercon	E12	h			12.0	39.4	113	1,217
Gaia Wind	11 kW	h			13.0	42.6	133	1,428
Furhländer	F30	h			13.0	42.6	133	1,428
Kano Rotor	30/13	h			13.0	42.6	133	1,428
Synergy Power Corp.	SLG	h			13.2	43.3	137	1,472
Pitchwind	20/14	h			14.0	45.9	154	1,656
Atlantic Orient	AOC 15/50	h			15.0	49.2	177	1,901
Vergnet	GEV 15/60	h			15.0	49.2	177	1,901
Cellpart	JBA-75	h			17.5	57.4	241	2,588
Lagerwey	LW 18/80	h			18.0	59.0	254	2,738

*Std. Power Rating= swept area x 200 W/m²

Key: nc=no control, h=horizontal furling, v=vertical furling, ab=air brake, tb=tip brake, p=pitch to feather, ps=pitch to stall,
GFRP=glass-fiber-reinforced polyester or fiberglass, wood lam.=wood laminate
Poly=glass-fiber-reinforced polypropylene PGFR=polycarbonate glass-fiber-reinforced

Std. Power Rating* (kW)	Mfg. Rated Power (kW)	Rated Wind Speed (m/s)	(mph)	Mfg. Perf. at Rated Power (%)	Spec. Rotor Loading at Rated Power (W/m²)	No. of of Blds.	Blade Material	Means of Control
4.3	3.0	10.0	22	0.23	139	3	GFRP	p
4.3	3.5	12.0	27	0.15	162	3	GFRP	p
4.3	5.0	9.0	20	0.52	231	3	GFRP	p
4.3	5.0	14.0	31	0.14	231	3	GFRP	p
4.6	5.0	12.0	27	0.21	218	3	GFRP	h
4.8	6.0	11.5	26	0.27	253	3	Poly	ps
5.3	2.3	9.0	20	0.19	87	3	GFRP	v
5.3	12.0	12.0	27	0.43	454	3	CFRE	v
5.7	5.0	14.0	31	0.11	177	2	wood lam.	ps
5.7	6.0	10.5	24	0.30	212	4	GFRP	p
7.1	7.5	10.5	24	0.30	213	3	GFRP	h
7.7	5.5				143	3	GFRP	pt
7.7	8.0	13.8	31	0.13	208	3	GFRP	h
7.7	8.0	12.5	28	0.17	208	3	GFRP	h
7.7	8.0	12.5	28	0.17	208	3	GFRP	h
7.7	10	13.8	31	0.16	260	3	GFRP	h
7.7	10	12.0	27	0.25	260	2	wood lam.	ps
7.7	10	13.8	31	0.16	260	3	GFRP	h
7.7	10	13.0	29	0.19	260	3	GFRP	h
7.7	10	14.0	31	0.15	260	3	GFRP	h
7.7	10	14.0	31	0.15	260	3	GFRP	h
7.7	10	11.6	26	0.27	259	3	wood lam.	p,h
7.7	13	12.5	28	0.27	324	3	wood lam.	p,h
9.9	15	11.6	26	0.32	304	3	wood lam.	p,h
9.9	18	11.6	26	0.37	355	3	wood lam.	p,h
10.1	15	10.5	24	0.42	298	3	GFRP	v
10.1	15	12.0	27	0.28	298	3	GFRP	p
10.1	12	12.0	27	0.23	239	3	GFRP	h
10.1	20	8.0	18	1.27	398	2		
12.3	20	11.6	26	0.34	326	3	GFRP	p,h
14.0	15	11.6	26	0.22	214	3	wood lam.	p,h
15.7	15	13.0	29	0.14	191	2	CFRE	ps
15.7	15	9.3	21	0.39	191	3	GFRP	mb
15.7	25	16.0	36	0.13	318	2	CFRE	ps
19.0	11	11.0	25	0.14	116	3	GFRP	s
22.6	15	9.0	20	0.30	133	2		
22.6	30	11.0	25	0.33	265	3	GFRE	p
26.5	11				83	2	GFRP	pt
26.5	30	12.0	27	0.21	226	3	GFRP	pt
26.5	30	12.0	27	0.21	226	3	wood	mb
27.4	30		0		219	3	GFRP	v
30.8	20	10.0	22	0.21	130	2	GFRP	p
35.3	50		0			3	wood lam.	tb
35.3	60	15.0	34	0.16	340	2	GFRP	ps
48.1	75	12.0	27	0.29	312	3	GFRP	s
50.9	80	12.0	27	0.30	314	2	GFRP	ps

pt=pitchable tip, mb=mechanical brake

CFRP=carbon-fiber-reinforced polyester CFRE=carbon-fiber-reinforced epoxy

GFRPoly=glass-fiber-reinforced polypropylene

Small Wind Turbine Manufacturers

There are more than 50 different manufacturers of small wind turbines worldwide. They produce more than 100 different models. The following list is the most comprehensive available, but it is not exhaustive.

Aerocraft
Moorkamp 38
D-27356 Rotenburg
Germany
phone: 49 42 61 96 00 34
fax: 49 42 61 96 00 35
info@aerocraft.de
www.aerocraft.de

African Windpower
Box 4533
Harare
Zimbabwe
phone: 263 4 77 15 81
fax: 263 4 77 15 80
www.power.co.zw/windpower/

AGROLUZ Aerogeneradores
Jose P. Varela 5377
Buenos Aires 1048
Argentina
phone: 54 11 4568 5934

Ampair
First Floor, The Doughty Building
Crow Arch Lane
Ringwood, Hants BH24 1NZ
United Kingdom
phone: 44 1425 480 780
fax: 44 1425 479 497
ampair@ampair.com
www.ampair.com

Atlantic Orient Corporation
PO Box 832
Charlottetown, PE
C1A 7L9
Canada
phone: 902 368 7171
fax: 902 368 7139
aoc@isn.net
www.aocwind.net

Atlantis Windkraftanlagen
Holzstrasse 10
DE-31556 Wölpinhausen
Germany
phone: 49 5037 98803
fax: 49 5037 98805
atlantis-windkraft@t-online.de
www.atlantis-windkraft.de

Bergey Windpower Co. Inc.
2001 Priestley Avenue
Norman, OK 73069
phone: 405 364 4212
fax: 405 364 2078
sales@bergey.com
www.bergey.com

Calorius Vindmøllen
Spånagervej 2
DK-4200 Slagelse
Denmark
phone: 45 58 26 80 60
fax: 45 58 26 80 60
calorius@calorius.dk
www.calorius.dk

Cataventos Kenya
Rodovia RS 130
KM 14, P.O. Box 111
Encantado RS
Brazil
phone: 55 51 751 1750
fax: 55 51 751 14 71
www.cataventoskenya.com.br

Cellpart Vindkraftverk
Sportvagen 6
S-89142 Ornskoldsvik
Sweden
phone: 46 660 850 00
fax: 46 660 850 03
www.cellpart.se

CSRI Elektropribor
30 Malaya Posadskaya Street
Saint Petersburg 197046
Russia
phone: 7 812 238 8199
fax: 7 812 232 3376
elprib@online.ru
www.elektropribor.spb.ru

Electromeccanica Salmini
Via Como, 5
Casorate Sempione
I-21011 Varese
Italy
phone: 39 03 31296729
info@salmini.it
www.salmini.it

Electrovent
663 Routhier Street
Sainte-Foy, Quebec G1X 3J8
Canada
phone: 418 654 1759
fax: 418 654 1759
www.electrovent.com

ENEL Green Power
Via G.B. Marini 7
I-00198 Roma
Italy
phone: 39 06 85097036
fax: 39 06 85092061
http://enelgreenpower.enel.it

Everfair Enterprises
1205 Elizabeth Street, A2
Punta Gorda, FL 33950
phone: 941 575 4404
fax: 941 575 4080
everfair@gate.net
www.charternet.com/fourwinds

Fortis
Botanicuslaan 14
NL-9751 AC Haren
Netherlands
phone: 31 50 534 0104
fax: 31 50 534 0104
fortis-windenergy@wxs.nl
www.fortiswindenergy.com

Fuhrländer
Auf der Hoehe 4
D-56477 Waigandshain Hessen
Germany
phone: 49 26 64 99 66 0
fax: 49 26 64 99 66 33
mail@fuhrlaender.de
www.fuhrlaender.de

Giacobone
Cerro Fitz Roy 1080
Rio Quarto Cordoba 5800
Argentina
phone: 54 358 463 4380
fax: 54 358 463 4379
info@giacobone.com
www.giacobone.com

GIAFA
Italia 3094
Cordoba 5009
Argentina
phone: 54 51 80 83 60
fax: 54 51 80 83 60
giafa@nt.com.ar
www.tecomnet.com.ar/giafa

Hamilton Ferris
P.O. Box 126
Ashland, MA 01721
phone: 508 881 4602
fax: 508 881 3846
sales@hamiltonferris.com
www.hamiltonferris.com

ICPE
313 Splaiul Unirii
74204 Bucuresti-3
Romania
phone: 40 1 3467229
fax: 40 1 3467268
icpe.sa@icpe.ro
www.icpe.ro

Inventus
Zum Frenser Feld - Halle 6
D-50127 Bergheim
Germany
phone: 49 2271 989190
fax: 49 2271 981042
info@inventusgmbh.de
www.inventusgmbh.de

J. Bornay Aerogeneradores
Paraje Ameraors, s/n
P.O. Box 116
ES-03420 Castalla, Alicante
Spain
phone: 34 965 560 025
fax: 34 965 560 752
bornay@bornay.com
www.bornay.com

LVM Ltd.
Old Oak Close
Arlesey, Bedfordshire SG15 6XD
United Kingdom
phone: 44 14 62 733 336
fax: 44 14 62 730 466
lvm.ltd@dial.pipex.com
www.lvm-ltd.com

Marlec Engineering Company
Rutland House
Trevithick Road
Corby, Northants NN17 1XY
United Kingdom
phone: 44 15 36 20 15 88
fax: 44 15 36 40 02 11
sales@marlec.co.uk
www.marlec.co.uk

Mecanix Sàrl
Rue Charpentier 6
CH-1880 Bex
Switzerland
phone: 41 24 463 43 09
mecanix@trianglevert.com
www.trianglevert.com/mecanix

Moratec
Immanuel-Kant Strasse 32a
D-84489 Burghausen
Germany
phone: 49 8677 9791 66
fax: 49 8677 9791 68
info@moratec.de
www.moratec.de

MV Frunze Arsenal Design Bureau
1/3 Komsomoa Street
Saint Petersburg 195009
Russia
phone: 7 812 542 29 73
fax: 7 812 542 20 60

Phoenix Windpower
P.O. Box 322
Somerville, Victoria 3912
Australia
phone: 414 303512
info@phoenixwp.com
www.phoenixwp.com

Proven Engineering Products
Moorfield Industrial Estates
Kilmarnock, Scotland KA2 0BA
United Kingdom
phone: 44 15 63 543 020
fax: 44 15 63 539 119
info@provenenergy.com
www.provenenergy.com

Ropatec
Via Siemens Str. 19
I-39100 Bozen-Bolzano
Italy
phone: 39 471 568180
fax: 39 471 568183
info@ropatec.com
www.ropatec.com

Solartechnik Geiger
Windener strasse 14
D-85051 Ingolstadt
Germany
phone: 49 84 50 73 90
fax: 49 84 50 73 90
windtechnik-geiger@t-online.de
www.windtechnik-geiger.de

Southwest Windpower
P.O. Box 2190
Flagstaff, AZ 86003-2190
phone: 520 779 9463
fax: 520 779 1485
info@windenergy.com
www.windenergy.com

Sunrise Solar (Soma)
49 Vista Avenue
Cococabana, New South Wales 2251
Australia
phone: 61 2 4381 1531
fax: 61 2 4382 1880
sunrise@dragon.net.au
www.somapower.com.au

Svensk Vinkdraft Industri AB (SVIAB)
Vettershaga
SE-7601 Bergshamr
Sweden
phone: 46 176 26 42 24
fax: 46 176 26 42 14
sviab@swipnet.se
www.sviab.com

Synergy Power Corporation
17/F The Strand
49 Bonham Strand
Hong Kong
China
phone: 852 2522-9000
fax: 852 28 10 04 78
spc@synergypowercorp.com
www.synergypowercorp.com

Travere Aerogenerateurs
341 Avenue Sainte Marguerite
F-06200 Nice
France
phone: 33 4 93 837 897
Fax: 33 4 93 723 766
contact@travere.com
www.travere.com

Vergnet S.A.
6 Rue Henri Dunant
F-45140 Ingre
France
phone: 33 2 38 22 75 00
fax: 33 2 38 22 75 22
eole@vergnet.fr
www.vergnet.fr

Westwind Turbines
Venco Products
29 Owen Road
Kelmscott, Western Australia 6111
Australia
phone: 61 8 93 99 52 65
fax: 61 8 94 97 13 35
venwest@iinet.net.au
www.westwind.com.au

Wind Generator Products
48 S.W. 4th St.
Homestead, FL 33030
phone: 305 247 2868
fax: 305 247 1808
moreinfo@windbugger.com
www.windbugger.com

Windside Oy
Niemenharjuntie 85
FIN-44800 Pihtipudas
Finland
phone: 358 20 8350 700
fax: 358 20 7350 700
finland@windside.com
www.windside.com

Wind Turbine Industries
Prior Lake Machine
16801 Industrial Circle, SE
Prior Lake, MN 55372
phone: 612 447 6064
fax: 612 447 6050
wtic@windturbine.net
www.windturbine.net

Characteristics of Selected Medium-Size Wind Turbines

The turbines in the table are ordered by swept area and not by rated power. Versions and rated power may vary depending upon the continent where the turbines will be used (50 or 60 Hz). Not all models are available on all continents. Turbines designed for offshore have been omitted.

Mfg. Perf. at Rated Power: Overall conversion efficiency of the wind turbine at rated wind speed derived from the rated output claimed by the manufacturer. This one of several tools useful in evaluating whether a manufacturers' performance estimates may be aggressive or conservative. Some manufacturer's are notoroiusly "aggressive" in estimating how much power their wind turbines will produce at any given wind speed. For example, if most manufacturers of medium-size wind turbines expect their turbines to convert 25 to 35 percent of the power in the wind at rated wind speed to electricity, and another expects more than 40 percent, the latter's claim may be exaggerated.

Spec. Rotor Loading at Rated Power: The power that a manufacturer expects its wind turbine to produce at its rated wind speed relative to the area swept by the rotor is an indication of how hard the blades have to work. It is also one indicator of potential noise. The higher the loading in W/m2, the higher the potential noise emissions.

Speed of Rotor: Medium-size wind turbine rotors can be operated at a relatively constant (c) speed with respect to wind speed, or rotor speed will vary (v) directly with wind speed. Some wind turbines use a hybrid approach (c/v) that varies rotor speed by up to 10 percent of nominal speed in response to changes in wind speed.

Rotor Control: There are two approaches to limiting rotor power in high winds: using aerodynamic stall (s), or by pitching the blades toward feather (p). Some wind turbines use a hybrid design. They pitch the blades toward stall (s+p).

Overspeed Control: Nearly all medium-size wind turbines and all large wind turbines include some aerodynamic means for preventing the rotor from overspeeding in an emergency. Most either pitch the blades toward feather or pitch only the blade tips toward feather. Some pitch the blades toward stall.

Characteristics of Selected Medium-Size Wind Turbines on the European and North American Markets

Manufacturer	Model	Rotor Dia. (m)	(ft)	Swept Area (m^2)	(ft^2)	Mfg. Rated Power (kW)	Rated Wind Speed (m/s)	Mfg. Perf. at Rated Power (%)	Spec. Rotor Loading (W/m^2)	No. of Blds.	(1) Speed of Rotor	(2) Rotor Control	Aerodynamic Overspeed Control
Fuhrländer	FL 100	21.0	69	346	3,700	100	13.0	0.21	0.29	3	c	s	pitchable tips
Vergnet	GEV 26/220	26.0	85	531	5,700	220	13.0	0.31	0.41	2	c	s+p	pitch to stall
Fuhrländer	FL 250	29.5	97	683	7,400	250	15.0	0.18	0.37	3	c	s	pitchable tips
Lagerwey	LW30/250	30.0	98	707	7,600	250	13.0	0.26	0.35	2	v	p	variable pitch
Enercon	E-30	30.0	98	707	7,600	300	12.0	0.40	0.42	3	v	p	variable pitch
Suzlon	S33	33.4	110	876	9,400	350	14.0	0.24	0.40	3	c	s	pitchable tips
Bonus	Mk IV	44.0	144	1,521	16,400	600	15.0	0.19	0.39	3	c	s	pitchable tips
Enercon	E40/6.44	44.0	144	1,521	16,400	600	12.0	0.37	0.39	3	v	p	variable pitch
Ecotecnia	44	44.0	144	1,521	16,400	640	14.5	0.23	0.42	3	c	s	pitchable tips
Mitsubishi	MWT 600	45.0	148	1,590	17,100	600	13.0	0.28	0.38	3	c	p	variable pitch
Vestas	V47	47.0	154	1,735	18,700	660	16.0	0.15	0.38	3	c/v	p	variable pitch
RePower	48/600	48.0	157	1,810	19,500	600	13.0	0.25	0.33	3	c	s	pitchable tips
DeWind	D4	48.0	157	1,810	19,500	600	11.5	0.36	0.33	3	v	p	variable pitch
Jeumont	J48	48.0	157	1,810	19,500	750	13.5	0.28	0.41	3	v	s	pitchable tips
RePower	48/750	48.0	157	1,810	19,500	750	14.5	0.22	0.41	3	c	s	pitchable tips
Ecotecnia	48	48.0	157	1,810	19,500	750	14.5	0.22	0.41	3	c	s	pitchable tips
Fuhrländer	FL 800	48.0	157	1,810	19,500	800	12.0	0.42	0.44	3	c	s	pitchable tips
Nordex	N50	50.0	164	1,963	21,100	800	15.0	0.20	0.41	3	c	s	pitchable tips
Lagerwey	LW52	52.0	171	2,124	22,900	750	12.0	0.33	0.35	3	v	p	variable pitch
Gamesa	G52	52.0	171	2,124	22,900	800	14.0	0.22	0.38	3	v	p	variable pitch
Vestas	V52	52.0	171	2,124	22,900	850	14.0	0.24	0.40	3	v	p	variable pitch
GE Wind	900	52.0	171	2,124	22,900	900	14.0	0.25	0.42	3	v	p	variable pitch
NEG-Micon	NM 52	52.2	171	2,140	23,000	900	15.0	0.20	0.42	3	c	s	pitchable tips
Fuhrländer	Fl 1000	54.0	177	2,290	24,600	1000	12.0	0.41	0.44	3	c	s	pitchable tips
Bonus		54.2	178	2,307	24,800	1000	15.0	0.21	0.43	3	c	s+p	pitch to stall
NEG-Micon	NM54	54.5	179	2,333	25,100	950	15.0	0.20	0.41	3	c	s	pitchable tips
GE Wind	900s	55.0	180	2,376	25,600	900	14.0	0.23	0.38	3	v	p	variable pitch
Winwind Oy	WWD-l	56.0	184	2,463	26,500	1000	12.5	0.34	0.41	3	v	p	variable pitch
Jeumont	J56	56.0	184	2,463	26,500	1000				3	v	s	variable pitch
GE Wind	900sl	57.0	187	2,552	27,500	900	14.0	0.21	0.35	3	v	p	variable pitch
RePower	57/100	57.0	187	2,552	27,500	1050	13.0	0.31	0.41	3	c	s	pitch to stall
Jeumont	J77	57.0	187	2,552	27,500	1500				3	v	s	variable pitch
Lagerwey	LW58	58.0	190	2,642	28,400	750	11.0	0.35	0.28	3	v	p	variable pitch
Fuhrländer	Fl 1000+	58.0	190	2,642	28,400	1250	14.0	0.28	0.47	3	c	s	pitchable tips
Enercon	E58	58.6	192	2,697	29,000	1000	12.0	0.35	0.37	3	v	p	variable pitch
Suzlon	S60	60.0	197	2,827	30,400	1000	13.0	0.26	0.35	3	c	p	variable pitch
Nordex	N60	60.0	197	2,827	30,400	1300	15.0	0.22	0.46	3	c	s	pitchable tips
Mitsubishi	MWT 1000	61.4	201	2,961	31,900	1000	12.5	0.28	0.34	3	c	p	variable pitch
Suzlon	S62	62.0	203	3,019	32,500	1000	12.0	0.31	0.33	3	c	p	variable pitch
Ecotecnia	62	62.0	203	3,019	32,500	1250	13.5	0.27	0.41	3	v	p	variable pitch
Nordex	N62	62.0	203	3,019	32,500	1300	15.0	0.21	0.43	3	c	s	pitchable tips
Bonus		62.0	203	3,019	32,500	1300	15.0	0.21	0.43	3	c	s+p	pitch to stall
Suzlon	S64	64.0	210	3,217	34,600	1000	11.0	0.38	0.31	3	c	p	variable pitch
DeWind	D6	64.0	210	3,217	34,600	1250	12.3	0.34	0.39	3	v	p	variable pitch
Suzlon	S66	66.0	217	3,421	36,800	1250	12.0	0.35	0.37	3	c	p	variable pitch
Vestas	V66	66.0	217	3,421	36,800	1750	16.0	0.20	0.51	3	c/v	p	variable pitch

Characteristics of Selected Medium-Size Wind Turbines on the European and North American Markets *continued*

Manufacturer	Model	Rotor Dia. (m)	(ft)	Swept Area (m^2)	(ft^2)	Mfg. Rated Power (kW)	Rated Wind Speed (m/s)	Mfg. Perf. at Rated Power (%)	Spec. Rotor Loading (W/m^2)	No. of Blds.	(1) Speed of Rotor	(2) Rotor Control	Aerodynamic Overspeed Control
RePower	MD 70	70.0	230	3,848	41,400	1500	12.5	0.33	0.39	3	v	p	variable pitch
Fuhrländer	MD 70	70.0	230	3,848	41,400	1500	12.5	0.33	0.39	3	v	p	variable pitch
Pfleiderer	PWE 1570	70.0	230	3,848	41,400	1500	12.0	0.37	0.39	3	v	p	variable pitch
Enercon	E66	70.0	230	3,848	41,400	1800	12.0	0.44	0.47	3	v	p	variable pitch
GE Wind	1.5s	70.5	231	3,904	42,000	1500	12.0	0.36	0.38	3	v	p	variable pitch
Lagerwey	LW72	71.2	234	3,982	42,800	2000	13.0	0.37	0.50	3	v	p	variable pitch
NEG-Micon	NM 72c	72.0	236	4,072	43,800	1500	15.0	0.18	0.37	3	c	s	pitchable tips
NEG-Micon	NM 72	72.0	236	4,072	43,800	1650	14.0	0.24	0.41	3	c	s	pitchable tips
Ecotecnia	74	74.0	243	4,301	46,300	1670	13.5	0.26	0.39	3	v	p	variable pitch
Bonus		76.0	249	4,536	48,800	2000	15.0	0.21	0.44	3	c	s+p	pitch to stall
GE Wind	1.5sl	77.0	253	4,657	50,100	1500	11.8	0.32	0.32	3	v	p	variable pitch
Fuhrländer	FL MD 77	77.0	253	4,657	50,100	1500	11.1	0.38	0.32	3	v	p	variable pitch
RePower	MD 77	77.0	253	4,657	50,100	1500	12.0	0.30	0.32	3	v	p	variable pitch
Ecotecnia	80	80.0	262	5,027	54,100	1670	13.5	0.22	0.33	3	v	p	variable pitch
Vestas	V80	80.0	262	5,027	54,100	1800	16.0	0.14	0.36	3	c/v	p	variable pitch
Gamesa	G80	80.0	262	5,027	54,100	1800	18.0	0.10	0.36	3	v	p	variable pitch
DeWind	D8	80.0	262	5,027	54,100	2000	13.5	0.26	0.40	3	v	p	variable pitch
Vestas	V80	80.0	262	5,027	54,100	2000	15.0	0.19	0.40	3	c/v	p	variable pitch
Nordex	N80	80.0	262	5,027	54,100	2500	14.0	0.30	0.50	3	v	p	pitch to stall
NEG-Micon	NM 82	82.0	269	5,281	56,800	1650	13.0	0.23	0.31	3	c	s	pitchable tips
Nordex	N90	90.0	295	6,362	68,400	2300	12.0	0.34	0.36	3	v	p	pitch to stall
Vestas	V90	90.0	295	6,362	68,400	3000	15.0	0.23	0.47	3	c/v	p	variable pitch

1. c=constant speed, v=variable speed

2. s=stall regulated, p=variable pitch, y=yaw regulated

Medium-Size Wind Turbine Manufacturers

There are numerous manufacturers that build reliable wind turbines for large-scale commercial applications such as wind power plants. Some of these models have also been installed as single turbines or in small clusters for farms, businesses, and cooperatives in Europe, notably in Germany and Denmark.

This is a dynamic industry. The number of manufacturers and their line of products change constantly. For current information on the models available, check the wind turbine market surveys mentioned elsewhere in these appendixes, or visit the manufacturers' Web sites. Many of the major manufacturers operate subsidiaries or joint ventures with other companies outside their national markets. For example, Bonus wind turbines are available in Germany from AN Windenergie.

Bonus Energy A/S
Fabriksvej 4
DK-7330 Brande
Denmark
phone: 45 97 18 11 22
fax: 45 97 18 30 86
bonus@bonus.dk
www.bonus.dk

DeWind Tecknik GmbH
Seelandstrasse 9
D-23569 Lübeck
Germany
phone: 49 451 39 09 771
fax: 49 451 39 09 778
dewind@dewind.de
www.dewind.de

Ecotecnia s. coop.
Amistat 23
ES-08005 Barcelona
Spain
phone: 34 93 225 7600
fax: 34 93 221 0939
ecotecnia@ecotecnia.com
www.ecotecnia.com

Enercon GmbH
Dreekamp 5
D-26605 Aurich
Germany
phone: 49 49 41 92 71 04
fax: 49 49 41 92 71 99
sales.international@enercon.de
www.enercon.de

Fuhrländer GmbH
Auf der Hoehe 4
D-56477 Waigandshain
Germany
phone: 49 26 64 99 66 0
fax: 49 26 64 99 66 33
info@fuhrlaender.de
www.fuhrlaender.de

Gamesa Eólica SA
Poligono Landaben, Calle E, s/n
E-31013 Pamplona
Spain
phone: 34 948 309010
fax: 34 948 309009
info@eolica.gamesa.es
www.gamesa.es

GE Wind
13000 Jameson Road
Tehachapi, CA 93561
phone: 661 823 6700
fax: 661 822 7880
windenergy.usa@ps.ge.com
www.gewindenergy.com

Jeumont Industrie
Boite Postal 189
F-59573 Jeumont
France
phone: 33 3 27 69 93 03
fax: 33 3 27 69 94 65
jicontact@jeumont-framatome.com
www.jeumont-framatome.com

Lagerwey Windturbine bv
P.O. 279
NL-3770 AG Barneveld
Netherlands
phone: 31 342 42 2724
fax: 31 342 42 2861
sales@lagerwey.nl
www.lagerwey.nl

Mitsubishi Heavy Industries
3-3-1, Minatomirai, Nishi-ku
Yokohama 220-8401
Japan
phone: 81 45 224 9537
fax: 81 45 224 9966
www.mhi.co.jp

NEG-Micon A/S
Alsvej 21
DK-8900 Randers
Denmark
phone: 45 87 10 50 00
fax: 45 87 10 50 01
mail@neg-micon.dk
www.neg-micon.dk

Nordex
Svindbaek
DK-7323 Give
Denmark
phone: 45 75 73 44 00
fax: 45 75 73 41 47
nordex@nordex.dk
www.nordex.dk

Pfleiderer Wind Energy
Ingolsädter Strasse 5l
D-92318 Neumark
Germany
phone: 49 91 81 28 8420
fax: 49 91 81 28 607
windenergy@pfleiderer.com
www.pfleiderer-wind.com

RePower Systems
Rödemis Hallig
D-25813 Husum
Germany
phone: 49 4841662 800
fax: 49 4841 662 888
info@repower.de
www.repower.de

Suzlon
5th Fl., Dodrej Millennium 9
Koregaon Park Road
Pune 411 001
India
phone: 91 20 4022000
fax: 91 20 4022100
asia@suzlon.com
www.suzlon.com

Vestas DWT
Smed Sorensens Vej 5
DK-6950 Ringkøbing
Denmark
phone: 45 96 75 25 75
fax: 45 97 75 24 36
vestas@vestas.dk
www.vestas.dk

Reconditioned and Used Wind Turbines

Used and reconditioned small wind turbines can be found in the classifieds of *Home Power* magazine or by contacting Lake Michigan Wind & Sun.

Abundant Renewable Energy
22700 NE Mountain Top Road
Newberg, OR 97132-6614
phone: 503 538 8292
fax: 503 538 8782
rwpreus@yahoo.com
www.abundantre.com

Lake Michigan Wind & Sun
1015 County U
Sturgeon Bay, WI 54235-8353
phone: 920 743 0456
fax: 920 743 0466
info@windandsun.com
www.windandsun.com

Ads for used medium-size wind turbines can be found in *Windpower Monthly* and in *New Energy* as well as the parent publication, *Neue Energie.* The following firms market used and reconditioned turbines.

Dansk Vindmølleformidling Aps
Oddesundvej 183, Visby
DK-7755 Bedsted
Denmark
phone: 45 97 95 16 58
fax: 45 97 95 16 02
dvf@nvn.dk

Energy Maintenance Services
P.O. Box 158
Gary, SD 57237
phone: 605 272 5398
fax: 605 272 5402
info@energyms.com
www.energyms.com

J. P. Sayler & Assoc.
P.O. Box 41515
Des Moines, IA 50311
phone: 515 255 4970
fax: 515 255 4925
windpowr@netins.net
www.windworkers.com

Mechanical Wind Pumps

For historical information on American water-pumping windmills, there are several museums worth visiting in North America. The largest is in Lubbock, Texas.

American Wind Power Center
1501 Canyon Lake Drive
Lubbock, TX 79403
phone: 806 747 8734
fax: 806 740 0668
www.windmill.com

Farm Windmill Manufacturers

The companies below manufacture farm windmills. All are back-geared and self-regulating. O'Brock and Topper offer catalogs full of windmills (both new and rebuilt), pumps, and paraphernalia. Airlift Technologies and Koenders offer airlift pumps.

AbaChem Engineering
Jessop Way
Newark, Nottinghamshire
NG24 2EH
United Kingdom
phone: 44 1636 76 483
fax: 44 1636 70 86 32

Aermotor Windmill Company
Box 5110
San Angelo, TX 76902
phone: 915 651 4951,
800 854 1656
fax: 915 651 4948
sales@aermotorwindmill.com
www.aermotorwindmill.com

Airlift Technologies
1540 Barton Rd., #263
Redlands, CA 92373
phone: 909 446 1780
sales@airliftech.com
www.airliftech.com

Bob Harries Engineering
Karamaini Estate
P.O. Box 40
Thika
Kenya
phone: 254 151 47 234
fax: 254 2 33 20 09

Cataventos Kenya
Rodovia RS 130
KM 14, P.O. Box 111
Encantado RS
Brazil
phone: 55 51 751 1750
fax: 55 51 751 14 71

Dempster Industries
P.O. Box 848
Beatrice, NE 68310
phone: 402 223 4026
fax: 402 228 4389
dempsterinc@beatricene.com

Ferguson Manufacturing
835 Old North Road, RD 2
Waimauku
New Zealand
phone: 64 9 411 8332
fax: 64 9 412 8897
ferman@windmills.co.nz
www.windmills.co.nz

FIASA (Fabrica de Implementos Agricolas SA)
Hortiguera 1882
Buenos Aires 1406
Argentina
phone: 54 11 4925 7066
fax: 54 11 4925 6747
fiasa@fiasa.com.ar
www.fiasa.com.ar

Koenders Windmills Inc.
P.O. Box 126
175 First Street E.
Englefeld Saskatchewan S0K 1N0
Canada
phone: 306 287 3720
fax: 306 287 3657
koenders.wind@sasktel.net
www.koenderswindmills.ca

Molins de Vent Tarrago
Raval Santa Anna, 30-32
ES-43400 Montblanc Tarragona
Spain
phone: 34 77 86 09 08
fax: 34 77 86 09 08
info@tarrago.es
www.tarrago.es

Place de la Victoire
Boite Postale 12
F-10380 Plancy l'Abbaye
France
phone: 33 25 37 40 15
fax: 33 25 37 43 72

Southern Cross Corporation
632 Ruthven Street
P.O. Box 109
Toowoomba, Queensland QLD 4350
Australia
phone: 61 76 38 4988
fax: 61 76 38 5898

Stewart and Lloyds
37 Leopold Takawira Street
P.O. Box 784
Harare
Zimbabwe
phone: 263 0 70 8 91
fax: 263 0 79 09 72

W. D. Moore Ltd.
3 Keegan Street
O'Connor, Western Australia 6163
Australia
phone: 61 9 337 47 66
fax: 61 9 314 13 06
www.wdmoore.com.au

Windtech International
P.O. Box 27
Bedford, NY 10506
phone: 914 232 2354
fax: 914 232 2356
info@windmillpower.com
www.windmillpower.com

Mechanical Wind Pump Dealers or Distributors

O'Brock Windmill Distributors
9435 12th Street
North Benton, OH 44449
phone: 330 584 4681
fax: 330 584 4682
windmill@cannet.com
www.obrockwindmills.com

Small Wind Turbine Towers

Tower manufacturers typically don't sell to individuals. They often deal directly with the manufacturer of the wind turbine. This assures the manufacturer that the tower is matched properly to the wind turbine. Rohn Industries is one of the few suppliers of guyed lattice and free-standing truss towers for small wind turbines in North America. NRG Systems builds a line of hinged anemometer masts suitable for selected micro turbines. Most wind machines greater than 10 meters (30 ft) in diameter are supplied with their own specially designed towers.

NRG Inc.
P.O. Box 509
Hinesburg, VT 05461
phone: 800 448 9463, 802 482 2255
fax: 802 482 2272
sales@nrgsystems.com
www.nrgsystems.com

Rohn Industries
6718 W. Plank Road
Peoria, IL 61656
phone: 309 697 4400
fax: 309 633 2695
mail@rohnnet.com
www.rohnnet.com/CommPro/Towers/Towers.htm

Rotor Blades

Many wind turbine manufacturers build their own rotor blades. There are, however, a few manufacturers of small wind turbine blades and several independent manufacturers of blades for medium-size and larger wind turbines.

Small Wind Turbines

Advanced Aero Technologies
920 N. Dale Street
St. Paul, MN 55103
phone: 651 487 9222
fax: 651 487 9222

Heliulm
F-46140 Cambayrac
France
phone: 33 56 53 69 290
fax: 33 56 53 69 738

Olsen Wings
Vads Møllevej 2
Sondrup
DK-8350 Hundslund
Denmark
phone: 45 86 55 0576
fax: 45 86 55 0160
olsen@olsenwings.dk
www.olsenwings.dk

Medium-Size Wind Turbines

Abeking & Rasmussen Rotec
Postfach 1229
Flughavenstrasse
D-27809 Lemwerder
Germany
phone: 49 421 67 33532
fax: 49 421 67 33115
info@abeking.com
www.abeking.com

Atout Vent
Actipole St. Charles
F-13710 Fuveau
France
phone: 33 42 29 14 62
fax: 33 42 29 14 61

LM Glasfiber A/S
Rolles Moellevej 1
DK-6640 Lunderskov
Denmark
phone: 45 75 58 51 22
fax: 45 75 58 62 02
info@lm.dk
www.lm.dk

MFG (Molded Fiberglass Companies)
P.O. Box 1459
Gainesville, TX 76241-1459
phone: 940 668 0302
fax: 940 668 0306
www.moldedfiberglass.com

Mail-Order Catalogs

Many micro and mini wind turbines can be purchased from mail-order companies. The turbines are small enough to be shipped via package delivery services. The following companies also sell the heavier small wind turbines, but they ship them by motor freight (truck or lorry). For wind turbines 1.5 kW and above, it's often best to buy from a local dealer who can both install and service the machine.

Kansas Wind Power
13569 214th Road
Holton, KS 66436
phone: 785 364 4407
fax: 785 364 5123
www.kansaswindpower.net

Real Goods Trading Co.
P.O. Box 593
Hopland, CA 95449
phone: 800 919 2400
fax: 707 462 4807
techs@realgoods.com
www.realgoods.com

Wind-Measuring Devices

These manufacturers build both wind-measuring instruments and sophisticated data recorders. These loggers require a computer with the requisite software to analyze the stored data.

Ammonit
Paul-Lincke-Ufer 41
D-10999 Berlin
Germany
phone: 49 30 612 79 54
fax: 49 30 618 30 60
ammonit@ammonit.de
www.ammonit.de

Campbell Scientific
815 West 1800 North
Logan, UT 84321-1784
phone: 435 750 9529
fax: 435 750 9596
www.campbellsci.com

Le Groupe Ohmega
Parc Industriael des Augustines
3, des Cerisiers
Gaspé, Quebec G4X 2M1
Canada
phone: 418 368 5425
fax: 418 368 7290
info@groupeohmega.com
www.groupeohmega.com

Met One Instruments
1600 Washington Boulevard
Grants Pass, OR 97526
phone: 541 471 7111
fax: 541 471 7116
sales@metone.com
www.metone.com

NRG Inc.
110 Commerce Street
P.O. Box 509
Hinesburg, VT 05461
phone: 800 448 9463,
802 482 2255
fax: 802 482 2272
sales@nrgsystems.com
www.nrgsystems.com

R. M. Young Company
2801 Aero Park Drive
Traverse City, MI 49684
phone: 616 946 3980
fax: 916 946 4772
met.sales@youngusa.com
www.youngusa.com

Second Wind
366 Summer Street
Somerville, MA 02144-3132
phone: 617 776 8520
fax: 617 776 0391
sales@secondwind.com
www.secondwind.com

Vector Instruments
115 Marsh Road
Rhyl, Denbighshire Wales LL18
2AB
United Kingdom
phone: 44 17 45 35 07 00
fax: 44 17 45 34 42 06
sales @windspeed.co.uk
www.windspeed.co.uk

Wind-Measuring Masts

Both NRG and Western Wind Power manufacture a line of hinged anemometer masts 10 to 40 meters (30 to 140 ft) tall designed for wind prospecting.

NRG Inc.
110 Commerce Street
P.O. Box 509
Hinesburg, VT 05461
phone: 800 448 9463,
802 482 2255
fax: 802 482 2272
sales@nrgsystems.com
www.nrgsystems.com

Western Windpower
Axiom House
Station Road
Stroud, Gloucestershire GL5 3AP
United Kingdom
phone: 44 1453 759 408
fax: 44 1453 756 222
info@western-windpower.com
www.western-windpower.com

Wind and Meteorological Data

For historical wind and weather data in the United States, you can search the records of the National Climatic Data Center for information on your area. The NCDC's information is available for a modest fee and sometimes can be downloaded directly from the Internet at http://lwf.ncdc.noaa.gov/oa/ncdc.html.

National Climatic Data Center
151 Patton Avenue, Room 120
Asheville, NC 28801-5001
phone: 828 CLI MATE, 828 271 4800
fax: 828 271 4876
orders@ncdc.noaa.gov
www.ncdc.noaa.gov

Some historical wind data for the United States can be found at National Oceanic and Atmospheric Administration's ClimVis Web site: http://www.ncdc.noaa.gov/onlineprod/drought/xmgr.html.

For an explanation of how air pressure—and, consequently, air density—changes with elevation and temperature, visit Shing Yoh's site at Kean University: http://hurri.kean.edu/~yoh/calculations. Yoh also provides a handy hypsometric calculator at http://hurri.kean.edu/~yoh/calculations/hydrostatic/home.html.

Links to numerous sites with weather information can be found at http://www.wrcc.sage.dri.edu/ams/wxsites.html#CLIMATE. This site includes links to numerous meteorological organizations outside the United States.

Periodicals

Windpower Monthly is the commercial wind industry's principal trade publication and is an unaffiliated observer of wind energy worldwide. *Home Power* is the principal magazine covering small wind turbine topics. *Home Power,* as the name implies, caters to homeowners using small off-the-grid power systems or small wind turbines interconnected with the utility. Both *WindStats,* a quarterly technical journal, and *Naturlig Energi,* a publication of the Danish Wind Turbine Owners' Association, are good sources for statistical data on wind turbine performance. The British and European wind energy associations jointly publish *WinDirections. Windenergie Aktuel* covers technical and business topics of wind energy in Germany. *Neue Energie* (in German) and its sister publication, *New Energy* (in English), are published by the German Wind Turbine Owners' Association and are good sources for developments in Europe. *Renewable Energy World* regularly features articles on wind development worldwide. *Energias Renovables* (in Spanish) regularly covers wind energy in Spain. *DEWI Magazin* is a quarterly technical publication of the German Wind Energy Institute. *DEWI Magazin* is written in German, but includes summaries in English. The *Journal of Wind Engineering* (electronic only) and *Wind Energy* are the wind industry's peer-reviewed technical journals. *Systémes Solaires* (in French) devotes at least one issue per year to wind energy developments in Francophone countries. *Home Energy* should not be confused with *Home Power* magazine. *Home Energy* deals with energy efficiency and conservation and is useful for tips on reducing energy consumption. *Windmillers' Gazette* focuses on water-pumping windmills with articles on both historical and contemporary technical issues.

DEWI Magazin
Eberstrasse 96
D-26382 Wilhelmshaven
Germany
phone: 49 44 21 48 080
fax: 49 44 21 48 0843
dewi@dewi.de
www.dewi.de

Energias Renovables
C/Miguel Yuste, 26
ES-28037 Madrid
Spain
phone: 34 91 327 79 50
fax: 34 91 327 26 80
info@energias-renovables.net
www.energias-renovables.net

***Home Energy* Magazine**
2124 Kittredge Street, #95
Berkeley, CA 94704-9942
phone: 510 524 5405
fax: 510 486 6996
contact@homeenergy.org
www.homeenergy.org

***Home Power* Magazine**
P.O. Box 520
Ashland, CA 97520
phone: 530 475 3179
fax: 530 475 0836
hp@homepower.org
www.homepower.org

Journal of Wind Engineering
AMSET Centre
Horninghold, Leicestershire
LE16 8DH
United Kingdom
phone: 44 1858 555 204
fax: 44 1858 555 504
amset@compuserve.com
www.multi-science.co.uk/windeng.htm

Neue Energie/New Energy
Bundesverband WindEnergie (BWE)
Herrenteichsstrasse 1
D-49074 Osnabrück
Germany
phone: 49 541 350 6031
fax: 49 541 350 6030
info@wind-energie.de
www.wind-energie.de/ne

ReNew
Australian Alternative Technology Association
P.O. Box 2001
Lygon Street North
Brunswick, East Victoria 3057
Australia
phone: 61 3 9650 7883
fax: 61 3 9650 8574
ata@ata.org.au
www.ata.org.au/

Renewable Energy World
35-37 William Road
London NW1 3ER
United Kingdom
phone: 44 171 387 8558
fax: 44 171 387 8998
james@jxj.com
www.jxj.com

Systémes Solaire
146, Rue de l'Universite
F-75007 Paris
France
phone: 33 1 44 18 00 80
fax: 33 1 44 18 00 36
systemes.solaires@energies-renouvelables.org
www.systemes-solaires.com

Wind Energie Aktuell
Querstrasse 31
D-30519 Hannover
Germany
phone: 49 511 844 1932
fax: 49 511 844 2576
SunMedia@compuserve
www.Energie-Online.de

***Wind Energy* (Journal)**
John Wiley & Sons, Inc.
605 Third Avenue
New York, NY 10158-0012
phone: 212 850 6645
fax: 212 850 6021
subinfo@wiley.com
www.interscience.wiley.com/jpages/1095-4244

WinDirections
European Wind Energy Association
26 Spring Street
London W2 1JA
United Kingdom
phone: 44 171 402 7102
fax: 44 171 402 7107
ewea@ewea.org
www.ewea.org/directions.htm

Windmillers' Gazette
P.O. Box 507
Rio Vista, TX 76093
phone: 254 582 2555, ext. 391
fax: 254 582 7591
www.windmillersgazette.com

Windpower Monthly
P.O. Box 100
DK-8420 Knebel
Denmark
phone: 45 86 36 59 00
fax: 45 86 36 56 26
mail@windpower-monthly.com
www.windpower-monthly.com

WindStats/Naturlig Energi
Havvej 32
Vrinners Hoved
DK-8420 Knebel
Denmark
phone: 45 86 36 59 00
fax: 45 86 36 56 26
windstats@forlaget-vistoft.dk
www.windstats.com

Wind Turbine Market Surveys

The German Wind Turbine Owners' Association (Bundesverband Windenergie) and the German magazine *Wind Energy Today (Wind Energie Aktuel)* both publish annual surveys of wind turbine manufacturers supplying machines to the German market. The surveys include small as well as large wind turbines.

Windkraftanlagen Marktübersicht
(Wind Turbine Market Survey)
Herrenteichsstrasse 1
D-49074 Osnabrück
Germany
phone: 49 541 350 6031
fax: 49 541 350 6030
info@wind-energie.de
www.wind-energie.de

Windkraftanlagen Markt
(Wind Turbine Market)
Querstrasse 31
D-30519 Hannover
Germany
phone: 49 511 844 1932
fax: 49511 844 2576
SunMedia@compuserve
www.wind-energie.de

Software

There are an increasing number of wind energy software suppliers. The software runs the gamut from full-fledged commercial siting programs (WindMap, WindPro, WindFarmer, and WindFarm) to multipage spreadsheets available for free downloads from the Web (RETScreen).

RETScreen
Natural Resources Canada
CANMET Energy Diversification
Research Laboratory
1615 Lionel-Boulet Boulevard
P.O. Box 4800
Varennes, Quebec J3X 1S6
Canada
phone: 450 652 4621
fax: 450 652 5177
http://retscreen.gc.ca/

WindFarm
Resoft Ltd.
7 Church Lane
Flitton
Bedford MK45 5EL
United Kingdom
phone: 44 1525 862616
fax: 44 1525 862616
www.resoft.co.uk

WindFarmer
St Vincent's Works
Silverthorne Lane
Bristol BS2 0QD
United Kingdom
phone: 44 117 972 9900
fax: 44 117 972 9901
windfarmerinfo@garradhassan.co.uk
www.garradhassan.com

WindMap
Brower & Company
154 Main Street
Andover, MA 01810
phone: 978 749 9591
fax: 978 749 9713
mbrower@truewind.com
www.browerco.com

WindPro
Energi og Miljødata
Niels Jernes Vej 10
DK-9220 Aalborg
Denmark
phone: 45 98 35 44 44
fax: 45 98 35 44 46
emd@emd.dk
www.emd.dk

Anchors and Guying Hardware

Both A. B. Chance and Joslyn manufacture screw anchors useful for installing guyed towers.

A. B. Chance Co.
210 N. Allen Street
Centralia, MO 65240
phone: 573 682 8414
fax: 573 682 8660
www.hubbell.com/abchance

Joslyn Manufacturing Company
9200 W. Fullerton Avenue
Franklin Park, IL 60131
phone: 773 625 1500
fax: 773 625 0090
jmcintl@joslynmfg.com
www.joslynmfg.com

Preformed Line Products
P.O. Box 91129
Cleveland OH 44101
phone: 440 461 5200
fax: 440 442 8816
inquiries@preformed.com
www.preformed.com

Work Belts and Fall-Arresting Systems

There are numerous manufacturers and dealers of fall safety equipment worldwide. The following are only three of the suppliers in the United States.

DBI/Sala
3965 Pepin Avenue
Red Wing, MN 55066-1837
phone: 651 388 8282
fax: 651 388 5065
solutions@dbisala.com
www.salagroup.com

Klein Tools Inc.
7200 McCormick Road
P.O. Box 599033
Chicago, IL 60659-9033
phone: 847 677 9500
fax: 847 677 0816
www.klein-tools.com

Rose Manufacturing
2250 South Tejon Street
Englewood, CO 80110
phone: 800 672 2222
fax: 800 967 0398
rose@msanet.com
www.msanet.com

Inverters

Advanced Energy Systems (Prime Power Systems)
14 Brodie Hall Drive
Bentley, Western Australia 6102
Australia
phone: 64 8 94 70 46 33
fax: 64 8 94 70 45 04
info@advancedenergy.com
www.advancedenergy.com

Exeltech Inc.
225 E Loop 820 North
Fort Worth, TX 76118-7101
phone: 817 595 4969, 800 886 4683
fax: 817 595 1290
info@exeltech.com
www.exeltech.com

Selectronic Australia
25 Holloway Drive
Bayswater, Victoria 3153
Australia
phone: 61 3 9762 4822
fax: 61 3 9762 9646
sales@selectronic.com.au
www.selectronic.com.au

SMA
Carl-Diem-Weg 13
D47803 Krefeld
Germany
phone: 49 2151 598989
fax: 49 2151 599531
info@sma.de
www.sma.de

Studer Solartechnik
Route des caserne 57
CH-1950 Sion
Switzerland
phone: 41 27 205 60 80
fax: 41 27 205 60 88
info@studer-inno.com
www.studer-inno.com

Vanner Power Group
4282 Reynolds Drive
Hilliard, OH 43026
phone: 614 771 2718
fax: 614 771 4904
pwrsales@vanner.com
www.vanner.com

Xantrex
8999 Nelson Way
Burnaby, BC V5A 4B5
Canada
phone: 604 422 8595
fax: 604 420 1591
customerservice@xantrex.com
www.xantrex.com

Batteries

Though there are numerous manufacturers of flooded lead-acid batteries in North America, most specialize in automotive starting batteries. Automotive batteries are not suited for the deep cycling typical of renewable power systems. The following companies manufacture batteries specially designed for off-the-grid power systems or traction (golf cart) applications.

C&D Technologies
Powercom Division
P.O. Box 3053
Blue Bell, PA 19422-0858
phone: 215 619 2700,
800 543 8630
fax: 215 619 7899
powercom@cdtechno.com
www.cdpowercom.com

Saft Nife
711 Industrial Boulevard
Valdosta, GA 31601
phone: 912 247 2331,
800 556 6764
fax: 912 247 8486
power@saftnife.com
www.saftnife.com

Surrette Battery Co. Ltd.
Rolls Battery Engineering
P.O. Box 671
Salem, MA 01970
phone: 978 745 3333,
800 681 9914
fax: 978 741 8956
sales@rollsbattery.com
www.rollsbattery.com

Trojan Battery
12380 Clark Street
Santa Fe Springs, CA 90670
phone: 310 946 8381,
800 423 6569
fax: 310 941 6038

U.S. Battery Manufacturing Company
1675 Sampson Avenue
Corona, CA 91719
phone: 909 371 8090,
800 695 0945
fax: 909 371 4671
info@usbattery.com
www.usbattery.com

Yuasa-Exide
P.O. Box 14145
Reading, PA 19612-4145
phone: 800 538 3627
fax: 610 372 8613
www.yuasa-exide.com

Plans for Small Wind Turbines

The Centre for Alternative Technology has plans for building a Cretan sail windmill. Picoturbine and R.A.M. Design offer a Canadianized version of Scoraig Wind Electric's plans for building a wind turbine from an automotive brake drum.

Centre for Alternative Technology
Llwyngwern Quarry
Machynlleth, Powys, Wales
United Kingdom
phone: 44 1654 702 400
fax: 44 1654 702 782
info@cat.org.uk
www.cat.org.uk

Kragten Design
Populierenlaan 51
NL5492 SG Sint-Oedenrode
The Netherlands
phone: 31 413 475770
fax: 31 413 75770

Picoturbine
146 Henderson Road
Stockholm, NJ 07460
phone: 973 208 0056
fax: 973 208 2478
www.picoturbine.com

R.A.M. Design
RR 2
Goderich, Ontario
Canada
phone: 519 482 5034
fax: 519 482 1593
windmill@windmill.on.ca
www.windmill.on.ca

Scoraig Wind Electric
Dundonnell, Scotland IV23 2RE
United Kingdom
phone: 44 18 54 633 286
fax: 44 18 54 633 286
hugh@scoraigwind.co.uk
http://www.scoraigwind.co.uk/

Griphoists

Tractel is a source of hand-operated hoists in a range of sizes suitable for raising and lowering small wind turbines on hinged towers. Tractel's hoists are also suited for a host of other industrial uses.

Tractel (World)
85-87, Rue Jean Lolive
F-93100 Montreuil
France
phone: 33 1 48 58 91 32
fax: 33 1 48 58 19 95

Tractel USA
Griphoist Division
P.O. Box 188
Canton, MA 02021
phone: 781 329 5650,
800 421 0246
fax: 781 828 3642
info@tractel.com
www.tractel.com

Videos

Many of the national wind energy trade associations offer promotional videos for a small fee. The videos are useful for seeing and hearing the wind turbines in operation. Mick Sagrillo and Scott Andrews have produced a video (in NTSC) on installing small wind turbines titled *An Introduction to Residential Wind Power with Mick Sagrillo.* The 63-minute video is available for $39.95 plus postage and handling.

Scott Andrews
P.O. Box 3027
Sausalilto, CA 94965
phone: 415 332 5191

Multilingual Beaufort Scale of Wind Force

Strength or Force	Mean Speed (knots)	(m/s)	Range of Speed (knots)	(mph)	(m/s)	Wind Pressure (lb/ft²)	(N/m²)	English
0	0	0	<1	<1	0–0.2	0.00	0	Calm
1	2	1	1–3	1–3	0.3–1.5	0.01	1	Light air
2	5	3	4–6	4–7	1.6–3.3	0.08	4	Light breeze
3	9	5	7–10	8–12	3.4–5.4	0.27	13	Gentle breeze
4	13	7	11–16	13–18	5.5–7.9	0.57	27	Moderate breeze
5	19	10	17–21	19–24	8–10.7	1.2	58	Fresh breeze
6	24	12	22–27	25–31	10.8–13.8	2.0	93	Strong breeze
7	30	15	28–33	32–38	13.9–17.1	3.0	146	Moderate gale
8	37	19	34–40	39–46	17.2–20.7	4.6	222	Fresh gale
9	44	23	41–47	47–54	20.8–24.4	6.6	313	Strong gale
10	52	27	48–55	55–63	24.5–28.4	9.2	438	Whole gale
11	60	31	56–63	64–72	28.5–32.6	12	583	Storm
12	68	35	64–71	73–82	32.7–36.9	16	748	Hurricane
13	76	39	72–80	83–92	37–41.4	20	935	
14	85	44	81–89	92–103	41.5–46.1	24	1,169	
15	94	48	90–99	104–114	46.2–50.9	30	1,430	
16	104	53	100–108	115–125	51–56	37	1,750	
17	114	59	109–118	126–136	56.1–61.2	44	2,103	

Source: Adapted from: *The Generation of Electricity by Wind Power* by E.W. Golding; *Kleine Windkraftanlagen* by Heinz Schulz; and *El Poder Del Viento* by J. Puig, C. Mesequer, and M. Cabre.

Claus Nybroe, Windmission, Denmark

Christof Stork and Daniele Niccolai, Riva Calzoni, Italy

Bernard Saulnier, Hydro-Québec Research Institute, Canada

Forian Schmidt, Fachhochschule Konstanz

General Description

Deutsch	Español	Français	Dansk	Italiano
Windstill	Calma	calme	Stille	Calma
leichter Wind	Aire ligero	souffle léger	Svag luftning	Bava di vento
leichte Brise	Brisa ligera	légère brise	Svag vind	Brezza leggera
schwache Brise	Brisa suave	petite brise	Let vind	Brezza tesa
mäßige Brise	Brisa moderada	jolie brise	Jævn vind	Vento moderato
frische Brise	Brisa fresca	bonne brise	Frisk vind	Vento teso
starker Wind	Brisa fuerte	vent frais	Kuling	Vento fresco
steifer Wind	Veinto moderado	grand frais	Stiv kuling	Vento forte
stürmisher Wind	Veinto fresca	coup de vent	Hård kuling	Burrasca moderata
Sturm	Veinto fuerte	fort coup de vent	Storm	Burrasca forte
schwerer Sturm	Veinto fortísimo	bourrasque	Stærk storm	Burrasca fortissima
orkanartiger Sturm	Tempestad	tempête	Orkanagtig storm	Fortunale
Orkan	Huracán	ouragan	Orkan	Uragano

Multilingual Lexicon

English	Dansk	Deutsch
Above ground (aerial) cable	Alslutning via luftledning	Freileitung
Aerodynamic brake	Luftbremse	Luftbremse
Aileron	Krængeror	Querruder
Air brake	Bremseklap	Bremsklappe
Air temperature	Lufttermperatur	Lufttemperatur
Anemometer	Anemometer	Anemometer
Angle iron	Vinkeljern	Winkeleisen
Annual average wind speed	Årlig middelvindhastiged	Jahresdurchschnittsgeswindgeschwindigkeit des Windes
Annual energy output (production)	Årsenergiproduktion	Jahrresenergieertrag
Availability (for operation)	Rådighedsfaktor	Verfügbarkeit
Bearing	Leje	Lager
Bearing, main	Hovedleje	Hauptlager
Blade	Vinge	Blatt
Blade angle	Indstillingsvinkel	Flügelanstellwinkel
Blade manufacturer	Vinge fabrikant	Flügelhersteller
Blade number	Antal blade	Blattzahl
Blade root	Vingerod	Flügelanfang
Blade tip	Vingespids, vingetip	Flügelspitze
Blade tip angle	Tipvinkel	Winkel der Flügelspitze
Brake	Bremse	Bremse
Brake, disk	Skivebremse	Scheibenbremse
Brake, emergency	Nødbremse	Notbremse
Capacity (kW)	Effekt	Leistung
Capacity factor (plant)	Kapacitetsfaktor	Ausnutzungsgrad der Nennleistung
Chord, blade	Korde	Blattsehne
Circuit	Kreds	Stromkreis
Circuit breaker	Motorværn	Abschalter
Coefficient of Performance	Virkningsgrad	Wirkungsgrad
Coning (of rotor)		Anstellwinkel (des Rotors)
Control	Kontrol	Regelung
Control system	Kontrolsystem	Regelungsystem
Converter	Omformer	Umrichter
Converter, frequency	Frekvensomformer	Frequenzumrichter
Cooling	Køling	Kühlung
Cooling surface	Overfladekøling	Kühleroberfläche
Current, alternating (AC)	Vekselstrøm	Wechselstrom
Current, direct (DC)	Jævnstrøm	Gleichstrom
Data sampling	Datasamling	Datenerfassung
Diameter (rotor)	Rotordiameter	Durchmesser
Direct drive	Direkte dreven	Direktantrieb
Direction of rotation (rotor)	Omdrejningsretning for rotor	Rotordrehrichtung
Distribution panel	Fordelertavle	Verteilerkasten
Downwind	Læside	Leeseite
Drag (aerodynamic)	Modstand	Luftwiderstand
Drive train		Antriebsstrang
Eddy	Hvirvel	Wirbel
Electrical	Elektrisk	Elektrische
Electricity	Elektricitet	Elektrizitat
Fantail	Sideror	Seitenrad
Feathering (of blade)	Drejning af vinge (pitch)	Blattverstellung in Richtung Fahne
Flap	Flap	Klappe
Foundation	Fundament	Fundament
Freewheeling (rotor)	Rotor der kører løbsk	Leerlauf

Español	Français	Italiano
Cable aereo	Ligne aérienne	Cavo aero
Freno aerodinámico	Frein aérodynamique	Freno earodinamico
Alerón	Aileron	
Freno de aire	Aérofrein	
Temperatura del aire	Température de l'air	Temperatura dell aria
Anemómetro	Anémomètre	Anemometro
Perfil angular	Profilé	
Velocidad promedio anual	Vitesse moyenne annuelle du vent	Velicitá media annua
Producción energetica annual	Rendement annuel (production annuelle) d'énergie	Produzione annua di energia
Disponibilidad	Disponibilité	
Rodamiento	Palier	Cuscinetto
Rodamiento, principal	Palier principal	
Aspa	Pale	Pala d'elica
Angulo del aspa	Angle de pale	Angolo di calettamento della pala
Fabricante de aspas	Fabricant de pales	Produttore di pale
Número de aspas	Nombre de pales	Numero di pale
Raíz de aspas	Base de pale	Radice della pala
Punta del aspa	Bout de pale	Estremitá della pala
Angulo de la punta del aspa	Angle de bout de pale	Angolo di calettamento periferico
Freno	Frein	Freno
Freno, disco	Frein à disque	
Freno de emergencia	Frein d'urgence	Freno di emergenza
Capacidad	Puissance	Potentza
Factor de capacidad	Facteur de charge	
Cuerda	Corde	Corda
Circuito	Circuit	Circuito
Interruptores del circuito	Disjoncteur	Interruttore
Coeficiente de potencia	Coefficient de puissance	
Conicidad	Cône de rotation	
Control	Réglage, contrôle, régulation	Controllo
Sistema de control	Système de réglage, de régulation	Sistema di controllo
Convertidor	Convertisseur	Convertitore
Convertidor de frecuencia	Convertisseur de fréquence	Convertitore di frequnza
Refrigeración	Refroidissement	Raffeddamento
Superficie de refrigeración	Surface de refroidissement	Superficie di Raffreddamento
Corriente alterna	Courant alternatif	Corrente alternata
Corriente continua	Courant continu	Corrente continua
Muestreo de datos	Echantillonage de données	Campionatura di dati
Diámetro (del rotor)	Diamètre (du rotor)	Diametro
Accionomiento directo	Transmission directe	Presa diretta
Dirección de rotación	Direction de rotation	Verso di rotatzione
Panel de distribución	Tableau de distribution	
Espaldas al viento	Sous le vent	
Arrastre (aerodinámico)	Traînée	
Tren de Potencia	Engrenages	
Remolino, vórtice	Tourbillon	Vortice
Eléctrico	Électrique	Elettrico
Electricidad	Électricité	Elettricitá
Hélice auxiliares	Gouvernail	
Poner en bandera	Mise en drapeau	
Flap	Volet de bord de fuite	Ipersostentatore
Base-Basamento	Fondations	Fondamenta
Rotanción libre	Rotation libre, roue libre	

Multilingual Lexicon *continued*

English	Dansk	Deutsch
Freestanding	Fritstående	Freistehend
Frequency	Frekvens	Frequenz
Furling (of rotor)	Dreje rotor ud af vinden	Aus dem Wind drehen
Fuse	Sikring	Sicherung
Gear ratio	Gearets omsætningsforhold	Getriebeübersetzung
Gearbox	Gearkasse	Getriebe
Gearbox, parallel shaft	1,2 eller 3-trins gear	Stirnradgetriebe
Gearbox, planetary	Planetgear	Planetengebtriebe
Gearless (direct-drive)	Uden gear	Getriebelos
Generator, doubly fed (wound rotor)		Doppeltgespeister Generator
Governor, fly-ball, or Watt	Regulatormekanisme	Fliekraftregler
Grid, network (Electricity)	Ledningsnet	Netz, Stromnetz
Ground (earth) wire	Jordledning	Erdleitung
Ground rod	Jordspyd	Erdungsstab
Gust	Vindbyge	Bö
Guy cable	Bardun	Abspannskabel
Height	Højde	Hohe
High voltage transmission	Højspændingsoverførsel	Hochspannungsleitung
Horizontal axis wind turbine	Vindmølle med vandret akse	Horizontalläufer, Windkraftanlagen mit horizontaler Achse
Hub	Møllenav	Nabe
Hub extender	Navforænger	Extender
Hub height	Navhøjde	Nabenhöhe
Imaginary power	Blindlast	Blindleistung
Induction generator	Asynkron generator	Asynchrongenerator
Inland	Indlands	Binnenland
Installation (erection)	Møllerejsning installation	Montage
Insulator	Isolator	Isolator
Integrated	Integreret	Integriertes
Interconnected (with the grid or mains)	Nettilsluttet	Netzparallelbetrieb
Kilowatt-hour meter	Elmåler	Stromzähler
Leading edge	Forkant	Vorderkante
Length	Længde	Länge
Lifetime	Levetid	Lebensdauer
Lift (aerodynamic)	Opdrift	Auftrieb
Lightning	Lyn	Blitz
Lightning arrestor	Lynafleder	Blitzableiter
Lightning protection	Beskyttelse mod lynnedslag	Blitzschutz
Low voltage distribution	Lavspændingsledning	Niederspannungsleitung
Main (power) cable	Hovedledning	Leistungskabel
Main bearing	Hovedleje	Hauptlager
Main disconnect	Hovedafbryder	Hauptschalter
Manufacturer	Fabrikant	Hersteller
Measurement height	Målehøjde	Messhöhe
Medium (distribution) voltage	Stærkstrømledning	Mittelspannungsleitung
Monitoring systems	Overvågningssystemer	Überwachungssysteme
Nacelle	Møllehat	Gondel
Nacelle cover	Inddækning	Gondelverkleidung
Neutral	Nul-ledning	Nulleiter
Noise emissions	Støj emmision	Geräuschemission
Nose cone, spinner	Spinner	Spinner
Pennant		Wimpel

Español	Français	Italiano
	Autoporteur	
Frecuencia	Fréquence	Frequenza
Puesta en bandera	Dispositif à empennage articulé	
Fusible	Fusible	fusibile
Relacion de transmisión	Ratio du multiplicateur	Rapporto di trasmissione
Reductor	Multiplicateur	
Reductor de ejes paralelos	Multiplicateur à trains parallèles	
Reductor planetario	Multiplicateur planétaire	
Sin reductor, acciomamento directo	Sans transmission	Pressa diretta
Generador de doble efecto	Générateur à double alimentation	
Regulador por contrapesos	Régulateur centrifuge	
Red eléctrica	Réseau (de distribution)	
Cable de tierra	Conducteur de terre	Filo di terra
Barra de Tierra	Prise de terre	Presa di terra
Racha	Rafale	Raffica
Cable tensor	Hauban	Cavo Tirante
Altura	Hauteur	Altezza
Linea de alto voltage	Transmission à haute tension	Transmissione di alta tensione
Turbina de eje horizontal	Eolienne à axe horizontal	Turbina eolica ad asse orrizontale
Hub, centro del rotor	Moyeu	Mozzo
Extensor del centro, prolongador de raiz de pala	Extension de base de pale	
Altura del hub	Hauteur de moyeu	Altezza del Mozzo
	Puissance réactive	Potenza Reattiva
Generador de inducción, generador asincrónico	Générateur asynchrone	Generatore a induzione
Interior	(À l') intérieur	All'interno
Instalación	Installation, montage	Messa in opera (erezione)
Aislador	Isolateur	Isolatore
Integrado	Intégré	Integrato
Conectado	Connecté au réseau	Collegato (con la rete o elettrodotto)
Metrocontador de kilowat-hora	Compteur d'électricité	Contatore
Borde de ataque	Bord d'attaque	Bordo d'attacco
Longitud	Longueur	Lunghezza
Tiempo de vida	Durée de vie	Vita
Sustentación	Portance, poussée	
Relampago/Rayo/Trueno	Foudre	Fulmine
Pararayos	Parafoudre	Scaricatore per sovratensioni di carattere atmosferico
Protección anti rayos	Protection contre la foudre	
Distribucción de vajo voltage	Ligne de distribution à basse tension	Distribuzione di bassa tensione
Cable principal, de potencia	Cable principal (d'alimentation)	Cavo principale
Rodamiento principal	Palier principal	Cuscinetto principale
Interruptor principal	Interrupteur principal	
Fabricante	Fabricant	Produttore
Altura de medición	Hauteur de measure	
Voltage medio	Ligne de distribution a moyenne tension	Distribuzione di media tensione
Sistema de monitoreo	Système de surveillance	system di monitoraggio
Cabina, nacelle	Nacelle	Gondola
Cuvierta de la cabina	Coque	
Neutral, neutro	Conducteur neutre	
Emisiones de ruido	Émissions sonores	Emissione di rumori
Cuvierta de la nariz	Nez conique, Nez du rotor	
Banderín	Fanion	

Multilingual Lexicon *continued*

English	Dansk	Deutsch
Permanent magnet generator	Permanent magnetisert generator	Permanentmagnetgenerator
Phase conductor	Fase ledning	Phase
Pitch control	Pitch kontrol	Rotorblattverstellung
Pitch mechanism	Pitch mekanisme	Pitchantrieb
Pitch, fixed	Fast pitch	Fester Pitchwinkel
Pitch, variable	Variabel pitch	Variabler Pitchwinkel
Plan (top) view	Plantegning	Draufsicht
Pole-switching generator	Polomkoppelbar generator	Polumschaltbarer Generator
Power (kW)	Effekt	Leistung
Power curve	Effektkurve	Leistungskennlinie
Power factor	Effektfaktor	Leistungsfaktor
Power plant	Kraftværk	Kraftwerk
Power regulation	Effektregulering	Leistungsregulung
Power, nominal or rated	Mærkeeffekt	Nennleistung
Profile	Profil	Profil
Protection	Beskyttelse mod lynnedslag	Sicherheit
Pulse width modulation	Pulsbreddemodulation	Pulsweitenmoduliert
Radiator	Køler	Kühler
Regulation	Regulering	Regelung
Renewable energy	Vedvarande energi	Regenerative Energie
Rib	Ribbe	Versteifungsrippe
Root, blade	Bladrod	Wurzel
Rotor	Rotor	Rotor
Rotor blade	Rotorblad	Rotorblatt
Rotor overspeed control	Løbskkørselskontrol	Rotor überdrehzahlkontrolle
Rotor speed (rpm)	Rotoromdrejningstal	Drehzahl (rpm)
Rotor speed control	Regulering af omdrejningstal	Drehzahlregulierung
Sail	Sejl	Segel
Sea level	Havoverfladen	Meereshöhe
Sea-level pressure	Lufttryk ved havoverfloden	Luftdruck in Meereshöhe
Service crane	Servicekran	Servicekran
Service drop	Elselshabets	Verteileischliiefe
Service entrance	Forbrugers eltilførselsledning	Wartungseingang
Shaft	Aksel	Welle
Shaft, main	Hovedaksel	Hauptwelle, Rotorwelle
Side view	Sidebillede	Seitenansicht
Slip (of asynchronous generators)	Slip	Schlupf
Slip rings	Slipringe	Schleifring
Sound	Lyd	Schall
Sound level	Lyd niveau	Schallpegel
Spar, blade	Bjælhe	Holm
Speed, constant	Konstant hastighed	konstante Geschwindigkeit
Speed, variable	Variabel hastighed	variable Drehzahl
Spinner, nose cone	Spinner	Propellerhaube
Spoiler	Spoiler	Störklappe or Spoiler
Stages (gear)	Trin	Getriebestufen
Stall regulated	Stallregulering	Stallregulierte
Start-up wind speed	Cut-in vindhastighed	Anlaufwindgeschwindigkeit
Streamer, pennant	Vimpel	Fahnenband
Strut	Stag	Vestrebung
Survival wind speed	Overlevelsesvindhastighed	Überlebeswindgeschwindigkeit
Swept area, rotor	Bestrøget rotor areal	Rotorfläche
Synchronous speed	Synkront omdrejningstal	Synchrondrehzahl

Español	Français	Italiano
Generador de imanes permanentes	Générateur à aimants permanents	Generatore a megnete permanente
Conductor de fase	Conducteur de phase	conduttore di fase
Control del paso	Régulation de pas	Controllo di beccheggio
Mecanismo del paso	Mécanisme de calage	Meccanisom di beccheggio
Paso fijo	Pas fixe	
Paso variable	Pas variable	
Vista superior	Vue de dessus	Vista in pianta dall'alto
Generador de inversión de polos	Génératrice à poles commutables	
Potencia	Puissance	Potenza
Curva de potencia	Courbe de puissance	Curva della potenza
Factor de potencia	Facteur de puissance	Fattore di potenza
Central eléctrica	Centrale	Impianto per la produzione di energia
Regulación de potencia	Régulation de puissance	Regoazione di potenza
Potencia nominal	Puissance nominale	Potenza nominale
Perfil	Profil	Profilo
Protección	Protection	Protezione
Modulación del ancho del pulso	Modulation à largeur d'impulsion (MLI)	Modulazione d'ampiezza
Radiador	Radiateur	Radiatore
Regulación	Régulation, réglage	Regolazione
Energías Renovables	Energies renouvelables	Energia da fonte rinnovabile
Costilla	Nervure d'aile	Centina (irrigidimento)
Raíz (base)	Tige de pale	Radice
Rotor, hélice	Rotor, hélice	Rotore
Aspas del rotor	Pale de rotor	Pala di rotore
Control de sobrevelocidad del rotor	Système de contrôle de sur-vitesse	
Velocidad del rotor	Vitesse du rotor, fréquence de rotation de l'hélice	Velocitá di rotazione del rotore
Control de velocidad del rotor	Contrôle de vitesse du rotor	Controllo della velocitá di rotazione del rotore
Vela	Voile	Vela
Nivel del mar	Niveau de mer	Livello del mare
Presion a nivel del mar	Pression au niveau de la mer	Pressione al livello del mare
Grua de servicio	Grue de service	Gru di assistenza
Conexión a red	Branchement de distribution	
Entrada de servicio	Branchement de l'abonne, éntree de service	
Eje	Arbre	Albero di trasmissione
Eje principal	Arbre principal	Albero Primario
Vista lateral	Vue de profil	Vista laterale
Deslizamiento de frecuencia	Glissement	
Anillos rozantes	Anneaux de glissement	Anelli di contatto
Sonido	Son, bruit	Suono
Nivel de sonido (ruido)	Niveau sonore	Livello di suono
Alma de una pala	Longeron	Longherone
Velocidad constante	Vitesse constant	Velocitá constante
Velocidad variable	Vitesse variable	Velocitá variablile
Cuvierta de la nariz		
Spoiler	Déporteur	Diruttore
Etapas	Étapes	
Stall controlado	Régulation par décrochement	Controllato in stallo
Velocidad del viento de arranque	Vitesse de démarrage	Velocitá di accensione
	Banderole	Banderuola
Contonearse	Entretoise	Montante
Velocidad del viento de supervivencia	Vitesse de survie	Velocitá di sopravvivenza
Area de barrido	Surface balayée	Area del disco d'elica
Velocidad sincrónica	Vitesse synchrone	

Multilingual Lexicon *continued*

English	Dansk	Deutsch
Tail vane	Haleror	Windfahne
Terminal	Terminal	Endableitung
Tip	Tip	Spitze
Tip speed ratio	Tishastighedsforhold	Schnellaufzahl
Tonal quality		Tonhaltigkeit
Torque	Moment	Drehmoment
Tower	Tårn	Turm
Tower, conical	Konisk tårn	Konischer Turm
Tower, lattice or truss	Gittermast	Gitterturm, Gittermast
Tower, steel tubular		Stahlrohrturm
Tower, tubular	Rørtåarn	Rohrturm
Trailing edge	Bagkant	Flügelhinterkante
Transformer	Transformator	Transformator
Transformer station	Transformatorstation	Trafostation
Type	Type	Bauart
Upwind	Luv	Luvseitig
Vertical axis wind turbine	Vindmølle med lodret akse	Windkraftanlagen mit vertikaler Achse
Voltage	Spænding	Spannung
Voltage flicker		Spannungsschwangkung
Wake	Skygge	Nachströmung
Water cooled	Vandkølet	Wassergekühlt
Weight	Vægt	Gewicht
Wind	Vind	Wind
Wind direction	Vindretning	Windrichtung
Wind energy	Vindenergi	Windenergie
Wind farms	Vindmøllepark	Windpark
Wind generator		Windgenerator
Wind power plants	Vindmøller	Windkraftwerk
Wind rose (of compass)	Vindrose	Windrose
Wind sock	Vindpose	Windsack
Wind speed	Vindhastighed	Windgeschwindigkeit
Wind speed, cut-out or shut down	Stop-eller cut-out vindhastighed	Abschaltwindgeschwindigkeit
Wind turbine	Vindmølle	Windkraftanlagen
Wind vane	Vindfane	Windfahne
Wind, prevailing	Fremherskende vindretning	Vorherrschende Windrichtung
Windmill (traditional European)	Vindmølle	Windmühle
Windmill, farm (American multiblade)	Vindrose	Western mill, Windrad
Wing	Vinge	Flügel
Winglet	Windglet	Winglet
Yaw bearing	Krøjeleje	Azimutlager
Yaw drive	Krøjeværk	Azimutgetriebe
Yaw orientation	Krøjeretning	Windnachfuhrung
Yaw rate	Krøjehastighed	Azimutgeschwindigkeit
Yaw system	Krøjesystem	Azimutsystem
Yaw, active (mechanical)	Aktiv krøjning (mekanisk hjulpet)	Aktive Windnachführung
Yaw, passive (free)	Passiv krøjning, fritkrøjende	Passive Windnachführung
Yawing	Krøjning	Das Drehen

Source: Staff and trainees at the Folkecenter for Vedvarende Energi, Denmark, especially Reinhard Löngsing, Giuseppe Leonardi, and Felix Varela. And the following: Túpac José Canosa Diaz, Universidad de Camagüey, Cuba; Rafael Oliva, Universidad Nacional de la Patagonia Austral, Argentina;Claus Nybroe, Windmission, Denmark; Klaus Kaiser, Germany; Florian Schmidt, Fachhochschule Konstanz, Germany; Bernard Saulnier, Hydro-Québec Research Institute, Canada;Charles Dugué, Cabinet Germa, France; and Charles Rosseel, France.

For terms not found here see www.windpower.dk/glossary.htm.

Español	Français	Italiano
Cola de orientación	Gouvernail	
Terminal	Borne	Terminale
Punta	Bout de pale	Estremitá
Relación de las velocidades de la punta	Quotient de vitesse périphéique	
	Timbre	
Torque	Couple moteur	
Torre	Tour, pylône	Torre
Torre cónica	Tour conique	Toree conica
Torres reticulares	Tour à treillis, pylône en treillis	
Torre tubular de acero	Tour tubulaire en acier	Torre cilindrica in acciaio
Torre tubular	Tour tubulaire	Torre cilindrica
Borde de salida	Bord de fuite	Bordo di fuga
Transformador	Transformateur	Trasformatore
Estación de transformación	Station de transformation	Stazione di trasformatzione
Tipo, clase	Type	Tipo
Cara al viento	En amont du vent	Sopravvento
Turbina de eje vertical	Eolienne à axe vertical	turbina eolica ad asse verticale
Voltaje	Voltage	Tensione
Picos de voltaje	Vacillement du voltage, flicker	
Estela	Sillage	Scia
Enfriado por agua	Refroidi par l'eau	Raffreddato ad acqua
Peso	Poids	Peso
Viento	Vent, éolien	Vento
Dirección del viento	Direction du vent	Direzione del vento
Energía eólica	Energie éolienne	Energia eolica
Parques eólicos	Ferme éolienne	
Generador eólico	Aérogenerateur	Generatore eolico
Plantas generadoras eólicas	Centrale éolienne	Impianti per la produzione di energia da foute eolica
Rosa de los vientos	Rose des vents	
	Manche á air	Colpo di vento
Velocidad del viento	Vitesse du vent	Velocitá del vento
Velocidad del viento de parada	Vitesse de coupure	Velocitá di spegnimento
Turbina eólica	Éolienne, aéogenerateur	Turbina eolica
Veleta	Girouette	Manica a vento
Viento prevaleciente	Vent dominant	
Molino de viento	Moulin à vent	Mulino a vento
Molino tipo americano (multipala)	Éolienne (à pales multiples)	
Ala	Aile	Ala
	Ailette	Aletta
Rodamiento de orientación	Palier d'orientation	Cuscinetto d'imbardata
Motor de orientación	Engrenage d'orientation	
Orientación	Orientation	
Relacion de transmision para la rotacion de orientación	Système d'orientation	Velocitá di imbardata
Sistema de orientación	Orientation	
Orientación, activa	Orientable	
Autoorientación	Auto-orientable	
Orientandose	Orientation	Imbardata

Bibliography

This bibliography includes entries for books on rigging of cables and wire rope, solar photovoltaics, and the general topic of wind energy.

Rigging

Carter, Paul. *Backstage Handbook: An Illustrated Almanac of Technical Information.* New York: Broadway Press, 1994. 310 pp. ISBN: 0-911747-39-7. Packed with useful information on tools, rigging, and an especially good section on wire rope.

Ellis, J. Nigel. *Introduction to Fall Protection.* Second edition. Des Plaines, Illinois: American Society of Safety Engineers, 1993. 228 pp. ISBN: 0-939874-97-0. An informative discussion of the development of fall safety standards for engineers and managers. Should be required reading for anyone working at heights or asking others to do so.

Kurtz, Edwin, and Thomas Shoemaker. *The Lineman's and Cableman's Handbook.* Ninth edition. New York: McGraw-Hill, 1997. 1,056 pp. ISBN: 0-0703601-1-1. An extremely useful handbook on setting poles, stringing conductors, and installing earth anchors.

Newberry, William G. *Handbook for Riggers.* Calgary, Alberta: W. G. Newberry, 1989 revision. ISBN: 0-9690154-1-0. An inexpensive booklet recommended by the Alternative Energy Institute's Ken Starcher for an introduction to working with wire rope.

Solar Photovoltaics

Davidson, Joel. *The New Solar Home Book: The Photovoltaic How-To Handbook.* Second edition. Ann Arbor, Michigan: Aatec Publications, 1987. 408 pp. ISBN: 0-937948-09-8.

Photovoltaic Design Assistance Center. *Stand-Alone Photovoltaics Systems: A Handbook of Recommended Design Practices.* Albuquerque, New Mexico: Sandia National Laboratories, 1990.

———. *The Design of Residential Photovoltaic Systems.* Albuquerque, New Mexico: Sandia National Laboratories, 1988.

Potts, Michael. *The Independent Home: Living Well with Power from the Sun, Wind, and Water.* Second edition. White River Junction, Vermont: Chelsea Green Publishing, 1999. 408 pp. ISBN: 1-890132-14-4. Explores the lives of those living and working with renewable energy.

Pratt, Doug. *The Real Goods Solar Living Sourcebook: The Complete Guide to Renewable Energy Technologies & Sustainable Living.* Tenth edition. White River Junction, Vermont: Chelsea Green Publishing, 1999. 562 pp. ISBN: 0-916571-3-3. The successor to Real Goods' popular *Alternative Energy Sourcebook.* Catalog and discussion of energy-efficient and renewable energy products, including compact fluorescent lights, photovoltaic panels, wind turbines, inverters, batteries, cables, and much more.

Strong, Steven, and William Scheller. *The Solar Electric House: Energy for the Environmentally-Responsive, Energy-Independent Home.* White River Junction, Vermont: Chelsea Green Publishing, 1993. 276 pp. ISBN: 0-9637383-2-1.

Wind Energy Technology in English

The following is only a sample of the books that have been published on wind energy in the past 50 years. There are hundreds of books on wind energy in various languages. Many are out of print and available only in large libraries or private collections. Each author makes a unique contribution to the subject. Reviews of some of these books, including their tables of contents, can be found on the World Wide Web.

Burton, Tony, David Sharpe, David Jenkins, and Ervin Bossanyi. *Wind Energy Handbook.* Chichester, United Kingdom: John Wiley & Sons, 2002. 617 pp. ISBN: 0-471-48997-2. Certain to become the English-language reference engineering text on wind turbine design The first such work to include chapters on design of wind farms, including a discussion of environmental impacts as well as cash flow.

Eggleston, David, and Forrest Stoddard. *Wind Turbine Engineering Design.* New York: Van Nostrand Reinhold, 1987. 352 pp. For professional wind turbine designers. Not for the faint of heart. The authors warn readers that a working knowledge of differential equations is essential to understanding the text, which includes an extensive discussion of aerodynamics and structural dynamics.

Freris, L. L., ed. *Wind Energy Conversion Systems.* Hemel Hempstead, United Kingdom: Prentice Hall, 1990. 388 pp. ISBN 0-13-960527-4. An engineering text by the leaders in British wind energy. A good academic reference book on contemporary wind technology.

Gipe, Paul. *Wind Energy Basics.* White River Junction, Vermont: Chelsea Green Publishing, 1999. 122 pp. ISBN: 1-890132-07-1. A cursory introduction to wind energy and small wind turbines.

———. *Wind Energy Comes of Age.* New York: John Wiley & Sons, 1995. 613 pages. ISBN: 0-471-10924-X. A chronicle of wind energy's progress from its rebirth during the oil crisis of the 1970s through a troubling adolescence in California's mountain passes in the 1980s to its maturation on the plains of northern Europe in the 1990s. Selected as one of the outstanding academic books published in 1995.

Golding, E. W. *The Generation of Electricity by Wind Power.* London: E. & F. N. Spon, 1955. 332 pp. Reprinted by John Wiley & Sons in 1976. Still a classic of English-language books on wind technology. Recounts British research on wind energy during the early 1950s.

Johnson, Gary L. *Wind Energy Systems.* Englewood Cliffs, New Jersey: Prentice Hall, 1985. 360 pp. An engineering textbook strong on the wind resources of Kansas and on electrical engineering.

Koeppl, Gerald W. *Putnam's Power from the Wind.* Second edition. New York: Van Nostrand Reinhold, 1982. 470 pp. The first half of the second edition is a reprint of Putnam's original book. The second half examines large wind turbine development programs in the United States and Europe during the 1970s.

Putnam, Palmer Cosslett. *Power from the Wind.* New York: Van Nostrand Reinhold, 1948. 224 pp. Reprinted in 1974. ISBN: 0-442-26650-2. The classic account of constructing the 1.250-megawatt Smith-Putnam turbine during the early 1940s. Like Golding's, many of Putnam's observations still apply.

Spera, David, ed. *Wind Turbine Technology: Fundamental Concepts of Wind Turbine Engineering.* New York: American Society of Mechanical Engineers, 1994. 650 pp. ISBN: 0-7918-1205-7. The book is noteworthy for a chapter on NASA/DoE's large turbine development program by the program's principal proponent, Lou Divone.

Walker, John F., and Nicholas Jenkins. *Wind Energy Technology.* Chichester, United Kingdom: John Wiley & Sons, 1997. 161 pp. ISBN: 0-471-96044-6. A handy introduction to wind energy for engineering students.

Modern Wind Energy History

Asmus, Peter. *Reaping the Wind.* Washington, D.C.: Island Press, 2001. 277 pp. ISBN: 1-55963-707-2. The book sings the praises of California's hitherto unsung energy heroes: the men and women who made the sometimes faulty technology in the state's windy passes work, and the eccentrics who made the pioneering days so colorful. A novelist would be hard-pressed to create the cast of characters that Asmus assembles: Zen Buddhists, Jungian dreamers, and a host of crooks, charlatans, and hucksters.

Baker, T. Lindsay. *A Field Guide to American Windmills.* Norman, Oklahoma: University of Oklahoma Press, 1985. 528 pp. ISBN: 0-8061-1901-2 The definitive history of the American farm windmill.

Heymann, Matthias. *Die Geschichte der Windenergienutzing 1890–1990.* Frankfurt am Main, Germany: Campus Verlag, 1995. 518 pp. ISBN: 3-539-35278-8. It's worth learning German to read this controversial book on the history of wind energy in Germany from the late 19th century to the early 1990s. It contains the most extensive discourse to date on the Third Reich's interest in wind energy and an unflattering, matter-of-fact description of attempts by some of the grand names in German wind energy to curry favor with the Nazis. This book is a powerful reminder that technology is not divorced from politics.

Righter, Robert. *Wind Energy in America: A History.* Norman, Oklahoma: University of Oklahoma Press, 1996. 366 pp. ISBN: 0-8061-2812-7. A thought-provoking account of the people and ideas behind the use of wind energy in the United States. The book emphasizes the conflict between centralization and distributed use of wind-generated electricity.

Torrey, Volta. *Wind-Catchers: American Windmills of Yesterday and Tomorrow.* Brattleboro, Vermont: Stephen Greene Press, 1976. 226 pp. ISBN: 0-8289-0292-5. This is an engagingly written history of wind energy in the United States with chapters on the Smith-Putnam turbine, the American water-pumping windmill, and 1970s pioneers of modern wind turbines. Righter's *Wind Energy in America: A History* adds his own original contributions and brings the story up to the 1990s.

van Est, Rinie. *Winds of Change: A Comparative Study of the Politics of Wind Energy Innovation in California and Denmark.* Utrecht, Netherlands: International Books, 1999. 368 pp. ISBN: 90-5727-027-7. This monumental work is destined to become a classic in its field. It paints a detailed, carefully researched picture of why the development of wind energy technology failed in the United States during the 1970s and 1980s, but succeeded in Denmark.

Wind Resources and Siting

Battelle Pacific Northwest Laboratory. March 1987. *Wind Energy Resource Atlas of the United States.* This is the updated version of Battelle's classic work mapping the U.S. wind resource. The update incorporates new data collected since the original atlas was published in 1980. DoE/CH 10094-4. The entire document is available on the Web at http://kepler.nrel.gov/wind/pubs/atlas.

Nelson, Vaughn, and Janardan Rohatgi. *Wind Characteristics: An Analysis for the Generation of Wind Power.* Canyon, Texas: West Texas A&M University, 1994. ISBN: 0-8087-1478-3.

Risø National Laboratory. *Wind Atlas Analysis and Application Program (WA^{S}P).* Published for the European Community Commission. 1992.

Troen, Ib, and Erik Lundt Petersen. Risø National Laboratory. *European Wind Atlas.* Published for the European Community. 1989. 656 pp.

Aesthetics and Noise

Nielsen, Frode Birk. *Wind Turbines & the Landscape: Architecture & Aesthetics.* Arhus, Denmark: Birk Nielsens Tegnestue, 1996. 63 pp. ISBN: 87-985801-1-6. A beautifully illustrated book showing how wind turbines can exist harmoniously on the landscape.

Pasqualletti, Martin, ed. *Wind Power in View: Energy Landscapes in a Crowded World.* San Diego: Academic Press, 2002. 234 pp. ISBN: 0-12-546334-0. An entire book devoted to how wind turbines appear on the landscape.

Wagner, Siegfried, Rainer Bareiß, and Gianfranco Guidati. *Wind Turbine Noise.* Berlin: Springer-Verlag, 1996. 205 pp. ISBN: 3-540-60592-4. The clearest yet most detailed book on wind turbine noise available.

Wolsink, Maarten. *Maatschappelijke acceptatie van windenergie: Houdingen en oordelen van de bevolking.* Amsterdam, 1990. 237 pp. A thesis on the public acceptance of wind energy (in Dutch).

Designing, Building, or Installing Your Own

For tinkerers and those wanting to learn more about installing small wind turbines, Michael Hackleman's book is hard to beat, but it's now also very hard to find. Jack Park's book is useful for experimenters wanting a broader background in the mechanics of wind turbines.

Hackleman, Michael. *The Home-Built, Wind-Generated Electricity Handbook.* Mariposa, California: Earthmind, 1975. Explains how to find, lower, and rebuild pre-REA wind generators.

Park, Jack. *The Wind Power Book.* Palo Alto, California: Cheshire Books, 1981. 253 pp. ISBN: 0-917352-06-8. One of the best overall books on the technology and design of small wind machines. A useful reference for experimenters with an aversion to metric units.

Piggott, Hugh. *It's a Breeze.* Machynlleth, Powys, Wales, United Kingdom: Centre for Alternative Technology, 1995. 36 pp. Straightforward advice on how to use small wind turbines.

———. *Windpower Workshop.* Machynlleth, Powys, Wales, United Kingdom: Centre for Alternative Technology, 1997. 160 pp. ISBN: 1-898049-13-0. Chock-full of tips on building your own wind-charger.

Water Pumping

Fraenkel, Peter, Roy Barlow, et al. *Windpumps: A Guide for Development Workers.* Intermediate Technology Publications, 1993. ISBN: 1-85339-1263 A thorough treatment of wind pumping by authorities on the subject.

Kentfield, J. A. C. *The Fundamentals of Wind-Driven Water Pumpers.* Amsterdam: Overseas Publishers Association, 1996. 286 pp. ISBN: 2-88449-239-9. This is a thorough engineering treatment of wind pumping and includes descriptions of the author's novel wind pump designs.

Wind Energy Technology in German

Gasch, Robert. *Windkraftanlagen.* 3rd Auflage. Stuttgart, Germany: B. G. Tuebner, 1996. 390 pp. ISBN: 3-519-26334-3. *Windkraftanlagen* is one of the most complete, and certainly the most up-to-date, engineering texts on wind energy available in any language. The book reflects the target audience of engineering students and Gasch's background in structural dynamics, but also includes chapters on wind pumps and on asynchronous generators. Also available in English (ISBN: 1-902916-37-3) from James & James, London.

Handschuh, Karl. *Windkraft gestern und heute: Geschichte der Windenergienutzung in Baden-Württemberg.* Staufen, Germany: Ökobuch, 1991. 115 pp. ISBN: 3-922964-33-8. The history of wind energy in the state of Baden-Württemberg in the southwestern corner of Germany. The book includes interesting chapters on the wartime development of wind turbines by Ulrich Hütter, as well as the manufacture of wind turbines in the 1950s by Allgaier.

Hau, Erich. *Windkraftanlagen: Grundlagen, Technik, Einsats, Wirschaftlichkeit.* Second edition. Berlin: Springer Verlag, 1996. 460 pp. ISBN: 3-540-57430-1. A thorough perspective on Germany's development of multimegawatt wind turbines during the early and mid-1980s. The second edition includes expanded coverage of medium-size wind turbines that were beginning to appear in large numbers on the German market in the mid-1990s.

Molly, Jens-Peter. *Windenergie: Theorie, Anwendung, Messung.* Karslruhe, Germany: Verlag C. F. Muller, 1990. 316 pp. ISBN: 3-78807269-5. German medium-size wind turbine technology development through the late 1980s. One of the most thorough and well-illustrated books on modern wind technology in any language. Includes useful German–English and English–German translation of common technical terms.

Schulz, Heinz. *Kleine Windkraftanlagen: Technik, Erfahrungen, Meßergebnisse.* Staufen, Germany: Ökobuch, 1991. 110 pp. ISBN: 3-922964-31-1. *Kleine Windkraftanlagen* offers an intriguing glimpse into the development of small wind turbines during the early 1990s in Germany, including test results on popular micro and mini wind turbines.

Wind Energy Technology in French

Cunty, Guy. *Éoliennes et aérogénérateurs: Guide de l'énergie éolienne.* Aix-en-Provence, France: Édisud, 2001. 167 pp. ISBN: 2-7449-0233-0. A dated introduction to wind energy in France, updated with photos of contempory wind turbines.

Guide de L'Énergie Éolienne: Les aérogénérateurs au service du développment durable. Quebec, Canada: Collection Études et filières, 1998. 161 pp. ISBN: 2-89481-004-0. A very good introduction to modern wind energy and how it can be used in France and French overseas territories, including technology, economics, and wind energy's environmental impacts as well as its benefits.

Le Gourières, Désiré. *Énergie Éolienne: Théorie, conception, et calcul pratique des installations.* Paris: Editions Eyrolles, 1980. 268 pp. (Published in English as *Wind Plants: Theory and Design.* Oxford, England: Pergamon Press, 1982.) At the time it was one of the first modern technical books on wind turbine design. It includes numerous sketches of small European wind turbines manufactured during the late 1970s, the most thorough documentation of early French wind turbines of any source, and a useful discussion of classic water-pumping windmill design.

Wind Energy Technology in Spanish

Gipe, Paul. *Energía Eólica Práctica.* Sevilla, Spain: Progensa (Promotor General de Estudios), 2000. 191 pp. ISBN: 84-86505-88-7. A Spanish translation of *Wind Energy Basics.*

Puig, Josep, Conrad Mesequer, and Miguel Cabre. *El Poder Del Viento: Manual práctico para conocer y aprovechar la fuerza del viento.* Barcelona, Spain: Ecotopia Ediciones, 1982. 154 pp. ISBN: 84-85813-09-X. A good description of wind technology development in Denmark, England, and Spain during the late 1970s and early 1980s. The book was produced by the Ecotecnia cooperative, now one of Spain's largest manufacturers of medium-size wind turbines.

Wind Energy Technology in Swedish

Wizelius, Tore. *Vind.* Täby, Sweden: Larsens Förlag, 1994. 158 pp. ISBN: 91-514-0283-1. Interesting photographs of Swedish wind turbines, including the construction of a home-built turbine based on a Danish design.

Glossary of Wind Energy Terminology

This glossary has been adapted from *Glossary of Wind Energy Terms* by Paul Gipe and Bill Canter (Forlaget Vistoft: Knebel, Denmark, 1997).

The wind energy vocabulary changes continuously as the technology evolves and as new issues arise to confront designers. Thus a glossary is not a static work; it shrinks and grows with the technology it attempts to describe.

A glossary is in part a window on a technology and the people who work with it. The development of wind energy requires the use of many divergent professions. As such this glossary includes terms from several different fields, not just those from mechanical engineering, aeronautics, and meteorology. For example, selected terms from the fields of biology, geography, aesthetics, and the electric utility industry are included. Moreover, unlike most glossaries that simply define terms, this glossary places the meaning of the term within its wind energy context.

1/7 power law: Empirically derived rule of thumb where the wind shear exponent, α, in the power law model is equal to one-seventh, representing a stable atmosphere above open grasslands typical of America's Great Plains. See also *Power law.*

Abandoned wind turbines: In U.S. usage, loosely applied to all wind turbines that stand inoperative for extended periods. Few inoperative wind turbines in California are legally abandoned, though the owners may not have made any substantive efforts to operate them for many years. Some wind turbines in California have stood idle for a decade or more. See also *Derelict wind turbines.*

Absorption, sound: Interception and attenuation of acoustic energy by use of sound-deadening materials, such as insulation and cladding, inside a wind turbine nacelle.

AC: See *Alternating current.*

Accumulator, hydraulic: Device for the storage of hydraulic pressure typically for use in braking systems. The accumulator is a tank divided in two parts. One side contains a fluid acting against a bladder or diaphragm; the other side contains compressed air.

Acoustic noise: Unwanted audible and subaudible sounds.

Acoustics: Study of sound.

Acuity, visual: Sharpness or keenness of visual perception.

ADEME: Agence de l'Environment et de la Maîtrise de l'Energie. French agency for environmental and energy management.

Adjusted availability: Fraction of time a wind turbine is available for operation after eliminating time and outages associated with activities that are nonstandard.

Advanced wind turbine (AWT): Term adopted by U.S. Department of Energy and the National Renewable Energy Laboratory to designate a series of wind turbines under development during the early 1990s. The designation was expropriated by one of DoE's grant recipients to designate its product line. The company and the turbine design are both defunct.

AEI: Alternative Energy Institute, West Texas A&M University in Canyon, Texas. Operates a wind turbine test center.

AEO: Annual energy output. North American usage for an estimate of the total energy a wind turbine can produce per year under standard conditions. Also annual energy production.

Aeolian Islands: Archipelago in the Tyrrhenian Sea, north of Sicily.

Aeolus: Greek god of the wind. See also *Éole.*

Aermotor: Trade name of the most widely recognized multiblade farm windmill built in North America. With its origin (a suburb of Chicago, Illinois) prominently displayed on its tail vane, the Aermotor became known to many outside North America as the "Chicago" mill.

Aeroelasticity: The study of the effect of aerodynamic forces on elastic bodies, such as wind turbine blades.

Aero-electric generator: Little-used term for "wind generator." Compound word with similar derivations as *hydroelectric generator,* which is commonly accepted. Technically more correct than many terms currently in vogue—for example, *wind generator.*

Aerogenerator: European expression for "wind generator." As with *wind generator,* a strict interpretation of the term would lead you to conclude that it's a machine that generates air (wind).

Aeroturbine: Similar derivations as *aerogenerator.*

Aesthetics: Philosophy that seeks to explain beauty. Metaphysical laws of perception. The study of the response to beauty and artistic expression. Perception of what is pleasing to the eye.

AID: Agency for International Development (U.S.).

Ailerons: Diminutive of the French *aile,* for "little wing." Movable control surfaces found on the outboard trailing edge of wings used for governing the roll of an airplane. Unlike wing flaps, ailerons move opposite one another. As the aileron on one wing moves up, the aileron on the other wing moves down. On airplanes, flaps are inboard of the ailerons, and flaps on either side of the fuselage move in the same direction, often down, to create drag for lowering air speed. The word *ailerons* has also been applied to movable flaps on the outboard trailing edge of a wind turbine blade for limiting power in high winds.

Air brake: Aerodynamic device, such as pivotable tips, spoilers, or parachutes, for slowing the speed of a wind turbine rotor. Flap or spoiler. Mechanism for destroying the aerodynamic performance of a wind turbine blade to limit rotor speed and torque in high winds. See also *Flap* and *Aileron.*

Air density: The mass of air relative to its volume. The density of air at sea level is 1.225 kg/m^3. Air density decreases with increasing altitude and temperature.

Airfoil: Curved surface, such as an airplane wing, designed to create lift. Airfoils for horizontal-axis or conventional wind turbines are asymmetrical, with the flat side toward the wind. Vertical-axis turbines use symmetrical airfoils. Some small wind turbines use airfoils with a single curved surface.

Air gap: The space between the rotor and stator in a generator. (Hugh Piggott)

Altamont Pass: Low saddle separating California's Livermore Valley from the San Joaquin Valley about 50 miles (80 km) east of San Francisco. Once one of the world's largest producers of wind-generated electricity. During the early 2000s, the area generated 0.8 to 1.1 TWh per year.

Alternating current (AC): Electric current that reverses its direction or polarity cyclically at 60 Hz in the Americas and 50 Hz elsewhere.

Alternator: Generator that produces alternating current.

Aluminum: Lightweight metal adapted unsuccessfully from the aircraft industry for use in wind turbine blades. The soft metal suffered fatigue failures in wind turbine applications and is no longer used in commercial wind turbine blades.

Ambient noise: Background sound level.

American farm windmill: Also called "classic" or "Chicago" windmill. Multiblade water-pumping windmill common across the breadth of the North American continent. The first wind turbine design indigenous to the New World. Multiblade wind pump commonly seen throughout North America, Australia, Argentina, and South Africa. A technology essential for European settlement on the American Great Plains, where "the cows chop the wood, and the wind pumps the water." Known as a classic wind pump among those who chafe at its American origins. Still used extensively around the world.

Ampere (amp): Unit of electric current equivalent to a flow of 1 coulomb per second, or to the steady current produced by 1 volt applied across a resistance of 1 ohm.

Amp-hour: Measure of energy storage potential in batteries. One amp-hour will deliver 1 amp of current for one hour at the battery's voltage. Amp-hours times battery voltage gives watt-hours of stored chemical energy.

Anemometer: An instrument for measuring and displaying wind speed or velocity. The term is often limited to the wind sensor, but correctly encompasses both the sensor and an indicating device.

Angle of attack: Angle between the chord line of an airfoil and the relative or apparent wind.

Annual energy output (AEO): See *AEO.*

Apparent wind: Relative wind. The wind seen by a wind turbine blade as it moves through the air. The resultant of the wind vectors acting on a wind turbine blade.

Armature: The current-carrying part of a generator where current is induced. The armature may be moving or stationary. In most wind turbine applications the armature is stationary and current is drawn off the stator.

Array effect: Sum of the interference of one wind turbine on those around it in a multiple-turbine array.

Array, wind turbine: Orderly grouping or arrangement of multiple wind turbines in relative proximity. Wind turbines are found in rectangular or geometric arrays on flat terrain and in linear arrays along dikes or breakwaters. Wind turbines can also be ordered vertically on towers of different heights in a wind wall.

Articulating blade: Airfoil on a straight-blade vertical-axis wind turbine that maintains its angle of attack regardless of the position on its path about the rotor axis. Any attachment of blades to their hub or central shaft through a hinge allowing out-of-plane or in-plane motions. In a VAWT, an airfoil that changes its angle of attack as the rotor revolves. Blades on ϕ-configuration Darrieus rotors, in contrast, are fixed in pitch. *Articulation* implies the coordinated movement of a jointed assembly. For example, the blades on a giromill or cycloturbine are mechanically articulated to maximize lift regardless of the position the blade, whether it is behind the tower, coming into the wind, or going downwind. This feature provides starting torque that is lacking in the conventional or ϕ-configuration Darrieus rotor.

Aspect ratio: In a vertical-axis wind turbine, the ratio of the rotor's height to its diameter. In a horizontal-axis wind turbine, the ratio of the square of rotor radius divided by the projected area of one blade.

Asynchronous generator: Generator where AC frequency is not exactly proportional to the speed of the generator rotor. The induction generators used in most wind turbines are asynchronous generators because they maintain a constant frequency as rotor speed varies 1 to –2 percent. Constant-frequency AC is produced by using an interconnection with a host utility to provide magnetization of the induction generator's field.

Atlantic Wind Test Site: Canadian center for testing wind turbines on Prince Edward Island.

Attenuation, noise: To reduce the intensity of acoustic energy by use of sound-deadening materials, baffling, or increase in distance between source and receptor.

Audible: Sounds that can be heard by most people with normal hearing.

Audubon: Short for National Audubon Society. North

American environmental group that emphasizes appreciation and protection of birds.

Augmentor: Device that increases the air flow through a wind turbine rotor. Called concentrators when used on the upwind side of a wind turbine and diffusers when used on the downwind side.

Autonomous system: See *Stand-alone power system.*

Availability: The quotient, expressed as a percent, of the total number of hours that a wind turbine is available for operation divided by the total number of hours in the period.

Average capacity: The power in kW of a wind turbine derived by multiplying the turbine's rated power times its capacity factor during a specific period.

Average wind speed: The mean wind speed during a specific period, often one year. One of the most common measures used to evaluate a wind resource. However, the average wind speed obscures the distribution of winds at various speeds during the period. The average power density in W/m^2 is a more accurate measure of the power in the wind. See *Mean wind speed.*

Avian mortality: Biological jargon for birds killed by wind turbines or their ancillary structures.

Avoided cost: Under federal law in the United States, the incremental costs to a utility of energy and capacity, which, but for the purchase from qualifying facilities, such utility would generate for itself or purchase from another source. The sum of fixed costs (such as power plant, transmission lines) and variable costs (such as fuel) avoided by a utility when a decentralized power generator contributes energy to the utility.

AWEA: American Wind Energy Association. Wind industry trade association in the United States.

A-weighted scale: Weighting network for sound level measurements that selectively discriminates against low and high frequencies to roughly approximate the response of human hearing.

AWT: See *Advanced wind turbine.*

Axial thrust: Force on a wind turbine rotor parallel to the windstream.

Azimuth: The angle about a vertical axis between a fixed heading and the direction of the object. Often applied to the orientation of a horizontal-axis wind turbine. See also *Yaw.*

Back-gearing: Term used to describe a transmission system developed for American farm windmills using gears to reduce the torque needed to lift the pump rod. In contrast to direct stroke, where the rotor lifts the pump rod for every revolution, back-geared windmills lift the pump rod once for every two or three revolutions.

Baffles, noise: Deflectors used to increase noise path and noise absorption inside a wind turbine nacelle as a means of attenuating acoustic energy.

Balance-of-station cost: The cost of components in a wind power plant other than the wind turbine, tower, and installation. This includes, but is not limited to, conductors, transformers, substations, and interconnection.

Baseload plant: Power plant that provides a constant portion of the demand for electricity throughout the day.

Battelle Pacific Northwest Laboratory: Private contractor to the U.S. Department of Energy, located in Richland, Washington, specializing in wind resource assessment of the United States. Battelle's wind resource functions have been assumed by the National Renewable Energy Laboratory's NWTC.

Batter: Slope or inclination caused by the taper of a wind turbine tower.

Battery-charging wind turbines: Mostly small wind turbines that generate DC directly or rectified DC for use in stand-alone power systems that include storage batteries.

Battery storage: The storage of electricity in the form of chemical energy. Batteries are often used in stand-alone wind power systems for providing electricity during periods of little or no wind.

Beard, of windmill: Ornate plaque beneath main shaft on the upwind side of many European smock mills displaying the name of the windmill or a proverb.

Bearingless rotor: Rotor in which the variable-pitch blades are attached to the hub without bearings. A variable-pitch rotor that depends on torsional flexure of the blade spar to change blade pitch.

Beaufort number: A arbitrary scale of wind strength or force created by British naval officer Sir Francis Beaufort (1774–1857). The scale uses common examples of the wind at work, such as the fluttering of flags or the movement of trees, to describe wind strength on a scale from 0 for calm to 12 for hurricane-force winds (74 mph). Also called Beaufort force.

Bed plate: Frame. A base plate on which generator, gearing, and other components of a wind turbine are fastened. Frame of a wind turbine's nacelle that supports the rotor and other components. Also called a strongback.

Bending stress: Stress created when a load causes a cantilevered beam to flex or bend. Wind thrust on a rotor blade causes bending stresses to occur at the juncture between the blade's root shaft and the hub.

Betz, Albert: German aerodynamicist of Göttingen, who found that an ideal wind machine could extract 16/27 (0.593) of the total power in the wind. Consequently the Betz limit is the maximum efficiency theoretically obtainable from a wind turbine.

Bicycle wheel turbine: Developed in the United States during the 1970s, a multiblade wind turbine that used many slender airfoils fixed between a metal ring (wheel) and a hub resembling a bicycle wheel.

Bin: Wind speed interval used for grouping wind speed data in the method-of-bins technique of wind measurement. The bins can be any width but are often 1 mph or 0.5 m/s. The 0–1 mph bin, for example, contains the occurrence of wind speeds from 0 to less than 1 mph.

Bin width: The size of a wind speed interval used in the method of bins (for instance, a bin having a span from 10 mph to 11 mph has a width of 1 mph).

Blade: The primary aerodynamic surface driving a wind turbine rotor. In a horizontal-axis wind turbine, blades are airfoils somewhat like the airfoils on aircraft propellers. In vertical-axis machines, blades may be vanes or buckets (as in Savonius rotors) or airfoil sections (as in Darrieus rotors). *Wing* in Danish usage (from *vinge*).

Blade-activated governor: Blade pitch-control mechanism used on some 1930s-era windchargers where centrifugal force acting on the blades themselves—rather than on separate weights, as in a traditional Watt governor—changes the pitch of the blades.

Blade area: Product of blade area per blade times the number of blades in the rotor.

Blade materials: Blade materials include the surface skin or layer, the core or body of the blade, and the spar or structural element.

Blade number: The number of blades in a wind turbine rotor.

Only one blade is needed. Most electricity-generating wind turbines use two or three blades. Traditional European windmills used four blades. Most mechanical wind pumps use multiblade rotors because of the high torque required to lift water.

Blade pitch: Blade angle. Angle between the chord line and the direction of motion.

Blade planform: Blade shape, including taper and twist. Plan view of a rotor blade as if looking down on a blade lying flat on the ground. Most blades on conventional wind turbines taper from hub to tip. Blades on vertical-axis wind turbines and on some conventional machines, especially those with extruded or pultruded blades, have a constant planform.

Blade root: Portion of a blade nearest the hub, or portion of a wing nearest the fuselage. Root shaft is a shaft connecting the blade to the hub.

Blade skin: Outer covering or surface of a blade.

Blade station: Airfoil section of a blade at a specified location along its quarter-chord line.

Blade tip: The portion of the blade farthest from the hub.

Blade twist: The difference between the blade angle at the root of a blade and that at the tip. Because the speed of the wind flowing over the blades varies continuously from root to tip (the tips travel much faster than the root), there is continuous variation in optimal blade angle. The blades on Denmark's Gedser mill were twisted from 13 degrees at the hub to 0 degrees at the tip.

BLM: Bureau of Land Management (U.S.). Federal agency administering public lands in the western United States. The BLM receives royalties from wind projects in the San Gorgonio and Tehachapi Passes.

BMFT: Bundesministerium Für Forschung und Technologie. German ministry for technology development. Original sponsor of the "250 MW" subsidy program now administered by Bundesministerium für Bildung, Wissenschaft, Forschung und Technologie (BMBF).

Bonneville Power Administration: Public power marketing agency serving the Pacific Northwest of the United States.

Booster mill: Water-pumping farm windmill added in line with another to move water higher or farther than can be done by a single farm windmill alone. Similar to Dutch windmill "gang" for successively lifting water out of deep polders.

Bora: Gravity squall caused by arctic air descending from the mountains of the former Yugoslavia into the Adriatic Sea.

Bottom-up technology development: Description applied by Danish scholar Peter Karnøe to a style of technology development characterized by myriad decisions and actions resulting in a gradual accretion of technology, as in Danish wind development. Contrasts with what Karnøe calls "top-down" development, characterized by central decision making and the search for technological "breakthroughs," as in the U.S. Department of Energy's wind program.

Boundary layer: The region between a fluid (air) that is moving relative to a surface where the fluid velocity is neither that of the fluid nor the surface, but between the two.

BPA: See *Bonneville Power Administration.*

Brake switch: Device used to safely short-circuit the conductors of a permanent-magnet alternators on small wind turbines. The load thus created can often stall the airfoils and slow the rotor to a safe speed. Best used when the rotor has been furled out of the wind. Not always a reliable means to stop the rotor in strong winds.

Breakwater: Civil engineering structure for sheltering a harbor. Many breakwaters are suitable for supporting wind turbines. The first "offshore" wind plant was built atop the harbor breakwater at Zeebrugge, Belgium.

British thermal unit (Btu): A measurement of heat content. The amount of energy needed to raise the temperature of 1 pound of water 1° Fahrenheit (F).

Broadband noise: Noise across the entire frequency spectrum.

Brushing: Deformation of the branches of trees and shrubs in areas of high winds where the "branches are bent to leeward like the hair in a pelt which has been brushed one way." Brushing is the most sensitive vegetational indicator of wind speeds and direction. Flagging and throwing indicate progressively greater wind speeds than those that cause brushing. (Putnam)

Brush windmill: In 1888 the successful manufacturer of arc lighting systems, Charles Brush, built a 56-foot (17 m) diameter multiblade wind turbine for charging batteries at his Cleveland, Ohio, estate. Brush's 12 kW wind-powered generating system rivaled that of banking magnate J. P. Morgan in his New York mansion, which operated on fossil fuel. The Brush windmill was the first American adaptation of wind energy to the generation of electricity and occurred at about the same time as Poul la Cour was beginning his experiments in Denmark.

Buffalo Ridge: Low-lying topographic feature trending southeastward from Watertown, South Dakota, across southwestern Minnesota to Spirit Lake, Iowa. There were sufficient wind resources along the Minnesota portion of Buffalo Ridge to supply 60 percent of the state's 1990 electricity consumption.

Bull pin: Forged rigger's tool with a tapered shaft on one end and an open-end wrench on the other. The tapered end is used to align bolt holes in metal tower parts.

Bull rope: Heavy rope used for temporary tower guys or raising and lowering heavy equipment from a tower.

Bundesverband Windenergie: German federal association of wind turbine owners. Largest association of its kind in the world. Sometimes translated as "German wind energy association," though it should not be confused with wind energy trade associations in Anglophone countries—the latter often represent only wind turbine manufacturers and commercial wind farm developers.

BuRec: Bureau of Reclamation (U.S.). Federal land management agency in the western United States responsible for the construction of major reservoirs in the Rocky Mountains. These reservoirs have often been eyed by federal planners for storage of wind-generated electricity. Constructors of Hamilton Standard's ill-fated WTS4, a 4 MW behemoth that has stood derelict near Medicine Bow, Montana, for many years.

Busbar costs: The cost of electricity at the generating plant. Busbar costs exclude the costs of transmission and distribution.

Buyback rate: The rate per kilowatt-hour a utility is willing to pay for excess energy fed into its lines by a small (relative to the grid) generator. In North America the buyback rate is usually lower than the retail rate charged by the utility for the power it sells. In Germany the buyback rate or "feed-in" tariff is set by the "electricity feed law."

BWE: See *Bundesverband Windenergie.*

BWEA: British Wind Energy Association. Wind industry trade association in Great Britain.

Calm: Period of no wind.

Camber: Measure of the curve in an airfoil section as the ratio of the distance from the peak of the curve to the chord line and the chord length.

Campbell plot: Graphic representation of a wind turbine's natural frequencies as a function of rotor speed.

Cantilever: A beam, pipe, or other structure supported or anchored at only one end. An unguyed, self-standing pipe or tubular tower is cantilevered. All modern horizontal-axis wind turbines use cantilevered blades. The blades are attached to the rotor at only one point, the blade root.

Cantilevered blade: Blade fixed at only one end. Self-supporting.

Cantilevered towers: Truss or tubular towers not dependent on guy cables for remaining upright.

CanWEA: Canadian Wind Energy Association. Canadian wind industry trade group.

Capacitor: Device that stores an electrical charge. Capacitors or condensers are used to boost the power factor of induction generators. Capacitors for wind turbine applications are often sized to compensate for the generators' volt-ampere-reactance at no-load.

Capacity credit: The amount of power that an intermittent resource such as a wind machine can be depended on to provide. Contrary to popular belief, a statistically determinable portion of a wind generator's capacity can be relied on to provide power when needed. This portion of a wind turbine's capacity can offset an equal amount of conventional generating capacity. Wind turbines therefore not only displace fuel but also displace some capacity. For most temperate wind regimes, the capacity credit is similar to the capacity factor.

Capacity factor: Measure of productivity. The quotient of the actual energy generated to that possible if the generator had operated at its rated capacity (power) over the time interval of interest, most often that of one year (8,760 hours). The capacity factor of wind turbines is dependent on reliability, performance, wind regime, and the rated power of the wind turbine.

Capacity, generating: Rated or peak power of a wind turbine or the sum of the rated or peak power of all wind turbines in an array in watts.

Capital-intensive: Technologies in which a large portion of total costs of production are associated with the cost of the equipment rather than with the costs of operation. Also used to differentiate from technologies that are labor-intensive or fuel-intensive. Like nuclear power, wind energy is capital-intensive.

Capture area: The area projected or swept by a wind turbine rotor. Also known as reference area or swept area.

Carbon dioxide: Product of fossil-fuel combustion. Non-regulated pollutant. Each kilowatt-hour of wind generation offsets the emission of approximately 1 pound (0.5 kg) of carbon dioxide from a natural gas-fired power plant, or 2 pounds (1 kg) of carbon dioxide from a coal-fired power plant.

Carbon-fiber reinforcement: Use of carbon fibers instead of glass fibers or wood to provide strength in composite materials.

Cascade effect: See *Slot effect.*

Case-hardened gears: Surface hardening of gear sets by the shallow infusion of carbon at high temperatures and subsequent quenching. Used to extend gear life.

CAT: Centre for Alternative Technology (Wales). Each year 60,000 to 90,000 tourists visit the exhibits of alternative energy at CAT.

Cathodic protection, of pipelines: During the 1930s small windchargers were used in North America to provide a DC charge to the surface of metal pipelines. This charge offset galvanic corrosion between the metal pipe and certain soils.

CEC: California Energy Commission, Sacramento. The world's second largest energy agency, after the U.S. Department of Energy. Responsible for siting power plants.

Centerline height, of rotor: Distance between the ground surface and the horizontal axis of a conventional wind turbine rotor.

Central-station power plant: As opposed to dispersed sources of electricity generation, one facility produces large quantities of electricity for an entire region.

Centrally directed research: Mechanism by which top-down technology development proceeds. Danish academic Peter Karnøe attributes the failure of wind energy research and development in the United States and Germany during the early 1980s to its centrally directed nature.

Centre for Alternative Technology: See *CAT.*

Cervantes: Author of *Don Quixote,* a novel about a delusional noble who tilts (jousts) at windmills on Spain's plains of La Mancha.

CFRE: Carbon-fiber-reinforced epoxy.

CFRP: Carbon-fiber-reinforced polyester.

Charge controller: See *Regulator, voltage.*

Chicago style: Multiblade mechanical wind pump used throughout the world. More than a million are still in use. Also known as the classic American water-pumping windmill, though the design has been widely copied by manufacturers in several countries. Called "Chicago" because of the dozens of factories that fabricated the familiar design in the late 19th century on the southern borders of Lake Michigan near Chicago. See *American farm windmill.*

Chinook: Warm, dry, powerful wind descending the eastern slopes of North America's Rocky Mountains.

Chord: Distance from the leading to trailing edge of an airfoil. A specific description of blade width in blades using airfoils.

Chord line: Straight line connecting the leading and trailing edges of an airfoil.

Clap-sail windmill: Translation of the Danish *klapsejlsmølle.* The type of windmill used by Poul la Cour in his experiments using the wind to generate electricity. *Clap-sail* is the Germanic description of the rotor control developed in England by Andrew Meikle in 1772. William Cubitt patented an improvement the Meikle's design in 1807. In England clap-sails are known as "patent sails" as a result. Clap-sails work like Venetian blinds or jalousie shutters.

Cloth sails: Traditional European or Dutch windmills used blades made from sail cloth draped over a wooden framework. Mediterranean windmills, including those of Spain, Portugal, and Greece, use sail cloth stretched from one spar to the next, much like a jib-rigged sail. Princeton University experimented with a two-blade, jib-rigged sail rotor during the 1970s.

Clusters: Of wind turbines. A group of turbines in the same vicinity.

Clutter, visual: Jumble or juxtaposition of multiple wind turbines of different types and sizes within a viewshed.

Co-ops: See *Cooperatives.*

COE: Cost of energy. Measure useful for comparing energy technologies of like characteristics. Commonly misused to compare the costs of fossil fuels to renewable resources because of the COE's inability to reflect future fuel price and environmental risk.

Coefficient of performance: C_p. The quotient of the power extracted by a wind turbine to the power in the wind. A measure of the rotor's aerodynamic efficiency. Rotors using modern airfoils operating at their design tip speed are capable of 0.35 to 0.45 C_p.

Cogging: Variation in speed of a generator due to variations in magnetic flux as rotor poles pass stator poles. Cogging in permanent-magnet alternators can hinder the start-up of small wind turbines at low wind speeds.

Collective pitch: Changing the pitch of all blades in a rotor simultaneously by the same amount in the same direction through a common linkage.

Columbia River Gorge: Deeply incised east–west valley of the Columbia River between the states of Washington and Oregon that funnels prevailing winds through the Cascade Range. During the 1970s the U.S. DoE installed three experimental Boeing Mod-2 wind turbines on a plateau overlooking the gorge near Goldendale, Washington. Though the turbines were removed in the 1980s, the area is again the scene of wind development.

Commutator: Segmented copper cylinder in a DC generator or motor that, in combination with brushes, provides an electrical connection between the coils of the rotating armature and stationary terminals.

Compliant drive train: Torsionally flexible drive train that absorbs peak torque loads by angular displacement of the gearbox or couplings.

Composite materials: Made up of more than one material. A composite blade may be made from a metal leading edge and blade spar with a fiberglass trailing edge. See *Fiberglass* and *Wood composites.*

Concentrator: Device that concentrates the wind stream striking a wind turbine rotor. See *Augmentor.* Concentrators have proven too cumbersome and costly for wind turbine applications. The same benefit provided by a concentrator can be obtained less expensively by extending the rotor swept area of a conventional wind turbine.

Conductors: Metal cables used to transmit electricity.

Conduit: Metal or plastic pipe for protecting conductors from physical damage.

Configuration, wind turbine: Type and number of blades, their location with respect to the tower, and the axis of rotation of a wind turbine rotor.

Coning: Sweeping the longitudinal axis of the blades on a rotor of a conventional wind turbine downwind. Coning permits both the shedding of some bending loads and, in downwind turbines, yawing without the need for a tail vane or mechanical yaw system. Coning is most accentuated on downwind turbines, but is also apparent on some upwind turbines. Coning is shallower on upwind turbines than on downwind turbines because of the risk that the blades will strike the tower. On upwind rotors of large wind turbines, the rotor is coned away from the tower.

Coning angle: The angle between a vertical plane and the blade axis on a conventional wind turbine. This description is from the appearance of the rotor on most downwind machines, which take the shape of a shallow cone swept by the rotor, with its apex at the hub.

Conspicuousness, visual: A measure of how prominent or easily observed a wind turbine or other object is on the landscape. Conspicuousness is not necessarily a measure of visual impact. An object that is visible on the landscape may not necessarily elicit objection.

Constant cost of energy: Cost excluding the effects of inflation. Also known as real cost of energy. Cost in constant currency is less than cost in nominal currency.

Constant speed operation: Asynchronous or induction generators operate at a relatively constant speed in comparison to true variable speed generators. Asynchronous generators operate within a few percentage points of their nominal speed, which is governed by the frequency of the utility line.

Constant-velocity test: Wind turbine test in which a steady airflow is maintained either in a wind tunnel or by moving the wind turbine relative to the ground. During the early 1980s DoE's Rocky Flats test center measured the performance of small wind turbines by attaching them to a railroad car and driving them along a track with a diesel locomotive. An obsolete form of testing.

Consumables: Supplies that are consumed during the normal operation and maintenance of a wind turbine (hydraulic fluid, lubricants, filters, and so forth).

Contactors: Electrical relays used for making "contact" between a wind generator and the utility grid or network.

Contrarotating: Wind turbine using dual rotors on the same shaft, one rotating clockwise, the other counterclockwise. The early German wind turbine NOAH used contrarotating rotors.

Control system: Wind turbine system that monitors the condition of the turbine and its environment. Depending on these conditions, the control system adjusts the operation of the wind turbine to protect it from damage or to optimize its performance.

Converter: See *Inverter.*

Cooperatives: Form of mutual ownership. Used frequently in Denmark, Germany, and the Netherlands for the purchase and operation of wind turbines.

Cornwall: County encompassing the windswept southwestern peninsula of England. Britain's first wind power plant was built in Cornwall near Camelford.

Cost of energy: The levelized cost of generating electricity during the life of a wind turbine or wind plant, including the installed cost, the cost of equity and debt, the cost of operations and maintenance, and the cost of fuel.

Cost of service: The cost to a utility for providing the generating capacity, distribution lines, and transformers. Usually expressed as a fixed charge in a utility bill regardless of the amount of power consumed. The cost of service must be recovered even if the customer uses no electricity at all, because the utility maintains the facilities whether used or not.

Counterweight: On one-blade turbines, the mass used to counterbalance both the mass and the thrust on the single blade.

Crane pads: Graded areas near a wind turbine foundation used for parking a crane in hilly terrain. Frequently access roads are used in place of separate crane pads.

Cretan sail: Identified with the island of Crete—though common throughout the Mediterranean—these jib sails use wooden poles for spars from which sail cloth is strung. Pictured in the Lasitti Valley—"the valley of 10,000 windmills."

Cross-arm of H-rotor: The structural member of an H-blade vertical-axis wind turbine that supports the blades.

Crosswind: Across the direction of the wind stream.

Cube factor: The quotient of the cube root of the mean cube of the wind speeds for a speed distribution divided by the mean wind speed. Equal to the cube root of the energy pattern factor. See also *Energy pattern factor.* (E. W. Golding)

Cube law: The power available in the wind increases with the cube of wind speed. Doubling the wind speed increases the power in the wind eight times:
2^3 or 2 x 2 x 2 x = 8

Cuffs: Highly tapered blade section near the hub. Results from the Glauert formula for the ideal rotor blade.

Cup anemometer: A vertical-axis drag device for measuring wind speed that uses cups mounted on radial arms. The most widely used form of wind sensor.

Current: Flow of electricity (I) measured in amperes.

Customer class: In the United States, the division of utility customers by rate schedule—for instance, R-1 Residential.

Cut-in wind speed: Wind speed at which a wind turbine begins to produce power. Not synonymous with *start-up windspeed,* except for wind turbines that motor the rotor to operating speed, or wind turbines that drive permanent-magnet alternators. See *Start-up windspeed.* Most analysts agree that lowering the cut-in speed from the common 4.5 m/s (7 to 9 mph) contributes little to total energy generation.

Cut-out wind speed: Wind speed at which the wind turbine ceases generating electricity. Most medium-size wind turbines have a cut-out speed of 30 m/s (65 mph). Some medium-size wind turbines have no cut-out speed. Many small wind turbines have no cut-out speed, but do furl or otherwise control the rotor instead. See *Furling speed.*

Cyclic pitch: Periodic change in blade pitch per rotor revolution of articulating, straight blades on vertical-axis wind turbines.

Cycloturbine: See *Giromill.*

Darrieus rotor: Sleek vertical-axis wind turbine developed by French inventor G. J. M. Darrieus in 1929. The (phi or ϕ) ϕ-configuration Darrieus rotor is often referred to as an eggbeater in North America because of its hoop shape. At one time Darrieus wind turbines represented about 5 percent of the installed wind capacity in California.

Data loggers: Recording instruments that store data electronically. Loggers are used most often to store wind data at remote sites.

dB: Decibel. Logarithmic scale used in electronics and acoustics to describe the difference in power or intensity between two levels. Defined as 10 times the common logarithm of the quotient of the two levels—that is, 10 log (P_1/P_2). In acoustics, the faintest audible sound, the threshold of hearing, is defined as 0 dB. The threshold of pain is 120 dB.

DC: See *Direct current.*

Decentralized systems: In wind systems, refers to establishment of autonomous units, such as households or neighborhoods, to provide electricity or heat.

Decibel. See *dB.*

Declared net capacity: British usage for the equivalent amount of conventional baseload capacity wind energy will offset. For example, if British nuclear plants operate at 70 percent capacity factor and wind turbines operate at 30 percent capacity factor, 1 MW DNC of wind plants is equal to 2.3 MW of installed wind capacity.

Decommissioning: Relative to wind energy, at a minimum refers to the removal of wind turbines and their towers. In some jurisdictions decommissioning also requires removal of all or a portion of the foundation and ancillary facilities.

DEFU: Danske Elvørker Forenings Udredningsafdeling. Research association of Danish utilities.

Delta connection: A three-phase connection where each phase is connected between two supply conductors. (Hugh Piggott)

Delta-3 blade hinge: Blade-to-rotor hub coupling that permits movement in and out of plane and about the pitch axis. On one- and two-blade teetered rotors, the d_3 hinge allows the blade to move out of plane as gusts strike the blade. As it does so the blade motion changes pitch slightly, reducing the blade's performance during the gust.

Demand charge: Charge by North American utilities based on the ratio of peak demand to total power consumed. This charge is part of rate schedules for commercial and industrial customers. The demand charge compensates the utility for maintaining sufficient spinning reserve to meet a customer's instantaneous demand.

Demand, for power: The amount of power required to satisfy the needs of a stated sector of the economy.

Density altitude: Aeronautical term for the combined change in air density with altitude and temperature.

Derelict wind turbines: Generic term for abandoned or nonoperating machines. Wind turbines that have been idle for an extended period regardless of wind speeds within the normal operating range. See also *Abandoned wind turbines.*

Design speed: Survival speed. The wind speed a wind turbine is designed to withstand without suffering catastrophic failure. Often 55 m/s (120 mph) or greater.

Design tip-speed ratio: Tip-speed ratio for constant speed-wind turbines at rated power.

DEWI: Deutches Windenergie Institut. German wind energy institute.

Diameter, rotor: Diameter of disk swept by a conventional horizontal-axis wind turbine. The single most important descriptor of conventional wind turbine size. Shorthand for the area swept by a conventional wind turbine. In ϕ-configuration Darrieus rotors, the width of the rotor at its equator. In H-rotors, the length of the rotor cross arm.

Differential billing: A method by which a utility can bill customers who provide a portion of their own electrical needs, such as with a wind generator. Two meters are used; one measures power consumed by the customer, the other measures excess power fed back into the utility's line, displacing the utility's own generation. In this way the utility can recover the cost of service and other charges by paying a lower rate for the power it buys from the customer than the rate it charges for power.

Diffuser: Downwind device that diffuses the wind stream through a rotor. See *Augmentor.*

Dike: Linear landscape feature common on the North German Plain that border polders. Wind turbines have been installed on or alongside dikes in the Netherlands and Germany.

Diode: An electronic check valve that allows alternating current to pass in only one direction. Diodes are used in automotive alternators to produce direct current for charging the battery. Diodes are also used for the same purpose in small wind turbines that use alternators.

Direct current (DC): Electric current that flows in one direction. See also *Alternating current.*

Direct-drive generator: Purpose-built generator that allows the low rotor speeds common in wind turbines to produce electricity without the use of a gearbox or transmission to increase rotor speed to that required to drive mass-produced generators.

Disco effect: Flash of bright light from the nacelle or blades of a wind turbine similar to that of a strobe light. As the rotor spins, the flash repeats with a rhythm akin to that of the flashing lights in a discotheque.

Disconnect switches: Manually operated, lockable switch—often located at the base of the wind turbine tower—that allows de-energizing the entire wind turbine circuit to isolate the wind turbine from the utility network and to facilitate safe maintenance of the wind turbine's electrical components.

Diseconomies-of-scale: Increasing costs with increasing size. For example, above a certain size large wind turbines cost more to service than medium-size wind turbines because of the specialized cranes necessary.

Disincentive: Something that deters an action. The initial cost of wind turbines is a disincentive to their use even though wind turbines have no fuel costs.

Disorder, visual: In reference to wind turbines, the confusing jumble of wind turbines of different types on towers of different types and differing heights in concentrated arrays.

Dispersed arrays: Arbitrarily, concentrations of wind turbines where the relative spacing across rows is greater than four diameters, and that between rows is greater than six diameters. In the dense arrays common in California, spacing of two diameters across the wind by six diameters downwind is not uncommon.

Distributed generation: Single or small clusters of wind turbines disseminated across the landscape in contrast to the concentration of wind turbines in large arrays or wind power plants.

Disturbance, noise: Annoyance caused by noise. A lesser impact than when noise interferes with some activity, such as sleep or communication.

Disturbance, visual: Annoyance caused by the intrusion of wind turbines on the landscape.

Diurnal variation of wind speed: Changes of wind speed between night and day.

Diversified load: A mix of different types of power-consuming devices. In residential use, various electrical appliances as opposed to space heating or water heating loads.

Diversity, of generating sources: A mix of technologies for the generation of electricity. As in the ecologist's credo "In diversity is stability," electrical networks using a mix of generating resources are more stable than those dependent on one plant or one fuel source.

DIY: British usage for "do it yourself." Often applied to plans for home-built windchargers.

DoE: Department of Energy (U.S.).

Downtime: Period when a wind turbine is not available for generation due to maintenance, repairs, or other causes.

Down tower: Usage by defunct wind company Kenetech Windpower (U.S. Windpower) for repairs requiring removal of the entire nacelle from the tower.

Downwind: Lee. The side away from the wind.

Downwind rotor configuration: A horizontal-axis wind turbine in which the rotor is oriented on the lee side of the tower. Downwind rotors employ prominent coning.

Drag: Resistance of a body moving through a fluid due to friction between the surface of the body and the fluid. A force that acts to retard the movement of a wind turbine blade through the air.

Drag device: One of two major classes of wind machines: lift devices and drag devices. Drag devices extract less energy from the wind than lift devices, and are much more material-intensive. The blades of drag devices move downwind at a speed slower than that of the wind. Horizontal-axis wind turbine blades move perpendicular to the wind and as such can never be true drag devices. See *Lift device.* (Hugh Piggott)

D-ring: Large metal rings in the shape of the letter *D* on work belts used for the attachment of lanyards.

Drive train: Portion of a wind turbine that transmits torque from the rotor to the generator.

Droop cable: Means for conducting electricity from the generator of a medium-size wind turbine to ground level through the use of extra-long conductors that are allowed to sag or hang down from the nacelle. The sag or droop in the cables permits the nacelle to yaw several revolutions before the conductors become twisted. This obviates the need for slip rings capable of transmitting high current.

Drop pipe: A galvanized pipe with a smooth interior in which the pump rod of a mechanical water-pumping windmill moves.

DSM: Demand side management.

DTI: Department of Trade and Industry (UK).

Dual-speed generator: Induction generator with dual windings. More specifically, an induction generator that excites the windings of six poles during low-power operation at which the synchronous speed is 1,000 rpm in Europe (1,200 in the Americas). For full-power operation, only four poles are energized, at which the synchronous speed is 1,500 rpm in Europe (1,800 rpm in the Americas).

Dual-speed operation: By using dual generators or dual-speed generators, constant-speed wind turbines can operate at two different speeds.

Dual-wound generator: See *Dual-speed generator.*

Ducted rotors: Wind turbine with a shroud to concentrate or diffuse the wind stream striking the rotor. See *Concentrator.*

Dump load: Electrical load lowest in priority in a series of loads supplied by a wind generator, a load used to absorb excess generation, a load used to maintain a desired voltage or frequency in an isolated network.

Dust, fugitive: Regulatory term in the United States for dust from road traffic and construction activities that crosses property boundaries. Fugitive dust can impair wind turbine operation by coating airfoil surfaces in arid regions, such as southern California, and can incur sanctions and fines against wind companies for use of unpaved roads.

Dwell time: The number of samples in a bin divided by the sampling rate.

Dynamic inducer: Airfoil transverse to the tip of a wind turbine blade used to induce airflow over the tip. Developed as an alternative to shrouded or ducted concentrators designed to improve energy capture. Like other forms of concentrators, it is usually less expensive and more reliable to simply extend the length of the rotor blade.

Dynamics: The branch of mechanics that examines forces and their effect on a moving object. In wind energy, the term refers to the interaction between forces on a rotor blade or other wind turbine components and their subsequent motion.

Dynamo: British usage for "DC generator."

Earthing: British for "grounding." See *Grounding.*

Easements, noise: Right-of-way registered with local government that grants permission to emit noise of a certain level, usually in excess of local ordinances, by the property owner.

Ebeltoft: Ferry terminal on the east coast of Denmark's Jutland

Peninsula where 16 turbines installed on the harbor breakwater became the world's second "offshore" wind plant. Revenues from the municipally owned turbines are used to pay for harbor improvements.

EC: European Community (now the European Union).

Eclipse windmill: Known as the railroad mill during the late 19th century on the American Great Plains.

ECN: Energieonderzoek Centrum Nederland. The Netherlands energy research foundation. Principal government center for wind energy research in the Netherlands.

Economies-of-scale: Reduction in cost derived from increases in size.

Eddies: Vortices or turbulence in a fluid acting contrary to the main current, often in the lee of an object, such as the vortices in lee of a rock in a mountain stream.

EdF: Electricité de France, French national electric utility.

Effective capacity: Dependable generating capacity calculated from the probability that generation would be available when needed.

Efficiency: Quotient of output divided by input.

Efficiency, conversion: Quotient of net output of a conversion device divided by the gross input required to produce the output.

Eggbeater: Colloquial term for f-configuration Darrieus rotor. See *Darrieus wind turbine.*

Eigen frequency: Natural frequency. The frequency at which an oscillating system, when excited with a constant force, will vibrate at constantly increasing amplitude. See *Resonance.* A critical design parameter for wind turbines rotors, drive trains, and towers. The Eigen frequency of wind turbines with stiff towers is greater than the Eigen frequency of the rotor. The Eigen frequency of soft towers is below that of the Eigen frequency of the rotor. To avoid destructive oscillations on a weakly dampened structure, a wind turbine rotor must avoid operating at the Eigen frequency of the tower for any extended period. (Klaus Kaiser, TU-Berlin)

Electrification, rural: Providing a regular supply of electricity to rural residents. Previously implied the extension of central-station power to rural areas by the construction of new power lines. Today the meaning includes the use of stand-alone or independent power systems.

Electrocution, as occupational hazard: Death caused by the passage of electricity through the body. An occupational hazard of working with wind-generated electricity.

Electrocution, of birds: A major cause of death among large birds of prey, especially raptors, when the bird comes into contact with a conductor and ground or two conductors of a high-voltage transmission line. Because birds of prey perch on tall objects, they frequently land on transmission poles or towers, where they come in contact with the energized conductors.

Electrolysis: The use of an electric current to split water into its constituent molecules of hydrogen and oxygen. Frequently mentioned as means for using wind-generated electricity to produce hydrogen as a transportation fuel.

Electrolyte: A chemical compound that conducts the flow of electrons.

Electromagnetic interference (EMI): The disruption of telecommunications by the magnetic field induced by the passage of electrons in a conductor.

Electromagnets: Magnets whose magnetism is provided by the passage of electrical current. Unlike permanent magnets, electromagnets can be controlled by controlling the current.

ELKRAFT: Electricity generation and transmission utility serving the eastern half of Denmark, including the island of Zealand.

ELSAM: Electricity generation and transmission utility serving Denmark's Jutland Peninsula.

Emissions, air pollutants: Relative to wind energy, the by-products of the mining, processing, transport, and consumption of fuels for the generation of electricity. The only emissions produced in the fuel cycle of wind-generated electricity are those from the processing of the raw materials used in the construction of the wind turbine and its ancillary structures and that of fugitive dust from traffic on service roads in arid regions. See *Dust, fugitive.*

Emissions, noise: The creation of audible and subaudible vibrations by a wind turbine and its components, especially that of the rotor and the drive train. See *Sound power level.*

Emissions offset: Relative to wind energy, the pollutants that would have otherwise been emitted by a fossil-fired power plant and the mining, processing, and transport of its fuel.

Endangered Species Act: U.S. law that prohibits "taking"—that is, killing—any species determined to be in danger of extinction. Often abbreviated as ESA.

Endesa: Empresa National de Electricidad, S.A. Spanish electric utility.

ENEL: Italian national electric utility.

Energy: The amount of work done over a given period of time. Wind energy is usually expressed in mechanical terms as horsepower-hours or electrically in kilowatt-hours (kWh).

Energy density: The ratio of energy per unit of the fuel's mass. The ratio of energy stored in a battery per unit of the battery's mass, often used to compare the abilities of batteries to store energy relative to the space they occupy.

Energy pattern factor: The quotient of the total energy available in the wind divided by the energy in the wind from the cube of the mean wind speed. (E. W. Golding)

Energy rose: A diagram that presents the direction and energy content of the wind from each point on the compass.

Enfield-Andreau: Experimental 24-meter (80 ft) diameter, 100 kW turbine designed by French inventor Andreau. Enfield Cables installed the prototype at St. Albans near London in 1957. The hollow rotor used centrifugal force to pull air through a turbine mounted vertically in the tower. The prototype was removed and reinstalled on Grand Vent in Algeria, then a French colony, by Electricité et Gaz d'Algérie.

Environmental costs: Monetized and nonmonetized costs due to the environmental impact of a technology. Environmental costs are often nonmonetized. For example, the costs associated with the climatic impact of global warming from carbon dioxide emissions of fossil fuels are not reflected in the price of fossil-fuel-derived energy. See also *Social costs* and *External costs.*

Environmental indicators: Plants or surface features that are associated with high wind speeds. Wind-flagged vegetation or wind-thrown trees are one environmental indicator of high winds; another is the shape, location, and movement of sand dunes.

Eole: Greek god of the wind. Also the sobriquet for the largest ϕ-configuration Darrieus turbine ever built.

EPA: Environmental Protection Agency (U.S.). Federal agency responsible for regulating air, water, and land pollution.

EPF: See *Energy pattern factor.*

EPRI: Electric Power Research Institute.

Equipment pod: Nacelle. Gondola. Enclosure at the top of the tower on a horizontal-axis wind turbine housing the generator and transmission.

Equity: The nondebt portion of financing. The holders of equity in an investment assume all investment risk for the potential of collecting all the profit.

Erosion: The process by which soil is removed and transported from one location to another through the abrasive action of wind or running water. Though erosion is a natural process, accelerated erosion from human activities causes water and air pollution, denudes land of topsoil, and alters the hydrology of stream courses, leading to flooding. Erosion due to improper wind development on steep terrain is a problem in California and Spain.

Erosion control: Techniques and structures used to minimize human-induced erosion. The principal erosion control practice for wind development is to minimize soil disturbance by minimizing roads and avoiding construction on steep slopes.

Escalation rate: A number that defines the annual increase in monetary value of a specified quantity.

ETSU: Energy Technology Support Unit at the United Kingdom's Harwell Laboratory. Once the principal government research center on wind energy in the U.K.

EU: European Union (formerly the European Community).

EWEA: European Wind Energy Association. Industry trade association for members of the European Union.

Exceedance levels: Noise. The percentage of the time that noise exceeds the given level. The level, L_{90}, indicates the noise level that will be exceeded 90 percent of the time.

Excitation: The current used to generate the field in an alternator or generator.

External costs: Nonmonetized costs attributed to a technology, such as the costs from extraction, processing, and combustion of a fossil fuel that are not reflected in its price. External costs include both social and environmental costs.

Externalities: See *External costs* and see also *Social costs* and *Environmental costs.*

Extrusion: Manufacturing process in which metal such as aluminum or soft steel is forced (pushed) through a die into a desired shape. Extrusion produces a product of constant planform and is most frequently used for the side rails of aluminum ladders and the frames of windows. Extruded aluminum has been used for f-configuration Darrieus wind turbine blades.

FAA: Federal Aviation Administration (U.S.).

Fabrication: Building an assembly with conventional manufacturing techniques, such as by stamping metal into component parts and fastening them together with welds, rivets, screws, or bolts.

Fail-safe braking system: When all powered systems are off, the brake is applied. Thus if an operating wind turbine loses power, the brake is automatically applied. Power to apply the brake may be provided by springs or by a hydraulic accumulator.

Fall-arresting systems: Combination of body harness, lanyard, and attachments that limit the length of fall should windsmiths lose their footing or handhold. One common fall-arresting system used in climbing towers includes a body harness with attached sleeve or clip that slides along a vertical steel cable or rail running the length of the tower. Fall-arresting systems are used only in emergencies and not for positioning.

Fall restraint: Combination of positioning belt, lanyards, and anchorages to restrain a windsmith from a fall. Should not be confused with a fall-arresting system. For proper fall protection, a fall restraint must also include a fall-arresting system.

Fan: Colloquial term for "rotor" on American farm windmills. See also *Windwheel.*

Fan tail: Small, multiblade rotor on late versions of the European windmill used to orient the main rotor into the wind. The fan tail's plane of rotation is at right angles to that of the main rotor so that off-axis winds will strike the fan tail and reorient the main rotor squarely into the wind. Early modern European wind turbines in Denmark and Germany, such as those built by Riisager, Windmatic, and MAN, used fan tails for orientation.

Fatigue cracks: Telltale signs that the material has become work-hardened and brittle where the amount of stress or the number of stress cycles exceeds the material's limits.

Fatigue life: The number of cycles of stress reversals that a material can withstand before it begins to show signs of brittleness and fatigue failure.

Fatigue, metal: The weakening and eventual failure of metal resulting from cyclic stress. All materials suffer from fatigue, though the failure of copper and aluminum is best known. Nearly everyone is familiar with the potential for the fatigue failure of the aluminum pull-tab on a can of soft drink. The pull-tab will eventually fail at its joint if worked successively back and forth. The soft aluminum at the joint eventually becomes work-hardened and brittle, then snaps off.

Fauna: Wildlife.

FDV: Foreningen af Danske Vindmøllefabrikanter. Association of Danish wind turbine manufacturers. Now known as Vindmølleindustrien.

Feathering: Rotating a blade about its longitudinal axis until its chord line is parallel to the wind. One method of controlling lift and torque on a wind turbine rotor. By decreasing the area of blade surface exposed to the wind, feathering reduces thrust and stress on the rotor and tower when the rotor is at rest.

FEE: See *France Énergie Éolienne,* French wind energy association.

Fehmarn: Island in the Baltic Sea between Germany and the Danish island of Lolland. The island has one of the largest concentrations of wind turbines in Germany; in 1994 this included more than fifty 500 kW wind turbines.

FERC: Federal Energy Regulatory Commission (U.S.). Federal agency responsible for regulating the interstate sales of electricity.

Fetch: The distance over which the wind blows with no obstructions.

Fiberglass: American usage for "glass-reinforced polyester composites, a composite material made of a polyester matrix reinforced with glass fibers."

Field magnet: An elctromagnet or a permanent magnet used to produce a magnetic field in an electric generator.

Field test: A performance test carried out under naturally occurring atmospheric conditions as opposed to tests conducted in a laboratory or wind tunnel.

Finite element: A system of analysis in structural and fluid mechanics where a mechanical system such as a wind turbine blade is divided into discrete elements connected to each other at discrete nodes or points.

Firm power: A term occasionally used to describe the contribution that a generator might make to the reliability of the overall power system.

Fixed charge rate: Multiplier used in engineering economics for determining the cost of energy. The multiplier includes the effects of inflation, the lifetime of the investment, and the cost of financing equity and debt.

Fixed coning: Describes a rotor with a rigid hub that does not permit out-of-plane or "flapping" deflection.

Fixed-pitch rotor: In contrast to a variable-pitch wind turbine, a rotor where the blades are fixed in pitch during operation. The blade pitch may be adjustable, but is not varied during operation.

Fixed-price contract: Payment for wind-generated electricity at a predetermined price for a fixed period of time. Germany's electricity feed law provides fixed prices for the first five years of the contract. See also *Tariff.*

Fjelds: Bald windswept plateaus or hilltops found above the Arctic Circle in Sweden, Norway, and Finland.

Flagging: Wind-deformed branches of trees and shrubs found most often on the lee side of coniferous trees in areas of high winds. Branches grow downwind on the leeward side of the trunk while the trunk remains bare on the windward side "like a flagpole carrying a banner flapping in a breeze." Flagging indicates higher winds than brushing. (Putnam)

Flapping: Blade motion in and out of the plane of rotation on a horizontal-axis rotor.

Flapping blade hinge: Connection between a rotor blade and hub on a conventional wind turbine that allows the blade to move in and out of the plane of rotation in response to gusts.

Flaps: Control surface usually located on the trailing edge of a wing designed to increase drag on the airfoil. Flaps are used in aircraft to reduce air speed during landing. Flaps can be used on the trailing edge of wind turbine blades to limit power in high winds by increasing drag. See *Air brake* and *Aileron.*

Flashing: Momentary but reoccurring bright reflections from blades or nacelles. These bright flashes of light draw attention to the wind turbine and have generated complaints of visual pollution from some residents near Palm Springs, California. See *Disco effect.* See also *Sparklies.*

Flettner rotor: The German inventor Anton Flettner crossed the Atlantic in 1925 on the *Baden Baden* using two upright spinning cylinders as the sole means of propulsion. These vertical-axis rotors depended on the Magnus effect for their forward thrust. Flettner also built a wind turbine that used four tapered rotating cylinders as blades. The rotor on this wind generator spanned 20 meters (60 ft) and stood atop a 33-meter (100 ft) tower. It was capable of generating 30 kilowatts in a 23 mph wind.

Flexure, blade: The downwind curve or bend in the blades of horizontal-axis wind turbines caused by the wind's thrust.

Flicker, shadow: The momentary but continuing passage of shadows across an occupied space caused by the blades of a wind turbine as the blades pass between the sun and the occupied space. Shadow flicker is more noticeable in the early-morning or late-evening hours, and in northern latitudes where the sun is low on the horizon and shadows cast are long. Some observers find shadow flicker annoying.

Floating price: Pool price.

Flora: Plants.

Flutter, blade: When wind turbine blades oscillate, beat, or flap rapidly with low amplitude through coupling of flapping and torsional deflections.

Flux: The lines of force due to a magnetic field.

Flyball governor: Watt governor used on 1930s-era windchargers in the United States to vary blade pitch with wind speed. Most commonly associated with windchargers build by the Jacobs Wind Electric Company. Often applied to any mechanical governor on the rotor of small wind turbines that uses centrifugal force acting on weights to alter blade pitch.

Flyway, migratory bird: Broad corridor through which migratory birds pass seasonally on their way to and from their breeding grounds.

Föhn: Warm dry wind rushing down from the Alps across the lowlands of Switzerland. Similar to the chinooks of western North America.

Forced outage: Termination of wind turbine operation due to an unplanned event such as equipment failure and unplanned maintenance. Forced outages are attributed to either the wind turbine or the utility system (for instance, network problems not associated with the wind turbine).

Foreground views: Close up. That part of a visual analysis of wind turbines where the wind turbines are nearest to and in front of the observer.

France Énergie Éolienne: French wind industry trade association. More closely resembles wind energy trade associations in Anglophone countries rather than the owners' association in Germany.

Freestanding tower: Tower not dependent on guy cables or stays to remain upright. See *Cantilevered towers.*

Freewheeling rotors: Wind turbine systems that use self-starting rotors and that permit the rotor to continue spinning in light winds whether or not the wind turbine is performing useful work.

Frequency, of alternating current: The number of cycles or hertz that alternating current reverses direction per second. The frequency in North America is 60 hertz; in Europe, 50 hertz.

Friends of the Earth: International environmental group with affiliates in most developed countries.

Frontal area: Area of a structure intercepting the wind. See *Swept area.*

Fuel-saver mode: Operation of a wind generator in which the power produced is used to displace conventional fuels at a utility power plant.

Fuel savings: Fuel not consumed or consumption offset by the production of wind-generated electricity. The use of wind energy provides both fuel savings and generating capacity avoided. Until the 1990s most of the value of attributed to wind energy by electric utilities was through fuel savings.

Full-span pitch control: In contrast to partial-span pitch control, motion of a wind turbine blade about its longitudinal axis along its entire length. In the United States, the Mod-0A controlled torque and power in high winds by varying the pitch of the entire blade. The Mod-2 and Mod-5B, however, controlled power by varying the pitch of only the outboard sections of the blades.

Furling: To roll up or take down a flag or sail. Form of overspeed protection used on traditional European windmills with both lattice or jib sails where the miller rolled up the sail. Also used in reference to a small wind turbine with a tail vane where either the tail swings toward the rotor or the rotor swings toward the tail so both are parallel. Furling prepares the wind machine for high winds or a period of inoperation.

Furling speed: The wind speed at which a wind turbine that uses furling to control rotor speed begins to furl. Many small wind turbines begin furling at 15 m/s or 30 mph.

Gearbox: Transmission. Mechanism that transfers the mechanical power of a wind turbine rotor to that of the load. Most commonly used to increase the speed of the main shaft to the speed required by a generator. In traditional wind pumps the gearbox is used to covert rotary motion of the windwheel to reciprocating motion of the pump rod.

Gear case: Commonly refers to the metal case surrounding a transmission or gearbox. See *Gearbox*.

Gear ratio: The ratio of the number of revolutions per minute between the main shaft and the output shaft of a wind turbine drive train.

Gedser: Ferry terminal and port on the southernmost extremity of the Danish island of Falster 150 kilometers (93 miles) south of Copenhagen. Gedser is located on an exposed peninsula jutting into the Baltic Sea. In 1956 Johannes Juul installed a wind turbine at Gedser with a three-blade, stall-regulated, upwind rotor that spanned 24 meters (80 ft). For overspeed protection, Juul devised a simple system for pitching the tips of each blade. The 200 kW Gedser mill operated in regular service from 1959 through 1967. During its lifetime the Gedser mill generated 2.2 million kWh and was capable of annual yields of 800 kWh/m^2 of rotor swept area. Juul's experimental turbine was so successful that it became the forerunner of all later Danish wind turbines. When Denmark looked again to wind energy for help in meeting the energy crisis of the 1970s, the country had a working model still standing at Gedser. The turbine has since been removed to a museum and a modern wind turbine installed atop its original concrete tower.

Geotextile mats: Woven mats of natural or synthetic fibers used as a underlayment for roads in boggy upland soils. The mats, developed by 19th-century engineers laying railroads across Britain's moorlands, distribute the loads imposed by road metal or fill material over a wide footprint. The mats can also be used to remove road material after the road is no longer needed.

German wind energy institute: See *DEWI*

GFRE: Glass-fiber-reinforced epoxy.

GFRP: Glass-fiber-reinforced polyester.

Gigawatt (GW): 1,000 kilowatts, 10^6 watts.

Gin pole: Pole derrick. Standing derrick. Vertical pole or pipe used to raise a wind generator and commonly associated with 1930s-era windchargers. When attached to the side of a tower and extending above the tower top, a gin pole acts like a portable crane and allows the generator to be raised into position and then lowered onto the tower top. Gin poles are also used to increase hoisting leverage when raising hinged towers from the ground.

Giromill: Cycloturbine. A vertical-axis H-configuration wind turbine with articulating straight blades. Rotors with articulating straight blades offer several advantages over conventional f-configuration Darrieus rotors because they are reliably self-starting and all lift along the blade is created at a maximum radius from the rotor axis, thus maximizing rotor torque. See also *Articulating blade*.

Golding: E. W. Golding was the technical secretary of the Wind Power Committee of Britain's Electrical Research Association during the 1950s and wrote what has become a classic in wind energy literature, *The Generation of Electricity by Wind Power*. As with Putnam before him, many of Golding's observations remain valid today.

Gondola: Nacelle. Equipment pod.

Governor: A mechanism that automatically controls the speed of a rotor, usually by changing the pitch of the blades. A mechanism for exerting control or governing some device. In wind energy it generally refers to a mechanism for controlling rotor speed, as in the flyball governor.

Grandpa's Knob: Long north–south ridge near Rutland, Vermont, where the 1,250 kW Smith-Putnam wind turbine operated intermittently from August 1941 to its failure in March 1945.

Greenpeace: International environmental organization with affiliates in most developed countries.

Grid: Network. Utility distribution system. Network of transmission and distribution lines carrying electricity from sources of generation to consumers.

Grid-connected wind turbines: Wind turbines that are physically linked or interconnected with a network or grid of an electricity distribution system. Such wind turbines use the network for field excitation or inverter commutation and are capable of feeding utility-compatible electricity into the grid or network.

Griggs-Putnam index: Named for biologist Robert F. Griggs of George Washington University and engineer Palmer C. Putnam. A system created in the mid-1940s that uses vegetational indicators of long-term average wind speeds. The techniques developed by Griggs were used in the siting of the 1,250 kW Smith-Putnam wind turbine on Grandpa's Knob near Rutland, Vermont, in 1941.

Grounding: Earthing. To connect an electrical circuit to the earth or to the ground.

Gullies: Form of erosion. Ditch or channel cut in the ground by erosion. Larger than rills, each gully may measure up to several meters across and several meters deep.

Gust: Sudden change in wind speed.

Guy anchor: Footing designed for anchoring the cables of a guyed tower.

Guy cables: Stays. Guy wire. In American usage, a wire or cable used in tension to support an object such as a tower or blade. Cable support. As in a guyed tower, the cables are equally spaced about the tower so that each is under tension.

Guyed towers: Towers that use guy cables and several far-flung anchors for remaining upright.

Habitat: The environment or surroundings where an organism normally lives. *Sensitive habitat* implies that the environment may either be easily disrupted by construction or use, or that the environment is home to a species of special concern.

Halladay, Daniel: Inventor who in 1854 built the first fully self-regulating wind pump. Until then millers had to manually turn the spinning rotor out of the wind or reef the sails during storms. Halladay constructed a multiblade rotor made up of several movable segments. Rather than attaching these segments to the hub directly, he pivoted them about a ring. In high winds, the segments would swing open into a hollow cylinder. Early models of Halladay's "rosette" or "umbrella" mill used a tail vane to point the rotor, or "wheel," windward. Later "vaneless" versions did away with the tail vane by orienting the wheel downwind of the tower.

Handline: Rope used for raising and lowering light material and tools from a tower.

Harmonic distortion: Undesired distortion of the sinusoidal voltage and current waveform of a utility's alternating current. Harmonics are of concern due to the damage they may cause utility and customer equipment. (Robert Putnam)

Harness, full-body: Unlike a work belt, which is solely an aid for positioning, a full-body harness is designed to maintain the wearer in an upright position after a fall and to facilitate extraction of a unconscious wearer from a confined space. Full-body harnesses include leg straps, belt, chest harness, and lanyard ring.

Hazards, bird: Structures that may injure or kill birds. In wind energy, exposed electrical conductors, guy cables, towers, and the moving rotor of a wind turbine are potential hazards to birds.

Hazards, occupational: Source of danger or risk to life and limb in the workplace. The principal occupational hazards of those working with wind energy are electrocution, falls, and becoming ensnared in moving machinery.

Head: Measure of the height that a pump lifts water.

Head, dynamic: The vertical distance a wind pump must lift water between the level of well drawdown and the level at the well outlet plus the friction head.

Head, friction: Loss of head caused by the resistance to flow of a fluid through a pipe. The friction head of water is a function of pipe diameter, length, and the number of elbows.

Head, pumping: The vertical distance a wind pump must lift water between the level of well drawdown and the level at the top of the well.

Heat pump: A reversible heating and cooling device that operates via a compressor and an evaporator, causing a liquid such as Freon to circulate in a closed loop. In one half of the cycle the compressor raises the pressure of the Freon, causing it to condense to liquid form and give off heat. In the other half the liquid passes to the evaporator, where lower pressure makes it expand and revert to a gas, absorbing heat. In the heat-pump process, heat can be either drawn off or delivered to a given space.

Heat rate: The amount of thermal input required to produce 1 kWh of electricity. The heat rate of conventional condensing cycle power plants is approximately 10,000 Btu (2,500 kcal) per kWh. The heat rate for a modern combined-cycle plant is about 7,000 Btu (1,750 kcal) per kWh.

Height-to-diameter ratio: For vertical-axis wind turbines, the ratio of the rotor's height to its width at the rotor's equator. Early f-configuration Darrieus rotors had height-to-diameter ratios near unity. Later designs are considerably more elliptical.

Hertz (Hz): Unit of frequency in cycles per second.

Hinged blades: On a horizontal-axis wind turbine, blades that are each separately hinged to the hub so as to allow some freedom of motion flapwise.

Histogram: Bar chart or graph.

Home-built: American usage for "do-it-yourself (DIY) project." Some homeowners have attempted to build their own wind turbines, often with little success.

Home light plant: Stand-alone or self-contained generating system originally used to provide lighting for homesteaders on the American Great Plains before the advent of the Rural Electrification Administration, circa 1930. Wind systems were sold during this period as "light plants" using glass batteries for storage and gasoline engine–generator combinations (such as the Delco Light Plant) as a backup power source.

Horizontal-axis wind turbine (HAWT): Conventional wind turbine. Wind turbine where the rotor spins about an axis near the horizontal as seen in the traditional European windmill and the American farm windmill.

Horsepower (hp): The power required in the English system to raise 550 pounds 1 foot in one second. One horsepower (hp) is equivalent to 746 watts.

H-rotor: Form or configuration of a vertical-axis wind turbine where in side or frontal views the straight vertical blades appear as the uprights of the letter *H* and the rotor cross-arm appears as the horizontal line of the *H*. The blades on an H-rotor may be fixed in pitch or articulate. See also *Giromill* and *Articulating blade.*

Hub: Center of wind turbine rotor. Component at which the rotor blades are attached to the main or drive shaft of a wind turbine.

Hub height: Centerline height. Distance from the center of a conventional wind turbine rotor to ground level.

Hütter, Ulrich: German engineer whose career spanned the half century from the 1930s through the 1980s and who postulated that the most cost-effective wind turbine will use long slender blades and lightweight components, and will operate at high tip speeds. Considered the father of high-speed, lightweight wind turbine design in Germany and the United States.

Hybrid power systems: A combination of renewable technologies (such as wind turbines or solar photovoltaics) and conventional technologies (such as a diesel generator) used to provide power at remote sites.

IEA: International Energy Agency (Vienna).

IGBT: Integrated gate bipolar thyristors. Electronic devices used in synchronous inverters.

Induction generator: An asynchronous electrical generator that must be supplied with magnetizing current. Power is produced only when the input shaft speed exceeds that necessary to generate an alternating current frequency synchronous with that of the magnetizing current.

Inertia, rotor: The resistance of a rotor at rest to being set in motion and the resistance of a rotor in motion to any change in its speed or direction.

Infrasound: Subaudible frequencies below 20 hertz.

Infrastructure: The basic facilities, structures, and services necessary for the functioning of a system such as electricity generation and distribution, transportation, communication, or finance.

In-plane: Direction or motion parallel with the rotor plane.

Installed capacity: The rated power of a wind turbine. Also called nameplate capacity. The installed or rated capacity differs from average capacity as used in the Pacific Northwest of the United States and the declared net capacity as used in Britain, because it's independent of actual generation or capacity factor.

Installed cost: More correctly, *installed price.* Total price for the turnkey installation of a wind turbine, including the price of the wind turbine, tower, foundation, installation, and any associated costs for interconnection.

Intercept area: Area of the wind stream swept by a wind turbine rotor. Swept area. Frontal area.

Interconnected wind turbine: A wind generator linked to the utility grid or network. Grid connected. Grid intertied.

Interest rate: A charge or cost for a loan expressed as the percent of the principal or amount loaned due during a specified period, usually per annum.

Intermediate plant: An electrical plant used to meet daily or seasonal variations in electrical load. The annual average use of this type of plant is less than that of baseload plants, but more than peak-load plants.

Intermittent resources: Energy resources, such as wind, hydro, and forms of solar energy, where the fuel or prime mover is not available 100 percent of the time. Wind is an intermittent resource because the wind does not blow all the time, and when it is blowing it may not be blowing at sufficient strength to drive a wind turbine at 100 percent of its rated capacity.

Intrusion, aural: Unwelcome or unwanted sound. Noise.

Intrusion, visual: Unwelcome or unwanted sight on a landscape.

Inversion, temperature: Meteorological condition in which air temperature increases with increasing altitude, or the inverse of the normal lapse rate where temperature decreases with increasing altitude. The inversion layer acts as a lid, preventing the dispersal of pollutants. Wind speeds are generally greater above an inversion layer than below.

Inverter: An electrical device to convert direct current (DC) to alternating current (AC).

Investor-owned utilities: Dubbed IOU in American usage. Private, for-profit utility company.

IRP: Integrated Resource Plan. System for determining how best to meet electrical demand over a specified planning horizon that includes improved energy efficiency and all sources of potential generation, including renewables.

ISET: Institut für Solare Energieversorgungstechnik, research group in Kassel, responsible for monitoring the performance of wind turbines under the German goverment's early research program. Its annual WMEP reports are a treasure trove of details on actual wind turbine performance over many years.

Isolation, noise: Component or device used for reducing noise emissions that prevents the interaction between components or the transmission of vibration between components.

Isovent: Lines of equal wind speed drawn on a geographic map of an area. Analogous to isobars (lines of equal barometric pressure) and elevation contours (lines of equal elevation).

Izaak Walton League: American environmental group active predominately in the Midwest.

Jacobs Wind Electric: 1930s-era manufacturer of windchargers. Jacobs built the self-styled "Cadillac" of windchargers.

John Brown: British manufacturer that developed 100 kW prototype wind turbine with a 15-meter (49 ft) downwind rotor circa 1955.

Joule: A unit of energy in the SI system. Work done when a force of 1 Newton is displaced through a distance of 1 meter.

Juul, Johannes: Widely credited as the father of the "Danish" configuration: three-blade, upwind, fixed-pitch, stall-regulated rotor. Juul trained in the Danish craft tradition by attending Poul la Cour's school for windmill electricians at the Askov *folkhøjskole.* Juul believed in incrementally increasing the size of wartime designs with which he was familiar. He emphasized low cost, simplicity, and the use of readily available materials. In 1956 he installed the 200 kW, 24-meter (80 ft) diameter turbine at Gedser, which became one of the period's most successful wind turbines. Juul's design approach has proven more successful than that of his German contemporary, Ulrich Hütter.

Kaiser-Wilhelm-Koog: Site of first German wind power plant and one of Germany's wind turbine test centers. Located in a former polder *(koog)* west of Marne in Schleswig-Holstein.

Kamakaze: Japanese for "sickle wind," a wind that cuts like a scythe.

Kamikaze: Divine wind. From the Japanese *kami* (divine) and *kaze* (wind). Named for a storm that destroyed a Mongol invasion fleet in 1281. Used in World War II for suicide pilots who were believed would save Japan from the invading Allies.

Khanasin: Arabic for "50," as in 50°C (122°F), a hot easterly wind from the desert in the Middle East.

Kilowatt: Unit of power. 1,000 watts.

Kilowatt-hour (kWh): Unit of energy equal to the use of 1,000 watts for one hour.

Kinderdijk: Site of world's oldest existing wind power plant where a gang of 19 windmills drained a polder southeast of Rotterdam until the 1950s.

Klapsejlmølle: Danish term for the traditional windmill on which the cloth sails and lattice work blades have been replaced with movable louvers, shutters, or jalousies that regulate the flow of air through the blade. In British usage *patent sails* derived from the inventions of Meikle and Cubitt. See *Patent sails.*

Knots: Nautical unit of speed. One knot equals 1 nautical mile per hour (1.15 mph, 0.515 m/s).

la Cour, Poul: Known as a the Danish Edison, he was one of the first Europeans to experiment with using wind energy to generate electricity. La Cour built experimental wind turbines using *klapsejlmøle* at Askov *folkhøjskole,* where he taught.

Lanchester-Betz limit: Anglophone academics argue that British engineer Frederick Lanchester should also be given credit for deriving the theoretical limits of a propeller turbine. He published his findings in 1915, several years prior to those of Albert Betz's widely quoted paper in 1920. Both reached similar conclusions. See *Betz.* (Karl Bergey)

Land–sea breeze: Onshore and offshore winds created by differential heating of land and large bodies of water. During daylight the land heats faster than the water, causing a breeze from sea to land (sea breeze). At night the land cools faster than the water, causing a breeze from land to sea.

Lanyard: Nylon cord or short nylon rope often with snap hooks at either end used to attach a work belt or body harness to a fall restraint system.

Lattice towers: Freestanding towers constructed of intersecting or crisscrossing metal bars, tubes, strips, or angles that resemble the latticework of a rose arbor. See also *Truss towers.*

Leading edge: Front of the airfoil. Region of an airfoil that first encounters the relative wind.

Learning curve: The decrease in cost of production as volume increases. One of the first demonstrations of this phenomenon occurred with the production of the Ford Model-T. During the 1970s the learning curve was often used to justify public expenditures on costly multimegawatt wind turbines. There were never enough large wind turbines built to gain economies from volume production, however, and thus the value of the learning curve for large wind turbines was never demonstrated.

Lease: A contract permitting use of property for a specified period in exchange for payment of rent or a fee. See also *Royalties.*

Lee: Downwind. On the side away from the wind. Snow accumulates in the lee of mountain summits.

Leeghwater, Jan Adriaenszoon: The Dutch father of polder drainage. Literally Jan "empty" water, so named for draining the Beemster polder with 40 windmills and the Schermer polder (near his birthplace in Noord Holland) in 1635 using 50 windmills.

Levelized cost of energy: A stream of equal costs whose present value is equivalent to a stream of both higher or lower costs throughout the life cycle of a generating plant.

Life-cycle cost: The all-inclusive cost of a technology from installation through dismantlement. Renewable technologies have higher initial costs than fossil-fuel power plants, but often lower life-cycle costs because of the ongoing costs for fuel during the life of a fossil-fuel plant.

Lifelines: See *Lanyard* and *Fall-arresting systems.*

Lift, aerodynamic: The force that pulls a wind turbine blade along, as opposed to drag.

Lift device: One of two major classes of wind machines. The blades move in a plane perpendicular to the wind. Lift devices are much more efficient at converting the energy in the wind than drag devices. The blades of lift devices can move at speeds greater than that of the wind. Fast-moving blades need airfoil shapes with good lift-to-drag ratios. High-speed lift devices have much lower solidity—that is, fewer and more slender blades—than drag devices. For example, lift devices can capture 50 times more power per unit of projected blade area than drag devices. (Paul Gipe and Hugh Piggott)

Lift-to-drag ratio: Ratio of the coefficient of lift to the coefficient of drag. Aircraft designers have long sought high ratios of lift to drag.

Lightning: An abrupt discharge of static electricity in the atmosphere. Because they're often the tallest objects on the landscape, wind turbines are susceptible to lightning.

Line commutated: Form of electronic inversion that uses the utility's alternating current waveform to trip electronic switches such as thyristors.

Live: In an electrical conductor or circuit, voltage is present and current can flow, producing a shock hazard. Hot.

Livestock watering: An important application for small wind turbines and small photovoltaic systems on the American Great Plains, Argentina's pampas, and the steppes of Central Asia. Water for livestock is essential for survival of livestock on these semiarid grasslands, and adequate supplies are critical for weight gain.

Livingston Bench: Upland on the south side of the Yellowstone River southeast of Livingston, Montana, long regarded as a potential site for wind development. By the mid-1990s, however, only four 65 kW wind turbines were in operation.

Load: A demand for electrical power.

Load duration: The amount of time during which the demand for power exists.

Load factor: In American usage, the coefficient of average electrical load divided by peak load. In British usage, capacity factor.

Load leveling: Technique for smoothing a power profile of a customer or an entire utility over a period of time, often entailing constraints on when power is used during the day.

Load-following capability: A measure of a utility system's response to change in load. As load increases the utility system responds by generating more from existing units or by bringing more units on line. When load decreases the utility system responds by dropping units or decreasing the generation from operating units.

Loads, standby: Secondary electrical loads that are supplied only when the demand of primary or essential loads is met. See also *Dump load.*

Logarithmic extrapolation: Method used by meteorologists to estimate changes in wind speed with height. It differs from the power law method, which was derived empirically.

$$V = [\ln (H/Z_0)]/[\ln(H_o/Z_0)]\ V_o$$

where V_o is the wind speed at the original height, V is the wind speed at the new height, H_o is the original height, H is the new height, and Z_0 is the roughness length. Roughness Class 1, representing "open" land, has a roughness length of 0.03 which is equivalent to the friction coefficient, in the power law equation of 0.13.

Lolly shaft: Pintle. Vertical shaft about which a conventional wind turbine yaws or turns to face the wind. American term commonly used when describing the turntable or the pintle of small 1930s-era wind generators, such as the Jacobs windcharger.

Loss-of-load probability: The statistical likelihood that a generating unit will not be available to meet demand. Utility engineers have traditionally raised the specter that wind generation is unreliable to meet demand because of its intermittency. Studies in the United States and Europe during the early 1990s showed that wind power plants exhibit a loss-of-load probability not unlike that of conventional sources.

Loudness, noise: A measure of the auditory response to sound, a measure of noise intensity. The human sensation of loudness depends not only on sound pressure, but also on frequency or pitch.

Low-frequency noise: Sounds below 100 hertz. Frequency range of troublesome noise most often associated with pulsating impulsive sound from downwind two-blade wind turbines. Includes audible and subaudible frequencies.

Lull: Calm. Period with no wind.

Lykkegaard wind turbines: Danish wind turbine patterned after Poul la Cour's *klapsejlsmølle* that was used during World War II.

Madaras rotor: J. Madaras used a slightly different approach than Flettner to harness the Magnus effect. He proposed that large cylinders rotating about a vertical axis be mounted on a circular track with an electrical generator in the wheels of each car or truck carrying a rotor. In 1933 a 90-foot (30 m) tall cylinder 28 feet (9 m) in diameter was erected in Burlington, New Jersey.

Magnus effect: Thrust created by the wind passing over a spinning cylinder. The spinning cylinder acts as an airfoil. The effect is comparable to that of baseball's curveball, where the pitcher imparts a spin to the ball as it leaves the hand, causing a curve in the ball's flight.

Main bearings: In a horizontal-axis wind turbine, the bearings supporting the main shaft, hub, and rotor. In a vertical-axis wind turbine, the bearings at the top and bottom of the rotor.

Mains: British usage for central-station power available from a network of transmission and distribution lines. Grid or network.

Main shaft: Spindle or axle connecting the rotor to the transmission and generator. In many conventional wind turbines the main shaft supports the rotor's mass and load while transferring its torque. In larger wind turbines the main shaft transmits torque but does not support the mass and thrust of the rotor and is sometimes called a quill shaft.

Marginal cost: The cost of the next unit beyond the margin or boundary of that needed.

Margin of safety: In engineering, an amount beyond that needed to meet design specifications.

Marineproof: Able to withstand the corrosive effects of the

marine environment—sea spray, salt-laden moisture, and so on.

Mark: British usage for "iteration, version, or model number," as in Supermarine Spitfire Mark IV.

Market-driven development: Technology development due to market pull or a demand for a product. As opposed to centrally directed technology development through research, development, and demonstration.

Maximum design wind speed: The maximum wind speed a wind turbine in automatic, unattended operation—but not necessarily generating—has been designed to sustain without damage to structural components or loss of its ability to function.

Maximum tested wind speed: The maximum wind speed a wind turbine in automatic, unattended operation—but not necessarily generating—has sustained without damage to structural components or the loss of its ability to function.

Mean-time-between failure (MTBF): A measure of reliability. Total operating hours divided by the total number of equipment failures.

Mean wind speed: Average wind speed. The value obtained by dividing the sum of a quantity of wind speeds by the number of wind speeds in the set. The mean wind speed obscures the distribution of wind speeds over time, but is the simplest measure of a wind resource.

Mechanical brakes: Spring-applied brakes.

Medium-size wind turbines: Arbitrary designation for wind turbines from 10 meters (30 ft) to 50 meters (150 ft) in diameter.

Megawatt (MW): Unit of power equal to 1,000,000 watts or 1,000 kilowatts. The Boeing Mod-5b is rated at 3.5 MW.

Megawatt-hour (MWh): Unit of energy equal to 1,000 kWh. The generation or use of 1,000,000 watts in one hour.

Mérida: City in Mexico on the north shore of the Yucatán Peninsula known for its multitude of water-pumping windmills, many mounted on the rooftops of multistory buildings.

Metal fatigue: Property characteristic of metals where strength decreases with use. More specifically, fatigue is a function of applied stress and the number of stress cycles. For example, the failure of a copper wire twisted and untwisted until it breaks is an example of metal fatigue. If the force or stress used in bending the wire is increased, the period of time before the wire breaks is shortened. The stress a material can withstand under cyclic loading is much less than it is under static loading. Flexure in metals is not completely reversible on the microscopic level, and cyclic loads cause strain hardening and a loss of ductility. Metal, in effect, becomes brittle with use. Metal rotor blades, for example, will eventually wear out and fail due to metal fatigue.

Meters per second (m/s): Unit of wind speed in the SI system. One m/s is equal to 2.24 mph and 1.94 knots.

Method of bins: A technique for collecting or analyzing a wind speed frequency distribution by grouping wind speed data into discrete intervals (bins). For example, wind speed data collected in m/s can be stored in bins of 0–1 m/s, 1–2 m/s, 2–3 m/s, and so on.

Microclimate: The wind regime or climate in the immediate neighborhood of the wind turbine.

Micrometeorology: The localized study of meteorological data, often used in planning wind farms.

Millwright: Skilled craft of building and maintaining a traditional European windmill. The term now also applies to anyone with the skills necessary to service any heavy machinery.

Missing-tooth syndrome: The visual impact of wind turbines on the landscape is accentuated by the human tendency to first notice objects unlike the others in a group of objects. People notice a black bean in a group of white beans or a white bean in a group of black beans. So it is with wind turbines. Observers first see a nonoperating wind turbine in a group of operating wind turbines.

Mistral: From Provençal, a cold, powerful, dry wind flowing down the Rhône Valley from the French Alps.

Mitigation: In American usage, the actions or measures taken to reduce the real or perceived environmental impacts from a project.

Mitigation, off-site: Actions or measures taken beyond the boundaries of a project site to compensate for a project's real or perceived environmental impacts.

Modified sine wave: Marketing sleight of hand for modified-square wave.

Modified-square wave: Waveform produced by some solid-state inverters that replicates, to varying degrees, the sinusoidal waveform produced by AC generators.

Modularity: Of or relating to discrete units that can be easily joined together to create a whole. Wind energy is modular because generating capacity can be added in units of individual wind turbines.

Moment of force: See *Torque.*

Mortality, bird: The number of dead birds from both natural and human causes. Relative to wind energy, the number of birds killed by wind turbines.

Mortality, human: Relative to wind energy, the number of people killed in accidents during the construction, operation, and removal of wind turbines.

Multiblade rotors: High-solidity rotors using numerous blades. The American farm windmill uses 10 or more blades that cover nearly the entire rotor disk. In contrast, most modern wind turbines use only two or three slender blades.

Multimegawatt wind turbines: Large wind turbines typically with rotors sweeping more than 60 meters (190 ft) in diameter.

Multiple rotors: Wind turbines using multiple rotors can take two forms: dual contrarotating rotors on the same axis or multiple rotors on the same towers.

Municipally owned utility: In American usage, an electric utility owned by a city government and ostensibly its citizens. This form of ownership contrasts with that of investor-owned utilities, whose owners may reside anywhere.

NACA: National Committee on Aeronautics (U.S.). Precursor to the National Aeronautics and Space Administration. No longer extant. Widely known for its series of standard airfoil configurations.

Nacelle: Machine cabin, gondola, equipment pod. The covering that houses the rotary components of a horizontal-axis wind turbine. Refers more specifically to the covering of the equipment pod.

Nacelle cover: Canopy. Shroud enclosing the equipment pod. Integral component for noise control and for protection of windsmiths during inclement weather in medium-size wind turbines.

Nameplate rating: Arbitrary rating of wind turbine in kilowatts. See *Rated power.*

NASA: National Aeronautics and Space Administration (U.S.).

National Research Council: NRC, Canada.

NATA: Network for Appropriate Technology and Technology Assessment (U.K.). Located at the Open University, Milton Keynes.

Natural frequency: The frequency with which a system vibrates absent any external force. See *Eigen frequency.*

NCAT: National Center for Appropriate Technology (U.S.). Located in Butte, Montana.

Necropsies, avian: Bird autopsies.

NEL: Formerly, National Engineering Laboratory (U.K.). Located southeast of Glasgow, Scotland, at East Kilbride.

NESA: Danish electric utility serving Copenhagen.

Net energy billing: One method by which a utility can compute the billing for providing electrical service to customers who use their own wind generator. One or two kilowatt-hour meters may be used, with billing by the difference between the two meters. Excess power from a wind generator, for example, runs the unratcheted meter. The utility, in effect, buys back energy at the same rate at which it sells energy to the consumer.

NEWIN: Dutch wind energy association.

NFFO: Non Fossil-Fuel Obligation (U.K.).

NIABY: Not In Anyone's Backyard. Bitter depiction of groups who oppose the use of certain technologies, such as wind energy. For example, Britain's so-called Country Guardians campaigns against wind energy, even where there is local support for the technology.

NIMBY: Not In My Backyard. Slightly derogatory acronym applied to those who oppose a land use in their vicinity, despite their support of or dependence on the same land use elsewhere. Relative to wind energy, a NIMBY response is characterized by statements such as "I support wind energy, just not here." Related to LULU or Locally Unwanted Land Use.

NIMSBY: Not In My Summer Backyard. Variant of NIMBY coined in Sweden surrounding controversy over installing wind turbines on islands frequented by summer tourists. NIMSBY implies or suggests a conflict between rural and urban inhabitants about the use of rural landscapes. While local year-round residents may support the use of wind turbines, urban dwellers who use the rural landscape only for recreation may object because of fears that the wind turbines will somehow disrupt their recreational use of the landscape. The same phenomenon is seen in Britain, Denmark, Germany, and North America.

Noise: Unwanted sound.

Noise ordinance: American usage for locally imposed restrictions on noise sources.

Nominal cost of energy: Includes the effects of both real cost escalation and inflation over time. Typically higher than costs expressed in real or constant dollars.

Noordoostpolder: One of the polders reclaimed from the Netherlands' former Zuiderzee by closure of the Afsluitdijk. Site of the world's longest linear array, in which 50 wind turbines border the Westmeerdijk north of Urk. The 6-kilometer (3-mile) string of wind turbines produces 12 percent of the polder's electricity.

Northwest, Pacific: American usage for the states of Washington and Oregon. Sometimes includes the states of Montana and Idaho.

NOVEM: Nederlandse Onderneming voor Energie en Milieu. Dutch ministry for energy and the environment.

NRC: National Research Council (Canada).

NRDC: Natural Resources Defense Council. U.S. environmental group.

NREL: National Renewable Energy Laboratory (U.S.). Located at Golden, Colorado. The wind division is located at the National Wind Technology Center on Rocky Flats near Boulder, Colorado.

Nuisance noise: In American usage, sound or noise that interferes with another's legal rights by causing damage, creating an annoyance, or causing inconvenience.

O&M: Operation and maintenance.

Obstructions: Obstacles in the wind's path that reduce its velocity and, hence, its power. Obstructions such as trees, buildings, and shelter belts retard the wind and cause turbulence damaging to wind turbines.

Off-peak: Refers to points on a load distribution curve or times of the day when electricity consumption is less the maximum.

Off-the-grid power systems: Stand-alone power systems. American usage for power systems independent of an electricity distribution network or grid.

Offshore: Siting or installation of wind turbines beyond the low-tide range of a coastline.

Omnidirectional: The ability to accept the wind from any direction. Unshrouded vertical-axis wind turbines are inherently omnidirectional.

One-off: British usage for preproduction prototype.

One-seventh power law: See *1/7 power law.*

Onshore: Siting or installation of wind turbines on land.

Operation and maintenance: Actions taken to operate or ensure the continuing operation of wind turbines and their ancillary equipment.

Opinion, public: Explicit or implicit expression of attitudes toward the use of wind energy by inhabitants of a physiographic or political locale, region, or state.

Orientation: Yaw. Azimuth. Bearing. Position of a conventional wind turbine about a vertical axis. The rotor on a conventional upwind turbine is oriented toward the north when the winds are from the north.

Out-of-plane: Direction of motion perpendicular to the rotor plane.

Overrunning clutch: A clutch that allows the driven shaft to turn freely under its own power when the speed of the driven shaft exceeds that of the driving shaft. The clutch on an automobile is a form of overrunning clutch that permits the engine to turn freely under its own power when its speed exceeds that of the starter. The overrunning clutch was used by the USDA at its Bushland, Texas, experiment station to match the mechanical characteristics of a Darrieus rotor to that of a high-volume, motor-driven irrigation pump. When sufficient wind was available, the Darrieus rotor drove the well pump mechanically. When winds were insufficient, the well pump was driven electrically.

Overspeed: Refers to excessive rotor speed. A speed where either fatigue will be accelerated or the rotor will fail catastrophically.

Overspeed control: Mechanism, device, or some combination of devices for limiting the maximum speed of the rotor to prevent it from self-destruction.

Packer head: A brass fitting installed over the top of a well pipe to prevent overflow. When used with a water-pumping windmill, it allows for pumping water uphill or into an aboveground storage tank.

Panemone: Simple wind machines using drag as the driving force where the effective surface or pane moves in the direction of the wind, as in the common cup anemometer. The maximum coefficient of power for simple panemones is 1/3. (Golding)

Parachutes: Method of overspeed control in which a parachute deploys at the tip of each blade during an overspeed event.

Parallel generation: Production of electricity in conjunction or parallel with that delivered by an electric utility network or grid. Wind power plants and individual wind turbines interconnected with the grid generate electricity in parallel with that of the local utility.

Parallel-shaft transmissions: Common arrangement of gears within a gearbox where the shafts of each gear parallel those of the others.

Partial-span pitch control: In contrast to full-span pitch control, where the rotation of a wind turbine blade about its longitudinal axis acts along its entire length, only the outboard sections of the blade are used. The inboard sections remain fixed in pitch. See *Full-span pitch control.*

Particulate pollution: Soot, smoke, and respirable particles. Includes fugitive dust in the United States.

Passerines: Songbirds. Order of small birds.

Patent sails: British usage for mechanical louvers, shutters, or jalousies that replaced traditional windmill sails. These self-regulating sails were developed by English millwright William Cubitt in 1807 following on Andrew Meikle's spring sails of 1772. Patent sails are found on later versions of windmills in Denmark, Germany, and Britain but are rarely seen on Dutch windmills. *Klapsejlmølle* in Danish.

Payback: Length of time before an investment is recovered. Time it takes for a wind turbine to pay for itself. Payback fails to incorporate benefits during the entire life cycle of an investment in renewable energy by excluding benefits after payback has occurred.

Peaking capacity: Generating units used to meet peak or maximum loads on a utility system. In the past, peaking units have had low capital costs but high running costs. Typically they used technologies, such as natural-gas-fired turbines, that permitted them to be started and ramped up to full output quickly.

Peak load: Maximum load or force on a structure, or maximum consumption in a system of supply.

Peak power: Maximum power output of a wind turbine. Peak power is often greater than rated power, especially among stall-regulated wind turbines. See *Rated power.*

Peak shaving: Method to reduce the need for standby generating capacity in a utility system by regulating or controlling when loads are brought on line.

Peak wind speed: Maximum instantaneous wind speed.

Penetration level: The relative amount of wind turbine generating capacity in an otherwise conventional electrical supply system. The quotient of wind-generated electricity divided by the system's total generation.

Pennines: The north–south hilly backbone of England that includes the Yorkshire dales. Small wind power plants operate in the mid-Pennines near Leeds-Bradford.

Perches, bird: Structures that enable birds to perch or roost. Lattice towers with horizontal cross-girts and nacelles with external work platforms offer numerous perches and are thought to contribute to collisions between birds and wind turbines in California's Altamont Pass.

Performance: Wind turbine productivity. Several units of measurement are in common use, including specific yield and capacity factor.

Performance curve: Graph of wind turbine power output relative to wind speed. See *Power curve.*

Permanent-magnet alternators: Generators using permanent magnets to create the field. Many small wind turbines use permanent-magnet alternators.

Photovoltaic cells: Solar cells. Laminated silicon wafers that convert sunlight directly into direct current.

Piers: Columns of steel or reinforced concrete used for the foundation of wind turbine towers. The legs of lattice towers typically rest upon piers for lateral stability.

Pilot vane: Method for overspeed control on American farm windmills that used a vane in the same plane as the rotor disk. This vane extended beyond the sweep of the rotor and was intended to furl the rotor during high winds. The same effect is accomplished today on farm windmills by offsetting the rotor axis slightly from the pivot axis (yaw axis). In both versions, the tail vane used to orient the rotor into the wind is not fixed but instead is spring-loaded. This allows the tail to furl when pressure on the rotor or pilot vane exceeds the counteracting pressure on the tail vane.

Pinned truss tower: Lattice tower that uses pinned or riveted members in a truss or triangular pattern; one example is NASA/DoE's tower for the MOD-0A series of wind turbines. Truss towers are rigid and unyielding in contrast to tubular towers.

Pintle: Lolly shaft. Vertical shaft about which a conventional wind turbine yaws. Term used by Palmer Putnam when describing the design of the 1.25 MW Smith-Putnam turbine.

Pitchable blade tip: Overspeed control mechanism common among early Danish wind turbines that swiveled a portion of the blade at the tip about its longitudinal axis during overspeed events. One of the characteristic features of Johannes Juul's Gedser turbine. The change in pitch of the tip reduced lift and increased drag sufficiently to prevent the rotor from destroying itself. Pitchable blade tips differ from tip brakes in that they form a uniform part of the blade prior to deployment, contributing to the blade's torque.

Pitch, blade: The angle between the chord line of a wind turbine blade and its direction of travel.

Pitch control: A system that varies blade pitch to control lift and hence rotor torque and power. Typically wind turbines using blade pitch control varied pitch from feather to the running position, though some designs vary pitch from the running position to stall.

Pitch-regulated rotor: Wind turbine rotor that varies blade pitch to regulate lift and, hence, torque and power.

Plan area: The area projected on a top view of an airfoil perpendicular to a section or profile of the airfoil.

Planetary transmissions: A gearbox consisting of a central gear, a coaxial internal or ring gear, and one or more pinions supported on a carrier. In contrast to parallel-shaft gearboxes, planetary transmission are lighter and more compact. In wind turbine applications, however, they have proven noisier than parallel-shaft gear boxes. Wind turbines in the 500 kW size class and larger have attempted to marry the advantages of the two technologies by using planetary gearing on the low-speed side and parallel gearing on the high-speed side where much of gearbox noise is generated.

Plant factor: British usage. Capacity factor. The energy produced by a wind turbine divided by the energy that would have been produced had the wind turbine operated at its rated power 100 percent of the time.

Pole shoe: Iron core of a field pole in a generator that faces the armature and is surrounded by the field coils.

Post mill: Earliest form of European windmill. The entire windmill—including the rotor, gearing, and millstones—rests upon and turns about a central post. Consequently the whole machine had to be turned to face the wind. A long pole (tail pole) extended from the post mill to ground level by which the miller could turn the mill about the post.

Power: The time rate of doing work or consuming or generating energy.

Power coefficient: A measure of a wind turbine rotor's aerodynamic performance. The quotient of the power captured by a wind turbine rotor divided by the total power in the wind stream.

Power conditioning: Treatment or conversion of electricity produced by a generator or delivered by a distribution system. See also *Synchronous inverter.* Typically variable-speed wind turbines generate electricity with variable frequency and voltage. In most applications this electricity must be conditioned or treated before it can be used by devices requiring constant frequency and constant voltage.

Power curve: Chart of a wind turbine's instantaneous power across a range of wind speeds. The same data is frequently presented as a table of the instantaneous power produced at wind speeds from cut in to cut out. Hypothetical or idealized power curves are often shown as a line on a chart with power on the vertical axis and wind speed on the horizontal axis. Actual measured power curves are shown as scatter diagrams. Power curves derived from field tests are based on the weighted average of measurements taken when the rotor is both speeding up and slowing down.

Power density: Measure of the strength of a wind resources in watts per square meter (W/m^2). The amount of power per unit area of the free wind stream. The energy flux in the wind stream.

Power duration curve: Graph of wind power versus time.

Power electronics: Electronic devices used in power modern conditioning equipment, including but not limited to thyristors and integrated-gate bipolar transistors.

Power factor: Relationship between true power and apparent or imaginary power. A measure of the degree with which current and voltage are out of phase with that from the utility's central generating station and the generator is consuming reactive power. See *Volt-ampere-reactance.*

Power form: Standard description of characteristics that describe the form in which power produced by a wind turbine is made deliverable to the load, such as voltage, phase, and form of current. Recommended by the American Wind Energy Association for inclusion in product literature.

Power in the wind: The ability of the wind to do work per unit of time. Often given in terms relative to a unit of cross-sectional area of the wind stream, for example in watts per square meter:

$$P/A = 1/2\ \rho V^3$$

where ρ is air density, and V is the velocity or speed of the wind.

Power law: Widely used empirical model for the increase in wind speed with increasing height aboveground:

$$V = V_o(H/H_o)^\alpha$$

where V is the wind speed, V_o is the wind speed at the reference height, H is the height above ground, H_o is the reference height, and α is a wind shear exponent derived from surface roughness. See also *Prandtl logarithmic model.*

Power quality: A subjective or qualitative comparison of the characteristics of voltage, frequency, and current harmonics (for alternating current) of electricity produced by a wind turbine relative to the norm or range of values acceptable to a load or receiving utility systems.

Power train: Drive train.

PPC: Greek national electric utility.

Prandtl logarithmic model: Formula for determining the variation in wind speed with height above the ground surface derived from the principles of fluid mechanics:

$$U(z) = (U^*/k)\ln(z/z_0)$$

where U(z) is the wind speed in m/s at height z in meters, U^* is the friction velocity, k is the von Karman constant (about 0.4), and z_0 is the roughness length in meters that characterizes the terrain. See also *Roughness length.*

Prevailing wind: Most frequently occurring wind. The prevailing wind may not necessarily represent the most wind energy at a site. Because of the cubic response of power to wind speed, less frequent but stronger winds may contain more wind energy. This distinction is important when siting a wind turbine near obstructions.

Price: Cost plus profit. The price of wind turbines is the value customers or buyers pay and not necessarily what it costs the manufacturer to build the wind turbine or construct and operate a wind power plant.

Projected area: Area of a structure intercepting the wind transverse to the direction of the wind. The area of the circle intercepting the wind represented by the spinning rotor of a conventional horizontal axis wind turbine. See also *Swept area.*

Propagation, noise: The spreading of sound waves from a source.

Propeller: Sometimes incorrectly used to describe the rotor of a wind turbine. The rotor of a wind turbine is propelled by the wind. The rotor does not propel the wind through the rotor or propel the wind turbine through the air.

PRS: The California Energy Commission's Performance Reporting System, which requires California's wind plant operators to report quarterly on the performance of their wind turbines.

PSC: American usage for "public service commission," a state regulatory agency that typically has jurisdiction over electricity and other publicly regulated monopolies. Also known as public utility commissions in some states.

Publicly owned utility: In American usage, an electric utility that is ostensibly owned by the people of a nation, state, or other geographic unit.

Public Utility Regulatory Policies Act (PURPA): Bans discrimination against small power producers in the United States and directs utility purchase of small power production at a rate commensurate with the utility's avoided (marginal) costs.

PUC: American usage for "pubic utility commission," a state regulatory agency with jurisdiction over electricity and other publicly regulated monopolies. Also known as public service commissions in some states.

Pulse width modulation: Electronic technique for adjusting the current to a load by rapid switching. (Hugh Piggott)

Pultruded blades: Wind turbine blades made from the pultrusion process.

Pultrusion: Manufacturing process in which fiberglass strands or mats are pulled through a resin-filled vat and then through a heated die, emerging into the desired shape, such as that of an airfoil, in nearly a finished form.

Pumped storage: Means of storing electrical energy off peak by pumping water uphill into a reservoir, where it is then released through a hydroelectric turbine to generate electricity on peak.

PURPA: See *Public Utility Regulatory Policies Act.*

Putnam, Palmer Cosslett: Noted American engineer and inventor who led the development of the 1.25 MW Smith-Putnam turbine. During World War II, Putnam was famed not for his wind turbine, but for his development of the DUKW and WEASEL amphibious vehicles.

PV: Photovoltaics or solar cells.

Quarter chord: The point along the chord line of an airfoil one-quarter of the chord length from the leading edge.

Quill shaft: A hollow mainshaft used on the Boeing Mod-2 to transmit rotor torque to the gearbox.

Quixote, Don: Protagonist in Cervantes's classic tale of self-delusion. *Quixotic* has become synonymous with foolhardy acts in response to a deluded sense of chivalry. Don Quixote attacked a colony of 38 "giants" (traditional windmills) on hills overlooking the plains of La Mancha in central Spain.

Radial station: Point or position along a blade as specified by its distance from the rotor axis measured perpendicular to the rotor axis.

RAL: Rutherford Appleton Laboratory (U.K.).

Raptorproof perches: Attempt to prevent perching by birds of prey on lattice towers in California's Altamont Pass by using various devices.

Raptors: Birds of prey such as hawks, eagles, falcons, and kestrels.

Rare-earth magnets: Permanent magnets fabricated from metallic elements with atomic numbers from 57 to 71, the so-called lanthanide series. Wind turbine designers sometimes use rare-earth magnets, such as those made from a neodymium-iron-boron compound, because of their greater magnetic density than the common ferric oxide magnets.

Rated capacity: The output power of a wind machine operating at the constant speed and output power corresponding to the rated wind speed.

Rated power: The power produced by a wind generator in kilowatts at its rated wind speed. Often confused with peak power, though on some small wind turbines rated power is also peak power. *Rated power* is a misleading descriptor of wind turbine size and harks back to the technology's formative years as a branch of electric utility research. A more reliable indicator of wind turbine size is rotor diameter.

Rated rotor speed: The nominal rotational speed of a wind turbine rotor when it is producing its rated power.

Rated wind speed: The wind speed at which a wind turbine produces its rated power. Rated wind speeds of different wind turbines vary from 8 m/s (18 mph) to 16 m/s (36 mph). The rated wind speed can indicate the wind regime for which the wind turbine was designed. Wind turbines with lower rated wind speeds are intended for low-wind-speed regimes.

Ratepayers: Utility customers.

Rate schedule: Published charges by each utility for cost of providing service, cost of generation, cost of maintaining reserve capacity, and cost of fuel to each customer class. See *Tariff.*

Rayleigh wind speed frequency distribution: Idealized distribution of wind speeds over time. The Rayleigh distribution is a special case of the Weibull distribution where the Weibull constant is 2 and the shape of the distribution depends on only mean wind speed.

$$f(V_R) = dV\,(\pi/2)\,(V/V_{avg}^2)\,\exp[-\pi/4\,(V/V_{avg})^2]$$

where dV is the width of the wind speed bin, V is the speed of the wind speed bin, and V_{avg} is the average wind speed.

Reactive power: Power necessary for the field excitation of induction generators. Reactive power can be inductive or capacitive.

Real cost of energy: Assumes no inflation. Also known as constant cost, it can include actual cost escalation over time. Typically lower than nominal cost, which includes the effects of inflation.

Rebar: Reinforcing bar or rod used in strengthening concrete.

Receptor: One that receives. Person or animal that may hear or sense a sound. Ordinances governing community noise are often described in terms of their effects on potential receptors.

Rectified alternating current: Direct current produced by feeding an alternating current through a diode, an electronic device that restricts current flow to one direction.

Rectifier: A electronic device, such as a diode, that converts alternating current (AC) to direct current (DC) by restricting current flow to one direction.

Rectifier bridge: A rectifier circuit using several diodes to produce a continuous DC output from an AC source. (Hugh Piggott)

Reefing: Furling. Bringing in, hauling down, or rolling up sails on a wind-driven machine that uses sails. Millers on traditional European windmills were constantly on the alert to reef or furl the mills' sails as needed.

Reflection: Light, sound, or other form of energy reflected or bounced off a wind turbine. Reflections of the harsh desert sunlight off metallic wind turbine nacelles or the glossy gel coat of wind turbine blades near Palm Springs, California, have caused complaints about "sparklies" from some neighbors. Similarly, there have been scattered incidents of television and radar signals being reflected from wind turbines in the United States and Europe.

Regulator, voltage: Device that controls or regulates battery voltage in a charging circuit by limiting the charge current. Charge controller.

Relative wind: Apparent wind. The wind seen by a blade as it moves through the air. The resultant of the wind and the wind created by the blade as it moves through the air.

Reliability: Dependability. The measure of reliability used in the wind industry is mechanical availability or the mean time between failure of a component sufficient to stop a wind turbine's operation. See *Availability.* Often incorrectly associated with intermittency. Wind turbines, though extremely reliable, are dependent on the wind. Wind turbines thus operate intermittently despite their dependable ability to generate electricity when the wind is present.

Repowering: Removing existing wind turbines and replacing them with newer, more cost-effective machines. Repowering existing sites in California's aging wind plants has produced striking improvements in productivity. At one site in the San Gorgonio Pass repowering increased installed capacity from 29 MW to 35 MW, more than doubling annual generation and specific yield. At a site in the Tehachapi Pass repowering increased annual generation five times and nearly tripled specific yield.

RERL: Renewable Energy Research Laboratory, located at the University of Massachusetts at Amherst.

Reserve margin: The difference between the total system dependable capacity and the actual or anticipated total system peak load for a specified period.

Resistance: In an electrical conductor, opposition to the flow of current. Measured in ohms.

Resonance: Response to a periodic force in which the resulting amplitude of oscillation becomes large when the frequency of the driving force approaches the natural frequency of a system or component. A common problem in wind turbine design is avoiding resonance between the frequency at which a blade passes the tower and the natural frequency of the wind turbine and its tower.

Retrofit: Space-age jargon for repairing or improving an existing structure or machine by adding new or modernized components.

Revegetation: To ensure that eroded or disturbed land grows a cover of vegetation. Postconstruction activity required to establish or reestablish vegetation on soils disturbed by grading for roads, ditching for buried power lines, and excavations for tower foundations.

Reynolds number: A dimensionless number describing the aerodynamic state of an operating airfoil. The number is used, along with angle of attack, to describe the limits of a particular airfoil's lift-to-drag ratio and the conditions at which stall occurs. Wind turbine airfoils typically operate in the range of Reynolds numbers from 0.5 million to 5 million. (Richard L. Hales)

Rib: Structural member that gives form to an airplane wing or shape to an airfoil used in a wind turbine blade.

Rigid hub: Attachment of wind turbine blades to the rotor hub where there are no degrees of freedom. A hub where there are no hinges permitting flapping or coning.

Ring generator: Generator or alternator where the generator diameter is large relative to the length of the generator case. In contrast to conventional generators—where the diameter is some fraction of the generator's length—the diameter of ring generators is several times the length or depth of the generator case. Ring generators are used in low-speed applications. For wind turbines, ring generators permit the use of direct drive and the elimination of step-up transmissions. For conventional wind turbines, the diameter of the ring generator approaches 10 percent of the wind turbine's rotor diameter.

Risø: Danish national laboratory near Roskilde on the island of Zealand. Famous for its early wind turbine test center.

Rocky Flats: Site of U.S. nuclear weapons plant near Boulder, Colorado, and now home to the National Renewable Energy Laboratory's Wind Division and its National Wind Technology Center.

Root, blade: Portion of a wind turbine blade where it attaches to the hub.

Root shaft: Nearly archaic term from early wind turbines for a bar or rod that connects a wind turbine blade with its hub. Most wind turbines today use some other form of blade attachment.

Rosette windmill: Umbrella mill. American water-pumping windmill where the vanes of the rotor were assembled in hinged segments. The segments flew open in high winds to regulate the speed and power of the rotor. Because they were more complicated and costly than the fixed rotors of their late 19th-century competitors, they eventually fell out of favor. Occasionally, one can be found at wayside tourist attractions on the Great Plains.

Rotor axis: Axis about which a wind turbine rotor revolves. On conventional wind turbines the rotor axis is nearly horizontal; on Darrieus wind turbines, vertical. The use of *rotor axis* to define the type of wind turbine in modern wind energy terminology is a departure from historical usage. Historians of technology describe the "plane" through which the rotor passes. Thus, historians call horizontal-axis wind turbines "vertical windmills"; they call vertical-axis wind turbines "horizontal windmills."

Rotor diameter: On conventional or horizontal-axis wind turbines, the diameter of a circle swept by the rotor perpendicular to the axis of rotation. On vertical-axis wind turbines, the maximum distance perpendicular to the axis of rotation. For an ϕ-configuration Darrieus turbine, this is the equator of the ellipse swept by the rotor. For an H-configuration turbine, it is approximately the length of the rotor cross-arm. A term commonly used to describe the size of a wind turbine. In general, the larger the rotor diameter, the greater the amount of energy captured from the wind because the area swept by the rotor of a conventional wind turbine increases with the square of the diameter.

Rotor, generator: The spinning component of an electrical generator.

Rotor plane: On conventional wind turbines, the plane perpendicular to the axis of rotation.

Rotor speed: Angular velocity of a rotor. Often given in revolutions per minute (rpm).

Rotor tilt: In a conventional wind turbine, the angle between the rotor's axis of rotation and the horizontal. For example, the axis of the 1.25 MW Smith-Putnam turbine was 12.5 degrees above the horizontal. The rotors on early post windmills were probably horizontal, but millers soon learned that tilting the wind shaft above the horizontal moved the rotor's weight and thrust more directly over the tower while enabling greater clearance between the blades and the tower. Both effects were advantageous. Some wind turbine manufacturers have erroneously believed or falsely claimed that their rotors were tilted above the horizontal to better catch the wind. This belief was possibly fostered by an early treatise on the aerodynamics of sails published in the mid-1920s, which illustrated the angle of clothes on a clothesline in high winds.

Rotor torque: The movement produced by a rotor about its axis. In a wind turbine, torque is a function of lift and the distance from the hub where the lift is produced.

Rotor, wind turbine: Fan, wheel, turbine. A term borrowed from the helicopter industry for a system of rotating airfoils, such as a helicopter rotor. The assembly of blades and hub on a wind turbine.

Roughness, blade surface: Imperfections or debris on the surface of an airfoil that disturb or disrupt the airflow. The performance of some airfoils is sensitive to surface roughness. When used on wind turbines in arid environments, the productivity of these airfoils can be cut by as much as 50 percent unless the blades are washed regularly.

Roughness, earth surface: The unevenness of the earth's surface due to topography, vegetation, and the built environment. This roughness disturbs the flow of wind within the boundary layer.

Roughness length: Measure in meters used in the Prandtl logarithmic model for estimating the variation of wind speed with height. Roughness length increases with increasing surface roughness. Roughness length varies from as little as 10^{-5} meters for surfaces with few or no obstructions, such as ice and open mudflats, to 4 meters in urban areas. See also *Prandtl logarithmic model.*

Royalties: A share or portion of the proceeds from the sale of wind-generated electricity paid to landowners for the use of their land. Royalties vary from country to country and depend on several factors, including the profitability of wind generation and competition for use of the land. Royalties in the United States range from 2 percent to more than 5 percent. See also *Lease.*

RSPB: Royal Society for the Protection of Birds in Britain. Equivalent to the National Audubon Society in the United States.

Runaway: See *Overspeed.*

Run of the wind: Distance the wind travels during a given period. It's measured by simple anemometers called wind-run odometers. Wind speed can be determined by dividing the wind run by the time period of measurement.

Rural Electrification Administration: REA. New Deal–era agency created by the Rural Electrification Act to bring central-station electricity to rural inhabitants of the United States. Expansion of the REA spelled doom for windcharger manufactures in the 1940s, though some areas did not receive power until the mid-1950s.

Safety, occupational: The risk or freedom from risk of injury or death on the job or in the workplace. The manufacture, installation, and operation of wind turbines entails a certain degree of risk to those who work with the technology. Twenty men were killed worldwide working with wind energy between 1976 and 2000.

Safety, public: The public's risk or freedom from risk of being injured or killed by the transport, installation, or operation of wind turbines. No passerby has ever been killed by the use of wind energy.

Sail: A section of fabric fitted to the spars and rigging of a vessel to use the power of the wind for propulsion. Because of their likely derivation from the sails of ships, the blades used to drive traditional European windmills are commonly called sails whether or not they use a fabric covering. The term is also applied to the individual blades of the American farm windmill rotor, and sometimes refers to the blades of modern wind turbines.

Sail wing: A wind turbine blade patterned after the sails of Mediterranean windmills, especially those found on Crete and in Portugal. The sail wing uses fabric sails strung between a spar and a taut cord. A sail wing uses a tubular spar for a leading edge, a taut cord for a trailing edge, and sail cloth strung between the two. The sail wing is recognized for its low cost and self-regulating features.

Sandia National Laboratory: Albuquerque, New Mexico. One of the national laboratories in the United States resulting from development of the atom bomb and contracted by the U.S. Department of Energy for research on Darrieus wind turbines.

San Gorgonio Pass: Near-sea-level gateway from southern California's coastal region to the Sonoran Desert about 90 miles (150 kilometers) east of Los Angeles. Bounded on the north by Mount San Gorgonio and on the south by Mount San Jacinto, the pass has some of the strongest winds in California. During the early 2000s, wind turbines in the pass generated about 600 million kWh per year.

Santa Ana: Powerful hot, dry wind that blows from the high deserts of southern California into the Los Angeles basin.

Savonius rotor: Simple drag device producing high starting torque developed by the Finnish inventor Sigurd J. Savonious. Sometimes called an S-rotor because of its appearance in plan view. Due to its ease of construction and high torque, it's well suited for water pumping in developing countries. For a drag device (similar to a cup anemometer with some flow recirculation), it's fairly efficient (31 percent), but like other drag devices is inefficient in its use of materials and, consequently, has had limited application.

Scavenging: The removal of dead and decaying animals from a locale by predators or by animals that feed on carrion. The rate at which dead birds are scavenged or removed from a site by carrion-eating animals, such as coyotes, vultures, and ravens, is an important factor in determining the total number of birds killed by wind turbines relative to those counted by an observer.

Scheduled outage: Termination of wind turbine operation due to a planned event such as scheduled (or planned) maintenance, training, tours, or the like. Scheduled outages are attributed to either the wind turbine, the utility system, or nonoperational wind-turbine-related events.

Schneefresser: Swiss wind known as the "snow-eater."

Sea breeze: Light onshore wind caused by differential heating of the land and water. The daytime wind associated with the land–sea breeze effect. See *Land–sea breeze.*

Self-excitation: When the field of a generator is excited internally as opposed to excitation provided by an external power source.

Self-furling: Form of rotor control used on conventional small wind turbines using tail vanes for orientation; the rotor automatically folds toward the tail in high winds. This can be accomplished by use of a pilot vane or by offsetting the rotor axis from the pivot or yaw axis. This moment or torque is counterbalanced by weights or springs in the American farm windmill or by careful design of the hinge between the tail vane and the nacelle in modern small wind turbines. See *Furling* and *Vertical furling.*

Self-regulating windmill: Term applied to American farm windmills that use self-furling to protect the rotor in high winds.

Self-starting: The ability of a wind turbine rotor to begin spinning solely using the power in the wind without the aid of an external power source. The manner of operating a wind turbine with this capability. Some wind turbines use rotors that are capable of self-starting, but these turbines motor the rotor up to operating speed to prevent the rotor from operating for any length of time at the rotor's resonant frequency.

Self-supporting tower: Freestanding or cantilevered tower. Tower that does not require guy wires to remain upright.

SEP: Samenwerkende Elecktriciteits-produktiebedrijven. Association of Dutch Electric Utilities.

SERI: Solar Energy Research Institute (U.S.). Renamed the National Renewable Energy Laboratory in the mid-1990s.

Serviceability: Ease of maintenance or repair.

Service hoist: Small winch sometimes located in the nacelle of medium-size and larger wind turbines for raising and lowering objects from the nacelle.

Servomotor: Electric or hydraulic motor used to control a device. The yaw motor used on a wind turbine to orient the turbine into the wind is a servomotor.

Setback, aesthetic: American usage for a buffer zone between wind turbines and a site's property line intended to compensate or ameliorate the visual intrusion of the wind turbines in a vista.

Setback, public safety: American usage for a buffer zone between wind turbines and a site's property line intended to protect the public from any real or perceived hazard from the wind turbines.

Shank: Shaft.

Shear: Force acting at right angles to the longitudinal axis of a bar, beam, or rod.

Short circuit: Accidental connection of low resistance established across a circuit causing an excessive or dangerous flow of current. Short.

Shroud: A structure surrounding a wind turbine rotor to concentrate or augment the wind stream in order to extract more energy from the wind. A structure to deflect the wind around an object, such as a wind turbine tower, to reduce drag or turbulence.

Shunt regulator: Electrical device that dumps or diverts current from a source to a ballast load, such as a resistance heater, for voltage control. (Hugh Piggott)

Shunt wound: Connecting the field windings of a generator in parallel with those of the armature. Field excitation is provided by current from the armature. Because it is self-exciting, shunt-wound generators are dependent on residual magnetism in the pole shoes of the field.

Sine wave: Waveform where the amplitude varies with the sine of a linear function of time as represented by a sinusoidal curve such as the voltage and current waveforms produced by AC generators.

Single phase: An alternating current supply with only one voltage component or waveform. Rural areas in North America are typically served by single-phase distribution lines. Most wind turbines designed for interconnection with the electricity network produce three-phase current.

Sirocco: Italian for a hot, humid southerly wind in southern Italy, Sicily, and the Mediterranean originating on the Sahara Desert. Also spelled *scirocco.*

Site survey: Assessing the potential for using wind energy at a specific location, including wind resources, access, and potential land-use conflicts.

Slewing: Danish for "yawing."

Slew ring: Danish for "yaw ring" or gear on wind turbines using active yaw.

Sling: Strap of nylon webbing or steel cable with loops in both ends, or a simple loop of nylon webbing used for cradling or hoisting heavy equipment with a crane. Used frequently when assembling and erecting a wind turbine.

Slip: Difference between synchronous speed and the operating speed of an induction generator. Given as percentage of synchronous speed. The slip increases with load on an induction generator. The slip of most induction generators is less than 3 percent—less than 50 rpm for a 1,500 rpm generator—but some induction generators can slip as much as 10 percent.

Slip rings: A series of metal rings used to transmit electric current or signals from a rotating shaft to a fixed shaft. Used on many small wind turbines to transfer electricity from the generator in the nacelle, which yaws in response to changes in wind direction, to fixed conductors on the tower. Medium-size wind turbines use droop cables instead of slip rings for conducting electricity from the generator to ground level. See also *Droop cable.*

Slot effect: The benefit derived from airflow through a gap between a slat or auxiliary airfoil and a wing or second airfoil. The slot effect was first noticed on jib-rigged sailboats where the foresail directs flow over the mainsail, accentuating the mainsail's performance.

Slow-speed shaft: See *Main shaft.* The shaft connecting a wind turbine's rotor and its drive train, so called because it spins at lower speed than the output or high-speed side of a step-up gearbox needed to drive conventional generators.

Smeaton, John (1724–1792): British civil engineer who rebuilt the Eddystone lighthouse. Recognized as one of the earliest professional engineers, Smeaton conducted the first scientific studies of wind turbine rotors and published his results—"On the Construction and Effects of Windmill Sails"—in 1759. One of his conclusions was that the maximum power of windmill sails was nearly proportional to the cube of wind speed.

Smidth, F. L.: Danish company that designed machinery for working with concrete and used this technology to build concrete silos and chimneys during the 1930 and 1940s. One of the world's first firms to marry the rapidly advancing field of aerodynamics to wind turbine manufacturing. During World War II the firm manufactured several 17.5-meter (57 ft) and 24-meter (80 ft) "Aeromotors" that used modern airfoils upwind of a concrete tower.

Smith-Putnam wind turbine: The world's largest wind turbine until the wind energy revival of the 1970s. Named for engineer Palmer Cosslett Putnam and the S. Morgan Smith Company, a manufacturer of hydroelectric turbines. Putnam assembled a talented team of engineers and academics to build a 1.25 MW wind turbine with a rotor 175 feet (53 m) in diameter. The turbine was installed atop Grandpa's Knob on Lincoln Ridge near Rutland, Vermont, in October 1941 and operated sporadically until it threw a blade in March 1945 and was dismantled. During its short lifetime the turbine generated only 62,000 kilowatt-hours.

Smock mill: Traditional European windmill in which the cap containing the rotor, main shaft, and gear assembly turns about the vertical axis of the tower to face the wind. So named because the flared wooden towers resemble a peasant's smock. In contrast to the smock mill, the entire tower and cap of post mills turn to face the wind.

Snap hooks: A hook closed on the open side by a spring-driven plate that is used on safety lanyards and fall restraint systems.

Social costs: Nonmonetary costs associated with a technology by its direct and indirect effect on human welfare. See also *External costs.*

Soft towers: Towers where the fundamental system frequency is less than the blade passage frequency. (Louis Divone)

Solidity: Quotient of total blade area divided by frontal or swept area of a wind turbine. In general, rotors with high solidity are less productive than those with low solidity. The three-bladed Gedser mill had a solidity of 0.09. The solidity of American-designed high-speed, two-bladed turbines, such as the Mod-O and the ESI-54, were 0.03.

Sound levels: See *Exceedance levels.*

Sound power level: L_w. A measure of acoustic power in decibels. The source or emission strength of wind turbine noise derived by field measurements of the sound pressure level:

$$L_w = L_p + 10\log(4\pi R^2)$$

where L_p is the measured sound pressure level and R is the slant distance from the nacelle. If the sound pressure level was measured on a reflective panel, 6 dB must be deducted from L_p. Sound power level is used in various noise models to project noise at various distances from a wind turbine. The sound power level of most commercial wind turbines varies from 95 dBA to more than 100 dBA.

Sound pressure level: L_p. A measure in decibels of pressure relative to a reference pressure of 20 micronewtons/m^2. Because sound pressure levels decrease with increasing distance from the source, location is always specified or implied. For most discrete sources, such as wind machines, the distance from the listener is just as important as the noise level of the source. For example, Danish wind turbine manufacturers estimate that the noise from a typical medium-size wind turbine will drop to 45 dBA within 150 meters (500 ft). Not to be confused with sound power level, though both measures use the same units: decibels.

Span: The overall length of the airfoil in the direction perpendicular to the cross section.

Spar, blade: A structural member running the length of an airplane wing or wind turbine blade.

Sparklies: American colloquialism. See *Reflection.* See also *Flashing.*

Specific capacity: Total energy generated in kilowatt-hours divided by the rated power of the wind turbine in kilowatts. At good sites medium-size wind turbines can generate more than 2,000 kWh/kW of rated capacity. At extremely energetic sites some wind turbines have produced in excess of 4,000 kWh/kW per year. Unlike specific yield, specific capacity is dependent on the arbitrary rating of the wind turbine in kilowatts.

Specific land use: Measure of the amount of land used by wind turbines relative to installed capacity, swept area, or annual generation. Open arrays, such as those found in Europe with an 8-rotor-diameter by 10-rotor-diameter spacing, occupy 100 m^2 of land per m^2 of swept area. Dense-packed arrays, such as those found in California, may use as little as 20 m^2 of land area per m^2 of swept area for an array with a 3-by-6-rotor-diameter spacing.

Specific tower head mass: A measure of a wind turbine's material intensity. The mass of the nacelle and rotor relative to the area swept by the rotor in kg/m^2. Lightweight high-speed turbines can have a specific mass as low as 10 kg/m^2, while large wind turbines have weighed in at more than 70 kg/m^2. Most medium-sized European wind turbines have a specific tower head mass of 20 to 30 kg/m^2.

Specific yield: A measure of wind turbine productivity within a specific wind resource. Net kilowatt-hour generation per square meter of rotor swept area per year (kWh/m^2/yr). Unlike capacity factor, specific yield is independent of arbitrary wind turbine ratings and is thus a more reliable indicator of performance.

Speed, generator: rpm or revolutions per minute.

Speed increaser: Transmission. Gearbox. Increases speed of low-speed or main shaft from turbine rotor to that needed to drive a generator or other device.

Speed, rotor: rpm or revolutions per minute of wind turbine's rotor.

Speed-up effect: Increase in wind speed induced by terrain features, such as the curvature of a hill.

Spinning reserves: The plant capacity, in excess of actual load, that can be called on to produce electricity at very short notice.

Spoilers, blade: A long narrow panel that, when raised on the upper surface of an airfoil, destroys or spoils lift and increases the drag of the airfoil. Method of overspeed control on wind turbine rotors.

Square wave: An oscillation of the voltage waveform between two extremes without any intermediary steps such as that produced by early solid-state inverters.

Squirrel cage rotors: The rotor of an induction generator where coils of copper bars or wire are arranged in slots on an iron core and are short-circuited on the ends of the rotor by rings.

Stage mill: Smock mill with a stage or platform from which the miller could work the sails.

Stall, aerodynamic: A condition of an airfoil in which an excessive angle of attack causes separation of the flow over the airfoil, resulting in a loss of lift and an increase in drag. This is a dangerous condition in aircraft, but it can be put to constructive use in limiting power from a wind turbine rotor.

Stall-controlled rotors: Wind turbine rotors that use stall to limit maximum power in high winds.

Stall regulation: Power control using stall. The blades of a fixed-pitch wind turbine operating at a constant rotor speed can be designed to stall in high winds to limit the wind turbine's maximum power.

Stand-alone power system: Electric power system independent of the network or grid often used in remote locations where the cost of stringing lines from large central power plants is prohibitive.

Star or Y-connection: A three-phase connection where each phase is connected between a supply conductor and a common or neutral point. (Hugh Piggott)

Start-up wind speed: Minimum wind speed at which a wind turbine rotor at rest will begin to spin. The start-up wind speed differs from the cut-in wind speed, when the wind turbine begins producing usable power, but in some small wind turbines the two are the same.

Stator, generator: That part of a generator that's stationary as opposed to the rotor that revolves.

Stay: Support or brace. Nautical expression for a guy rope or cable used to support a mast or spar. Early Danish wind turbines, such as the Windmatic 14S, were patterned after the Gedser mill, which used struts and stays to brace the rotor.

Stiff towers: Towers where the lowest natural frequency of the system is higher than the blade passage frequency. (Louis Divone)

Stock: The spar used to support the sail of a traditional European windmill. Most Dutch windmills used stocks from a single continuous piece of timber. German, Danish, and British windmills used one-piece, two-piece, and three-piece stocks made from either timber or metal.

Stock watering: Providing ample water for cattle to accelerate weight gain as well as to ensure their survival on the plains of North America during the droughts of late summer. An important application for wind pumps.

Storage batteries: The ability to transform kinetic energy to potential energy in the form of chemical reactions in batteries.

Storage, compressed-air: The ability to transform kinetic energy to potential energy in the form of a compressed-air reservoir. In practice surplus electricity is used to drive a compressor. When energy is needed, compressed air is released through a turbine, where it's used to burn a fuel under pressure. In this way the compressed-air reservoir replaces the compressor side of the gas turbine.

Storage, pumped: The ability to transform kinetic energy to potential energy in the form of water in a reservoir. Surplus electricity is used to pump water uphill, where it's stored in a reservoir until needed. When electricity is needed, water is released to drive hydroelectric turbines.

Straight-blade VAWT: As the name implies, the airfoils are straight as opposed to the curved (troposkein) shape of the blades on a ϕ-configuration (eggbeater) Darrieus. H-rotors use straight blades.

Stress-relieved: Heat treatment for metals that allows internal stresses to dissipate.

Strip chart recorders: Obsolete form of recording device that uses long sheets or strips of paper on which data is recorded.

Struts: Rods or bars used to brace a tower or wind turbine blade. The Gedser mill used struts and stays to strengthen the rotor.

Sucker rod: In North America, the wooden or metal pole connecting the windwheel to the pump cylinder in a water-pumping farm windmill.

Surge protection: Electronic device designed to dissipate rapid increases in voltage or current.

Survival wind speed: Maximum sustained wind speed a wind turbine can withstand without catastrophic damage. See *Design speed.*

SWECS: Archaic jargon for "small wind energy conversion system," or small wind turbine.

Swept area: The area of the wind stream swept by a wind turbine rotor. For a conventional wind turbine, the swept area is the area of a circle:

$$A = \pi R^2$$

where R is the rotor's radius. For H-configuration VAWTs the swept area is:

$$A = HD$$

where H is the length or height of the blades and D is the diameter of the rotor. For ϕ-configuration Darrieus, swept area is approximately the area of an ellipse. See also *Intercept area* and *Capture area.*

Synchronized stop of two-blade rotors: A technique thought to ameliorate the visual intrusion of two-blade wind turbines where the rotors of all units in an array are parked in the same position, preferably horizontal. First proposed by landscape architects at the University of California Polytechnic in Pomona during the mid-1980s.

Synchronous generator: Generator where AC frequency is directly proportional to the speed of the generator's rotor. Most wind turbines interconnected with the utility network use asynchronous generators because of the generator's simplicity. Those wind turbines that use synchronous generators produce variable-frequency current at variable voltages.

Synchronous inverter: Power conditioning device for synchronizing wind generator frequency and voltage with that from the grid or network and, where necessary, inverting DC to AC.

Tag line: A lightweight rope used to steady unwieldy loads being hoisted by a crane. Tag lines are frequently used to prevent damage of a wind turbine rotor as it is lifted to the nacelle.

Tail vane: As in the weather vane, the vertical surface that aligns itself parallel to the wind. The tail vane of a small conventional upwind turbine keeps the rotor facing into the wind.

Taper, blade: Narrowing of chord from root to tip. Taper can be linear or nonlinear. High-performance rotor blades are tapered for much the same reason the blades are twisted: The tangential velocity of a point on the blade decreases the closer the point is to the hub. Consequently the tip-speed ratio declines toward the hub, and a greater solidity is required to maintain an optimal coefficient of performance. This greater solidity can be provided by tapering the blade to a much wider chord at the root.

Tariff: Rate schedule. A schedule of prices or fees for the use of electricity.

Tax credit: Under U.S. tax law, a credit directly reduces tax liability. In contrast, a tax deduction reduces net taxable income and only indirectly reduces tax liability. One dollar of tax credits is worth $1 in after-tax value to a U.S. taxpayer.

Tax deduction: Under U.S. tax law, a deduction reduces net taxable income. This reduces taxable liability in proportion to the level of taxation. One dollar in deductions is worth $0.30 in after-tax value for someone in the 30 percent tax bracket.

Teetered hub: Rotor hub with a pivot that permits two-blade rotors of conventional wind turbines to move as a rigid body several degrees perpendicular to the plane of rotation (±10 degrees). So called because the motion of the rotor resembles that of a playground teeter-totter or seesaw. Teetering is a passive means for reducing the aerodynamic loading on a wind turbine rotor and the cyclic loads on the wind turbine's drive train.

Tehachapi Pass: The only all-weather route through California's Sierra Nevada. The pass follows the trace of the Garlock fault through the east–west trending Tehachapi Mountains. The Tehachapi wind resource area is one of the world's largest producers of wind-generated electricity. During the early 2000s wind turbines in Tehachapi Pass generated 1.3 to 1.4 Terawatt-hours per year.

Terrain enhancement: Increase in wind speed induced by terrain features, such as the curvature of a hill. Terrain features, such as a narrow mountain pass, that increase wind speed.

Three phase: An alternating current supply where each of three voltage components or waveforms differ in phase by one-third of a cycle, or 120 degrees. All medium-size and nearly all wind turbines designed for interconnection with the grid produce three-phase current. The simplest three-phase system uses three conductors. Each conductor serves as the return path for the other two.

Throwing: Deformation of trees and shrubs from high winds, where the trunk as well as the branches lean or bend to leeward. Throwing indicates higher winds than flagging. (Putnam)

Throw line: Light rope thrown over a limb, cross-arm, or tower to raise a heavier rope.

Thrust, on rotor: Drag in the streamwise direction that acts to force the blades of an upwind rotor toward the tower. Overturning force on a wind turbine and tower.

Thrust, on tower: Drag in the streamwise direction that acts to force the tower to bend downwind. Overturning force.

Thyristors: High-speed electronic switches used in electronic inverters or power conditioning equipment.

Tip, blade: That portion of a wind turbine blade opposite the hub or root end.

Tip brake: Means of overspeed control incorporating a plate located at the end of a wind turbine blade such that when deployed the plate dramatically increases drag. The term is sometimes incorrectly applied to pitchable blade tips. Unlike pitchable blade tips, tip brakes do not contribute to the blade's torque when undeployed. Tip brakes create parasitic drag and turbulence that accentuates tip noise.

Tip loss: The loss of lift associated with airflow around the tip of a blade instead of across it. Both Aerovironment in the United States and the Technical University of Delft in the Netherlands have attempted to design blades with devices that prevent tip losses. However, commercial blade manufacturers have found that it's cheaper and structurally more sound to extend the blade's span than to add tip augmentors.

Tip speed: Speed at the tip of a rotor blade as it moves through the air, often given in feet or meters per second. The tip speed of "slow-running" medium-size wind turbines ranges from 50 m/s to 60 m/s. The tip speed of "fast-running" medium-size wind turbines ranges from 70 m/s to 120 m/s. Aerodynamic noise is nearly proportional to tip speed.

Tip-speed ratio: The ratio of blade tip speed to wind speed. Modern wind turbines use high-speed airfoils that operate at tip-speed ratios above 5. Lift devices use airfoils that operate at tip-speed ratios of more than 1. Drag devices run at tip-speed ratios of less than 1.

Tip vortices: Power-robbing, noise-creating turbulence found at the end of wind turbine blades. The swirling spirals of air are similar to eddies downstream of a rock in a swift mountain stream.

Tændpibe-Velling Mærsk: A geometric array of 100 wind turbines on Denmark's Jutland Peninsula situated southeast of Ringkøbing on the east side of Ringkøbing fjord. One of the first wind power plants built in Europe.

Top-down technology development: Command and control of technology development by a central organization, such as a government agency or other large institution.

Tornado vortex: American boondoggle. Wind generation concept funded by the U.S. Department of Energy through the Solar Energy Research Institute's Advanced and Innovative Concepts program during the 1970s. The concept conceived of a large cylindrical tower resembling a natural-draft cooling tower with slotted openings and guide vanes that would direct incoming air to create a tornado vortex. This vortex would draw in further air at the base of the tower. The rising column of air within the tower would drive a high-speed turbine. No prototype was never built.

Toroidal accelerator rotor platform (TARP): Another candidate in U.S. Department of Energy's Advanced and Innovative Concepts program during the 1970s. In the TARP concept, a structure or augmentor is constructed in the shape of a hollow doughnut (toroid); dual high-speed turbines are mounted tangentially on either side of the core. Accelerated flow around the toroid drives the turbines. No prototype was never built.

Torque: Turning or twisting force. A force that produces a rotating motion or the force created by a rotating motion. When a wrench is used to tighten a bolt, torque (force times the moment arm) is applied to the bolt. For the same amount of force, torque is increased by increasing the moment arm. Power is the product of torque and shaft speed or rpm. To transmit an equal amount of power, a rotating shaft can operate either at high speeds and low torque or low speeds and high torque. As wind turbine rotors increase in size, their rotational speed decreases. Consequently to transmit the power available, torque must increase. As torque increases, so does shaft size. At higher torques, a larger-diameter shaft is needed to resist the force acting to twist the shaft in two.

Torque ripple: Effect on drive train from two-blade vertical-axis wind turbines. The lift and hence torque produced by the blades of a vertical wind turbine varies with the position of the blades as they move about around the tower. Lift, and hence torque, is greatest when the blades are moving across the wind, and least when they're moving with or against the wind. Thus, the torque varies from a maximum to a minimum several times per rotor revolution.

Tower: A component of conventional wind turbines necessary to elevate the rotor above the ground.

Tower mill: Similar to the smock mill except that the tower is usually made of stone or brick instead of wood.

Tower shadow: Wake created by the wind stream in the lee of a tower. Each blade on a downwind rotor must pass through the wake once per revolution, creating the characteristic *whop-whop-whop* of downwind rotors.

Trailing edge: Opposite the leading edge of an airfoil.

Tramontana: Provençal for a strong cold wind flowing down the northeastern flank of the Pyrenees toward the Mediterranean Sea. Not as well known as the mistral but equally as powerful.

Tramontano: Italian for a cold north wind.

Transfer switch: Electrical switch for disconnecting utility customers from the utility network while connecting them to an independent power system. Frequently used in rural America to provide power for emergency loads during outages of the utility distribution system.

Transient: A pulse or other temporary spike in voltage or current in an electrical distribution system. Transients can be caused by lightning, by the opening and closing of switch gear, or by operation of certain loads.

Translation: Change in physical position. Used to describe the motion of a wind machine's blades and rotor, as in a translating airfoil. Drag translators are devices that convert the energy in the wind into the motion of a blade by means of aerodynamic drag.

Transmission: Device for transmitting the power of the rotor to that of the load, most commonly used for converting the low speed of the main shaft to the high speed required by a generator. See also *Gearbox.*

Transmission and distribution: The combination of power lines, transformers, switch gear, and other electrical equipment used by an electric utility to transfer electric power from a source of generation to the utility's loads.

Troposkein: Greek for a shape that approximates that of a rope fixed at both ends and sagging due to gravity. The shape of a spinning "skip rope" in American parlance. Many ϕ-configuration Darrieus rotors used blades curved to approximate the troposkein shape to eliminate bending stresses.

Truss towers: Freestanding towers constructed from a rigid triangular framework of metal beams, bars, tubes, or angles. See also *Lattice towers.*

Tubular towers: Towers made from rolled steel plate, generally unguyed though sometimes with stays near ground level.

Turbine: Rotor. Sometimes used in reference to a rotor with a large number of blades—for example, the "bicycle wheel turbine."

Turbulence: Sudden changes in both wind speed and direction associated with airflow about an obstruction. Turbulence is undesirable because it decreases harnessable wind power and increases wear and tear on a wind turbine.

Turbulence intensity: Quotient of instantaneous wind speed divided by the mean wind speed for a given period.

Turntable: Revolving platform, as in a phonograph turntable. In American usage it refers to an assembly of a shaft or pintle and a platform that allows a conventional wind generator to yaw in response to changes in wind direction.

Tvind: Between 1976 and 1978 students at the Tvind school near Ulfborg on Jutland's North Sea coast assembled one of the world's largest wind turbines. Though the 54-meter (180 ft) wind turbine never was permitted to operate at its rated power of 2 MW, it did generate about 1 million kilowatt-hours per year to heat the school. Tvind operated the turbine more than 50,000 hours before it broke a blade in 1993. The rotor was replaced and the turbine is still operating.

Twist, blade: Variation of blade angle or pitch from root to tip. The pitch is greatest at the root and smallest at the tip, where the blade is parallel, or nearly so, to its direction of travel. Johannes Juul varied the pitch of the blades on the Gedser mill from 13 degrees at the hub to 0 degrees at the tip.

Two-speed generators: Asynchronous or induction generators with dual windings. One set of windings energizes the field of six poles. The second set of windings energizes the field of four poles. Under normal wind conditions the controller energizes only four poles, enabling the generator to reach nominal synchronous speed at 1,500 rpm in Europe or 1,800 rpm in the Americas. During low winds the wind turbine controller energizes six poles, enabling the generator to reach synchronous speed at 1,000 rpm in Europe or 1,200 rpm in the Americas. This permits operation of the wind turbine rotor at two speeds.

Two-speed operation: The functioning of a constant-speed wind turbine rotor at dual speeds by use of a two-speed induction generator. Operation at two constant speeds enlarges the envelope of wind speeds over which the constant-speed wind turbine rotor is most aerodynamically efficient.

Two-stage gearbox: A series of gears in a transmission that are arranged in two steps.

UCS: Union of Concerned Scientists. Environmental group in the United States that has aggressively encouraged energy efficiency and renewable energy development.

Umbrella mill: Rosette mill. Daniel Halladay's first version of the self-regulating, water-pumping windmill. Halladay hinged the wooden slats (commonly used for blades at the time) around a metal ring. The blades were counterbalanced so that under normal operation they would form the familiar disk of the farm windmill. During high winds, however, the force on the blades counteracted the weights and the wind swung the blades into a hollow cylinder. This allowed the wind to pass through the rotor unimpeded. Also dubbed the rosette, the umbrella mill was supplanted by the more popular fixed rotors that were the direct forebears of the American farm windmills seen today.

Uniformity, visual: Resemblance in size, color, shape, and configuration. Method proposed by landscape architects in California to reduce the visual intrusion of wind turbines on a landscape by minimizing contrasts between like machines. Objects need not be identical, but merely appear similar, to elicit the positive human response to uniformity.

Upper Midwest: The states of Iowa, Minnesota, and Wisconsin and portions of the states of North and South Dakota. Vast reaches of open land and moderate-to-good wind resources have long enticed wind developers. By the mid-1990s only Buffalo Ridge in southwestern Minnesota had seen any extensive activity.

Upwind: Windward. The side facing the wind.

Upwind rotor configuration: A horizontal-axis wind turbine in which the rotor is oriented on the windward side of the tower. Upwind turbines employ some means for orienting the rotor into the wind. Small wind turbines use a tail vane. Medium-size wind turbines mechanically orient the rotor into the wind.

USDA: U.S. Department of Agriculture. Administers an experiment station at Bushland in the Texas Panhandle, where it tests wind turbines for rural use.

Usefulness: Concept by California landscape architects that visual intrusion of wind turbines on the landscape is deemed acceptable by many observers when the observer perceives that the wind turbines are being used—that is, are operating—and are useful, or producing electricity.

Utility-compatible generation: Electrical generation of alternating current that meets an electric utility's requirements for voltage and frequency.

Utility interconnection: The point of electrical connection between a wind turbine and an electric utility's distribution or transmission network. In contrast, a stand-alone power system is not connected to an electric utility.

Vane: Wooden or metal blade on the American farm windmill.

Vaneless mill: Umbrella or rosette mill. Early American farm windmill that did not use a tail vane (hence *vaneless*) to orient the rotor with changes in wind direction. The rotor on vaneless mills operated downwind of the tower.

Variable-geometry VAWT: Nonarticulating (fixed-pitch), straight-blade, vertical-axis wind turbine that reduces its intercept or frontal area to limit power in high winds. Invented by Peter Musgrove at Reading University, England, and pursued by Sir Robert McAlpine in an unsuccessful attempt to commercialize the concept. Up to its rated speed, the rotor has an H-shape with the rotor blades in a vertical position. Above its rated speed, the geometry of the rotor varies as the blades are flung from the vertical about a hinge where they are attached to the rotor cross-arms. This movement reduces the rotor intercept area, limiting the amount of power the wind turbine will capture and preventing the rotor from destroying itself.

Variable pitch: A method of controlling rotor torque and power by changing blade pitch.

Variable-speed generator: Any of several kinds of generators and accompanying power conditioning equipment that enables a generator to operate at variable speeds while generating utility-compatible electricity.

Variable-speed operation: The operation of a wind turbine where rotor speed is proportional to wind speed. Operating the turbine at variable speed enables the rotor to maintain an aerodynamically optimal relationship between tip speed and wind speed, and it permits the rotor to store the energy in gusty winds as inertia rather than forcing the drive train to instantaneously absorb the increased torque.

Variable-stroke wind pump: Mechanical wind pump that automatically varies the pump's stroke to optimize the relationship between rotor torque and the load.

Velocity: A vector of speed and direction.

Velocity duration curve: Graph of velocity or wind speed and the amount of time or duration the wind occurs at a range of wind speeds.

Vertical-axis wind turbine (VAWT): Wind turbine whose rotor spins about a vertical axis—examples include Darrieus rotors and cup anemometers. Not to be confused with the term *vertical wind turbines* used by historians and industrial archaeologists to refer to the rotor plane, not the axis of rotation.

Vertical furling: Form of rotor control on conventional small wind turbines where the rotor tips vertically toward the tail vane in high winds. This movement is counterbalanced by weights, springs, or careful design of the hinge axis. See also *Furling* and *Self-furling.*

Vertical wind speed profile: Graphical representation of the change in wind speed with height.

Village electrification: Supply of electricity to rural communities currently without power. With modern hybrid wind and solar power systems, it's now possible to provide electricity without extending the distribution system from central stations.

Visibility: The state of being visible or of being easily observed. Visibility is distinct from visual intrusion or visual impact. An object, such as a wind turbine, that is visible is not necessarily an intrusion on a landscape. See also *Conspicuousness, visual.*

Volt (V): The unit of electric potential or electromotive force equal to the difference between two points where the current flow of one amp dissipates one watt of power.

Volt-ampere-reactance (VAR): The reactive power consumed when magnetizing the field in a generator equal to the product of current, voltage, and the sine of the phase difference between current and voltage. A measure of reactive power.

Vortex generators: Auxiliary aerodynamic devices attached to the low-pressure surface of airfoils that delay flow separation, increasing the airfoil's aerodynamic performance. Small vanes normal to the low-pressure surface of the airfoil are set at an angle to the incoming flow which creates vortices on the wing surface. The vortices increase lift, but also increase drag.

Vortices: Eddies. A whirling mass of water or air.

Wake: Turbulence downstream of an obstruction in a moving fluid.

Water churn: A vessel in which water is agitated to generate heat.

Water-pumping windmills: Wind pumps. Most often applied to farm windmills that mechanically pump water.

Watt: Unit of power equal to 1 amp at 1 volt.

Watt governor: Flyball governor. Named for James Watt, the Scottish inventor of the condensing-cycle steam engine (1769). The revolving weights or flyballs, acted upon by centrifugal force, limit shaft speed through a mechanical linkage. Though often credited to Watt, this early feedback mechanism was used on traditional European windmills long before the age of steam.

Watt-hour (Wh): Unit of electrical energy equal to the generation or use of 1 watt for one hour.

Weibull wind speed distribution: A mathematical idealization of the distribution of wind speeds over time: the amount of time the wind blows at a given wind speed.

$$f(V) = k/C\ (V/C)^{k-1} \exp[-(V/C)^{k}]$$

where f(V) is the frequency of occurrence for the wind speed (V) in a frequency distribution, (exp) is the base e exponential function, C is the empirical Weibull scale factor in m/s, and k is the empirical Weibull shape factor.

Weldment: A multipiece metal assembly fastened together by welds rather than bolts or other means.

Well casing: Pipe installed in the well bore to prevent the well wall from collapsing.

Wheel: Rotor on American farm windmill. Colloquial expression for rotor of multiblade wind turbine.

Wincharger: Brand of windcharger or battery-charging wind turbine manufactured in the United States from the 1930s into the 1970s.

Wind assist: Innovative approach to using wind energy where the wind turbine provides all the power to the load when the wind is sufficient, and where it assists a conventional power source when the winds are inadequate to deliver all the power needed. When the wind is not blowing, the conventional source picks up the load as the wind turbine stands by.

Wind-assisted irrigation: The application of wind energy as an aid or supplement or high-volume irrigation applications. Concept developed by the U.S. Department of Agriculture's Bushland, Texas, experiment station to help farmers on the high plains.

Wind atlas: Collection of wind resource maps for a region.

Windcharger: American usage for battery-charging wind turbines from the pre-REA era (1930–1950).

Wind-dependent availability: The amount of time a wind turbine is available for operation relative to the amount of time wind speeds are within the turbine's operating range.

Wind–diesel systems: Hybrid power systems using wind turbines in parallel with diesel generator sets for remote villages.

Wind-driven generator: Wind generator. More accurate but less commonly used description of a wind-electric turbine.

Wind dynamo: Wind turbine. Wind-electric generator. From Volta Torrey's description in *Wind Catchers* (1976) for a machine that converts the wind's mechanical energy into electrical energy.

Wind-electric generator: Wind generator, as in hydroelectric generator.

Wind energy conversion system (WECS): Archaic jargon for wind machines. A generic term that includes wind turbines, wind generators, and windmills that was used by the U.S. Department of Energy and its contractors during the 1970s and 1980s.

Wind engine: American farm windmill adapted for mechanical tasks other than pumping water.

Wind farm: Wind power plant. Array. An array of multiple wind machines sited within one geographic area and operated as one unit. The term was derived from the literary association between farming and harvesting wind energy: Wind generation and farming depend on seasonal cycles, the turbines are planted in rows like fields of corn (maize), and both are practices predominantly found in rural areas.

Wind furnace: Wind turbines designed strictly for heating. Advocates of the wind furnace concept believe that the same winter winds that rob a house of its warmth could be used for heating. Most attempts at manufacturing wind turbines solely for home heating have failed commercially.

Wind generator: Wind machine used to generate electricity.

Wind machine: Generic expression for any device that translates wind energy into motion, whether it be motion in one direction (as in a sailboat) or motion around an axis (as in a wind turbine).

Windmill: Generic term for wind machine. Specifically refers to a wind machine used for grinding grain. However, the familiar Dutch or European windmill was used not only to grind grain but also to cut timber, pump water, shred tobacco, and perform a multitude of other industrial tasks. In North American usage, the term *windmill* most often refers to a multiblade water-pumping wind machine: the "American farm windmill."

Windmill, Dutch: Windmill commonly associated with the Netherlands, although it's found in nearly all European countries.

Windmill, water-pumping: Variously known as farm, Chicago, or classic American farm windmill. Wind pump.

Wind park: See *Wind power plant.* Incorrect colloquial usage. In American and British English *park* connotes recreational enclaves or sylvan settings publicly protected from development. Arrays or clusters of multiple wind turbines are more correctly described as wind power plants. During the 1970s utilities in North America envisioned building "energy parks" of multiple nuclear reactors. Propagandists for the utilities deliberately chose the term *parks* to mislead public debate. Continued use of the term *wind park* by the wind industry may be construed as a similar attempt to deceive the public.

Wind plant: See *Wind power plant;* see also *Wind turbine.* In both American and European usage the term can mean individual wind turbines. In North America, Marcellus Jacobs used the term to describe individual wind turbines during the 1930s and again during the 1970s.

Windplant: Kenetech Windpower trademark for a wind power plant built by the defunct Livermore, California, manufacturer. The same company patented variable-speed operation despite documented widespread prior use.

Windpower generator. Wind generator. Wind turbine.

Wind power plant: Array of multiple wind turbines in one locale operated as one entity, much like the multiple steam turbines at a typical coal-fired power plant. Any cluster of wind turbines used for the bulk generation of electricity.

Wind power station: Electric Power Research Institute (U.S.) description of wind power plant.

Wind pump: Wind turbine used to raise water. The equivalent of more than 100 MW of wind pumps are in use worldwide. See also *American farm windmill.*

Wind regime: Sum of regional wind characteristics.

Wind rose: Circular bar graph of wind speed, direction, and sometimes frequency of occurrence. It gets its name from the radiating bars that represent the principal points of the compass.

Wind run: Distance the wind travels during a given period. When used over a specific time period, an average measure of wind speed can be found. For example, if an anemometer records that 150 miles of wind passed through in a 10-hour period, the wind speed averaged 15 mph during the period.

Wind-run odometer: Simple device used for deriving average wind speed that measures and records the number of miles or kilometers that pass the anemometer.

Wind shear: Change in wind speed with height above the ground. See also *Wind speed gradient.*

Wind shear exponent: A number derived empirically from surface roughness and used in the power law model for estimating changes in wind speed with changes in height. See *Power law.* The wind shear exponent for low-grass prairies such as the American Great Plains is approximately one-seventh.

Windsmiths: Modern millwrights. Those who operate and maintain wind turbines.

Wind speed: Wind velocity without the directional component of the vector. Colloquial for "wind velocity."

Wind speed duration curve: Graphical representation of the amount of time the wind equals or exceeds a given speed with time on the horizontal axis and wind speeds on the vertical axis.

Wind speed frequency curve: Graphical presentation of the wind speed frequency distribution.

Wind speed frequency distribution: The number of occurrences or the amount of time the wind occurs at discrete wind speeds. Presented in both tabular and graphical form. In graphical form wind speed is shown on the horizontal axis, and the number of occurrences or time is shown on the vertical axis.

Wind speed gradient: Variation of wind speed with height above the ground. Normally wind speed increases with height. The rate of increase varies with the roughness of the terrain. The rougher the terrain, the greater the rate at which wind speed increases with height, because of greater frictional effects on airflow in the air nearest the ground. See *Wind shear.*

Wind speed profile: Graphical representation of changes in wind speed with height above the ground.

Wind structure: Details of wind flow over an area. Includes variations in direction with height as well as speed, gustiness, and the nonuniform nature of the wind.

Wind turbine: Often used as a generic term for any wind machine. Correctly used, it refers only to a rotor made up of high-speed airfoils. Wind generator.

Wind turbine generator (WTG): Jargon to describe a wind turbine driving a generator. Often used by attorneys (solicitors) and brokers in California to impart a sense of technical sophistication.

Wind vane: Device for indicating wind direction. Weather vane.

Wind velocity: *Wind speed* in common parlance.

Wind wall: An array of wind turbines arrayed vertically on towers of different but uniform heights as well as arrayed horizontally across the prevailing wind.

Windward: Upwind. On the side facing the wind. Snow is swept clear from the windward side of mountain summits.

Windwheel: North American usage for the rotor on the water-pumping farm windmill. The revolving elements in a wind machine: blades, vanes, hub, and shaft.

Windy City: Chicago, Illinois. Named for the blowhards (boosters) who promoted the city's hosting of the Columbian Exposition in 1883. No connection with wind or wind energy.

Wing: Translation of Danish *vinge,* for wind turbine blade or vane.

Wood composites: Assemblage of wooden blocks or veneer and polyester or epoxy resins for use in wind turbine blades.

Work belt: Positioning device. Waist belt of wide leather or woven nylon straps used with a lanyard for freeing the hands while working in an elevated position. Often incorrectly called a safety belt. Not part of a fall-arresting system.

WMEP: ISET's annual report, Wissenschaftliches Meß und Evaluierungsprogramm, on wind turbine performance in Germany derived from actual field data.

Yaw: Rotation about a vertical axis of a horizontal-axis wind turbine nacelle to maintain alignment with the wind direction.

Yaw, active: A horizontal-axis wind turbine yaw control system that uses an electrical or hydraulic servomechanism to orient the rotor with the wind in response to a signal from a wind vane.

Yaw angle: Angle about a vertical axis between the wind direction and the rotor axis of a horizontal-axis wind turbine.

Yaw axis: Vertical axis about which a horizontal-axis wind turbine nacelle changes its orientation in response to changes in wind direction.

Yaw control: Means for orienting the rotor into or out of the wind or for limiting yaw movement. Wind turbine yaw control is subdivided into active and passive mechanisms. Passive yaw control uses forces in the wind itself to orient the nacelle with changes in wind direction, whereas active yaw uses a signal from a sensor to direct the nacelle into or out of the wind.

Yaw dampening: Means for diminishing or retarding yaw movement.

Yaw, passive: A horizontal-axis wind turbine yaw control system that uses the natural aerodynamic forces of the wind to orient the nacelle and rotor in proper alignment with the wind. There are three common forms of passive yaw: coning on downwind rotors, tail vanes on small upwind rotors, and the fan tail on either upwind or downwind turbines.

Yaw rate: The speed at which a horizontal-axis wind turbine reorients the nacelle to changes in wind direction. Small wind turbines using tail vanes have high yaw rates compared to medium-size wind turbines using active yaw control.

Yield, specific energy: Quotient of annual energy output divided by rotor swept area and given in $kWh/m^2/yr$. The most reliable measure of overall wind turbine performance relative to a specific wind regime.

Zaan district: Site on the Zaan River in the Dutch province of Noord Holland with the largest documented concentration of windmills during the Netherlands' golden age. Historians attribute the birth of the Industrial Revolution to the more than 700 windmills used along the Zaan in the 17th century.

Zaanse Schans: Open-air museum of operating windmills adjoining Zaandam in the Dutch province of Noord Holland.

Zeebrugge: Belgian port with breakwater. Site of the world's first "offshore" wind power plant.

Zephyr: Greek god of the west wind.

INDEX

the politics and practice of sustainable living

CHELSEA GREEN PUBLISHING

CHELSEA GREEN sees publishing as a tool for effecting cultural change. Our trendsetting books explore the many facets of sustainable living, including organic gardening, green building, renewable energy, and whole and artisan foods. Our published works, while intensely practical, are also entertaining and inspirational, demonstrating that an ecological approach to life is consistent with producing beautiful, eloquent, and useful books.

We invite you to learn more at **www.chelseagreen.c**
To place an order or to r
catalog, call toll-free (800
or write to us at P.O.
White River Junc
Vermont 0500

Gardening

The Apple Grower: A Guide for the Organic Orchardist, Revised and Expanded Edition
Michael Phillips
ISBN 1-931498-91-1 | $40

Food

Full Moon Feast: Food and the Hunger for Connection
Jessica Prentice, foreword by Deborah Madison
ISBN 1-933392-00-2 | $25

lanet

to Growth: The 30-Year Update
Donella Meadows, Jorgen Randers, and Dennis Meadows
ISBN 1-931498-58-X | $22.50

Edens Lost & Found: How Ordinary Citizens Are Restoring Our Great American Cities
Harry Wiland and Dale Bell, with Joseph D'Agnese
ISBN 1-933392-26-6 | $25

CHELSEA GREEN PUBLISHING
the politics and practice of sustainable living